Radiology, Pathology, and Immunology of Bones and Joints

Radiology, Pathology, and Immunology of Bones and Joints
A Review of Current Concepts

Edited by Frieda Feldman, M.D.

Appleton-Century-Crofts
New York

Copyright ©1978 by APPLETON-CENTURY-CROFTS
A Publishing Division of Prentice-Hall, Inc.

All rights reserved. This book, or any parts thereof, may not be used or reproduced in any manner without written permission. For information, address Appleton-Century-Crofts, 292 Madison Avenue, New York, N.Y. 10017.

78 79 80 81 82 / 10 9 8 7 6 5 4 3 2 1

Prentice-Hall International, Inc., London
Prentice-Hall of Australia, Pty. Ltd., Sydney
Prentice-Hall of India Private Limited, New Delhi
Prentice-Hall of Japan, Inc., Tokyo
Prentice-Hall of Southeast Asia (Pte.) Ltd., Singapore
Whitehall Books Ltd., Wellington, New Zealand

Main entry under title:

Radiology, pathology, and immunology of bones and joints.

Includes index.
1. Bones—Diseases. 2. Joints—Diseases.
3. Bones—Diseases—Immunological aspects.
4. Joints—Diseases—Immunological aspects.
5. Bones—Radiography. 6. Joints—Radiography.
I. Feldman, Frieda. [DNLM: 1. Bone and bones—
Radiography. 2. Bone diseases—Immunology.
3. Bone diseases—Pathology. 3. Joints—
Radiography. 4. Joint diseases—Immunology.
5. Joint diseases—Pathology. WE225 R129]
RC930.5R3   616.7'1'07   78-10340
ISBN 0-8385-8254-0
ISBN 0-8385-8253-2 pbk.

Design: Edmée Froment

PRINTED IN THE UNITED STATES OF AMERICA

Contributors

GEORGE F. ASCHERL, JR., M.D.
Assistant Professor of Radiology
Department of Radiology (Neuroradiology)
Neurologic Institute
Columbia Presbyterian Medical Center
New York, New York

DAVID H. BAKER, M.D.
Professor and Director
Pediatric Radiology
College of Physicians and Surgeons of Columbia
  University
New York, New York

WALTER E. BERDON, M.D.
Professor of Radiology
College of Physicians and Surgeons of Columbia
  University
New York, New York

ROBERT E. CANFIELD, M.D.
Professor of Medicine
Department of Medicine
College of Physicians and Surgeons of Columbia
  University
New York, New York

WILLIAM J. CASARELLA, M.D.
Professor of Radiology
College of Physicians and Surgeons of Columbia
  University
New York, New York

DAVID C. DAHLIN, M.D.
Professor of Pathology
Mayo Medical School;
Department of Surgical Pathology
Mayo Clinic and Mayo Foundation
Rochester, Minnesota

MURRAY K. DALINKA, M.D.
Professor of Radiology
Department of Radiology
University of Pennsylvania
Philadelphia, Pennsylvania

ROBERT H. DE BELLIS, M.D.
Assistant Professor of Clinical Medicine
Department of Medicine
College of Physicians and Surgeons of Columbia
  University
New York, New York

JOHN R. DENTON, M.D.
Associate Orthopedic Surgeon
Department of Orthopedic Surgery
Columbia Presbyterian Medical Center
New York, New York

HAROLD M. DICK, M.D.
Associate Clinical Professor of Orthopedic Surgery;
Chief, Children's Orthopedic Surgery
Department of Orthopedic Surgery
College of Physicians and Surgeons of Columbia
  University
New York, New York

JACK EDEIKEN, M.D.
Professor and Chairman
Department of Radiology
Thomas Jefferson University Hospital
Philadelphia, Pennsylvania

RASHID A. FAWWAZ, M.D., Ph.D.
Assistant Professor
Department of Radiology
College of Physicians and Surgeons of Columbia
  University
New York, New York

FRIEDA FELDMAN, M.D.
Professor of Radiology
College of Physicians and Surgeons of Columbia
  University;
Attending Radiologist
Columbia Presbyterian Medical Center
New York, New York

J. WILLIAM FIELDING, M.D.
Clinical Professor of Orthopedic Surgery
College of Physicians and Surgeons of Columbia
    University
New York, New York

WILLIAM R. FRANCIS, M.D.
Fellow, Orthopedic Surgery
College of Physicians and Surgeons of Columbia
    University
New York, New York

ROBERT FREIBERGER, M.D.
Professor of Radiology
Cornell University Medical School;
Director of Radiology
Hospital for Special Surgery
New York, New York

HARRY K. GENANT, M.D.
Associate Professor
Chief, Skeletal Section
Department of Radiology
University of California
San Francisco, California

RICHARD J. HAWKINS, M.D.
London, Ontario, Canada

HARVEY L. HECHT, M.S.
Assistant Clinical Professor of Radiology
Department of Radiology
College of Physicians and Surgeons of Columbia
    University
New York, New York

SADEK K. HILAL, M.D.
Professor of Radiology
The Neurologic Institute
New York, New York

HAROLD G. JACOBSON, M.D.
Professor and Chairman
Department of Radiology
Albert Einstein College of Medicine;
Director of Radiology
Montefiore Hospital
New York, New York

PHILIP M. JOHNSON, M.D.
Professor of Radiology
Director of Nuclear Medicine
Department of Radiology
College of Physicians and Surgeons of Columbia
    University
New York, New York

PETER M. JOSEPH, Ph.D.
Assistant Professor of Clinical Radiology
Department of Radiology
College of Physicians and Surgeons of Columbia
    University
New York, New York

HOWARD A. KIERNAN, JR., M.D.
Associate in Orthopedic Surgery
Department of Orthopedics
College of Physicians and Surgeons of Columbia
    University
New York, New York

ELLIOT F. OSSERMAN, M.D.
Professor of Medicine
American Cancer Society;
Associate Director
Institute of Cancer Research;
Department of Medicine
College of Physicians and Surgeons of Columbia
    University
New York, New York

WILLIAM B. SEAMAN, M.D.
Professor and Chairman
Department of Radiology
College of Physicians and Surgeons of Columbia
    University
New York, New York

STANLEY S. SIEGELMAN, M.D.
Director of Diagnostic Radiology
Department of Radiology and Radiological Sciences
Johns Hopkins University School of Medicine
Baltimore, Maryland

ETHEL S. SIRIS, M.D.
Assistant Professor of Medicine
Department of Medicine
College of Physicians and Surgeons of Columbia
    University
New York, New York

BARBARA N. WEISSMAN, M.D.
Assistant Professor of Radiology
Department of Radiology
Peter Bent Brigham Hospital;
Clinical Associate (Radiology)
Robert B. Brigham Hospital
Boston, Massachusetts

JOSEPH P. WHALEN, M.D.
Professor and Chairman
Department of Radiology
The New York Hospital-Cornell Medical Center
New York, New York

Contents

Preface xi

A VARIETY OF PEDIATRIC OSTEOPATHIES

1. Abnormal Structure, Modeling, and Density of Bone 1
   Joseph P. Whalen, M.D.

2. Angiomatosis and the Skeleton 7
   Frieda Feldman, M.D.

3. Skeletal Dysplasias: Osteogenesis Imperfecta, Osteopetrosis, Hyperphosphatasemia 31
   Joseph P. Whalen, M.D.

4. Fibromatosis and the Skeleton 35
   Frieda Feldman, M.D.

5. The Mucopolysaccharidoses: Lipid and Non-Lipid 47
   David H. Baker, M.D.

6. Hematologic and Storage Diseases 53
   Walter E. Berdon, M.D.

7. Marrow Imaging with Radioiron: Clinical Application 57
   Rashid A. Fawwaz, M.D.

THE SKULL AND SPINE

8. The Pediatric Spine as a Clue to Other Diseases 71
   Walter E. Berdon, M.D.

9. Role of Myelography 75
   George F. Ascherl, jr., M.D.

10. Computed Tomography of the Skull and Spine 83
    Sadek K. Hilal, M.D., Ph.D.

THE ARTHRITIDES

11. Juvenile Rheumatoid Arthritis 89
    Barbara N. Weissman, M.D.

12. Arthritides: Complications and Unusual Features 99
    Jack Edeiken, M.D.

13. Systemic Lupus Erythematosus, Dermatomyositis, Progressive Systemic Sclerosis, and Mixed Connective Tissue Disease 101
    Barbara N. Weissman, M.D.

14. HLA-B27 Associated Arthritis 109
    Barbara N. Weissman, M.D.

15. Osteoarthrosis (Osteoarthritis, Degenerative Joint Disease) 121
    Jack Edeiken, M.D.

16. Erosive Osteoarthritis (Degenerative Joint Disease) 125
    Jack Edeiken, M.D.

17. Neurotrophic Arthropathy 127
    Jack Edeiken, M.D.

18. Infections of Bones and Joints 129
    Jack Edeiken, M.D.

19. Clinical Applications of Magnification in Skeletal Radiology   133
Harry K. Genant, M.D.

METABOLIC BONE DISEASES AND RELATED CONDITIONS

20. Paget's Disease of Bone: Clinical and Therapeutic Considerations   139
Robert E. Canfield, M.D. and Ethel S. Siris, M.D.

21. Osteolytic Paget's Disease   147
William B. Seaman, M.D.

22. Osteoporosis, Osteomalacia, and Hyperparathyroidism   153
Stanley S. Siegelman, M.D.

23. Radioimmunoassay: Parathyroid Hormone and Calcitonin   171
Robert E. Canfield, M.D. and Ethel S. Siris, M.D.

24. Bone Mineral Determinations: Methods and Techniques to Date   175
Peter M. Joseph, Ph.D.

25. Quantitative Bone Mineral Analysis Using Computed Tomography   183
Harry K. Genant, M.D.

MALIGNANT BONE TUMORS

CLINICAL RADIOLOGIC AND PATHOLOGIC CONSIDERATIONS

26. Plasma Cell Dyscrasias: Multiple Myeloma and Related Conditions   189
Elliot F. Osserman, M.D.

27. The Radiology of Plasma Cell Dyscrasias   201
Stanley S. Siegelman, M.D.

28. Round Cell Lesions: Pathology and Newer Classifications   211
David C. Dahlin, M.D.

29. Round Cell Lesions: Radiology   215
Harold G. Jacobson, M.D.

30. Osteosarcoma and Chondrosarcoma: Newer Variants and Pathologic Subclasses   221
David C. Dahlin, M.D.

31. Osteosarcoma and Chondrosarcoma: Radiology   227
Harold G. Jacobson, M.D.

DIAGNOSIS AND MANAGEMENT AND SPECIAL STUDIES

32. Chemotherapy of Malignant Bone Tumors   231
Robert H. DeBellis, M.D.

33. Bone Tumors: Role of Nuclear Medicine   237
Philip M. Johnson, M.D.

34. Angiography in the Management of Bone Tumors   243
William J. Casarella, M.D.

35. Limb-Saving Resections of Malignant Bone Tumors with Autograft/Allograft Replacement   247
Harold M. Dick, M.D.

36. Musculoskeletal Applications of Computed Tomography   253
Harry K. Genant, M.D.

TUMORS AND TUMOR-LIKE LESIONS

37. Histiocytosis X: Pathology   257
David C. Dahlin, M.D.

38. Histiocytosis X: Radiology   259
Frieda Feldman, M.D.

39. Giant Cell Tumor and Its Variants   275
David C. Dahlin, M.D.

40. Skeletal Lesions Simulating Malignancy   279
Harold G. Jacobson, M.D.

41. Cartilaginous Tumors and Tumor-Like Conditions   281
Frieda Feldman, M.D.

TRAUMA

42. The Upper Cervical Spine   303
J. William Fielding, M.D., William R. Francis, M.D., and Richard J. Hawkins, M.D.

43. Fractures and Dislocation about the Shoulder   313
Murray K. Dalinka, M.D.

44. Angiography of Trauma: Diagnostic and Therapeutic   321
    William J. Casarella, M.D.

45. Ankle Fractures   329
    Murray K. Dalinka, M.D.

46. Bone Trauma: Assessment by Scintigraphy   339
    Rashid A. Fawwaz, M.D.

47. Orthopedic Considerations of Trauma to the Growing Skeleton   345
    John R. Denton, M.D.

ARTHROGRAPHY, ARTHROSCOPY, AND PROSTHETICS

48. Arthrography of the Ankle, Elbow, Wrist   351
    Harvey L. Hecht, M.S.

49. Shoulder Arthrography   363
    Murray K. Dalinka, M.D.

50. Arthrography of the Knee   371
    Robert Freiberger, M.D.

51. Total Knee Replacement   373
    Howard A. Kiernan, M.D.

52. Arthrography of the Postsurgical Hip   377
    Robert Frieberger, M.D.

Index   381

Preface

This compendium represents an enlarged version of topics covered during the course of a symposium on the radiology, pathology, and immunology of bones and joints sponsored by the College of Physicians and Surgeons and held in New York City October 17–20, 1978.

This is not a formal text. Rather, it represents a record and an extension of what transpired at the meeting. It also represents a "cross section of the art" as applied to particular sub-specialties in the broad field of musculo-skeletal health and disease as seen by several of its outstanding specialists.

I am grateful to all of them for their participation and for the "particular brand" of enthusiasm and knowledge with which they seem to have been blessed in abundance.

Frieda Feldman, M.D.

Abnormal Structure, Modeling, and Density of Bone

Joseph P. Whalen

Before abnormality of structure can be appreciated, an understanding of the mechanism by which bone grows is required. Growing bone, of necessity, must change both in size and shape. Prior to the discovery of the importance of osteocytes in resorbing bone, it was concluded that the mechanism by which bone is resorbed in this modeling process was exclusively osteoclastic. The studies by Belanger demonstrate the importance of the osteocyte in bone resorption.[1]

In bone there are three cell types:—the osteoblast, whose function it is to form bone; the osteoclast, a surface cell that resorbs surfaces of bone; and the osteocyte, which has the capability of both resorbing and producing bone.[2] Long bones grow in length by enchondral bone formation, which occurs at the physis when chondrocytes lay down cartilaginous matrix, which is then calcified. On this matrix, osteoblasts lay down osteoid, which is then calcified. Bone is then produced on a calcified cartilaginous core. Eventually the cartilage disappears as primary trabeculae become secondary trabeculae. The cartilage deep to the envelope of bone is removed by the osteocyte adjacent to the cartilage (osteocytic chondrolysis)[3] (Figs. 1 and 2). As the osteocyte removes the cartilage, "holes" do not appear because of the osteoblasts or osteocytes, which form bone after removal of bone by the osteocyte. Bone then increases in length and can be shaped by this process. On surfaces of bone where marked angulation is required, osteoclasts are important in this remodeling process.

Bone increases in diameter in its diaphyseal portion by osteoblasts situated on the surface in the periosteum. As the diameter of bone increases, the medullary cavity increases. Bone is formed on the periosteal surface and removed on the endosteal surface (Figs. 3–5). In the normal process in human bone, there is transition from woven bone to lamellar-type bone. Woven bone is disorganized, reflecting rapid turnover. There is gradual transition from woven to lamellar bone, and by the age of eleven years, all woven bone is replaced by lamellar bone.[4] This transition does not occur by virtue of osteoclasts "boring holes" through the cortex and then "filling in" by concentric osteoblastic activity; rather it occurs by the budding of the osteoblasts on the surface with elevation of the periosteum. The hiatus is then eventually filled in (Fig. 5).[5]

Alteration of this normal transition from woven bone to lamellar bone occurs when turnover of bone is too active. Radiolucent bone results either from excessive resorption or decreased production. Radio-dense bone results from either decreased resorption or increased production.

Apart from this, various pathologic processes select areas where alteration is occurring, and a diagnostic radiographic picture may appear. If, for instance, one ingests heavy metal or if experimental animals, such as the rat, are given a heavy metal, the toxic effect of the heavy metal is known to inhibit resorption. This results in a growing bone that does not completely model. The metaphysis does not taper and it appears dense. This results in thickened trabeculae, as seen histologically, with retention of the cartilaginous core, as demonstrated in Fig. 6.

Another illustration of histologic–radiologic correlation is that of excess calcitonin given to growing animals. A similar failure of modeling occurs, again secondary to the direct effect of calcitonin in preventing resorption (Fig. 7).[6]

It is difficult to categorize the many dysplasias of

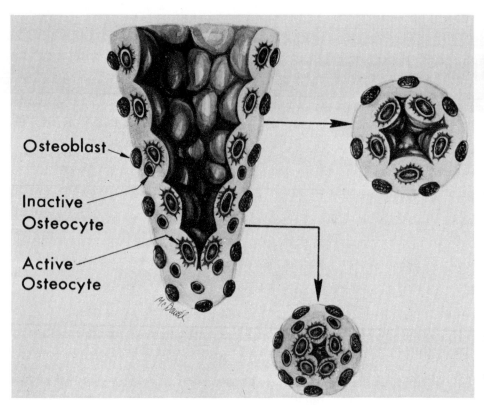

FIG. 1. Primary spongiosa. The cartilaginous core (dark shading) is enveloped by bony tissue. Osteoblasts are on the surface actively laying down bone. Inactive osteocytes are deep to the surface "entrapped" by bony tissue. Deeper, adjacent to cartilaginous core, ost ocytes are actively removing cartilage to be replaced by trabecular bone. Cross sections to the right at different levels illustrate central position of core, its relationship to deep-seated osteocytes, and its eventual disappearance without cavity formation in the progression to secondary spongiosa. (From Whalen JP et al: Some metabolic considerations in bone disease. In Potchen EJ (ed): Current Concepts in Radiology, Vol. 3. St. Louis, Mosby, 1977, pp 143–92. Courtesy of the C.V. Mosby Company.)

FIG. 2. Metaphyseal trabeculae in a growing animal showing the surface of the trabeculae lined by osteoblasts (black arrowheads). The central cartilaginous core (dark staining) (white arrowheads) and active osteocytes resorbing the cartilaginous core (arrows). The bone is assuming a lamellar-type appearance with the normal removal of cartilage and replacement by lamellar bone.

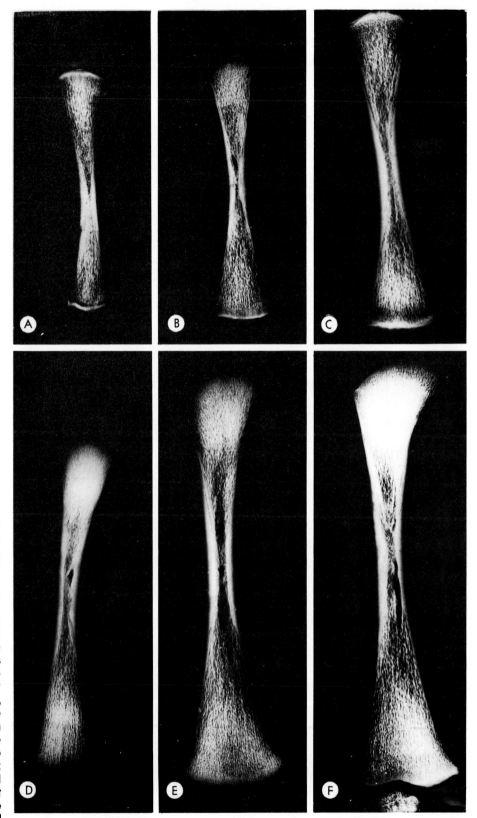

FIG. 3. Roentgenograms of the midsagittal slices of femurs, necropsy specimens of fetuses weighing up to (A) 500, (B) 1000, (C) 1500, (D) 2000, (E) 2500, and (F) 3000 g, respectively. Progressive widening of the medullary cavity at the mid diaphysis and progressive thickening of the cortices is seen through this series. Note that the cortices are actually thinnest in the 500-g fetus (A), but the close proximity of the cortices gives an impression of sclerosis. The progressive increase in length is also noted. (From Whalen JP et al: Neonatal transplacental rubella syndrome: its effect on normal maturation of the diaphysis. Am J Roentgenol Ther Nucl Med 121:166, 1974.)

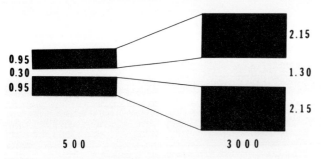

FIG. 4. Schematic illustration of relative increases in width at the mid diaphysis of cortex and medullary cavity in millimeters from 500-g to 3000-g fetuses. The width of the cavity increases more than four times, the total diameter about 2.5 times. (From Whalen JP et al: Neonatal transplacental rubella syndrome: its effect on normal maturation of the diaphysis. Am J Roentgenol Ther Nucl Med 121:166, 1974.)

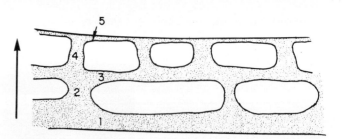

FIG. 5. Schematic drawing of growth in the diameter of the normal diaphysis. The drawing is of cortex with the endosteal surface to the bottom and periosteal surface to the top. Growth occurs in the direction of the arrow. The oldest bone formed is labeled 1; 2 represents a bud (area 2), which then elevates the periosteum. Area 3 is bone formation beneath the periosteum elevated by the bud (area 2). Number 4 represents a bud occurring from the periosteal surface of 3, elevating the periosteum, which then in turn forms bone in area 5. The speckled areas represent formation of bone, with the spaces in between occurring because of distances between buttresses. These spaces eventually fill in and form lamellar bone. The spaces are not the result of osteoclastic activity but merely represent separation of buttresses elevating the periosteum and awaiting filling in by osteoblastic activity.

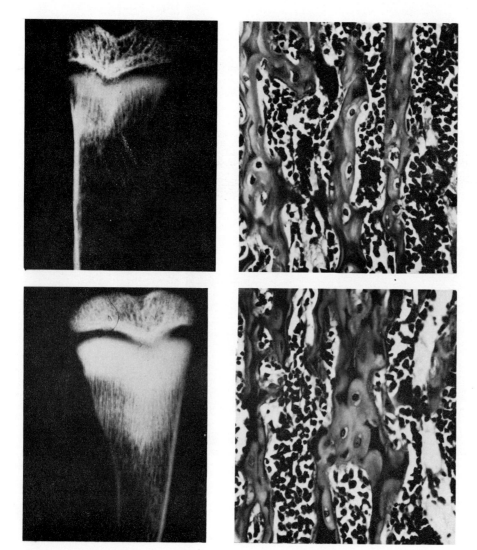

FIG. 6. Top left: Radiograph of a normal proximal tibia of a normal rat 40 days of age. Top right: Histologic section of normal metaphyseal trabeculae (corresponding to top left) showing slender trabeculae with relatively little chondroid core (H&E, x300). Bottom left: Radiograph of proximal tibia of 40-day-old rat after 16 days of yellow phosphorus ingestion. The metaphysis is sclerotic and broadened. Bottom right: Histologic section (corresponding to bottom left) of metaphyseal trabeculae showing the thickened trabeculae with persistence of the cartilaginous core (arrested osteocytic chondrolysis). (Left-hand illustrations from: Whalen JP et al: Pathogenesis of abnormal remodeling of bone: effects of yellow phosphorus in the growing rat. Anat Rec 177: 15, 1973. Courtesy of *Anatomic Record*)

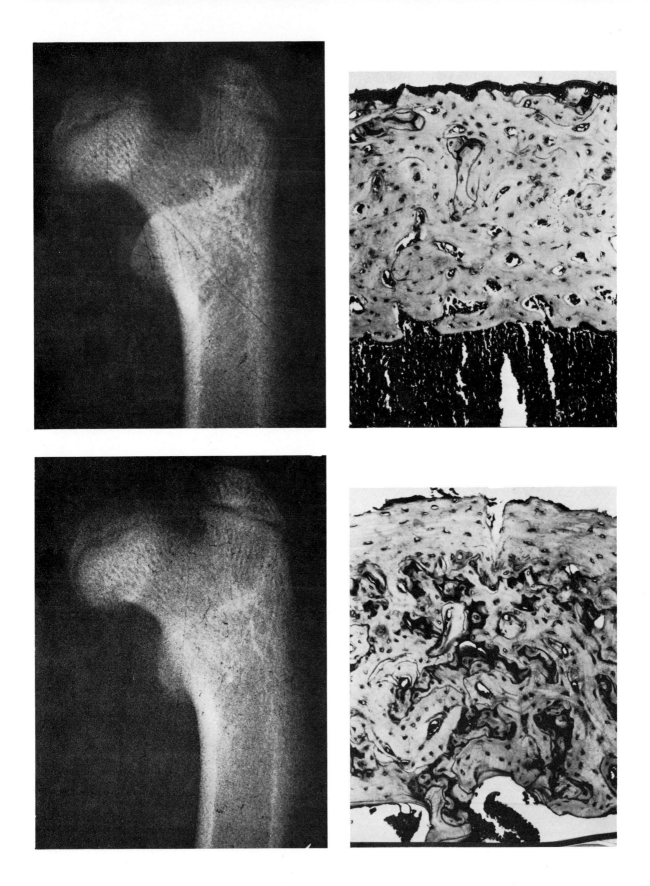

FIG. 7. Top left: Radiograph of normal 46-day-old rat showing normal tapering of metaphysis. Top right: Caudomedial aspect of femur mid diaphysis in transverse section of normal 46-day-old rat. Bottom left: Radiograph of calcitonin-injected rat showing broadening of the metaphysis. Bottom right: Caudomedial aspect of femur mid diaphysis in transverse section of 46-day-old rat with retention of cartilage core (arrested osteocytic chondrolysis). (Left-hand Figures from Whalen JP et al: Calcitonin, parathyroidectomy and modeling of bones in the growing rat. J Endocr 66:207, 1975. Courtesy of *Journal of Endocrinology*.)

bone in terms of the precise mechanism of alteration and perhaps potential treatment, but at least three such entities can be explained by this concept of modeling—osteogenesis imperfecta, osteopetrosis, and hyperphosphatasemia.

References

1. Belanger LF et al: Resorption without osteoclasts (osteolysis). In Siggnaes Mechanisms of Hard Tissue Destruction. Washington, D.C., American Association for the Advancement of Science, 1963, p 531
2. Whalen JP et al: Growing bone—the role of internal resorption and its control. In Progress in Pediatric Radiology, Vol 4, Intrinsic Diseases of Bones. Basel, Karger, 1973, p 45
3. Whalen JP et al: Pathogenesis of abnormal remodeling of bones: effects of yellow phosphorus in the growing rat. Anat Rec 177:15, September 1973
4. Whalen JP, Krook L, Nunez EA: Some metabolic considerations in bone disease. In Potchen EJ (ed): Current Concepts in Radiology, Vol 3, Chap 7. St. Louis, Mosby, 1977, p 143
5. Volberg FM Jr et al: Lamellated periosteal reactions; a radiologic and histologic investigation. Am J Roentgenol 128:85, 1977
6. Whalen JP et al: Calcitonin, parathyroidectomy and modeling of bones in the growing rat. J Endocrinol 66:207, 1975

Angiomatosis and the Skeleton

Frieda Feldman

Hemangiomatosis and Lymphangiomatosis

Hemangiomas in bone are composed of blood vascular channels that are either cavernous or capillary in type. The former are more common in bone and consist of large vessels and sinuses interspersed among trabeculae or embedded in a connective tissue matrix. The latter are composed of fine capillary loops and occur less frequently in bone,[41] although in some series this pattern was most commonly found in vertebra hemangiomas.[10] The existence of intramedullary lymphatics has been questioned; although the periosteum contains abundant lymphatic channels, intraosseous lymphatics have been rarely identified. They have, however, been demonstrated by lymphangiography under certain circumstances, albeit abnormal, such as lymphedema.[65] It has been postulated that they may gain entry into the medullary cavity by accompanying the nutrient vessels that traverse the cortex.

Hemangiomas and lymphangiomas, in both bone and soft tissue, are represented by small or dilated cysts or cavities, the contents of which are often the deciding factor in their classification as either hemangiomas or lymphangiomas. A particular lesion may have the features of both or they may coexist in a particular patient, with a lymphangioma occurring at one or more sites and a "typical" hemangioma occurring at another. Therefore, the general term *angioma* or *angiomatosis* for multiple lesions has been recommended to include both hemangiomas and lymphangiomas, and since soft tissues and other organs as well as the skeleton may be involved, the condition has also been referred to as *systemic angiomatosis*.

Classification

The variations in size and distribution of these angiomas and their association with other abnormalities have resulted in certain specific clinical syndromes as well as in diversified clinical presentations. Some of the latter may not be readily recognized as part of an associated or concomitant angiomatosis. It has therefore been clinically advantageous to group these lesions in terms of the number of sites and kinds of sites involved. Such a framework serves to facilitate the consideration of and search for concomitant lesions. It also serves as a basis for prognostic evaluations (Table 1).

Pathogenesis

In addition to their frequent histologic nonspecificity, which has prompted their being designated by the broad term *angiomas*, the question as to whether hemangiomas and lymphangiomas represent congenital vascular malformations or inborn errors of tissue development, i.e., angiohamartomas, rather than congenital benign vascular tumors (true neoplasms), remains unanswered. The distinction is not always possible on the basis of morphology. In favor of their being vascular tumors is their ability to grow or increase in size out of proportion to somatic growth, their local invasiveness, and their multicentric distribution. The last has caused them to be alluded to as *benign metastasizing hemangiomas*. On the other hand, their indolent behavior, the failure to detect a primary lesion, and the limited number of metastatic malignancies producing well-defined ex-

Table 1 Classification of Benign Vascular Lesions Involving Bone

Number of sites involved
    Solitary
    Multiple
        Focal
        Diffuse
Specific sites involved
    Skeleton alone
    Skeleton + viscera
    Skeleton + skin and/or somatic soft tissues
    Skeleton + skin and/or somatic soft tissues + viscera

panding bone lesions, often with sclerotic margins, are features that militate against their designation as tumors or metastases and particularly against such designations as *benign metastasizing angiomas*. The frequent delay in their appearance, which may suggest a tumor arising de novo, has been compared to the late presentation of an inborn error of metabolism. Most favor the classification of the angiomas (hemangiomas and/or lymphangiomas) as hamartomas along with the fibromatoses.[68]

Pathology

Gross Pathology. Partial excision of a segment of involved bone is preferable to needle or open biopsies, which frequently yield inadequate specimens.[10,58,65] The fundamental gross finding is that of a cystic lesion, which may appear multilocular and dark red when sectioned. Thickened trabeculae that are not of neoplastic origin and are continuous with the adjacent osseous tissue may be concentrated around the periphery of the soft bloody area or may traverse it. Depending on their distribution, bony trabeculae may be responsible for distinctive roentgenographic patterns variously described as *soap bubble, sunray, honeycomb, corduroy, paint brush,* and *spoke wheel* that are associated with angiomas at particular sites. The surrounding medullary cavity may be grossly involved, while the cortex may be thickened, thinned, or eroded either internally or externally. When the neighboring soft tissue harbors a similar lesion, a decision as to the point of origin becomes difficult if not impossible. The degree of bone resorption or proliferation depends on the location of the lesions.

Microscopic Pathology Hemangiomas and lymphangiomas, in both bone and soft tissues, are represented by small or dilated intercommunicating vascular cavities or channels lined by a single layer of flattened endothelium and separated from one another by loose connective tissue or bony trabeculae (Fig. 1). Smooth muscle is seldom found. Conspicuous elastic fibers may be present. Identification is usually dictated empirically by the contents of the cavities, i.e., red blood cells and/or a proteinaceous material. They may, on occasion, be empty. Blood or lymph may also have entered into the cavities during manipulation.

Solitary Primary Hemangioma of Bone

Solitary primary hemangiomas of bone are most commonly found in the axial rather than in the appendicular skeleton. The majority occur in the spine. The first vertebral column hemangiomata were described at autopsy by Virchow in 1863,[64] while the first report referring to their roentgenographic appearance was by Hitzrot.[29] The spine is the locus for more hemangiomas than all other bones of the body combined.[54] Among extravertebral sites, the skull is most frequently involved. In Sherman and Wilner's series,[54] more than half of 45 solitary hemangiomas were in the skull and spine, and in that of Dorfman et al, 17 of 24 were similarly located.[10] If one excludes these sites, primary hemangioma of bone becomes an uncommon condition constituting less than 1 percent of bone tumors in several large series[9,10,54] (Fig. 2).

Incidence

A wide disparity exists between the anatomic and roentgenographic incidence of vertebral hemangiomas. Spinal column angiomas were noted in 10.7 percent of 3829 autopsies studied by Schmorl.[52,53] They were found in 8.9 percent of 1948 men and 12.5 percent of 1881 women. One vertebral angioma was demonstrated in 66.5 percent of the cases, 2 to 5 angiomata in 32.8 percent, and in 0.7 percent more than 5 vertebral angiomata were present. They were located in the thoracic, lumbar, and cervical vertebrae and in the sacrum, in descending order of frequency. The favored vertebral sites in descending order of frequency were the twelfth thoracic, the fourth lumbar, the first lumbar, the second lumbar, and the third lumbar vertebrae. They may, therefore, occur in any vertebra, be located in different portions of the vertebra, and vary in size.

Topfer,[61] a pupil of Schmorl, found an approximately 12 percent incidence of vertebral angiomas, i.e., 267 in 2154 consecutive autopsies. They were

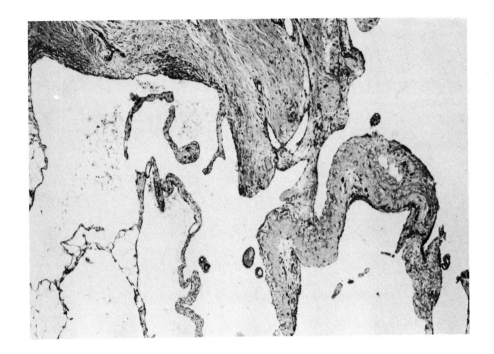

FIG. 1. Congenital hemangiolymphangiomatosis of the upper extremity. This patient had extensive, diffuse involvement of soft tissues as well as of bone, leading to pathologic fractures and extensive bone absorption. This illustration shows intercommunicating vascular spaces lined by inconspicuous endothelial cells from an area just outside the periosteum of a phalanx of a finger. (H&E X 84.) Courtesy of Dr. Raffaele Lattes.

multiple in one-third of cases. The majority were incidental findings at routine necropsy and were not demonstrable roentgenographically. The frequency of angiomas among men and women increased with advancing age. In Schmorl's autopsy series[52,53] only 3 to 4 percent of young persons had vertebral angiomata, while the percentage increased to 12 percent of men over 60 and 60 percent of women over 60. However, clinically there is a higher incidence of symptoms among younger patients, probably indicating an earlier or more extensive angiomatous change or more rapid progression in size. Ages ranged from 8 to 76 years, but the majority of patients were 60 years of age or older.

Clinical Data

Solitary hemangiomas have been observed in all age groups but are most commonly diagnosed in the middle decades of life. Most studies show no familial tendency and no predilection as to race. In some series, a female preponderance of about 2 to 1 has been noted as far as vertebral hemangiomas are concerned. Of 39 hemangiomas studied at the Mayo Clinic,[17] 69 percent were in females. Schmorl noted a 57.7 percent autopsy incidence in women.

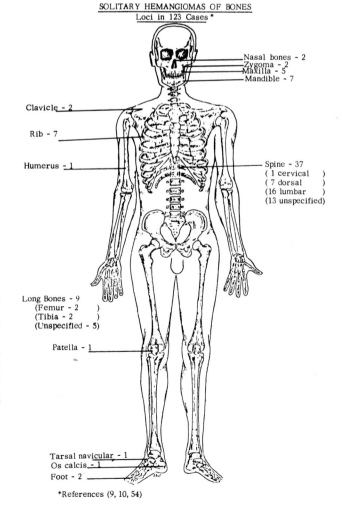

FIG. 2. Solitary hemangioma of bone—loci in 123 cases.

Although most are clinically insignificant and are incidentally discovered on routine roentgenograms or at autopsy, they may be responsible for local pain, tenderness, muscle spasm, spinal column rigidity, or neurologic symptoms, particularly in the case of vertebral lesions.[15,17] Hypoesthesias, radiculitis, and complete transverse myelitis have been reported. A relationship has been noted between pregnancy and the development of clinical signs of spinal cord compression by vertebral hemangiomas. It has been suggested that this may be influenced by increased venous pressure and blood volume during pregnancy and by direct hormonal effects on the endothelium.[10] Intraosseous hemangiomas at other sites and particularly those associated with soft-tissue components may become manifest by soft-tissue masses, chylous or hemorrhagic effusions into body cavities, hemoptysis, anemia, and thrombocytopenia in addition to bone pain and pathologic fracture. Serum calcium and phosphorus are usually normal despite bone involvement.

Four clinical categories based on symptomatology as well as on physical and roentgen findings have been associated with vertebral hemangiomas:

1. *Incidental finding on roentgenograms.* Classic roentgen pattern. Normal vertebral body contour. No subjective or objective symptoms or signs of spinal cord or nerve injury. These lesions are clinically indolent and show little tendency toward further growth.
2. *No objective roentgen findings.* Subjective symptoms of local pain or tenderness; radicular pain. No objective signs of nerve or spinal cord compression.
3. *Classic roentgen pattern with no changes in vertebral body contour or myelographic abnormalities.* Subjective symptoms. Objective signs of spinal cord and/or nerve compression without paraplegia.
4. *Change in classic roentgen pattern.* Loss of normal vertebral body contour, i.e., fracture, compression, ballooning, or expansion, with or without accompanying myelographic or angiographic abnormalities. Subjective symptoms. Objective signs of nerve or spinal cord compression.

The ballooning or outward extension of the vertebral body as seen roentgenographically could be a fifth category, which may not be associated with symptoms. The latter have been related to thrombosis, inflammation, and subsequent swelling and edema.[17]

Symptoms produced by vertebral hemangiomas may mimic intraspinal tumors, except that they develop more rapidly than those associated with benign neoplasms of the cord. They more closely resemble those produced by metastases to the spinal column and may even progress more rapidly, particularly as regards compression fracture. With increased compression, the spinal canal may be narrowed, with a resultant block of the subarachnoid space. However, a localized hemangioma may involve only a part of the vertebra, so that solely the body, a pedicle, a lamina, or a spinous process may be affected. The spinal cord and nerve roots may

Table 2 Hemangiomas of the Spine—Clinical–Roentgenographic Correlation

Subjective symptoms radicular or local pain, tenderness	Objective signs—nerve or spinal cord compression, paraplegia	Roentgenographic findings	Roentgenogram
−	−	+	Classic hemangioma Incidental finding Normal vertebral contour
−	−	++	Change in classic roentgenogram Ballooning or convex vertebral body
+	−	−	No objective finding
+	+	+	Classic hemangioma No change in vertebral contour or on myelogram
+	+	+++	Loss of normal vertebral contour (fracture, compression, expansion) $\bar{c}$ or $\bar{s}$ angiographic and/or myelographic abnormalities

therefore escape injury. Conversely, nerve roots may be compromised by involvement of a single pedicle. The microscopic findings are those of a capillary or cavernous hemangioma. In Dorfman's series, the capillary pattern was most frequent.[10] Most commonly, the lesions are confined to the cancellous portion of the vertebral body.

Hemangiomas of the skull have been noted over a wide age range, from the newborn to over 70 years of age.[69] A superficial localized mass may occasionally be palpated. No alteration of hair growth is noted. Neurologic symptoms are unusual, since they are most commonly associated with external rather than internal expansion. Intracranial pressure is rarely elevated and CSF chemistries remain normal.

Roentgenographic Findings

Although any portion of the vertebra may be affected, involvement of the body is most often identified roentgenographically. It is usually recognized by its generalized osteopenia. However, the most distinctive roentgenographic feature of a vertebral hemangioma is its vertically oriented arrangement of coarse trabeculae separated by intervening areas of reduced density (Fig. 3A). The latter make the bony columns stand out in bold relief, creating a "corduroy cloth" or "paint brush" pattern, which is particularly prominent on the lateral projection. Narrow, more irregularly spaced horizontal bony bands may occasionally be seen traversing or connecting the more prominent vertically oriented columns creating a fine reticulated criss-cross effect. This cross-striation is most often evident on the anteroposterior projection (Fig. 4A). The pedicles and posterior elements may participate in this pattern. Occasionally, small radiolucent vacuoles may be noted. The normal vertebral body configuration and cortex is most commonly retained, but occasionally its contour may become straightened or convex, presumably pressed outward by pathologic tissue (Fig. 4). The cortex may, additionally, become ill defined, although its outer border is usually discernible.

All these classic roentgen findings may be eliminated by a compression fracture, which may be acute or gradual. When gradual, the upper and lower vertebral body endplates may assume a markedly concave appearance simulating osteomalacia. Differentiation is based on the fact that involvement, rather than being universal as in a metabolic process, is most commonly confined to a single vertebra and less often to several vertebrae. The pedicles and posterior elements may likewise, when involved, either retain their shape or become deformed. The intervertebral spaces are generally intact. Metastases, lymphoma, and myeloma may be difficult to differentiate when the posterior elements and particularly the pedicles are affected. However, involvement of pedicles is not an early manifestation of myeloma. Paravertebral soft-tissue masses, relatively large, uniform areas of osteolysis with no internal trabecular pattern, reactive new bone formation, and multiple sites of involvement are several roentgenographic features that may help to separate these conditions, since they are not usually associated with vertebral hemangiomas. Paget's disease most commonly exhibits a coarser, more heavily accentuated trabecular and cortical pattern, which results in a generalized appearance of increased bone density. The vertebral body contours are particularly dense due to marginal bony condensation and cortical hypertrophy. A resultant thick, sclerotic border is commonly seen to rim a relatively radiolucent medullary center prompting the classic comparison with a picture frame. The posterior elements may be similarly affected so that structures involved by Paget's disease may be distinguished by virtue of their larger size when compared to neighboring segments of the spinal column.

Solitary hemangiomas of the skull are most common in the parietal and frontal bones, where they usually appear as a single, round, or oval area of rarefaction.[69] The rarefied area usually has an internal pattern, which en face appears reticulated or honeycombed. Occasionally, a rounded lesion will have a "spoke wheel" pattern with trabeculae arranged as striae radiating from a common center (Fig. 5). The interstices between the bony trabeculae vary in size and may occasionally be large. On tangential views, the lesion may have a spiculated configuration, since the same striae may be seen radiating externally in a manner likened to a sunburst. Therefore, in profile, hemangiomas commonly extend outward beyond the normal skull confines. External bulging is accompanied by erosion of the inner aspect of the outer table with preservation of the integrity of the inner table, and therefore, signs and symptoms of an intracranial space occupying lesion rarely occur. There are no dilated vascular channels or diploic lakes in relation to the lesion.

Smaller lesions may be entirely lytic with no discernible internal pattern but with sharp, well-delineated margins. The borders of the rarefied area may be irregular, but they are never serpiginous. Peripheral bony condensation or reactive sclerosis is uncommon and serves to differentiate meningiomas from hemangiomas. The former commonly produce marked hyperostotic changes in related portions of the cranium. Striations when noted in meningiomas tend to be vertical and parallel with one another, rather than being radially disposed as in heman-

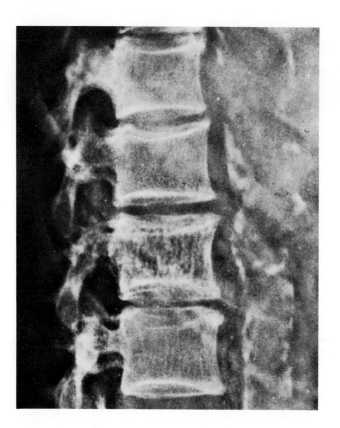

FIG. 3. Hemangioma of the lumbar spine—lateral view. The vertebral body appears striped by vertically oriented coarsened trabeculae separated by fairly regularly spaced areas of decreased density. A "corduroy cloth" pattern is simulated. Note that the vertebral body is not enlarged and that the normal concave contours have been maintained. The process appears to involve the pedicle.

giomas. Osteomyelitis produces rarefied areas that are diffusely mottled, ragged, irregular, and serpiginous in outline and not uncommonly multiple. The lesions of histiocytosis are commonly multiple; have sharply demarcated, commonly beveled edges; are centrally clear with no internal pattern and may be associated with a central island or spicule of bone mimicking a sequestrum. Certain anemias may be accompanied by the development of calvarial striations. However, these tend to be diffuse, with a generalized increase in diploic space size with no discrete visible or palpable lumps. Osteomas are

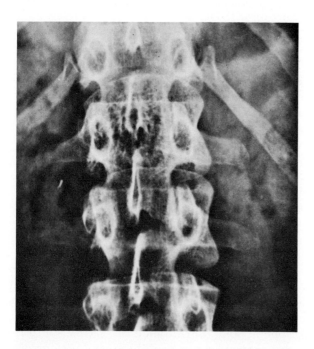 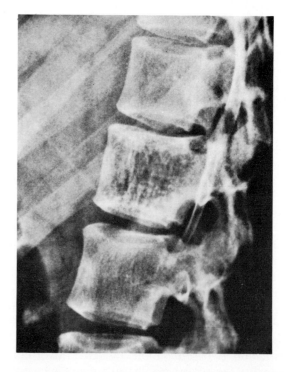

FIG. 4. Hemangioma of the lumbar spine. Left: Frontal view. The L1 vertebral body appears relatively osteopenic, but the trabeculae now seem to have a criss-cross or netlike arrangement. The normal height and contour appear to be preserved. Right: Lateral view. Same vertebral body as on left now has a vertically oriented trabecular pattern. There is a loss of the normal anterior concavity, with slight bulging of the anterior cortex as compared to the neighboring vertebrae. However, the overall normal size of the affected vertebral body is maintained.

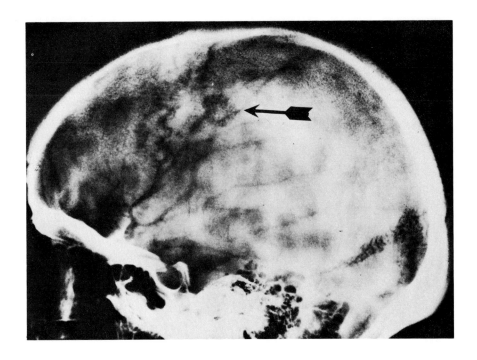

FIG. 5. Hemangioma of the skull—lateral view. The lesion appears as a rounded, relatively radiolucent area within the parietal bone. Note the fine internal trabeculae, which appear to originate from a common center—the so-called spoke wheel pattern. Although well defined, the lucent lesion has no associated peripheral bony condensation or sclerotic border.

commonly uniformly dense and sclerotic both in profile and en face.

Epidermoids, though radiolucent, are outlined by sclerotic borders, are commonly serpiginous in contour, and lack a reticulated or honeycombed internal pattern. Metastases and myeloma present as osteolytic lesions; however, they are commonly multiple, while metastases are generally poorly defined. Therefore, roentgen characteristics most commonly associated with primary calvarial hemangiomas include the following features:

1. Solitary, well-circumscribed, round to oval lytic lesions.
2. Usually located in the frontal or parietal region.
3. Have a reticulated, honeycombed or spoke wheel appearance en face.
4. Have a spiculated or sunburst appearance on profile with trabeculae oriented at right angles to the internal table.
5. Project beyond the normal contour of the skull in profile.
6. Expansion is more marked externally than internally and is associated with erosion of the inner aspect of the external table.
7. Preservation of inner table integrity.
8. Absence of reactive sclerosis or hyperostosis.
9. Absence of vascular changes in the neighboring bone, i.e., dilated vascular channels or diploic lakes in relation to the lesion.

Occasionally, frontal bone hemangiomas may encroach on the frontal sinus, producing deformity of its walls. The orbital roof may rarely be depressed, with resulting venous congestion, unilateral papilledema, and proptosis. The hemangiomas involving the skull bones usually show a capillary or cavernous histologic pattern with more abundant intervening connective tissue than in the vertebral hemangiomas. The most striking feature is the presence of abundant trabecular bone between the engorged vascular elements.

In addition to an overall coarsening, the trabeculae or spongiosa of other involved bones may exhibit a latticelike or weblike arrangement. The internal structure of the web is well defined, with a sharp line of demarcation between individual cells as well as with neighboring normal bone (Fig. 6). The cortex may be thin or bulge externally, but it is most commonly intact with no evidence of a periosteal reaction unless accompanied by fracture. A soft-tissue component is a rarely appreciated accompaniment of solitary hemangioma of bone. Larger irregular areas of destruction are occasionally noted. A hemangioma of a rib or clavicle commonly appears as a lytic expansile lesion. Profound reactive new bone production is unusual but may be noted without accompanying fracture (Fig. 7). The new bone is externally smooth, may have an ivorylike quality, and may involve a long segment of an affected tubular bone. The honeycomb and sunburst patterns that are fairly regularly encountered in the flat bones of the skull may also be associated with pelvic lesions and less commonly with centrally situated long-bone lesions. In some instances, the sunburst pattern may suggest a

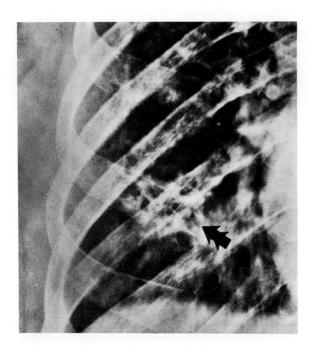

FIG. 6. Hemangioma of the rib. An asymptomatic lytic slightly expansile lesion was noted on the routine chest film of a 40-year-old female. The lesion is anteriorly situated within the right fifth rib. It has a latticelike or meshlike structure. Trabeculae traverse the area sharply demarcating the individual lucent "chambers" or "cells." There is no periosteal reaction or soft-tissue component.

malignant tumor, particularly when seen in narrow tubular bones such as the ribs and clavicle. Hemangiomas of the small bones of the hands and feet are represented by lytic lesions whose rarefied areas may have a completely vacuolated (Fig. 8) or more solid soap bubble appearance, or may have fine delicate septa noted on a tangential view. They may be ballooning or expansile lesions.

Solitary hemangiomas or multiple focal hemangiomas of long bones are uncommonly encountered. They are usually centrally situated, often in close proximity to the nutrient artery, and may produce fusiform enlargement or expansion of the shaft. The epiphysis is usually spared. The trabeculae in the area of the lesion are coarser and denser than those in the neighboring normal cancellous bone. Occasionally, coarse trabeculae or osseous plates appear to arise and spread radially from one center, again producing a sunburst appearance. The cortex remains intact, though it may become extremely thin and/or hypertrophied.

Occasionally, lytic lesions may undergo spontaneous sclerotic reaction. Unlike the pathognomonic roentgenographic appearance associated with solitary hemangiomas of the vertebrae or skull, those in the long bones and flat bones such as the pelvis or scapula can have unusual roentgenographic features that require biopsy for accurate diagnosis. The histology for the most part is that of a cavernous type of hemangioma. They are rarely associated with clinical symptoms.

Multiple hemangiomas of bone focally distributed, i.e., involving a single area or two or more bones of an extremity or an extremity girdle, occur less frequently (Fig. 9A). The roentgenographic findings are similar to those of solitary hemangiomas. They are commonly radiolucent lesions that deform the contour or diameter of the involved bone. Expansion to the point of extreme cortical thinning or dissolution may occur. Cortical thickening, an undulating periosteal reaction, as well as soft-tissue swelling may be associated. Solitary or focally distributed osseous hemangiomas (Fig. 9B, C) may be associated with either superficial or deep neighboring soft-tissue hemangiomas (Fig. 9D). The latter are also commonly composed of cavernous vascular channels, which may reach large proportions without involving bone. They may be identified by means of single or multiple rounded discrete calcifications, which commonly have radiolucent centers and represent phleboliths. Phleboliths are not found in primary hemangiomas of bone. A bone in the vicinity of a soft-tissue hemangioma either may be independently affected or may be secondarily involved in a variety of ways, since soft-tissue hemangiomas may exert secondary effects by means of traction, mechanical pressure, or infiltration with resulting local deformities. Localized cortical thickening or thinning or smooth, saucerlike indentations or erosions of the external cortical margin may result (Fig. 9).

Pathologic fractures in the structurally weakened bones are a common presenting and recurrent clinical feature. Abnormalities in the overall size and shape of the extremity may occur. An overgrowth of the soft tissues of the extremity or affected part may occur with either an associated increase or decrease in the size of the neighboring skeletal parts (Fig. 9A, B). Complete destruction or disappearance of bone may also occur.[18]

Skeletal enlargement or hypertrophy has been associated with stimulation of the neighboring growth plate, which in turn has been related to the increased local blood supply or hemostasis. The same localized hyperemia has also been held responsible for an overall decrease in size due to early growth plate closure (Fig. 9). Other growth aberrations, including

FIG. 7. Hemangioma of the rib. This middle-aged female complained of right upper quadrant discomfort. A gall bladder series was requested and radiographs incidentally revealed an expanded ivory dense lesion of the anterior segment of the right seventh rib with an inhomogeneous reticulated relatively radiolucent central area. The upper and lower cortices were thickened but externally smooth. The boundary of the lesion with the proximal normal bone is fairly well defined.

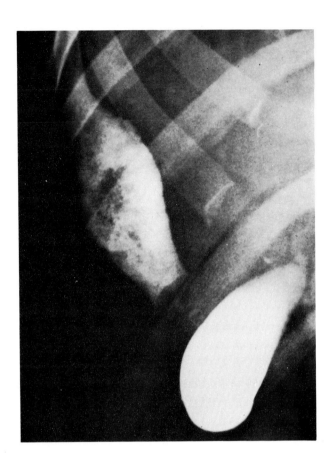

bowing and/or modeling deformities, have been attributed to asymmetric as well as early growth plate closure. Local hyperemia has also been held responsible for cortical hypertrophy and localized periosteal new bone formation, as well as increased bone resorption radiographically reflected as severe osteopenia.

Treatment

Solitary skeletal angiomas of the long bones when involved with a pathologic fracture are treated with corrective orthopedic procedures and/or with surgical extirpation when indicated in accessible sites. Local radiation has been most frequently used in association with symptomatic large or inaccessible hemangiomas with varied success. Recently, hemangiomas of bone as well as those in the soft tissues have been treated by means of selective intravascular embolization techniques.[13]

Prognosis

The overall prognosis of solitary and/or localized multifocal osseous hemangiomas is excellent. Most are stationary, although some show indolent growth and may progress slowly in size. There are no reports of malignant degeneration. Most of these lesions eventually stabilize with or without treatment. With radiation therapy, hemangiomas will occasionally undergo at least partial sclerosis. However, in some instances, sclerosis and/or regression has occurred spontaneously.[23,58,65]

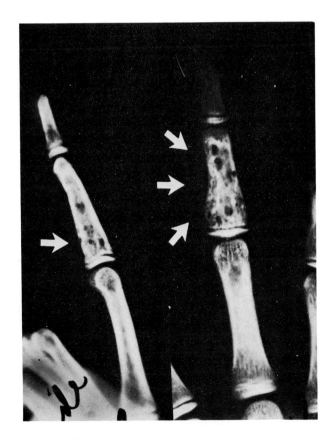

FIG. 8. Hemangioma of the finger. Numerous rounded osteolytic lesions create a "Swiss cheese" or vacuolated appearance in the middle phalanx of the middle finger of an 11-year-old female with a prominent localized radial soft-tissue mass. Left: Lateral view. Note small cortical lesions (arrow). Right: Posteroanterior view. The soft tissues are locally but asymmetrically prominent, with a focal radial bulge (arrows).

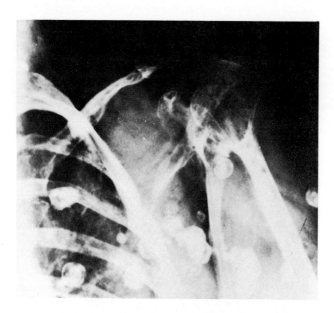
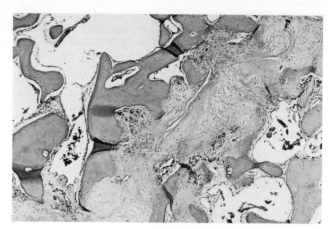

FIG. 9. Multiple (focal) hemangiomatoses of bones and soft tissues. This 42-year-old female was admitted with a flail left upper extremity, which she could not abduct or extend. The overlying skin and soft tissues had a reddish blue discoloration. Top left: Frontal view of the left shoulder girdle. The bones are abnormal in size and shape. Numerous, small, radiolucent defects are seen within the proximal humeral shaft and clavicle, which has an undulating contour. The cavernous hemangioma of the neighboring soft tissues may be diagnosed on the basis of multiple large, calcified phleboliths, several of which have radiolucent centers. Bottom left: Anteroposterior view of the left hand. The cavernous hemangioma involved the entire left upper extremity. Note the lumpy appearance of the soft tissues about the digits. Large phleboliths are noted proximally. An arterogram showed no opacification of the abnormal vascular channels. There was no evidence of arteriovenous communications. Note the "undergrowth" of several of the metacarpal bones, the extreme overtubulation, the associated fracture (arrow), and the numerous intraosseous radiolucencies Top right: Hemangiolymphangiomatous upper extremity. Note extension of the lesion in the marrow spaces between bony trabeculae. The cartilage is from an interphalangeal joint.(H&E,X100.) Courtesy Dr. Raffaele Lattes. Bottom right: Surgical specimen. Injection of preserved atretic radial artery revealed no abnormal arteriovenous communications. The abnormal angiomatous network failed to opacify.

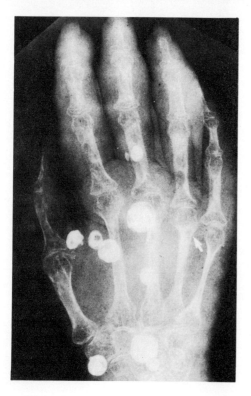
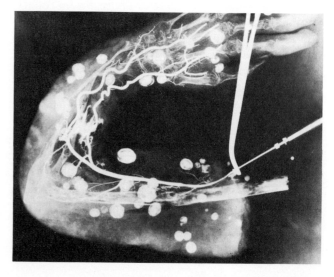

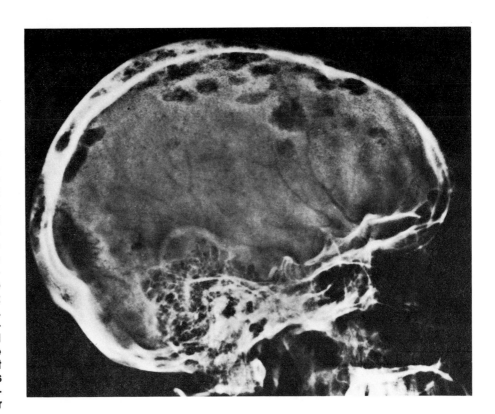

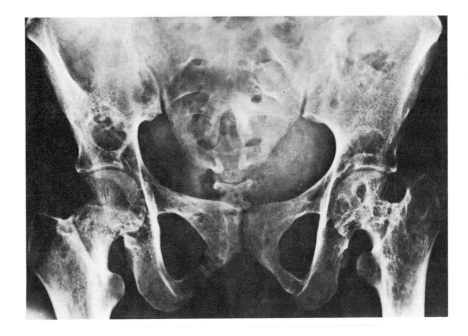

FIG. 10. Multiple diffuse skeletal hemangiomatosis. The patient was a 23-year-old female who was admitted with acute abdominal pain. She had a hemoperitoneum related to an associated hemangiomatosis of the small and large bowel. A skeletal survey revealed numerous skeletal hemangiomas. Top: Lateral skull. Multiple rounded, well-circumscribed radiolucencies represent intraosseous hemangiomas in the frontal, parietal, and occipital bones. Note irregularly widened diploic space and the cortical thinning of the outer table of the occipital bone. Bottom: Frontal view of the pelvis. Sharply demarcated osteolytic defects are scattered throughout the pelvis. Note the lytic defect in the left sacrum, and the numerous defects within the left ischium, which has an irregular cortical contour. The left femoral head, neck, and acetabulum are deformed and smaller than their counterparts on the right. Several of the lytic lesions are rimmed by sclerotic borders.

Multiple Angiomatosis

Solitary skeletal hemangiomas particularly of the spine and skull are not uncommon, while multiple (focal) skeletal hemangiomas and diffuse soft-tissue (Fig. 9) and visceral hemangiomas, though less common, are not rare. However, in a survey by Wallis et al,[67] only 24 cases of diffuse hemangiomatosis involving the skeleton alone were reported in the world literature to which they added two cases of their own. Diffuse skeletal hemangiomatosis may be distinguished by a wider distribution of its lesions (Fig.10), as opposed to multiple focal hemangiomatosis of bone in which the lesions are commonly localized to two or more bones of an extremity or an extremity girdle.

Clinical Data

More than 95 percent of angiomas of the skin and subcutaneous tissue are visible at birth or appear within the first 6 months of life. Those in deeper tissues or viscera may not present until childhood or even adult life. They may also exhibit the phenomenon of delayed appearance. Therefore, cited age distributions relate to the site of the lesion. Generalized diffuse skeletal involvement is most commonly recognized in children and young adults. The age of recognition has ranged from 1 year to 72 years. A greater incidence is noted in males in some reviews. There is no evidence to suggest a familial occurrence. Approximately 50 to 78 percent of patients with multiple skeletal hemangiomas have some associated skin, soft-tissue, or visceral angioma if diligently sought.[58] This figure is probably low, since the absence of signs, symptoms, or obvious clinical stigmata do not exclude the existence of subclinical extraskeletal lesions. Their discovery is of more than academic interest, since lesions limited to the skeleton are characterized by a benign course with slow progression over many years and eventual stabilization. No patient with diffuse hemangiomatosis limited to the skeleton has died of the disease. This excellent outlook is in contrast to that which prevails when diffuse skeletal involvement is associated with extensive or with particular types of visceral involvement. Wallis et al[67] noted that 8 of 12 cases with associated visceral involvement were dead 1 to 12 years after onset. The 4 patients who survived had limited visceral involvement. Therefore, the presence, site, and extent of visceral involvement adversely affects the clinical course. Conversely, the viscera, skin, and somatic tissues, as well as the skeleton, may be solely affected (Table 1). Skeletal lesions may be extensive, almost always involving the spine, skull, ribs, and pelvis as well as the long bones (Fig. 10). The spleen and liver are the viscera most often involved (Fig. 11). However, the lungs, kidneys, mediastinum, thymus, and muscles as well as subcutaneous tissues and lymph nodes may be affected. The spleen, lungs, liver, and pleura are visceral sites that have been most frequently detected. Any part of the skeleton, including the small bones of the hands and feet, may be affected but may not necessarily be visualized roentgenographically. Symptoms of respiratory involvement are of serious significance. Whenever intrathoracic structures such as pleura or lungs were grossly involved, the clinical course usually terminated in death through recurrent effusions, pneumonias, and empyemas. Hepatosplenomegaly, ascites, weight loss, cachexia, and mesenteric venous thrombosis have all been noted.

Cervical or mediastinal angiomas, some of which have been noted to be mesenteric cystic lesions filled with chylous fluid, i.e., lymphangiomas, may be associated with chylothorax, hemothorax, hemoperitoneum, or chylous ascites. Chylous or sanguinous pericardial effusions due to angiomatous mediastinal involvement have been described.[11,19,42] They may represent one extreme of the spectrum of cystic angiomatosis.[19]

Disturbances in fat and protein metabolism with hyperglobulinemia have been reported.[36] The association of severe thrombocytopenia with hemangiomas is a rare but well-known finding.[38,57,60] These patients may have a normal platelet count initially. Hypochromic anemia may also be associated, but particularly in patients with visceral lesions. Two-thirds of patients with histologically benign lesions may have grave prognoses despite the presence of a basically benign condition.

Pathogenesis

The question of congenital versus postnatal development as well as hamartoma versus true neoplasm has been argued in connection with diffuse as well as with solitary angiomas. Some have thought that visceral foci represent multiple hematogenous or lymphatic emboli disseminated by a tumor that originated in bone marrow or bone and have called the entity *benign metastasizing hemangiomatosis*. Others contend that they represent multiple independent foci comparable to and contemporary with the bony lesions.[34,68]

Factors given in favor of their being neoplastic include the fact that they may increase in size and number, that they may appear histologically benign and yet behave in a malignant fashion, and that lesions are diffusely scattered throughout the skeleton and other tissues (Fig. 12). However, the spleen, which is the most common extraskeletal organ involved with cystic angiomatosis (Fig. 11 Top), is an infrequent site, along with skeletal muscle, of metastatic involvement.

Pathology

Gross Pathology. Biopsies are generally unrewarding. The lesions are no different from those seen in localized hemangiomas of bone. They are most commonly of the cavernous type. At times, they appear to be lymphangiomatous in origin.

FIG. 11. Diffuse skeletal and visceral hemangiomatosis. This middle-aged man had a 26-year history of anemia, a palpable left upper quadrant mass, and roentgenographically documented lytic and blastic lesions in the spine and pelvis. He had refused surgical exploration. He had intermittent left upper quadrant pain and melana in the 2 years prior to admission. Esophagoscopy and gastroscopy had disclosed numerous reddish dilated gastric vessels. He developed acute abdominal symptoms on admission. Top: Erect view of the abdomen. The stomach and small bowel are markedly displaced to the right by a massively enlarged spleen. Note multiple air fluid levels in the distended small bowel. Free air outlines the medial aspect of the spleen. Note the blastic lesions in the right sixth rib (arrow) and within multiple vertebrae. A previous rib and iliac crest biopsy had been diagnosed as showing "multifocal sinusoidal telangiectasis." Exploratory laparotomy revealed a 1500-ml abscess cavity in the left abdomen and a perforated gastric fundus. Biopsies of the spleen, liver, and left twelfth rib revealed visceral and skeletal angiomatoses. Bottom: Frontal views of the right hip. Several areas of ill-defined osteolysis are present. However, the visualized skeletal lesions are now predominantly osteoblastic. Note large ischial lesions as well as smaller blastic lesions within the right femoral neck. Several surround and appear to partially reconstitute a previously lytic area (arrow). The patient expired with a long-delayed diagnosis of diffuse skeletal hemangiomatosis. This case was the basis of a report by Waldron and Zecker.[65]

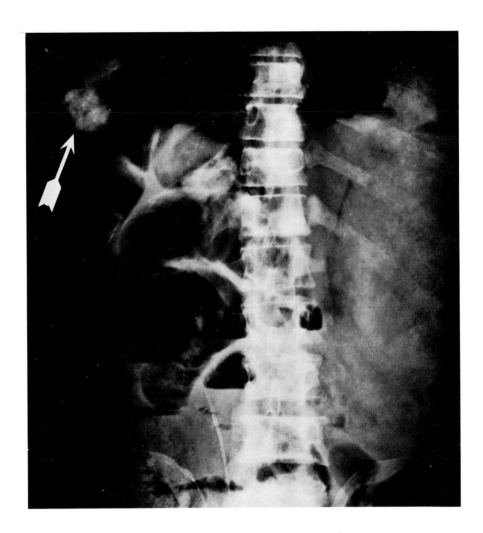

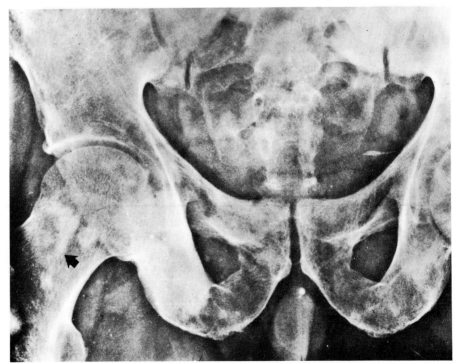

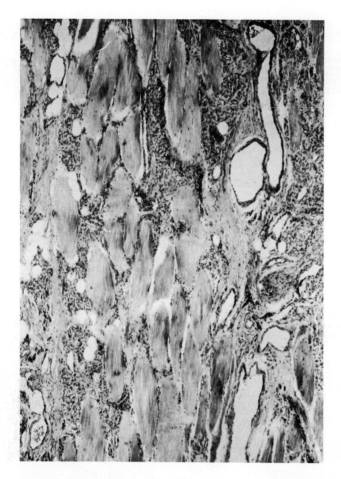 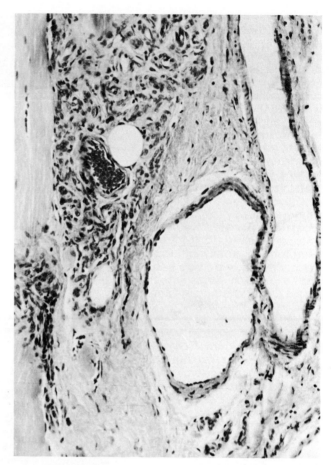

FIG. 12. Diffuse benign angiomatosis. Left: Note the extensive involvement of the skeletal muscle of an extremity. The psuedo-infiltrating pattern of this benign vascular lesion can lead to the erroneous diagnosis of angiosarcoma. (H&E, X84.) Right: Same case at a higher magnification. Note the two larger vessels of arterial type, suggesting a vascular malformation associated with proliferation of smaller blood vessels, with a cytologically benign, orderly arranged endothelial lining. (H&E, X270.) Courtesy Dr. Raffaele Lattes.

Laboratory Findings. There are no characteristic laboratory findings and no consistent abnormalities in blood calcium, phosphorus, or alkaline phosphatase despite widespread skeletal and/or visceral lesions.

Roentgenographic Features The majority of lesions are lytic, may be expansile, and vary in size from a few millimeters to a few centimeters (Fig. 10). The areas of osteolysis are well defined, occasionally with a thin sclerotic border (Fig. 10B). Less commonly, lesions may be entirely sclerotic or blastic (Fig. 11A, B). Lytic lesions were found in all cases reviewed by Wallis et al,[67] while sclerosis was noted in half of them. The propensity of vascular tumors to cause bone destruction and/or bone production is not understood.

Although in most cases the lesions are seen to be static or indolent, they may be seen to be slowly progressive on serial roentgenograms. A minority rapidly become larger. They eventually stabilize, however, while others may be seen to regress. Some may be sclerotic de novo or become so spontaneously or following surgical procedures such as curettage. Two cases of angiomatosis involving the small bones of the hands and feet were followed for 32 years and for 12 years by Spjut and Lindbom.[58] In one case, the lesions were slowly progressive, while in the other they became either gradually or rapidly larger. Several regressed and others stabilized.

Roentgenographically, lesions are most commonly seen in the medullary cavity. When progressive, they may thin or erode the endosteum with gradual resultant expansion of the cortical contour (Fig. 10). Expansion is most evident in the small tubular bones, where the lucent lesions often have a lacy or soap bubble internal trabecular pattern. They may have an irregular ill-defined border. Layered or laminated periosteal reactions are infrequent without antecedent trauma.

The diagnosis of angiomatosis is suggested by the generalized distribution of the lesions and their indolent behavior. Its roentgen appearance differs from that classically associated with solitary skeletal angiomas, particularly in the vertebral column. Vertebrae involved in association with diffuse angiomatosis usually do not have the characteristically vertically oriented striated appearance associated with solitary angiomas.

Vertebral lesions are usually round or oval areas of osteolysis. Occasionally a "vertebra plana-like" appearance is seen, with the vertebral body height reduced to wafer thinness.

The presence of multiple osteolytic and occasional blastic lesions may suggest metastatic disease, myeloma, or lymphoma in the older age group, while lymphoma, leukemia, and metastatic seminoma and melanoma are considerations in the young. Predominantly sclerotic lesions, in addition to metastases, may simulate bone islands, myelosclerosis, and mastocytosis particularly in view of frequently associated hepatosplenomegaly (Fig. 11A). However, the failure to detect a primary lesion, the limited number of metastatic deposits that expand bone and respect cortical integrity, the degree of indolence, and the sharp definition of most angiomatous lesions argue against a roentgenographic diagnosis of malignancy or metastasis. Neuroblastoma and leukemia in children are most commonly associated with diffusely distributed, permeative, ill-defined areas of medullary and cortical destruction, as well as with aggressive periosteal new bone formation. Dystrophic changes in the form of metaphyseal radiolucent bands are another feature.

Occasionally, biopsy of osteolytic lesions has yielded so-called foam cells[48] with no other specific features. This finding in association with hepatosplenomegaly has led to the diagnosis of histiocytosis X. However, skeletal lesions of histiocytosis X may be distinguished by their frequently associated periosteal and endosteal reactions particularly when cortical involvement is noted, their bony sequestralike spicules within the area of lysis, and a predilection for the diaphysis of long bones and for the skull. Pulmonary parenchymal changes and skin lesions supply other unique features.

Polyostotic fibrous dysplasia, enchondromatosis, congenital generalized fibromatosis, and hyperparathyroidism may be associated with osteolytic lesions that have distinctive features. Those of fibrous dysplasia may be hazy, often have mineral within them, may be predominantly unilateral and segmental in distribution, and are often associated with distinctive modeling deformities. Distinctive skull lesions and deformities are commonly present. Enchondromas commonly contain stippled calcifications within them indicative of cartilage, and often have an hour glass configuration and other associated modeling deformities. Generalized fibromatoses may be distinguished by the younger age of the patient at discovery and by associated skin and soft-tissue lesions that do not contain phleboliths as often seen within soft-tissue cavernous angiomas. The osteolytic lesions of hyperparathyroidism are often accompanied by pathognomonic subperiosteal bone resorption at multiple sites as well as by characteristic clinical and laboratory features.

Roentgenographically detectable soft-tissue involvement includes hepatosplenomegaly, ascites, pleural or pericardial effusion, and mediastinal enlargement indicative of an angioma, lymphangioma, and/or hygroma.[47] Skeletal surveys in such patients (those which primarily present with cystic hygromas, visceral hemangiomas, or chylous or sanguinous effusions) would probably yield additional cases of multiple skeletal angiomatosis.

Prognosis and Treatment

Although vascular lesions may be malignant de novo, i.e., of the 17 cases with visceral involvement tabluated by Wallis et al, 13 were benign and 4 were reported to be malignant;[67] malignant transformation has not been reported in skeletal angiomatosis. Some of the bony lesions remain stationary, others may be slowly progressive, and others appear to undergo sclerosis. No patient with diffuse hemangiomatosis limited to the skeleton appears to have died of the disease, and therefore, despite its widespread osteolytic and blastic roentgenographic appearance, it is basically a benign condition. However, its association with visceral involvement considerably alters the prognosis. Despite the underlying histologic benignity, two-thirds of patients with combined skeletal and visceral involvement die of the disease.[67]

Therapy

Radiotherapy, alkylating agents, sclerosing solutions, and selective vascular embolization have been utilized to treat soft-tissue tumors that compress vital structures, for cosmetic reasons, and when they are responsible for hemorrhagic episodes. Skeletal lesions have been managed by curettage, bone grafting, and radiotherapy as indicated.

Skeletal angiomas may, therefore, be solitary, multiple and focal, multiple and diffuse, or associated with systemic lesions. In this last instance, they may give rise to a number of separate syndromes depending on the anatomic location of the lesions. A variety of other hemangiomatous disorders have also been associated with several other[49] and more familiar syndromes, which include hereditary, multiple, hemorrhagic telangiectasis (Rendu–Osler–Weber disease); encephalofacial hemangiomatosis (Sturge–Weber disease); coexistent hemangiomas of the retina and central nervous system plus frequently associated cystic lesions of lung, pancreas and kidney (Von Hipple–Lindau disease); and coexistant hemangiomas of the skin and enchondromatosis of bone (Maffucci's syndrome).[50]

Skeletal Lymphangiomatosis

Primary lymphangiomas of bone are exceedingly rare. The histologic features may be similar or identical to lymphangiomas of skin or soft tissue as well as to skeletal hemangiomas, and in some cases both entities coexist. Some reported cases may actually represent mixed lesions or hemangiomatosis.[2,1,12,27,28,39,45,46,51,55,59,62]

Soft-tissue lymphangiomas are more commonly encountered than the skeletal lesions and presumably originate directly from mesodermal rests or from lymphatic tissue normally present in an area (Fig. 13). Although lymphatics are normally present in the periosteum, lymphangiomas of bone originating in this location would be expected to erode or destroy the periosteal surface in order to secondarily involve the medullary cavity. However, this is not borne out roentgenographically, since many of the osteolytic foci appear internally within the medullary cavity. The presence of intraosseous lymphatics have been demonstrated with lymphangiomas but predominantly in abnormal situations such as lymphedema. It is therefore generally believed that intraosseous lesions result from mechanical pressure atrophy, from neighboring ectatic endothelial lined channels that are variously held to originate from lymphatic tissue (lymphangiomatosis), blood vascular tissue (hemangiomatosis), or from mesodermal rests within bone.

The pathogenesis of lymphangiomatosis of bone disease is, therefore, as poorly understood as that of hemangiomatosis. Inflammation, lymphatic obstruction, and neoplasia have been suggested, but congenital malformation is felt to be the most likely cause.

Small biopsy specimens fail to demonstrate any definite abnormality, and microscopic differentiation of the cystic lesions similarly lined by endothelium is often difficult if not impossible to differentiate from hemangiomatosis. The course and prognosis vary. Reported lymphangiomas in bones, as well as in the soft tissues, have also produced a number of syndromes, depending on their location.[16] Chylous pericardial, pleural, and peritoneal effusions have been associated with visceral and mediastinal lymphangiomas or angiomas. Massive pericardial and pleural effusions may reaccumulate rapidly in relatively asymptomatic patients, and an increased mortality has been noted despite a misleading paucity of clinical and physical findings. Patients with bone involvement alone are generally asymptomatic, but pathologic fractures are a frequent complication.

Bone lesions are most commonly radiolucent and often have a fine, thin, sclerotic margin. Intramedullary lesions may be large and ill defined, with cortical thinning and expansion of the normal bony contour. Widespread abnormalities of both the cortex and spongiosa may exist and, apparently, entirely intracortical radiolucencies may be noted. All bones, including the skull, flat bones, and small bones of the extremities, may be involved.

Bone lesions, as in hemangiomatosis, may be slowly progressive, and additional bone lesions may develop

FIG. 13. Lymphangiomatosis. A 46-year-old male complained of increasing swelling of both lower extremities since birth. The swelling was first confined to the ankles and feet, but during childhood and adolescence it spread proximally. Lymphangiomatosis was diagnosed on the basis of biopsy at age 11. The patient suffered from pruritus and developed infections easily. A paternal grandfather was noted to have a similar condition. Top: Frontal views of right and left legs. Note bilateral marked prominence of the soft tissues. Small curvilinear, calcific deposits are seen medial to the distal left tibia (arrows). The right fibula has a wavy external contour. The bones did not otherwise appear to be involved. Middle and Bottom: Lateral views of right and left feet. Physical examination revealed scarring and discoloration of the skin, as well as vegetative lesions, which are obvious on the radiographs (white arrows). The soft tissues appear edematous and are peppered with linear and punctate calcifications (black arrows), which do not have the appearance of phleboliths.

Angiomatosis and the Skeleton

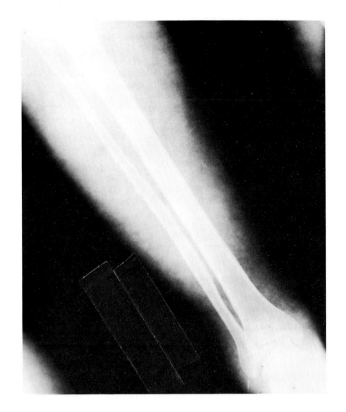
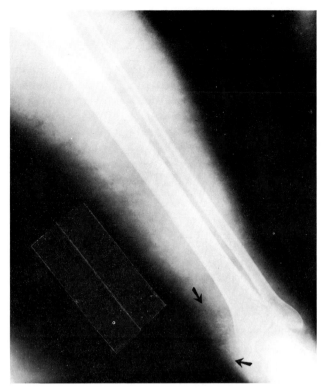
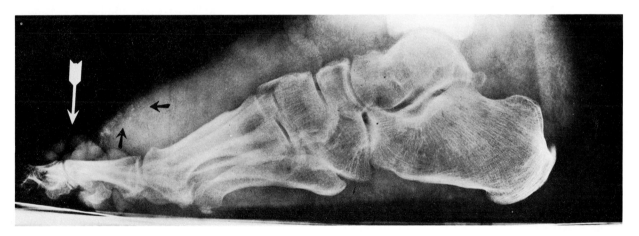
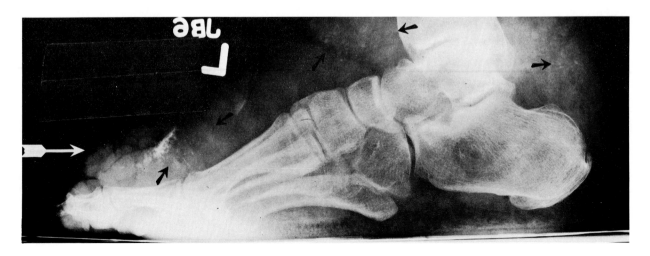

during observation. Patients with bone involvement alone are generally asymptomatic, but pathologic fractures are a frequent complication.

The definition of lesions by means of contrast material has been infrequently attempted or reported. Bilateral lymphangiography in Nixon's case,[46] which involved bone, demonstrated contrast material within a femoral cortical lesion on a delayed 48-hour film. Widespread abnormalities in soft-tissue lymphatics included dilatation, delayed drainage, and collateral drainage. In Najman's case,[45] contrast was injected directly into a cystic region in the skull that communicated with adjacent multilocular areas. Lymphangiography performed in four other cases showed no contrast medium within the bone.[39,45,46,59] Late filming may be of value, since abnormal lymphatics that are invariably present make delayed drainage a factor.

Lymphangiography done in a case with chylopericardium failed to show opacification until a 24-hour upright film revealed puddling of the opaque medium within the pericardium.[8] A superior mediastinal lymphangioma was removed at thoracotomy. Chylothorax or chylopericardium, with or without apparent lymphatic bone disease, may be indicative of a widespread abnormality of the lymphatic system that should be systematically searched for. It is conceivable that skeletal angiomas may play an additional role in the etiology of heart failure in particular patients.

Disappearing Bone Disease (Massive Osteolysis)

Massive osteolysis is a rare phenomenon that may result in extensive disappearance or complete destruction of an entire bone or several bones within a particular area. It has been associated with hemangiolymphangiomatosis of bone in the literature. However, the extent of bone destruction in massive osteolysis serves as a point of difference. The phenomenon of marked destruction is also distinct from that occurring in association with disuse atrophy or with atrophy resulting from trauma or infection. These processes, though leading to the disappearance of a sizable segment of an affected bone, allow a considerable portion to remain. Massive osteolysis differs from that caused by other familiar causes of bone destruction, including tumors, certain central nervous system diseases, and hyperparathyroidism.

Since the first case of spontaneous bone absorption recorded by Jackson in 1838,[30,31] approximately 60 cases have been reported under a variety of names, including *phantom bone, disappearing bone, spontaneous resorption of bone, spontaneous absorption of bone, progressive atrophy of bone, progressive lysis of bone,* and *Gorham's disease.* Gorham and Stout[20] first noted the histologic similarities of several scattered cases variously described in the literature as showing "connective tissue and capillaries, fibrous tissue rich in blood vessels, marrow replaced by fibrous vascular tissue, abundant vascular channels, fibrocystic disease." They suggested that progressive osteolysis was associated with angiomatosis of blood vessels and sometimes of lymphatics. They, as well as subsequent observers, were unable to state whether osteolysis occurred as a direct result of active hyperemia, mechanical causes, changes in pH or other eventualities.

Some contend that no real increase in the number of vessels exists but that, rather, only a distention or expansion of those already present has taken place.[6] In other words, the hypervascularity may not be the primary change and the basic abnormality lies in the bone absorption itself. Since the marrow has such a tremendous capillary network, the capillaries tend to distend by default. It has in fact been noted that vascular changes are not as marked in cases actually undergoing resorption and that histologic evidence of an almost angiomatous appearance has been most frequently found in late cases after the bone has been completely resorbed. Therefore, the marrow hypervascularity does not exactly duplicate that noted in an angiomatous malformation nor does it have neoplastic characteristics.

Gorman and Stout[20] noted no osteoclasts in tissues they examined. However, according to Bullough,[5] osteoclasts may be seen in one photomicrograph of the first case reported by Gorham et al.[21] Osteoclasts were also noted by several other authors, but resorption has not been uniformly associated with increased osteoclastic activity.[6,43]

In general, alterations in bone vascularity have resulted in either bone deposition or resorption, with arterial hyperoxygenation promoting bone resorption and venous stasis promoting bone formation. A lower level of oxygen favors bone formation.

To date, though most authors agree on the presence of an increased vascularity in the affected bone or area, the point of contention as to whether this represents a dilatation of the existing vascular bed or an actual vascular proliferation has not been settled. The pathogenesis of massive osteolysis is therefore still unknown.

Clinical Data

The condition is most common in children and adolescents with the age of onset ranging from 1½ to 58 years. The majority of patients are under 30. No sex-related or sex-associated systemic disease has been noted. Although generally not held to be hereditary, a study of several families indicated that this may be a syndrome with autosomal dominant inheritance.[5]

Destruction is most commonly confined to a single bone or two or more bones centered around a joint. A history of trauma, usually of mild degree, is often elicited. The onset, however, is usually insidious, with complaints ranging from a dull aching pain to progressive weakness. The first symptom may be related to pathologic fracture.

The course is usually protracted. Often, after a number of years, the disease stabilizes itself. Progression usually, but unpredictably, ceases, with residual degrees of deformity.[14,24,37] Severe crippling deformities, if they occur, do so after a number of years. There is no evidence of reossification. Fatal termination may occur particularly in those with thoracic involvement. Severe deformities of the thorax associated with chylous or serosanguinous pleural effusions that recur rapidly after thoracentesis and secondary infection are common features.

Pathology

Gross Pathology. The resorbed segment of bone may be replaced by such dense fibrous tissue that not only is normal function maintained, but normal weight may be borne. Autopsy and amputation specimens have described the bones as thin and tapered and the bone tissue as soft, weak, and spongy in texture. The soft-tissue component, when present, may be larger and more vascular than the intraosseous component. A fibrous band replaced the original bone in most cases. In others, no such excess of fibrous tissue was noted in the position where the bone might have been. It is not known whether the lesion begins within or outside the bone in such instances.

The maximum increase in dilated vascular spaces may be seen where only traces of bone remain. However, fibrous tissue may entirely replace the marrow spaces.

Roentgenographic Features. A subcortical or medullary radiolucency may be initially noted that subsequently develops into a larger area of osteolysis, which may eventually involve the entire bone, as well as multiple contiguous bones (Fig. 14). The process commonly involves apposing bones about a joint space. A characteristic concentric shrinking of a tubular bone is roentgenographically demonstrated as a uniform decrease in the caliber of the bone and a tapering of its ends, giving it a "sucked candy cane" appearance. This is a feature often associated with neuropathic osteoarthropathy. The cone-shaped spicules of bone at the edges of the lesion may be related to removal of the medullary support by intraosseous involvement or to an extraosseous process that brings external pressure to bear on the bone.[5] If the process continues, almost complete resorption of the involved skeleton may occur. There is no associated coarsened trabecular pattern, no reactive bone formation, and no neoplastic osteogenesis of any kind. No evidence of phleboliths or soft-tissue calcifications is noted.

Contrast studies, including arteriography, venography, and lymphangiography, have afforded little additional information. Hambach et al[25,26] attempted phlebography in a case involving the shoulder girdle by injecting contrast into a cubital vein but failed to visualize the vessels in the region of lysis. Contrast was then injected into the area formerly occupied by the scapula. Opacified vessels in the distal part of this region were interpreted as outlining the general appearance of a scapula and were felt to represent indirect evidence of an angiomatous lesion. Johnson and McClure[35] failed to produce opacification with aortography but achieved partial success with osseous angiography. Vascular opacification occurred in some areas of bone resorption. However, their patient had had radiotherapy and it was postulated that some channels had thereby become obliterated. Milner and Baker[43] showed no opacification with arteriography. In most instances where arteriography was done, hypervascularity was not demonstrated.

Treatment

There is no known effect of various treatments, including radiation therapy and surgery. Resection has not always been successful, since the remainder of the bone or other bones may become involved after the affected part has been removed. Results of treatment are additionally difficult to assess in view of the frequent spontaneous arrest of the disease.

The ordinary, focal, nonagressive hemangioma of bone such as is classically seen in the vertebrae and skull exhibit markedly different behavior and rarely, if ever, are associated with the syndrome of pro-

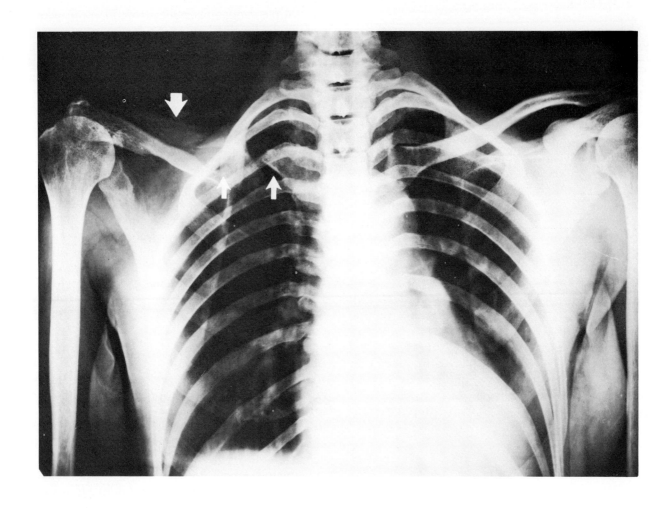
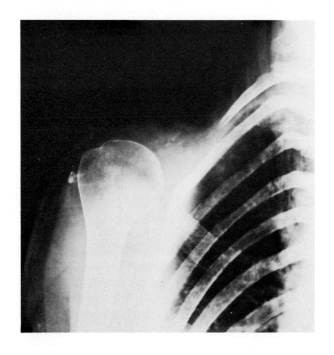
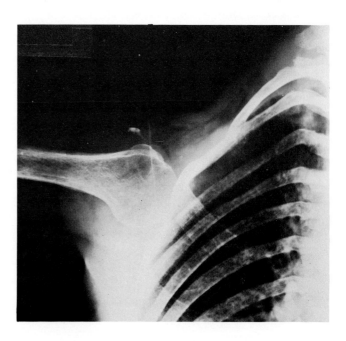

FIG. 14. "Disappearing bone disease" (massive osteolysis). This 42-year-old woman gave the following history. While lifting a child 5 years prior, she dropped it for "no reason." Two weeks later, while lifting a cake of ice, something "snapped" in the right shoulder. A fracture of the right clavicle was diagnosed on a roentgenogram taken elsewhere. The scapula was reputedly normal. She was treated with a "splint." She had occasional intermittent pain at the fracture site for 2½ years prior to admission, which was relieved by heat and rest. Radiographs done elsewhere 2 years prior to admission reputedly showed absorption of the inner half of the right clavicle. Top: Posteroanterior view of the chest. Note the large gap between the medial and lateral fragments of the right clavicle (arrows). The intervening bone is completely absent. Note tapered or "sucked candy" appearance of the medial end of the lateral fragment. A large osteolytic area is also seen within the body of the scapula (large white arrow). Bottom left: Frontal view of the right shoulder 1 year after Top. The right clavicle and scapula have been completely resorbed. Only a small fragment of bone is noted lateral to the proximal right humerus. The right first rib has been partially destroyed and the superior cortex of the right second rib appears depressed but smooth. Bottom right: Frontal view of the right shoulder with arm abducted (same time as Bottom left). Note that despite the "massive osteolysis" and severe deformity, function has not been interfered with. There was no associated pain, tenderness, or mass. Severe muscular atrophy was noted both clinically and roentgenographically. Blood chemistries were all normal.

gressive osteolysis. Malignant vascular tumors of bone may be diagnosed by biopsy, although some of these show multicentric involvement.

With the exception of the solitary hemangioma of the skull and vertebral column, primary intraosseous lesions of vascular origin are rare. A mountain of misleading terminology has added to the difficulty in comparing material from various sources. In addition, the great variation in clinical manifestations of the angiomatoses further explains the numerous names that have been given to them. Their proper recognition and evaluation as a possible part of a larger syndrome or syndromes is often based on the patient's initial presentation and the combined efforts of several physicians representing various disciplines at various times.

Acknowledgment

I am grateful to Dr. Raffaele Lattes, Chief of the Department of Surgical Pathology of the Columbia–Presbyterian Medical Center, for the pathologic sections and descriptions contained herein and for the time and energy he devoted in the review of this manuscript.

References

1. Ackerman AJ, Hart MS: Multiple primary hemangioma of the bones of the extremity. Am J Roentgenol 48:47, 1942
2. Bickel WH, Broders AC: Primary lymphangioma of the ilium: report of a case. J Bone Joint Surg 29:517, 1947
3. Brower AC, Culver JE Jr, Keats TE: Diffuse cystic angiomatosis of bone. Report of two cases. Am J Roentgenol 118:456, 1973
4. Bucy PC, Capp CS: Primary hemangioma of bone with special reference to roentgenologic diagnosis. Am J Roentgenol 23:1, 1930
5. Bullough PG: Massive osteolysis. NY State J Med October 1, 1971, p 2267
6. Case Records of Massachusetts General Hospital. Case 16. N Engl J Med 270:731, 1964
7. Cohen J, Craig JM: Multiple lymphangiectases of bone. J Bone Joint Surg 37A: 585, 1955
8. Collard M, Fievez M, Godart S, Toussaint JP: Contribution of lymphangiography in study of diffuse lymphangiomyomatosis: report of case with anatomic observations. Am J Roentgenol 102:466, 1968
9. Dahlin DC: Bone Tumors, 2nd ed. Springfield, Illinois, Thomas, 1967
10. Dorfman HD, Steiner GC, Jaffe HL: Vascular tumors of bone. Hum Pathol 2:349, 1971
11. Effler DD: Primary chylopericardium. N Engl J Med 250:20, 1954
12. Falkmer S, Tilling G: Primary lymphangioma of bone. Acta Orthop Scand 28:99, 1956
13. Feldman F, Casarella WJ, Dick HM, Hollander VA: Selective Intra-arterial embolization of bone tumors. Am J Roentgenol 123:130, 1975
14. Fornasier VL: Haemangiomatosis with massive osteolysis. J Bone Joint Surg 52B:444, 1970
15. Foster DB, Heublein GW: Hemangioma of vertebra associated with spinal cord compression. Am J Roentgenol 57:556, 1947
16. Frank J, Piper PG: Congenital pulmonary cystic lymphangiectasis. JAMA 171:1094, 1959
17. Ghormley RK, Adson AW: Hemangioma of vertebrae. J Bone Joint Surg 23:887, 1941
18. Goidanich IF, Campanacci M: Vascular hamartoma and infantile angioectatic osteohyperplasia of the extremities. J Bone Joint Surg 44A:815, 1962

19. Goldstein MR, Benchimol A, Cornell W, Long DR: Chylopericardium with multiple lymphangioma of bone. N Engl J Med 289:1034, 1969
20. Gorham LW, Stout AP: Massive osteolysis (acute spontaneous absorption of bone, phantom bone, disappearing bone). Its relation to hemangiomatosis. J Bone Joint Surg 37A:985, 1955
21. Gorham LW, Wright AW, Shultz HH, Maxon FC: Disappearing bones: a rare form of massive osteolysis. Reports of 2 cases, one with autopsy findings. Am J Med 17:674, 1954
22. Gramiak R, Ruiz G, Campetti FL: Cystic angiomatosis of bone. Radiology 69:347, 1957
23. Gutierrez RM, Spjut HJ: Skeletal angiomatosis. Report of 3 cases and review of the literature. Clin Orthop 85:82, 1972
24. Halliday DR, Dahlin DC, Pugh DG, Young HH: Massive osteolysis and angiomatosis. Radiology 82:637, 1964
25. Hambach R, Hendrich F: Ein Fall von generalisierter Haemangiomatose. Zentralbl Allg Pathol 100:236, 1959
26. Hambach R, Pujman J, Maly V: Massive osteolysis due to hemangiomatosis. Report of a case of Gorham's disease with autopsy. Am J Roentgenol 71:43, 1958
27. Harris R, Prandoni AG: Generalized primary lymphangiomas of bone; report of a case associated with congenital lymphedema of the forearm. Ann Intern Med 33:1302, 1950
28. Hayes JT, Brody GL: Cystic lymphangiectasis of bone. J Bone Joint Surg 43A:107, 1961
29. Hitzrot JM: Hemangioma cavernosum of bone. Ann Surg 65:476, 1917
30. Jackson JBS: Boneless arm. Boston Med Surg J 8:368, 1838
31. Jackson JBS: Absorption of humerus after fracture. Boston Med Surg J 10:245, 1872
32. Jacobs JE, Kimmelstiel P: Cystic angiomatosis of skeletal system. J Bone Joint Surg 35A:409, 1953
33. Jaffe HL: Tumors and Tumorous Conditions of the Bones and Joints. Philadelphia, Lea and Febiger, 1958, p 224
34. Johnson LC: General theory of bone tumors. Bull NY Acad Med 29:164, 1953
35. Johnson PM, McClure JG: Observations on massive osteolysis. A review of the literature and report of a case. Radiology 71:28, 1958
36. Jones DV, Landing BH, Wittgenstein E: Hyperglobulinemia secondary to massive lymphangioma. Am J Dis Child 101:510, 1963
37. Jones GB, Midgley RL, Smith GS: Massive osteolysis—disappearing bones. J Bone Joint Surg 40B:494, 1958
38. Kasabach HH, Merritt KK: Capillary hemangioma with extensive purpura. Report of a case. Am J Dis Child 59:1063, 1940
39. Kittredge RD, Finby N: Many facets of lymphangioma. Am J Roentgenol 95:56, 1965
40. Lichtenstein L: Bone Tumors. St. Louis, Mosby, 1972
41. Lindholm SO, Lindbom A, Spjut HJ: Multiple capillary hemangiomas of the bones of the foot. Acta Pathol Microbiol Scand 51:9, 1961
42. Miller SV, Purett HJ: Fatal chylopericardium caused by hamartomatous lymphangiomatosis. Am J Med 26:951, 1959
43. Milner SM, Baker SL: Disappearing bone. J Bone Joint Surg 40B:502, 1958
44. Moseley JE, Starobin SG: Cystic angiomatosis of bone: manifestation of hamartomatous disease entity. Am J Roentgenol 91:1114, 1964
45. Najman E, Fabecic-Sabadi V, Temmer B: Lymphangioma in inguinal region with cystic lymphangiomatosis of bone. J Pediatr 71:561, 1961
46. Nixon GW: Lymphangiomatosis of bone demonstrated by lymphangiography. Am J Roentgenol 110:582, 1970
47. Pachter MR, Lattes R: Mesenchymal tumors of mediastinum. III. Tumors of lymph vascular origin. Cancer 16:108, 1963
48. Parsons LG, Ebbs JH: Generalized angiomatosis presenting the clinical characteristics of storage reticulosis, with some observations on reticulo-endothelioses. Arch Dis Child 15:129, 1940
49. Pochaczevsky R, Sussman R, Stoopack J: Arteriovenous fistulas of the maxillofacial region. J Can Assoc Radiol 23:201, 1972
50. Rabinovich I: Maffucci's syndrome (dyschondroplasia associated with hemangiomas). Review of the literature. Khirurgia 38:139, 1962
51. Rosenquist CJ, Wolfe DC: Lymphangioma of bone. J Bone Joint Surg 50A:158, 1968
52. Schmorl G: Die pathologische Anatomie der Wirbelsäule. Verh Dtsch Orthop Ges Kong 21:3, 1927
53. Schmorl G: Die Gesunde und Kranke Wirbelsäule im Roentgenbild. Thieme, Leipzig, 1932, p 74
54. Sherman RS, Wilner D: The roentgen diagnosis of hemangioma of bone. Am J Roentgenol 86:1146, 1961
55. Shopfner CE, Allen RP: Lymphangioma of bone. Radiology 76:449, 1961
56. Silver KH, Aggeler PM, Crane JF: Haemangioma with thromboctyopenic purpura. Am J Dis Child 76:513, 1948
57. Southard SC, DeSanctis AG, Waldron RJ: Hemangioma associated with thrombocytopenic purpura. Report of a case and review of the literature. J Pediatr 38:732, 1951
58. Spjut HJ, Lindbom A: Skeletal angiomatosis. Report of 2 cases. Acta Pathol 55:49, 1961
59. Steiner GM, Farman J, Lawson JP: Lymphangiomatosis of bone. Radiology 93:1093, 1969
60. Sutherland DA, Clark H: Hemangioma associated with thrombocytopenia. Am J Med 33:150, 1962
61. Töpfer DII: Über ein infiltrierend wachsendes Hamangiom der haut und multiple Kapillarektasien der haut und inneren Organe. II. Zur Kenntnes der Wirbelangiome. Frankfurt Z Pathol 36:337, 1928
62. Tucker SM: Bilateral chylothorax with multiple osteolytic lesions: generalized abnormality of lymphatic system. Proc R Soc Med 60:17, 1967
63. Unni KK, Ivins JC, Beabout JW, Dahlin DC: Heman-

gioma, hemangiopericytoma and hemangioendothelioma (angiosarcoma) of bone. Cancer 28:1403, 1971
64. Virchow R: Die krankhaften Geschwilste 111, 373 Berlin, Hirshwald, 1863, 1867
65. Waldron RL, Zeller JA: Diffuse skeletal hemangiomatosis with visceral involvement. J Can Assoc Radiol 20:119, 1969
66. Wallace S: Dynamics of normal and abnormal lymphatic studies with contrast media. Cancer Chemother Rep 52:31, 1968
67. Wallis LA, Asch T, Masel BW: Diffuse skeletal hemangiomatosis. Report of 2 cases and review of literature. Am J Med 37:545, 1964
68. Willis RA: The borderland of embryology and pathology, 2nd ed., London, Butterworths, 1962, p 351
69. Wyke BD: Primary hemangioma of skull: rare cranial tumor. Am J Roentgenol 61:302, 1949

Skeletal Dysplasias: Osteogenesis Imperfecta, Osteopetrosis, Hyperphosphatasemia

Joseph P. Whalen

Osteogenesis Imperfecta

The radiographic findings of osteogenesis imperfecta are those of decreased bone density varying in severity depending upon the clinical classification. It is apparent the lucency is either the result of poor production of bone or excessive resorption. Some authors favor the former,[1] and state that there is impaired obsteoblastic activity with the formation of abnormal osteoid matrix, which resembles immature collagen. This is felt to be a part of the widespread disorder of abnormal connective tissue. In this disease, the serum calcium, phosphorus, phosphatase, and ascorbic acid on urinalysis are usually within normal limits. However, the serum alkaline phosphatase is sometimes elevated as a result of fracture. On the other hand, others feel the subnormal accumulation of bone mass in osteogenesis imperfecta is caused by overactive bone resorption rather than decreased bone formation.[2]

Osteopetrosis

Osteopetrosis is a disease of universal failure of modeling. All bones, whether enchondral or membranous, fail to model, although the most striking changes occur in bones preformed in cartilage. The hallmarks of this disease are the accumulation and persistence of calcified cartilage that normally would be resorbed in the course of modeling. It has been stated in the past that this is due to metaplasia of cartilage. It has been our feeling that this is rather due to failure of osteocytic resorption of the calcified cartilage, thus explaining both the histologic and radiologic findings.[3] The findings of osteopetrosis histologically are very similar to those associated with excess calcitonin or of heavy-metal poisoning, i.e., persistence of cartilaginous core, inactive osteocytes, decreased number of osteoclasts (in proportion to diminished osteocytic resorption), and nonlamellar bone in areas where lamellar bone should be present. This results radiologically in bone that is denser than normal, particularly in the growing areas of the bony skeleton. Fig. 1 illustrates human osteopetrosis with increased density in the growing portion of the vertebral bodies. The increased density in osteopetrosis represents calcified cartilaginous core. It is interesting to note the similarities of x-ray changes in other diseases that involve the growing ends of bone, particularly in the vertebral bodies as illustrated in Fig. 2. In Fig. 2, there are two cases of osteomalacia in which treatment resulted in the development of sclerosis of the growing ends (a failure of modeling). In this instance, however, the density represents calcified osteoid. It is important for the radiologist, then, to correlate the clinical and biochemical findings and not merely to compare similar radiographic appearances. The role of calcitonin in this disease is not known, but it is attractive to think that the disease may represent a calcitonin excess state.

Hereditary Bone Dysplasia with Hyperphosphatasemia

This disease is a rare bone disease well described by Dr. Caffey.[4] The basic defect in this disease appears to be excessive resorption of bone with compensatory increased production. This results in rapid bone turnover without the sequential transition of woven bone to mature bone. The histologic appearance of this disease is one of woven bone in areas where lamellar bone should be found.

Since this disease is clearly one of excessive resorption, therapy with calcitonin appears to be a logical approach. Its response to calcitonin has been dramatic clinically, radiologically, histologically, and biochemically[5,6] (Fig. 3). Biochemically, this disease is characterized by those parameters that reflect excessive resorption and excessive production, i.e., increased alkaline phosphatase serum levels and increased urinary hydroxproline levels.[6]

Summary

An appreciation of the role of the osteocyte in bone modeling is required in the understanding of the pathogenesis of the diseases listed above. Specifically, the osteocyte resorbs cartilage (osteocytic chondrolysis), which then "makes room" for osteoblasts to produce bone in a normal, orderly fashion. If excessive resorption occurs, the cartilage disappears and woven bone results. If resorption is limited, the cartilaginous core persists and modeling does not occur.

References

1. Currarino G, Brooksaler F: Osteogenesis imperfecta. Prog Pediatr Radiol 4:346, 1973
2. Spranger J: The biochemical basis of bone dysplasias. Prog Pediatr Radiol 4:29, 1973
3. Whalen JP, Krook L, Nunez EA: Some metabolic considerations in bone disease. In Potchen EJ (ed): Current Concepts in Radiology, Vol. 3, Chap. 7. St. Louis, Mosby, 1977, p 143
4. Caffey J: Familiar hyperphosphatasemia with ateliosis and hypermetabolism of growing membranous bone; review of the clinical, radiographic and chemical features. Prog Pediatr Radiol 4:438, 1973
5. Whalen JP, et al: Calcitonin treatment in hereditary bone dysplasia with hyperphosphatasemia: A radiographic and histologic study of bone: Am J Roentgenol 129:29, 1977
6. Horwith M, et al: Hereditary bone dysplasia with hyperphosphatasemia: Response to synthetic human calcitonin. Clin Endocrinol 5(Suppl):341s, 1976

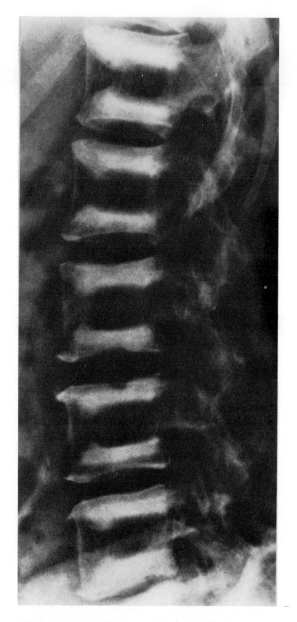

FIG. 1 Radiograph of lumbar spine in a patient with human osteopetrosis. Note increased density in subchondral area of vertebral body. This corresponds to nonresorbed calcified cartilage. This radiographic appearance is reminiscent of healing rickets and sclerotic renal osteodystrophy. (From Whalen JP et al: Some metabolic considerations in bone disease. In Potchen EJ (ed): Current Concepts in Radiology, Vol. 3, St. Louis, Mosby, 1977, p 143–92. Courtesy of the C.V. Mosby Company.)

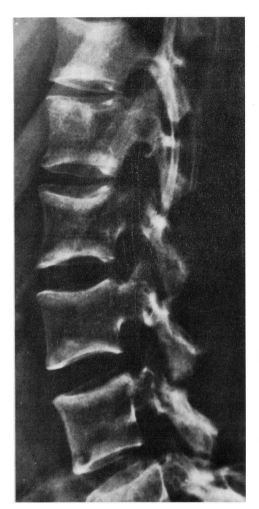
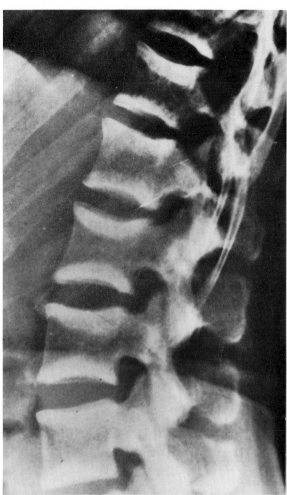
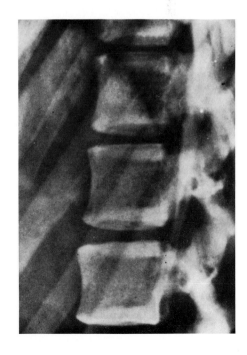

FIG. 2. Above left: Osteomalacia of growing spine showing osteopenia but with some minimal subchondral sclerosis. Above right: Same patient as in adjacent illustration, after a month of therapy with vitamin D and calcium. There is marked sclerosis of the subchondral surface of vertebral bodies. These areas represent calcified osteoid and nonlamellar bone and are reminiscent of the rugger jersey spine seen in renal disease. Opposite: Patient with osteomalacia on dilantin therapy for seizures. One month prior to this radiograph, the patient had been treated for his osteomalacia. This radiograph demonstrates the sclerotic subchondral bone secondary to calcified osteoid. (From Whalen et al: Some metabolic considerations in bone disease. In Potchen EJ (ed): Current Concepts in Radiology. Vol. 3, St. Louis, Mosby, 1977, p 143–92. Courtesy of the C.V. Mosby Company.)

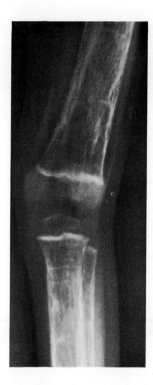 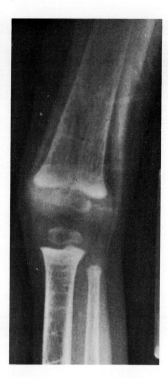

FIG. 3. Radiographs of the distal femur and proximal tibia and fibula in a patient with hyperphosphatasemia. The radiograph (left) is prior to treatment and (right) 4½ months after calcitonin therapy. Discrete cortices can be seen after treatment in areas of preexisting meshlike bone. (From Horwith M et al: Hereditary bone dysplasia with hyperphosphatasemia: response to synthetic human calcitonin. Clin Endocrinol 5 (Suppl); 341, 1976. Courtesy of *Clinical Endocrinology*.)

Fibromatosis and the Skeleton

Frieda Feldman

Fibromatosis is a local, invasive, but nonmetastasizing growth of moderately well-differentiated fiboblastic tissue. However, these growths may be multiple and confined to a particular site or tissue, or multiple and diffusely distributed. Fibrous proliferations can be divided into those that occur both in children and adults and that are similar in morphology, location, and behavior in all age groups and those that occur principally during infancy, have no clinical or morphologic counterpart in adult life, and are relatively rare.

This chapter will emphasize the latter, since they frequently pose a special problem in diagnosis, not only clinically but histologically. Multiple fibromatosis, particularly in infants, is often erroneously associated with poor prognosis, since their microscopic appearance often fails to reflect their true biologic behavior, and such features as cellularity, immaturity, and rapid growth may be mistaken for evidence of malignancy.[2,11,35,36] However, survival with regression and eventual disappearance of all lesions is a common occurrence if vital viscera are not involved. Although considered rare, the number of reported cases does not reflect the true incidence of the disorder. Many instances have gone unrecognized or unreported, while others have been improperly classified, most commonly as neurofibromatosis or congenital fibrosarcoma.

Until Stout in 1954[35] reviewed the previous heterogeneous group of juvenile fibromatoses and suggested a more clinically correlated classification, it had been difficult to compare the various forms of fibrous lesions. Among the 44 cases reviewed, two were defined as congenital generalized fibromatosis, which is a specific clinical and pathologic entity within the broad category of the juvenile fibromatoses characterized by the presence of nodules of well-differentiated fibrous tissue disseminated throughout the body. One of these patients died on the first day of life with autopsy documentation of multiple fibrous nodules in the skin, most muscles (including those of the tongue, myocardium, and gastrointestinal tract), the lung, peritoneum, lymph nodes, and femur. The other died on the twenty-second day of life similarly involved with subcutaneous and visceral nodules composed of well-differentiated fibrous tissue that for the most part resembled other benign juvenile fibromas histologically. Stout contended that the condition was one of multiple independent fibroblastic tumors rather than widespread metastases from a single primary source.[35]

In 1965, Kaufman and Stout[19] divided the congenital fibromatoses into two major groups—solitary and multiple or generalized. Since then this classification has been further modified and subdivided into 3 groups:

1. Localized congenital fibromatosis—The most common type of congenital fibrous tumor, which may consist of a solitary mass or several fibrous masses confined to one site or area. It can occur in the soft tissues and in nearly every organ, including bone, but arises chiefly in the subcutaneous soft tissues or muscles of the head, neck, and proximal extremities.
2. Multiple congenital fibromatosis—Same as 1, but involves either more than one soft-tissue site or soft tissue and bone (Fig.1).
3. Generalized or disseminated congenital fibromatosis—Multiple sites of involvement occur in the viscera as well as within soft tissues and bone.

Only a small percentage of cases have the third type, or full-blown form, of the disorder. Lesser degrees of expression are common in the form of a

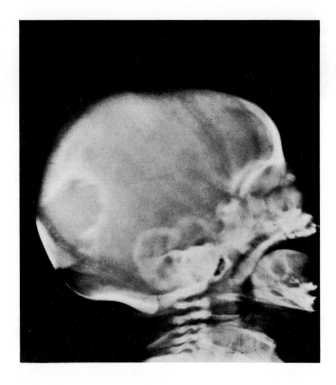
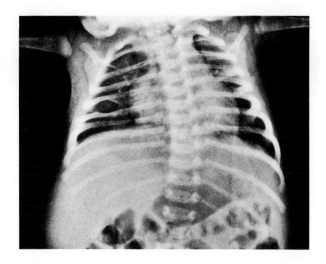
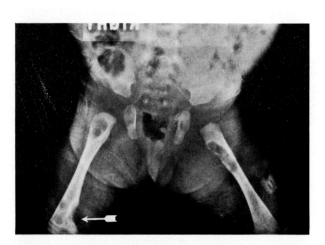
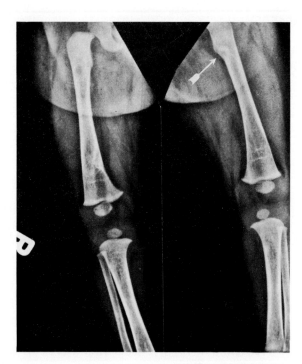

FIG. 1. Congenital multiple fibromatosis. On physical examination at 5 days of age, this 2551-g spontaneously delivered female was noted to have a hard, 3 x 3 cm mass in the right posterior parietal region of the skull and a second firm 2 x 2 cm mass in the distal left thigh. Top left: Lateral skull at 5 weeks. A 4 x 2.5 cm lytic lesion with a sclerotic lobulated but smooth border is seen in the right posterior parietal region. Biopsy revealed interwoven strands of maturing fibroblasts. Top right: Posteroanterior view of the chest at 5 weeks. The right sixth rib appears irregular in contour. A sharply bordered, superiorly concave depression is noted in its midposterior portion. The pulmonary parenchyma appear normal. Bottom left: Frontal view of the pelvis and femora at 5 weeks. Additional well-circumscribed lesions with thin sclerotic borders involved both proximal femora as well as the distal right femur. The latter lesion is eccentrically situated within the metaphysis and involves the cortex (arrows). Biopsy revealed tissue similar to that obtained at the site of the skull lesion. Neither lesion was encapsulated, and several areas showed invasion of muscle bundles. No mitotic figures were noted. Bottom right: Frontal view of the lower extremities at 5 1/2 months. Minimal residual defects are present. The proximal L femoral lesion is almost completely remineralized (arrow). Two of the four lesions in the right femur had concomitantly remineralized, while the other two had disappeared. The cranial defect had also decreased in size and was similarly surrounded by roentgenographic evidence of reparative sclerosis. This case is the subject of a report by Baer and Radkowski.[4]

solitary fibrous nodule or mass, which may be situated in various tissues including voluntary and smooth muscles, viscera, skin, lymph nodes, or bone.[4,9,20-22,27,32,34,38] An interesting parallel is provided by hemangiolymphangiomas and neurofibromas that may be associated with solitary or multiple lesions.

The extent and site of involvement seriously influences the prognosis of the lesion. The prognosis of the first two types is uniformly good (Fig. 1) except for local complications such as pathologic fracture of intrinsic bone lesions, invasion of bone from neighboring soft-tissue lesions, osteomyelitis, or post surgical complications including local recurrence. However, in type 3 (generalized or diffuse fibromatosis)[26] prognosis may be grave. Mortality has approached 80 to 90 percent in some series, particularly when vital organs are heavily involved.[30,34]

Clinical Data

Although the fibromas have their inception in utero, some affected newborns appear normal.[28] Most commonly, however, the lesions are obvious at birth or within 2 to 4 months of age. It had been stated that no cases were noted in which the clinical picture developed in a previously healthy child after the first few months. This has been disproven.[10]

Geshickter described in an adult an unusually aggressive case that was not congenital. He thought, however, that it represented a latent form of fibromatosis, since the lesion became active at a more advanced age. Numerous organ systems were involved and resulted in the death of the patient. There are, in fact, adolescent and adult forms of progressive and aggressive fibromatosis, the so-called extraabdominal desmoids, which infrequently affect the skeleton[18,35,36] (Fig. 2). They may, however, erode neighboring bones or compress local nerves and vessels.[3]

Juvenile fibromatoses in general affect both sexes, with some preference for males. Subcutaneous nodules may arise from or invade underlying skeletal muscles or fascia, and involvement may be limited to a single mass (Fig. 3). Bony and visceral involvement may also be single or multiple (Fig. 2), but the latter in particular may be small, asymptomatic, difficult to identify, and may remain unsuspected. The lesions may resolve spontaneously, remain stationary, or grow progressively but do so slowly. Soft-tissue lesions in particular may ulcerate and become secondarily infected.

Etiology

The etiology of juvenile fibromatoses and of the congenital fibromatoses in particular is unknown. Several hypotheses have been advanced for the latter, including an increased estrogen level, which is implicated by the following observations. Multiple fibrous lesions have been experimentally produced by the administration of large doses of estrogen to guinea pigs.[23,29] Abdominal desmoids are more common in pregnant women. Plantar fibromatoses are commonly associated with liver disease, which, in turn, may be associated with raised estrogen levels. Increased gonadotropin has been found in these tumors.[15] Although the mothers of some patients[30] had been treated with estrogens during pregnancy this was not a universal finding; others took no medication at all.

Its occasional generalized character has suggested a systemic disturbance of interstitial fibrous tissue. An intrauterine metabolic disorder, toxic state, or infection during the last weeks of pregnancy has also been postulated. Maternal syphilis has been implicated in some cases, but such a history is not uniformly elicited. A viral cause has been suggested based on inclusion bodies described in some lesions.[7] Others have failed to produce evidence for a viral etiology.

Trauma has been implicated in some of the congenital fibromatoses, such as fibromatosis colli, i.e., the sternomastoid "tumor" of the newborn producing wry neck, as well as in other lesions, with resultant scar formation. Early in development, teratogens are known to produce lesions with hemorrhagic necrosis, which are said to heal and atrophy with scar tissue formation before circulating or tissue macrophages are available. Scar formation related to other intrauterine episodes have been documented, i.e., gliosis in toxoplasmosis gondii infections and cicatrization after intrauterine volvulus or after mesenteric occlusion. Many fibromatous nodules occur in otherwise well-formed organs, suggesting that these nodules developed after major organogenesis was complete. Another etiologic mechanism considered was that of a temporary suspension of normal growth, with a form of fibrogenesis presumably replacing normal enchondral bone formation. The lesions' location in relation to growth plates suggests that the abnormal process ceased at or shortly after birth, as there is a distance of 0.5 to 1 cm between almost all intraosseous fibromas and the adjacent epiphyseal plate. Against this is the fact that in some cases lesions have progressed after birth, with multilating and even fatal results.

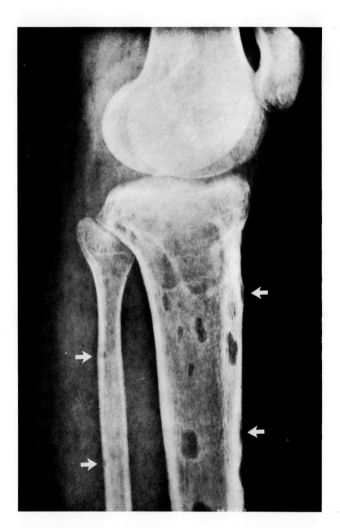 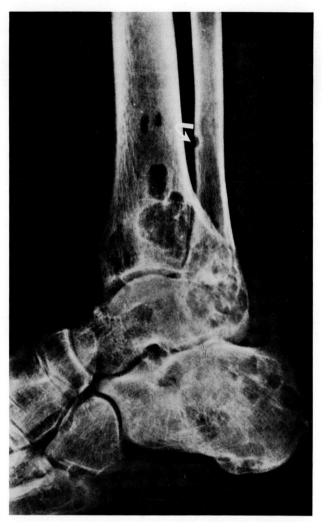

FIG. 2. Multiple skeletal fibromatoses. This 20-year-old male had a 1-year history of L leg pain. Radiographs were said to reveal "abnormal bones." Additional investigation elsewhere included an "inconclusive" angiogram and a "nondiagnostic" biopsy. He was admitted due to continued discomfort. Left: Lateral view of the proximal tibia 7 fibula. Numerous round and oval well-defined radiolucencies are seen scattered throughout the shafts of the visualized bones. Some have sclerotic rims. A larger area of osteolysis is seen within the medullary cavity of the proximal tibial metaphysis. Note the small intracortical lesions within both bones (arrows) that have expanded and thinned but not destroyed the cortex. Right: Oblique view of the left lower leg and ankle. Note the small osteolytic lesion that occupies the entire fibular cortex and deforms its endosteal surface. Its outer cortical margin cannot be definitely identified. Note numerous varying-sized intramedullary osteolytic lesions. All are sharply defined, while some are bordered by a thin sclerotic rim. Note several lesions within the tarsal and metatarsal bones.

Since some lesions continue to enlarge or proliferate at a faster rate, at the time of birth, a dominant fibrous mass may give the impression of a primary malignant tumor, while the smaller lesions scattered throughout the body may mimic metastases. The problem of whether multiple, diffusely distributed, histologically benign lesions are of independent multifocal origin has also been debated with reference to other lesions, such as multiple hemangiomatosis, lymphangiomatosis, lipomatosis, and neurofibromatosis. The weight of opinion has favored their being of independent multicentric origin. In addition, many of the multicentric fibrous lesions occur in smooth or striated muscle or viscera, such as the spleen, which are rarely the sites of metastases. Surprisingly, predominantly fibrous structures such as tendons and fascia are only infrequently involved.

A relationship of self-limiting fibromatosis to

FIG. 3. Juvenile fibromatosis. This 13-year-old white male noted a painless mass in his mid right thigh 1 month prior to admission. It was firm and nontender when palpated and had no associated discoloration. Opposite: Frontal view of the right thigh. A well-defined, elliptical, nonmineralized soft-tissue mass is seen posterolateral to the midshaft of the right femur (arrows). A radiolucent halo gives a false impression of encapsulation. An angiogram revealed the mass to be relatively avascular. Below: CAT scan of both thighs. The mass lies in the posterior lateral aspect of the thigh (arrow) and is completely separate from the femur. However, it cannot be divorced from neighboring muscle bundles, which are discretely defined on the contralateral side. No mineral is distinguished within the lesion. At surgery, several lobulated tan–pink masses having a whorled appearance were found lateral to the biceps femoris muscle. The lesion encompassed the sciatic nerve and was adherent to the surrounding posterior compartment muscles and soft tissues in a portion of its course. Microscopically, proliferating spindle shaped cells at the periphery of the mass compressed and entrapped fascial structures, adipose tissue, and skeletal muscle fibers that had become secondarily atrophic. Fibromatosis was diagnosed.

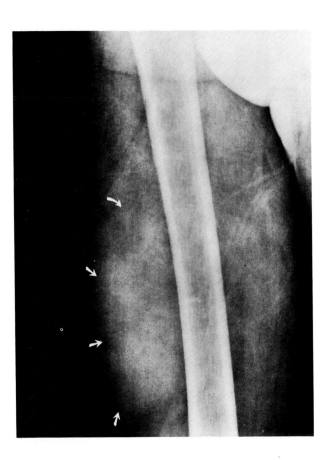

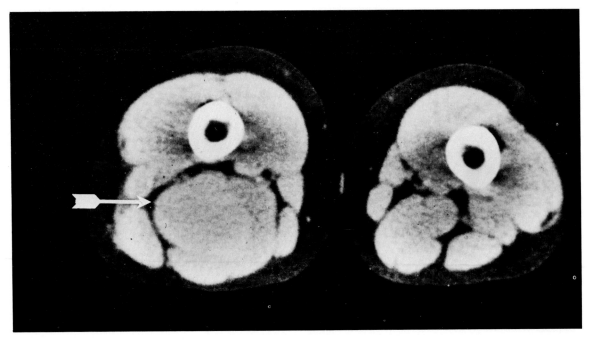

spontaneously involuting neoplasms such as the in situ neuroblastoma and Wilms tumor of early infancy has also been considered. However, not all observers believe that the fibromatoses are true tumors.

It has been suggested that, except for teratomas, practically all masses at birth described as benign neoplasms might be classified as hamartomas.[34] Hamartomas are defined as tumorlike malformations that consist of an abnormal mixture of the normal constituents of an organ. The abnormality may be one of quantity, arrangement, or degree of differentiation of components or of all three. Congenital generalized fibromatosis could therefore be placed in the hamartoma category, particularly in view of its observed "smooth muscle-like" component[6,13,14] as well as on the basis of a peculiar admixture of vascular, fatty, and fibroblastic components observed in some lesions.

Therefore, it is believed that congenital generalized fibromatosis may appear during intrauterine life as multiple, locally infiltrative, fibroblastic nodules that may involve several organ systems. Each nodule is an independent focus of benign proliferation rather than a metastasis from a primary tumor, and some investigators believe that the histologic appearance of this benign proliferation is more suggestive of a tissue malformation or a hamartoma than of a neoplasm.[6,8] Metastases seem unlikely, since (1) lesions appear to arise de novo in interstitial fibrous tissue and bear no relation to transmitters of metastases such as blood vessels or lymph channels; (2) they are composed of well-differentiated fibrocellular tissue that, if it metastasizes at all, does so with extreme rarity; (3) lesions occur in organs that are rarely the sites of metastases, i.e., voluntary muscles, heart, tongue, and muscle layers of the gastrointestinal tract and spleen.

Genetics

A hereditary influence has been suggested[5,10,17] as playing a possible role in the development of fibromatosis. Six sets of siblings have been described, and a question of an autosomal recessive inheritance has been raised by Baird and Worth.[5] Although congenital generalized fibromatosis is listed by McCusick as a possible autosomal dominant,[24] Baird and Worth have pointed out that it has never been described in a parent of an affected child and note that consanguinity plus the familial nature in their cases supports an autosomal recessive type of inheritance.

It is of interest that Baird and Worth's five cases of congenital generalized fibromatoses were originally diagnosed as neurofibromatosis and occurred in a family who were given a poor prognosis in terms of life expectancy and were told of a high genetic risk of recurrence in future siblings. Two of the cases were brothers and two others were their first cousins. The consanguine parents had no unusual physical findings, cafe au lait spots, or subcutaneous nodules. Neurofibromatosis with sarcomatous degeneration was diagnosed in two of their newborns, and amputation was considered in one child. Both boys were expected to die in infancy but came to medical attention years later when the parents inquired about the advisability of sterilizing them so that the disorder would not be passed on. They had been unaware of the same lesions in their relatives.

Drescher[10] described two siblings with multiple fibromatosis that became manifest at 2 years of age and that seemed to regress over a period of 7 years.

Histopathology

Gross Features

The lesions are rubbery, firm, spherical-to-ovoid unencapsulated nodules, which, on section, may bulge slightly and display a finely trabeculated glistening pinkish-gray surface and which vary from 0.2 to 2.5 cm in size. Some lesions, though not encapsulated, may be well circumscribed and shell out easily. Others may compress, invade and destroy adjacent tissues, or merge with adjacent muscle fibers and bony structures and may be recognizable only on palpation as areas of increased firmness and induration. They may be widely distributed over the entire body in skin and subcutaneous tissues, skeletal and smooth muscles, the gastrointestinal tract, bladder, viscera, peritoneum, heart, and occasionally lymph nodes and bone. Nodules in skeletal muscle always occupy the belly of the muscle rather than its fibrous site of origin or tendinous insertion. None is usually associated with synovial tendon sheaths, deep fascia, intermuscular septa, or paraosteal fibrous tissue. Those in skeletal muscles favor areas in which the larger masses of voluntary muscles are located and tend to predilect the trunk and proximal portions of the extremities, i.e., the shoulder girdle, upper extremities, low back, gluteal regions, and thighs. Their exact distribution and number vary from case to case.

Juvenile and localized forms of fibromatosis may represent a lesser degree of expression of the same abnormality and are characterized by solitary fibromatous nodules or a single mass (Fig. 3) microscopically identical to the generalized form. In addition to soft tissues and viscera, bone marrow may be largely replaced by white dense fibrous tissue.

Microscopic Features

Microscopically, the lesion consists of a cellular fibrous tissue composed of spindle cells arranged in both loosely and tightly packed interlacing fasciculi

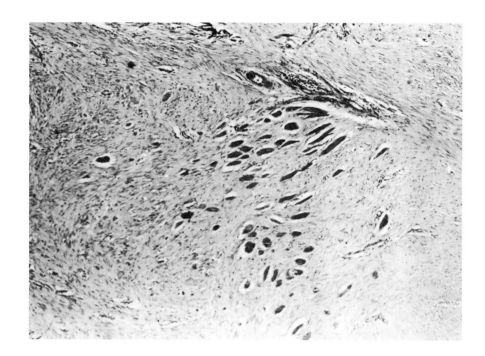

Fig. 4. Typical fibromatosis (so-called extraabdominal desmoid). Note the invasive pattern of an otherwise mature fibrous tissue that has led to entrapment of skeletal muscle fibers. (H&E, X 84). Courtesy of Dr. Raffael Lattes.

generally associated with abundant collagen ground substance. The cellular population consists of maturing fibroblasts of uniform appearance with oval or fusiform nuclei. Mitoses, anaplasia, and cellular atypism are generally lacking (Fig. 4).

Some lesions have a hemorrhagic appearance, and degenerative changes with calcification are occasionally noted. Degenerating nodules may be yellowish brown in color with hemorrhagic foci and areas of myxomatous change. Foci of degeneration and calcification are occasionally noted. However, there is no associated inflammatory reaction.

Most lesions are remarkably similar, exhibiting a fascicular growth pattern with loose edematous stroma. However, compact cellular areas displaying mitotic activity, a vascular background, or unusual vascular proliferations have rendered interpretation difficult in some cases.

The cellular elements of the nodule may be arranged predominantly around blood vessels and vascular channels, which are abundant in most lesions. Occasional small, thin-walled venous channels may show focal circumferential abnormal fibromuscular proliferation. Therefore, this feature superimposed on the basic pattern, which may include irregular, thin-walled cavernous sinusoidal or capillary channels lined by a single layer of flattened endothelium may suggest a tissue malformation rather than a neoplasm.

Although several cases in the literature were initially diagnosed as neurofibromas,[1] the cases do not have their distinctive histologic features.[25,33,39]

Occasionally, because of variations in cellular density, fibromatosis may be alternatively diagnosed as well-differentiated fibrosarcoma. However, although the cellular morphology of these fibrous lesions may be suggestive of malignancy, their natural history indicates that they are benign lesions that may completely regress provided they are situated in nonvital locations.[4,16,31,37]

In interpreting infant and juvenile lesions, it is important to remember that neither cellular proliferation nor mitotic activity nor continued aggressive local recurrence constitutes true biologic malignancy.[2,35,36] Recent electron microscopic studies[28] have documented smooth muscle-like elements as well as fibroblastic elements within these lesions, supporting the concept that they are hamartomas rather than atypical fibromas,[6,8] and the term *generalized hamartomatosis* has been proposed as better suited to this entity. Morettin additionally felt that the stipulation "congenital" should be reserved for those lesions noted at birth or during the first week of life.[28] He, too, divided generalized hamartomatosis into two types on the basis of its anatomic distribution:

1. Somatic—involving subcutaneous and muscular tissues with or without skeletal involvement.
2. Somaticovisceral—involving viscera as well as soft tissue and/or skeleton.

The somatic form has a benign course leading to complete resolution of both cutaneous and skeletal lesions in many instances. The combined somaticovisceral form, however, has a poor prognosis, with the majority of cases dying in the first few months of life. Therefore, the above classification is also generally based on the distribution of the lesions, which, in turn, may influence clinical evolution and prognosis.

Roentgenographic Findings

The number and location of bone lesions in congenital fibromatosis vary widely. They are most commonly radiolucent, less than 0.5 cm in diameter and usually well defined, but most often without sclerotic borders. Calcification within the areas of lysis is not a feature.

Generally, the defects are eccentrically situated when found in the long bones, where they tend to be metaphyseal in location and symmetrically distributed. The lesions appear to originate or grow eccentrically, destroying the medial more than the lateral cortical border (Fig. 1 bottom left). It has been suggested that the lesions arise from the periosteum and extend to erode bone. This could explain the fact that, in some cases, the early changes are limited to cortical erosion of bone (Fig. 2). Therefore, the cortex as well as the medullary cavity may be involved, and cortical expansion and thinning with occasional destruction and extension of the intramedullary lesion into the soft tissues may be a feature. Conversely, invasion of bone by a neighboring soft-tissue lesion may occur.

It is not uncommon to observe expansion or an increase in size of the intraosseous lesions before spontaneous regression begins. Regression commonly results in complete disappearance of bony abnormalities and has been noted to occur without sequelae within months (Fig. 1 bottom). In Heiple's case of a newborn boy with congenital generalized fibromatosis limited to the skeleton, only faint bone scars remained at the age of 22 months to indicate the sites of 100 spontaneously regressing fibrous lesions of bone.[16] The areas of bone destruction may not always be apparent at birth and may develop rapidly in a period of days or weeks. Pathologic fractures are a frequent complication. However, they tend to heal rapidly and may also produce abundant exuberant callus.

Lesions may involve the calvarium (Fig. 1 Top left) and may be due to intrinsic osseous foci or to erosion by fibrous nodules arising from the adjacent dura. This is an important differential point, since, as with the localized form of fibromatosis, there may be considerable destruction of bone associated with a neighboring mass due to extrinsic mechanical pressure. This may be particularly pronounced when the mass is vascular or has undergone secondary infection. Periosteal reaction has been noted that may occasionally be quite extensive, may be diaphyseal, and may involve the long bones of all extremities. However, this is distinctly unusual. A recent case of congenital generalized fibromatosis of Familusi et al[12] was therefore of particular interest, since in addition to skin and visceral lesions, which included gingival hypertrophy and lymphatic dilatation of the ileal villi, diffuse bony involvement also included the rare manifestations of skeletal hyperostosis and ankylosis of numerous joints.

Pulmonary and other visceral lesions may not be roentgenographically apparent either at birth or subsequently and may be revealed only at autopsy (Fig. 1 Top right). The roentgenographic pattern of pulmonary involvement is nonspecific and resembles interstitial fibrosis or generalized bronchopneumonia, so that they rarely have roentgenographic expression.

Although nodules concomitantly occur in other viscera and are important to prognosis, they are usually small and generally not roentgenographically identifiable. Fibromas of the gastrointestinal tract are common; however, gastrointestinal tract nodules are primarily located within the muscular coat or project from the bowel contours in the form of pedunculated subserosal lesions, so that they rarely have roentgenographic expression. The largest gastrointestinal tract lesion was 1.1 cm in diameter situated under the serosa.[34] Fibromas may be directed toward the intestinal lumen as sessile polyps producing a moderate degree of narrowing with occasional ulceration of the overlying mucosa. Neonatal or early postnatal demise is most commonly due to complications arising from visceral involvement as well as to debility and intercurrent infections.

Differential diagnosis includes a number of congenital disorders linked by subcutaneous tumor nodules appearing at birth or shortly thereafter, as well as a number of other systemic hamartomas and tumors. These include neurofibromatosis, hemangiomatosis, lymphangiomatosis, lipomatosis, enchondromatosis, the reticuloendothelioses, fibrous dysplasia, congenital syphilis, and neuroblastoma.

Among these, reticuloendotheliosis, or histiocytosis X, is the most important differential consideration, particularly since it may similarly involve infants and very young children as in Letterer–Siwe Disease. However, the latter shows a predilection for visceral involvement in the form of hepatosplenomegaly and lymphadenopathy and is associated with typical skin lesions. In the reticuloendothelioses in general, lytic lesions predilect the skull but may involve the entire skeleton without a metaphyseal preference in the long bones as in congenital generalized fibromatosis.

Focal lytic areas commonly have irregular albeit well-demarcated borders with beveled edges, while periosteal new bone formation is frequently associated with long-bone lesions.

Fibrous dysplasia, particularly in its polyostotic form, is associated with marrow cavity expansion and with thinning and/or thickening of the cortex resulting in modeling deformities. Most lesions are bordered by a prominent sclerotic rim or rind, while their central portions have a hazy glazed appearance with evidence of matrix calcification frequently visible on roentgenograms. The osseous lesions of fibrous dysplasia are rarely manifest in the neonatal periods as the lesions of congenital generalized fibromatosis commonly are.

Neurofibromatosis, in distinction to congenital generalized fibromatosis, is rarely manifest in the newborn, is often hereditary, has its major distribution along nerve trunks, and may be associated with cutaneous pigmentation and other neuroectodermal disturbances. Primary intraosseous long-bone lesions are uncommon. The skull and spine are frequently involved, with typical stigmata including enlarged optic foramina, absence or dysplasia of predilected areas such as the left side of the calvarium and the sphenoid bone, and dysplastic ribs and spine. Enchondromatosis is invariably associated with modeling deformities due to disturbed enchondral bone formation with resultant expanded, hour glass- or dumbbell-shaped diametaphyseal lesions. These are often seeded with stippled calcific deposits. Endosteal scalloping and cortical thickening and buttressing are frequently present. In addition, enchondromas are uncommonly appreciated before the age of 2 years. Osteomyelitis is most often associated with one or two rather than multiple sites and is rarely symmetrically distributed. Affected bones display a more diffuse and permeative pattern of destruction accompanied by periosteal reactions that may be associated with profound involucrum and/or sequestra formation.

Congenital syphilis often produces multiple metaphyseal bone lesions. However, additional widespread skeletal stigmata include deformed, frayed, and fractured growth plates, as well as symmetric periosteal reactions. Neuroblastoma, when it metastasizes to bone, is a more aggressive process in terms of its distribution within individual bones and its dissemination throughout the skeleton. The roentgenographic findings are more typically those of a malignant process that diffusely permeates bone without a metaphyseal predilection as in congenital generalized fibromatosis. Periosteal reactions of an aggressive laminated variety are common. Skull involvement may be associated with diastasis of the sutures.

Hemangiomatosis and lymphangiomatosis frequently produce alterations in the size and shape of the bones and/or tissues of the affected part. The latter may harbor laminar calcifications or phleboliths in hemangiomatosis and linear calcifications in lymphangiomatosis. An increase or decrease in the overall size of the extremity may result from early or asymmetric growth plate closure. Multiple intraosseous lipomas are rare. However, associated soft-tissue lesions may be typically radiolucent or may calcify.

Congenital generalized fibromatosis should be suspected in a newborn or infant with single or multiple masses or nodules that may be roentgenographically apparent and that arise from muscles or subcutaneous tissues, particularly about the trunk, head, and upper extremities. However, the incidence of fibromatosis in neonates is second only to that of vascular tumors.[19] In addition, multiple, localized, well-defined lytic skeletal lesions that when present are symmetric and often eccentrically situated within the medial metaphysis should aid in differentiating congenital generalized fibromatosis. Congenital generalized fibromatosis when entirely limited to the subcutaneous and muscular tissues without osseous or visceral involvement may be difficult to distinguish roentgenographically. Larger superficial lesions may become infected, with secondary involvement of neighboring bones. The resultant osteomyelitis and accompanying periosteal reaction may make diagnosis difficult. Pathologic fracture is a common form of initial presentation. Periosteal elevation and new bone formation are infrequently observed. Serial roentgenograms may reveal resolution (Fig. 1C, D) or progression of established lesions, as well as the development of new, well-circumscribed areas of osteolysis. Involvement of the lungs and other viscera should aid in differentiation as well as in determining prognosis. Some cases may go on to complete regression[36] despite visceral involvement. It is probable that nonsymptomatic visceral lesions have also been present in patients who have had complete regression.

Localized juvenile fibromatosis may likewise be deforming or crippling because of anatomic sites involved, but the majority are not a threat to life. In untreated patients, growths may cease and fibromatosis remain inactive or even disappear.

Summary

A peculiar characteristic that the angiomatoses and generalized fibromatoses share is that, despite their being benign lesions, they may be associated with significant morbidity and mortality. All these lesions may occur diffusely and may invade, compromise, or destroy adjacent structures as well as involve multiple organs.

However, despite this behavior, they most commonly follow a benign course, particularly when they are limited to the skin and deeper somatic tissues with or without bone involvement. Bony lesions, although multiple and sizable, are also capable of complete regression particularly in congenital generalized fibromatosis. Many cases with congenital generalized fibromatosis involving bone alone, skin alone, or bone and skin alone with no major visceral involvement have, as a rule, had complete clearing without complications or residua. Patients with minor visceral involvement that had had a benign outcome have also been documented. However, most patients with generalized fibromatosis as well as those with angiomatosis who have generalized visceral and bone involvement or diffuse visceral involvement alone die. Those with congenital generalized fibromatosis die within the first 4 months of life 80 to 90 percent of the time. Neonatal or early demise is most commonly associated with debility and intercurrent infections as well as from complications arising from visceral involvement. Therefore, prognosis in both types of lesions chiefly depends on the number and sites of associated visceral lesions. When these are few or are situated in nonstrategic loci where they do not interfere with function, a benign course is most common and, particularly in the case of congenital fibromatoses, even complete regression is possible.

Recognition of all these lesions is important so that falsely pessimistic prognoses can be avoided and so that unnecessary radical surgery and/or radiotherapy is not performed in the belief that a malignancy is present. It is important in terms of giving accurate genetic counseling. The most frequent sources of misinterpretation arise in relation to the histologic confusion of congenital fibromatosis with well-differentiated fibrosarcoma and neurofibromatosis. Although rare, congenital fibromatosis should be a major consideration in the infant in the proper clinical context.

Acknowledgment

I am grateful to Dr. Raffaele Lattes, Chief of the Department of Surgical Pathology of the Columbia–Presbyterian Medical Center, for the pathologic sections and descriptions contained herein and for the time and energy he devoted in the review of this manuscript.

References

1. Agneesens A: Been ongewoon geval van neurofibromatose bij een zuigeling. Manndschr Kindergeneesk 18:347, 1950
2. Anderson DH: Tumours in infancy and childhood. I. Survey of those seen in pathology laboratory of the Babies Hospital during years 1939–1950. Cancer 4:890, 1951
3. Arlen M, Koven L, Frieder M: Juvenile fascial fibromatosis of the forearm with osseous involvement. Report of a case. J Bone Joint Surg 51A:591, 1969
4. Baer JW, Radkowski MA: Congenital multiple fibromatosis. A case report with review of the world literature. Am J Roentgenol 118:200, 1973
5. Baird PA, Worth AJ: Congenital generalized fibromatosis: an autosomal recessive condition? Clin Genet 9:488, 1976
6. Bartlett RC, Otis RD, Laakso OA: Multiple congenital neoplasms of soft tissues. Report for four cases in one family. Cancer 14:913, 1961
7. Battifora H, Hines JR: Recurrent digital fibromas of childhood. Cancer 27:1530, 1971
8. Beatty EC: Congenital generalized fibromatosis in infancy. Am J Dis Child 103:128, 1962
9. Condon VR, Allen RP: Congenital generalised fibromatosis: case report with roentgen manifestations. Radiology 76:444, 1961
10. Drescher E, Woyke S, Markiewicz C, Tegi S: Juvenile fibromatosis in siblings. J Pediatr Surg 2:427, 1967
11. Enzinger FM: Fibrous hamartoma of infancy. Cancer 18:241, 1965
12. Familusi JB, Nottidge VA, Antia AU, Attah EB: Congenital generalized fibromatosis. An African case with gingival hypertrophy and other unusual features. Am J Dis Child 130:1215, 1976
13. Feiner H, Kay GI: Ultrastructural evidence of myofibroblasts in circumscribed fibromatosis. Arch Pathol Lab Med 100:265, 1976
14. Gabbiani G, Majno G: Dupuytren's contracture: fibroblast contraction? An ultrastructural study. Am J Pathol 66:131, 1972
15. Geshickter CF: Desmofibromatosis and viscerofibromatosis. Bull Georgetown Univ Med Center 8:200, 1955
16. Heiple KG, Perrin E, Aikawa M: Congenital generalized fibromatosis. A case limited to osseous lesions. J Bone Joint Surg 54A:663, 1972
17. Hower J, Gobel FM, Ruttner JR, Surster K: Familaire kongenitale generalisierte Fibromatose bei zwei Halbschwestern. Schweiz Med Wochenschr 101:1381, 1971
18. Jaffe HL: Tumors and Tumorous conditions of the Bones and Joints. Philadelphia, Lea and Febiger, 1958, p 249
19. Kauffman SL, Stout AP: Congenital mesenchymal tumours. Cancer 18:460, 1965
20. Kingsbury GH, Perrin E, Atikawa M: Congenital generalized fibromatosis. J Bone Joint Surg 54A:663, 1972
21. Larregue M, Poitou JP, Bressieux P, DeGiacomoni P, Vant F: Fibromatose congénitale généralisée. Ann Dermatol Vénéreol (Paris) 104:349, 1977

22. Lin JJ, Svoboda DJ: Multiple congenital mesenchymal tumors. Multiple vascular leiomyomas in several organs of a newborn. Cancer 28:1046, 1971
23. Lipschutz A, Vargas L: Structure and origin of uterine and extra genital fibroids induced experimentally in guinea pig by prolonged administration of oestrogens. Cancer Res 1:236, 1941
24. McKusick VA: Mendelian Inheritance in Man. Catalogs of Autosomal Dominant, Autosomal Recessive and X-Linked Phenotypes. Baltimore, Johns Hopkins Univ. Press, 1975
25. Mamelle JC, Salle B, Daudet M, Rosenberg D, Brunat M, Monnet P: Fibrosarcome et fibromatose congénitaux. Ann Pédiatr Paris 16:255, 1969
26. Mande R, Hannequet A, Loubry P, Cloup M, Marie J: Fibromatose congénitale diffuse de nouveau-né à évolution régressive. Ann Pédiatr Paris 12:692, 1965
27. Mauclaire P, Legry H: Fibromatose généralisée de nature congénitale. Bull Mem Soc Chir Paris 41:1156, 1915
28. Morettin LB, Müller E, Schreiber M: Generalised hamartomatosis (congenital generalised fibromatosis). Am J Roentgenol 104:722, 1972
29. Nadel EM: Histopathology of oestrogen induced tumours in guinea pigs. J Natl Cancer Inst 10:1043, 1950
30. Plaschkes J: Congenital fibromatosis: localized and generalized forms. J Pediatr Surg 9:95, 1974
31. Schaffzin EA, Chung SMK, Kaye R: Congenital generalized fibromatosis with complete spontaneous regression. A case report. J Bone Joint Surg 54A:657, 1972
32. Schlangen JT: Congenital generalized fibromatosis. A case report with roentgen manifestations of the skeleton. Radiologia Clin 45:18, 1976
33. Schlösser W, Rotzler A: Zur Kasuistik angeborener Geschwülste; spindelzilleges Sarkim mit ausgedehnter Metastasierung. Zentralbl Allg Pathol 88:161, 1952
34. Schnitka TK, Asp DM, Horner RH: Congenital generalized fibromatoses. Cancer 7:953, 1958
35. Stout AP: Juvenile fibromatosis. Cancer 7:935, 1954
36. Stout AP, Lattes R: Armed Forces Institute of Pathology. Atlas of Tumour Pathology. Vol 1, 2nd ser, 1967
37. Teng P, Warden MJ, Cohn WL: Congenital generalized fibromatosis (renal and skeletal) with complete spontaneous regression. J Pediatr 62:748, 1963
38. Ts'o TO, Teoh TB: Fibromatosis in an infant. J Pathol Bacteriol 85:521, 1963
39. Williams JO, Schrum D: Congenital fibrosarcoma. Report of a case in a newborn infant. Arch Pathol 51:548, 1951

The Mucopolysaccharidoses—Lipid and Nonlipid

David H. Baker

"Mucopolysaccharidosis" (MPS) is a term introduced by Brante in 1952 that is now used to define a group of heritable diseases having a common etiology. The affected patients show abnormal metabolism of mucopolysaccharides (Fig. 1). Intracellular storage and/or abnormal excretion of urinary mucopolysaccharides, together with certain physical findings, characterize this group of diseases.

The term "Hunter–Hurler syndrome" results from Major Ian Hunter's report in 1917 and Gertrude Hurler's report in 1919, where in one instance an X-linked form of the disease was seen and in the other an unknown, presumably autosomal recessive form, was seen. The term "gargoylism" was introduced by Ellis in 1936 at the proceedings of the Royal Society of Medicine. It is an unfortunate term that unfortunately has stuck—unfortunate because of the unpleasant connotation for the families and unfortunate because of the fact that "gargoyle" really refers to the throat and was used to describe decorative rain spouts on Gothic buildings.

By 1965, McKusick and colleagues had differentiated six mucopolysaccharidoses according to clinical findings, genetic pattern of inheritance, and the pattern of excretion of mucopolysaccharides in urine. These were designated as MPS I through VI. MPS I was the contraction for the Hurler syndrome, MPS II for the Hunter syndrome, MPS III for the Sanfilippo syndrome, MPS IV for Morquio syndrome, MPS V for Scheie syndrome, and MPS VI for the Maroteaux–Lamy syndrome. Since that time, further classification has suggested an MPS I-H for Hurler and MPS I-S for Scheie syndromes, so that currently there is no V classification, but rather MPS I-H and MPS I-S.

The more we know about the biochemistry of these diseases, the less important the radiographic distinctions become. However, the major radiographic features of the various forms of mucopolysaccharidoses are listed in Table 1. MPS I-H, MPS II A, MPS IV, and MPS VI A have significant roentgen features. The main clinical features of MPS I-H consist of skeletal abnormalities which include stiff joints, dwarfism, marked and progressive mental retardation, and corneal clouding. Most cases do not live to the age of 10 years. This disease is usually diagnosed in early childhood rather than at birth and the children are usually considered normal in the first few months. The circumstances that cause the child to be brought to medical attention include musculoskeletal system problems, with stiff joints, lumbar gibbous, and inability to walk at the usual time. The patients also show excessive nasal production of mucus along with noisy breathing. Progressive medical and physical deterioration is a striking characteristic. Stunted growth, coarsening of the facial features, and clouding of the corneas appear in the first years of life. Limitation of joint motion becomes progressive, as does mental impairment and hepatosplenomegaly. Anatomic changes in the coronary arteries and heart valves are also progressive and result in death due to cardiac failure, usually before 10 years.

This component of the disease is frequently complicated by recurrent pulmonary infections. A good clinical description of the gargoyle comes from Jackson's article in the *American Journal of Medicine* and states in part "these ugly, blind, weak, pasty, hoarse, big-eared, bushy eyebrowed, hyperteloric, snub-nosed, open-mouthed, prognathic, hairy, claw-fingered, pot-bellied, knock-kneed, flat-footed, knobby-jointed, dwarfed idiots." That is not par-

TABLE 1. Mucopolysaccharidoses

Designation	Name	Clinical Features
MPS I-H*	Hurler syndrome	Early clouding of cornea; grave manifestations; death usually before age 10 years
MPS I-S	Scheie syndrome	Stiff joints; cloudy cornea; aortic regurgitation; normal intelligence; (?) normal life span
MPS I-H/S	Hurler–Scheie compound	Phenotype intermediate between Hurler and Scheie
MPS II A*	Hunter syndrome, severe	No clouding of cornea; milder course than in MPS I-H, but death usually before age 15 years
MPS II B	Hunter syndrome, mild	Survival to 30s to 50s; fair intelligence
MPS III A	Sanfilippo syndrome A	Identical phenotype; mild somatic; severe central nervous system effects
MPS III B	Sanfilippo syndrome B	
MPS IV*	Morquio syndrome	Severe bone changes of distinctive type; cloudy cornea; aortic regurgitation
MPS V	Vacant (formerly Scheie syndrome)	—
MPS VI A*	Maroteaux–Lamy syndrome, classic form	Severe osseous and corneal change; normal intellect
MPS VI B	Maroteaux–Lamy syndrome, mild form	Mild osseous and corneal change; normal intellect
MPS VII	β-glucuronidase deficiency (more than one allelic form)	Hepatosplenomegaly; dysostosis multiplex; white cell inclusions; mental retardation

*Significant roentgenographic features.

ticularly kind, but it is an accurate clinical description of a patient with Hurler's syndrome (Fig. 1).

The radiographic aspects closely parallel the clinical findings. There are several features that, taken in combination, make the diagnosis in the patient of the proper age quite specific. These include the hands, where the proximal ends of the metacarpals become pointed and the phalanges become bullet-shaped; at the same time, the radius and ulna grow together, with a rather characteristic growth deformity. Posterior ribs are thin and the anterior lateral ribs wide, giving them a spatula appearance. There is genu valgum characteristically of the proximal femoral necks and thickening and apparent shortening of the clavicles. The skull, which is originally normal, shows progressive elongation and eventual evidence of increased intracranial pressure in the coronal and lambdoid sutures, while the sagittal suture has become solidly fused (Fig. 2). This is the one example we know of where the postnatal fusion of the sagittal suture results in a deformity. The sella turcica enlarges, and there may be large deposits of mucopolysaccharide around the base of the brain and also within the mandible, causing large radiolucent defects in that bone.

A keel-shaped or hypoplastic L1 is another usual accompaniment. This has caused gargoylism to be confused with a number of other conditions, principally cretinism, where a similar change may also be seen, and by the widest stretch of imagination, achondroplasia.

The inheritance of Hurler's syndrome is felt to be

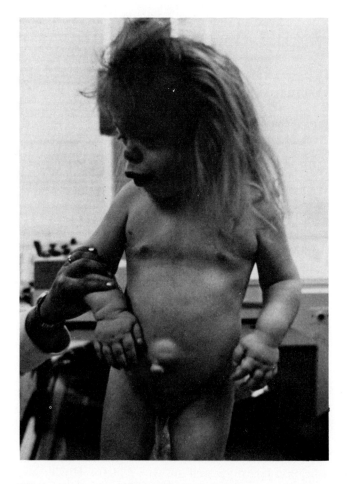

FIG. 1. Typical patient with Hurler's disease, showing the coarse facial features, the pot belly, and the claw hand.

The Mucopolysaccharidoses

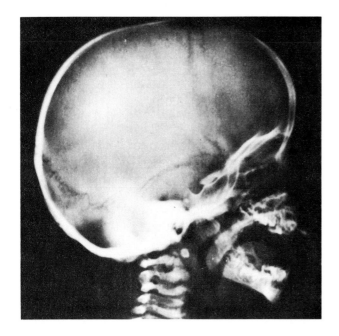
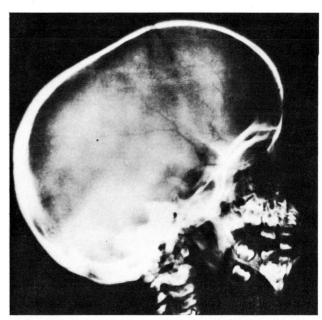
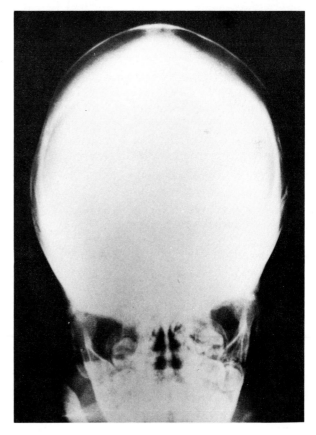

FIG. 2. Above left: Normal skull in lateral projection Above right: The same patient 4 years later; A dolicocephalic configuration with prominence of the coronal sutures. Opposite: The skull in frontal projection showing solid bony fusion of the sagittal suture.

autosomal recessive. MPS I-S, the Scheie syndrome, will not be discussed at length, because distinctive radiographic features are essentially absent.

MPS II—Hunter Syndrome

This mucopolysaccharidosis has many points in common with MPS I-H. It can be differentiated by less severe mental impairment and X-linked pattern of inheritance and a lack of corneal clouding.

Currently there are felt to be two different phenotypic components to MPS II, one a much more severe type than the other, with the patients usually succumbing before the age of 15 years with mental retardation, and the other a milder form of the disease with better intelligence and considerably longer survival. The roentgen changes and the physical features are quite similar to those of MPS I-H, although not as severe. Corneal clouding cannot be found in the majority of patients. The sex-linked inheritance has been quite well worked out in this disorder, and there is one report of a patient with progeny whose daughter subsequently had two of three male children born with the condition.

MPS III—Sanfilippo Syndrome

These patients show fewer somatic manifestations than those with the other mucopolysaccharidoses. The most marked manifestation is severe and progressive mental retardation. Radiographs are not usually helpful in this group.

MPS IV—Morquio Syndrome

This is the other severe radiographically involved disorder. The confusion over the names and who described the condition and who should get credit for it is beyond the scope of this paper. Needless to say, the finding of keratin sulfate in the urine helps to confirm and classify this condition as an abnormal mucopolysaccharidosis and to allow a more definitive description.

The major clinical findings are short-trunked dwarfism with protrusion of the sternum, genu valgum (as in Hurler syndrome), pes planus, loose ligaments and skin, mild protrusion of the lower portion of the face, and possible fine corneal opacities. Hearing is impaired and there may be hepatomegaly. Most patients are not diagnosed at birth, and psychomotor development is normal for the first year. Patients are frequently brought to medical attention because of delay in standing and walking or because of a waddling gait or flat feet, protrusion of the sternum, or thoracolumbar kyphosis. As the child grows older, the pes planus and genu valgum worsen, there is increased protrusion of the sternum, the patients stand in flexion, and dwarfism becomes evident. A good clinical description also appears in Jackson's article as follows: "The changes produce short necked, round shouldered, hump backed, pidgeon chested, knock-kneed, flat-footed dwarfs, waddling like ducks with head thrust forward, lower ribs flared upward, hips and knees flexed and with hanging arms which are too long for their stunted bodies." This is a good clinical description that mirrors the x-rays (Fig. 3). Another finding is hypoplasia of the odontoid process, which may lead to pyramidal signs and actually sudden death with acute atlantooccipital subluxation. Roentgen changes include those that are quite similar but not identical to those of MPS I-H, and here the spine goes through striking changes with a much more universal vertebral plana. Coxa valga is always present and helps serve to distinguish this type from the major non-mucopolysaccharidoses with which it is confused, namely, spondyloepiphyseal dysplasia. Odontoid hy-

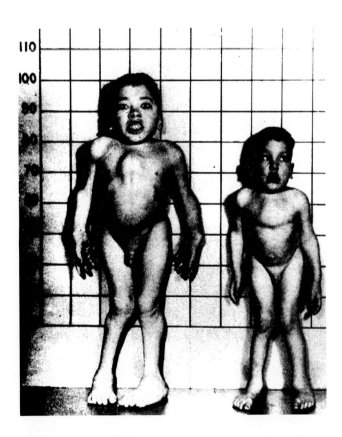

FIG. 3. Frontal photograph of two brothers (From McKusick VA: Heritable Disorders of Connective Tissue, 3rd ed. St. Louis, Mosby, 1966.)

poplasia was felt for a while to be a distinguishing feature, but unfortunately, both conditions may show that particular abnormality.

MPS VI—Maroteaux–Lamy Syndrome

This condition is very similar radiographically to that of patients with MPS I-H. These patients maintain normal or relatively normal mental development, and in our experience this is a much rarer condition. We have not been in the position of confusing this with the more commonly seen Hurler syndrome, but the radiographs, particularly any single radiograph (e.g., wrist or pelvis), might be very confusing.

Mucolipidosis GM I and I Cell Disease

The conditions which can occasionally be confused with the mucopolysaccharidoses such as the mucolipidoses GM I, for instance, are seen in early infancy, and the osseous changes here may be similar to those in MPS I, but much more severe and with a much earlier onset, or they may merely consist of a keel-shaped vertebral body in the upper lumbar area with the rest of the spine and skeleton appearing normal. GM I, generalized infantile gangliosidosis, or neural visceral lipoidosis, can be easily confused with I cell disease, particularly in early infancy. Cherry red maculae have not been seen in I cell disease, while they are the rule of GM I.

Bibliography

Gardner LI: Endocrine and Genetic Disease in Childhood and Adolescence, 2nd ed. Philadelphia, Saunders, 1975

Neuhauser EBC, Griscom TN, Gilles FH, Crocker AC: Arachnoid cyst in Hurler Hunter syndrome. Arch Radiol Dis 11:453, 1968

Spranger JW, Langer LO Jr, Widemann HR: Bone Dysplasia. An Atlas of Constitutional Disorders of Skeletal Development, 1st ed. Philadelphia, Saunders, 1974

Hematologic and Storage Diseases

Walter E. Berdon

The common childhood hemolytic anemias that have x-ray changes are Cooley's anemia (thalassemia major) and sickle-cell anemia and its variants, SC and S-Thal.

Sickle-Cell Anemia

Sickle-cell anemia has a wide spectrum of findings, primarily related to ischemia and infarction rather than marrow hyperplasia. The findings are age related. The infant's bone changes are in the small bones of the hands and feet (hand–foot syndrome). With increasing age, extremity destructive patterns are seen that are similar in appearance to those noted in association with malignancies such as leukemia (Fig. 1).

The older child, adolescent, or adult finally shows the changes of aseptic necrosis as seen in the femoral head and humeral head, as well as in the spine, in the form of depressions of the vertebral centra (codfish vertebra, Fig. 2). Infection is common in sicklers, may occur on the basis of salmonella or hemophilus influenza, and is often probably superimposed on infarction. We cannot differentiate infection from infarction solely on the basis of radiographs.

Cooley's Anemia

The well-known changes of marrow hyperplasia account for the hair-on-end skull changes and widened medullary cavities in the long bones and vertebral bodies (Fig. 3), which can narrow the space for the spinal cord so as to produce a narrow canal syndrome in the adult. Premature closure of part of the humeral epiphyseal plate may produce a growth

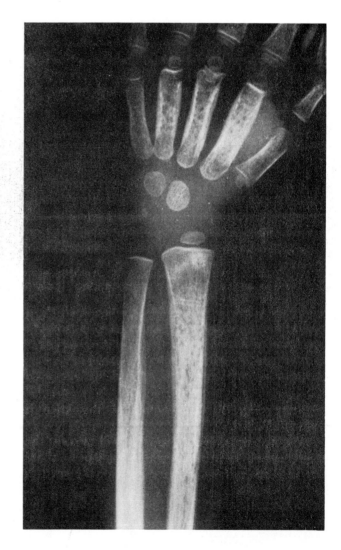

FIG. 1. Diffuse destructive pattern in hands and forearm bones of a 3-year-old sickle-cell patient who was febrile. Blood cultures were negative, though the patient was treated with penicillin for a possible superimposed infection.

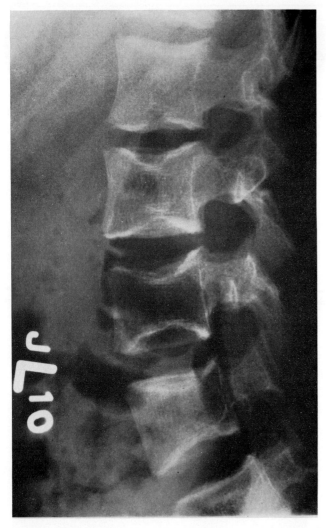

FIG. 2. Typical central depression in an older sickler's lumbar spine probably representing growth disturbance.

necrosis, moth-eaten bones with fractures, and modeling deformities) (Fig. 5) that resemble sickle-cell anemia in some ways.

The modeling deformity of expanded femoral shafts is also seen in Cooley's anemia. Unlike the case in Cooley's patients, the skull is rarely involved. The "infantile" form of Gaucher's disease is one of rapid clinical course with central nervous system involvement leading to early death. Other stigmata of Gaucher's disease seen in the pediatric age group are similar to those seen in adults.

deformity (Fig. 4); less commonly, the knee is similarly affected.

Early high transfusion treatment will prevent the bone changes. Infarctive changes similar to those seen in sickle-cell anemia are occasionally seen in Cooley's anemia.

"Adult" Gaucher's Disease

Gaucher's disease is a storage disease affecting primarily Jewish people and other people of Mediterranean descent. It produces bone changes (aseptic

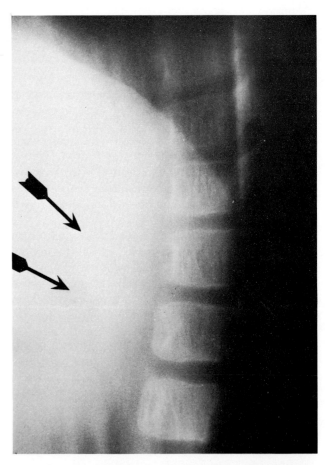

FIG. 3. Coarse reticular pattern in the spine of a 10-year-old with thalassemia major. The huge dense liver, laden with iron from many high transfusions, casts a white image over the abdomen. Dense lymph nodes also laden with iron are dimly seen (arrows).

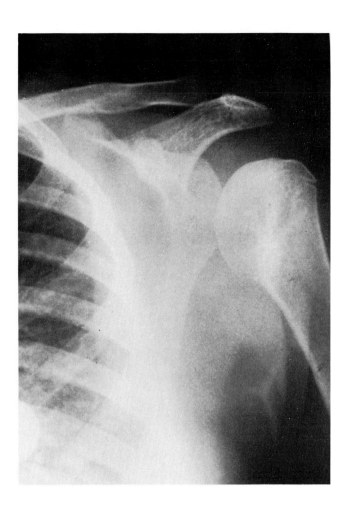

FIG. 4. Sloping humeral epiphyseal plate in older thalassemic patient. This condition is of unknown etiology and appears late in the first decade.

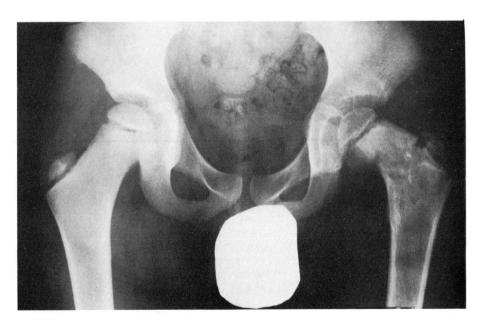

FIG. 5. Pathologic fracture of left femoral neck in an 8-year-old with Gaucher's disease. Both diaphyseal and epiphyseal involvement are seen.

Marrow Imaging with Radioiron: Clinical Application

Rashid A. Fawwaz

Marrow imaging is rarely done, due in large part to the physical and biologic shortcomings of the radioactive compounds used.[1,2] Marrow imaging should detect extramedullary as well as medullary hematopoietic sites. This requirement severely limits the usefulness of radiocolloids and radioindium, since they accumulate in normal liver and spleen. In contrast, radioiron is more specific and identifies sites of erythropoiesis and iron storage. The Anger whole-body scanner Mark II[3] allows visualization of the high-energy gamma-emitting radionuclides of iron ^{52}Fe and ^{59}Fe.

Diagnosis by pattern recognition of the marrow image without correlation with the clinical problem can be erroneous. We will present examples of useful information that can be obtained when radioiron images are interpreted in conjunction with the hematologic status of the patients, based on our experience in over 150 studies.

Methods

Iron-52 as ferrous citrate was used alone in the first eleven cases presented. Imaging was performed 16 hours after an intravenous dose ranging from 50 to 200 microcuries. In the last two cases, both ^{52}Fe and ^{111}In were used. The latter radionuclide was used as indium chloride in a dose of 3 millicuries, with scanning performed 72 hours after intravenous administration and within 1 week of the ^{52}Fe scan. The Anger whole-body scanner Mark II was used as the imaging device. Whole-body scans were performed in the anterior and posterior views; each view required 22 minutes for ^{52}Fe. Occasionally, spot scintiphotos were taken with an Anger positron camera.

When a scan is obtained approximately 1 day following the administration of iron, splenic localization represents erythropoiesis; hepatic localization reflects erythropoiesis and/or storage of iron diverted to the liver due to an overall decrease in erythropoietic activity.

Case Reports and Results

Staging of Polycythemia Vera

Case 1. At age 14, this patient was found to have polycythemia vera with painful splenomegaly, erythrocytosis, thromboleukocytosis, and hyperplastic marrow. Intermittent courses of busulfan were given throughout a 2-year period. At age 16, he was well, with no splenomegaly and a normal blood count. At age 26, he had an asymptomatic relapse with slight splenomegaly and marked thrombocytosis. He was appropriately concerned about his prognosis. A ^{52}Fe scan (Fig. 1) was done and showed normal central erythropoiesis. Intermittent courses of busulfan were again given for the next 2 years. Presently, at age 31, he is still well, with a normal blood count and no splenomegaly.

Our experience has confirmed the initial study of Pollycove et al[4] in that marrow imaging shows a changing pattern over the life span of patients with polycythemia vera. In the early stage, the scan is normal; later, it shows peripheral extension; still later, it shows splenic erythropoiesis, central marrow failure, and possibly greater peripheral extension.

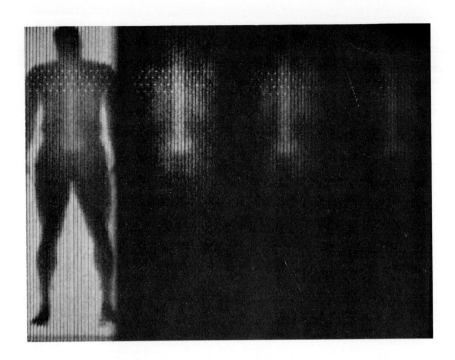

FIG. 1. Whole-body scan (posterior view) demonstrates normal central erythropoiesis in a patient with polycythemia vera. Unless indicated, this and subsequent figures show varying-intensity images of a single-pass ^{52}Fe body scan.

The normal scan in this patient was clinically useful as a favorable prognostic indicator.

Evaluation for Splenectomy in Myeloid Metaplasia

Case 2. At age 52, this man developed polycythemia vera, subsequently controlled with ^{32}P and phlebotomy therapy. At age 64, he was anemic and had painful splenomegaly, with early satiety and marked weight loss. The sternal marrow was still hyperplastic. The ^{52}Fe scan (Fig. 2) showed a third of his erythropoiesis in the spleen and two-thirds in the central and peripheral marrow. Acquisition of quantitative information was obtained by utilizing computer-assisted delineation of areas of interest.[5] Because a major component of erythropoiesis was extrasplenic, splenectomy was performed. The spleen showed marked myeloid metaplasia. He did well postoperatively, with regain of weight and disappearance of the anemia; the latter recurred 2 years

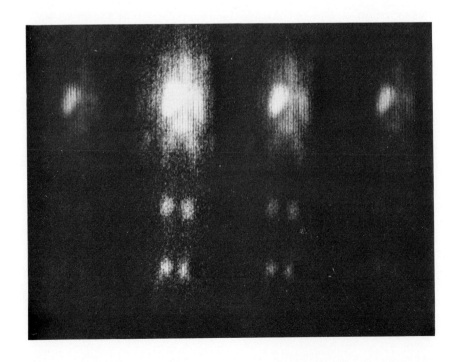

FIG. 2. Whole-body scan (posterior view) demonstrates splenic myeloid metaplasia, with the major component of erythropoiesis extrasplenic, in a patient with polycythemia vera.

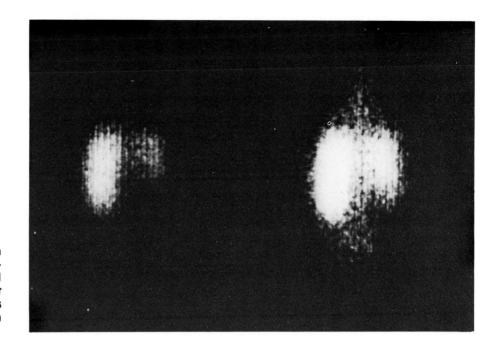

FIG. 3. Whole-body scan (posterior view) demonstrates splenic myeloid metaplasia, with the major component of erythropoiesis intrasplenic, in a patient with polycythemia vera.

later. Presently aged 67, he is being treated with testosterone and has had no further progression of the anemia.

We wish to emphasize that diagnosis based purely on the appearance of the scan—pattern recognition—is deplored. We have seen a patient with a scan identical to the one discussed above due to DiGuglielmo's erythroleukemia.

Case 3. At age 70, this woman developed polycythemia vera requiring ^{32}P and phlebotomy therapy. By age 74, splenomegaly had progressed to 18 cm below the costal margin, with a 5-kg weight loss. Although she was not anemic, marrow biopsy showed significant myelofibrosis. The ^{52}Fe scan (Fig. 3) showed that most of her erythropoiesis was in the spleen and liver, with only a minimal amount in her marrow. Splenectomy was not done. Instead, cautious intermittent courses of busulfan were given, resulting in cessation of weight loss and moderate shrinkage of the splenomegaly. She is still alive at age 77, with a normal hemoglobin.

Myeloid metaplasia may occur *de novo* or subsequent to a previous myeloproliferative disorder, typically polycythemia vera. The spleen may be massive and painful. Associated findings may include weight loss (due to early satiety and/or hypermetabolic state), and thrombopenia (due to increased splenic trapping). Anemia is common, due to inadequate erythropoiesis (related to associated myelofibrosis), splenic hemolysis, and/or splenomegalic hypervolemia with a large plasma volume. Though splenectomy may help such patients, the decision to perform major surgery on elderly people who are chronically ill is difficult. A ^{52}Fe scan is useful in that quantitatively significant splenic erythropoiesis may be considered a contraindication to splenectomy.

Diagnosis of Myelofibrosis with Myeloid Metaplasia (MMM)

Case 4. At age 53, this man developed a mild thromboleukocytosis with occasional peripheral myelocytes and normoblasts. Bone marrow examination was refused. At age 60 he had an acute myocardial infarction with a concurrent cerebral arterial thrombosis associated with a marked thrombocytosis; the latter responded to a course of busulfan. At age 61 he had a more marked leukoerythroblastic blood picture with anemia and "teardrop" red blood cells; he developed pain and tenderness of his shoulders, thighs, and knees. Radiographs of the affected bones were normal. A ^{52}Fe scan (Fig. 4) showed splenic erythropoiesis, diminished central marrow erythropoiesis, hepatic iron uptake, and peripheral marrow extension corresponding to the areas of bone pain.

The scan confirmed the clinical diagnosis of MMM, even though no bone marrow biopsy was done. Metastatic carcinoma of the bones may cause a leukoerythroblastic picture similar to that seen in MMM but does not have a radioactive iron scan like this.

Skeletal pains may occur in patients with diseased bone marrow in the absence of radiographic changes. "Bone-seeking" radioisotopes may show abnor-

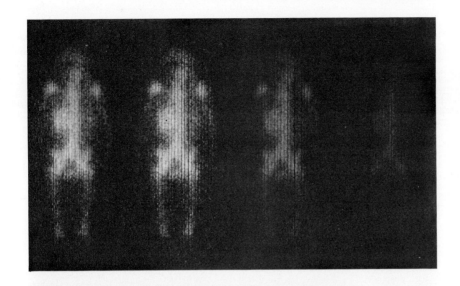

FIG. 4. Whole-body scan (posterior view) shows peripheral marrow expansion, corresponding to areas of bone pain, in a patient with polycythemia vera.

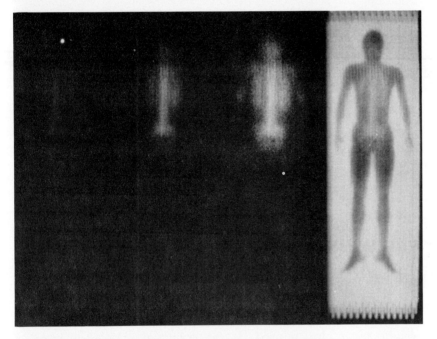

FIG. 5. Whole-body scan (posterior view) of a 9-year-old girl with hypoplastic anemia shows normal central marrow activity but lack of peripheral marrow extension expected for this age group.

FIG. 6. Whole-body scan (posterior view) of an 11-year-old male with aplastic anemia demonstrates absent marrow erythropoiesis; only hepatic iron storage is seen.

malities in such patients due to abnormal bone uptake and/or blood flow. Radioiron scanning is also useful in patients whose disease involves the marrow erythroid cells by showing increased uptake, as in the above case, or decreased uptake, as occurs in acute bone infarction (e.g., sickle-cell disease) or tumor (see Case 9).

Evaluation of Aplastic Anemia

Case 5. At the age of 9, this girl developed aplastic anemia. (Two siblings had died under the age of 10.) She had a hemoglobin of 6.7 g/100 ml, reticulocytes 0.2 percent, a platelet count of 35,000/cu mm with 14 percent neutrophils. An iliac marrow biopsy showed marked hypoplasia. A ^{52}Fe scan (Fig. 5) showed marrow activity still present but diminished in intensity and also restricted in extent (at this age much more peripheral activity should be present). Androgen therapy was begun with good response, so that at age 12, the hemoglobin was 13.9 g/100 ml, the platelet count was 87,000/cu mm, and the white cell count was 5,700/cu mm. A repeat ^{52}Fe scan showed normal central marrow uptake. The patient then refused further androgen therapy and her hemoglobin subsequently fell to 4.0 g/100 ml.

It is of interest that patients with aplastic anemia who respond to androgen therapy did not show marrow extension, suggesting that response occurs in hypoactive central marrow and does not involve conversion of peripheral fatty marrow. We have seen some patients with MMM who responded to androgen therapy with improvement of anemia and the concurrent development of brisk peripheral erythropoiesis, which had not been present before therapy. Finally, we have studied a man who developed erythrocytosis after about 1 year of testosterone injections, given for diminished libido; his scan was normal.

Case 6. The scan (Fig. 6) of this 11-year-old boy with idiopathic aplastic anemia shows absent marrow erythropoiesis; only hepatic iron storage is seen. An identical pattern can occasionally be seen in patients with complete replacement of erythropoiesis due to an acute leukemia. There was no response of his disease to 6 months of high-dosage androgen therapy, and he died within 1 year.

It is important to note that some patients with aplastic anemia show small nonhomogeneous foci of residual erythropoiesis in the marrow spaces. This is demonstrated in the spot scintiphoto of the pelvis of another patient with aplastic anemia (Fig. 7). In this case, routine bone marrow examination was not adequate for complete evaluation of the marrow status.

Finch et al state: "Limited quantitative data are available concerning the critical level of marrow function in aplastic anemia. The success of marrow

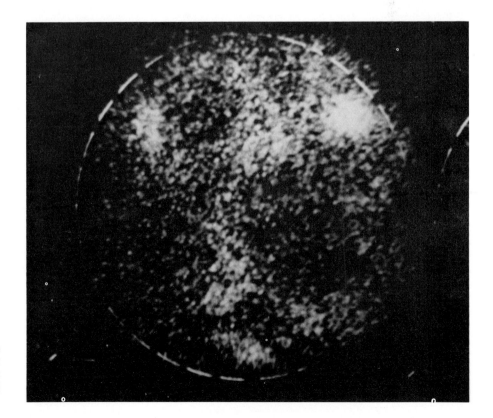

FIG. 7. Scintiphotos of the right hemipelvis of a patient with aplastic anemia. There is nonhomogeneous residual erythropoiesis.

transplantation in severely affected patients makes it essential to distinguish those patients in whom the degree of marrow damaged will be lethal."[6] We suggest that a radioiron scan may be useful; the presence or absence of residual erythropoietic tissue may be of importance in regard to the clinical decision to initiate androgen therapy or to attempt early marrow transplantation.

Differentiation of Benign from Malignant Eosinophilia

Case 7. At age 29 this man was found to have eosinophilia. Splenomegaly developed 1 year later. Extensive tests and biopsies were not contributory except for marked eosinophilic hyperplasia of the marrow and large numbers of eosinophils in a splenic aspirate. When first seen by us at age 33, he was asymptomatic, with splenomegaly, a mild anemia, and a white cell count of 23,000/cu mm (77 percent eosinophils). The marrow was hypercellular due to a mature eosinophilic hyperplasia, with a normal chromosomal pattern. Serum B_{12} was slightly elevated, and neutrophil alkaline phosphatase activity was slightly low. A ^{52}Fe scan (Fig. 8) showed significant peripheral marrow extension; hepatic iron storage was present. This is indicative of compromised erythropoiesis and gave support to the diagnosis of eosinophilic leukemia. At age 36 he developed bone pain and tenderness. At age 38 peripheral microembolization of his fingers and brain

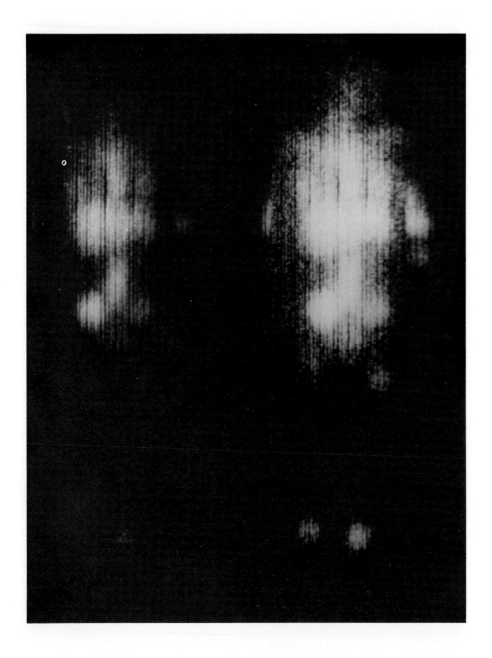

FIG. 8. Whole-body scan (anterior view) demonstrates peripheral marrow extension and hepatic iron storage in a patient with eosinophilic leukemia.

occurred; chemotherapy was started, but congestive heart failure and angiopathic hemolysis led to his death at age 39. Autopsy confirmed the clinical diagnosis of endomyocardial fibrosis. Eosinophilic leukemia was still present in his marrow and spleen.

We have studied three other patients with marked blood and marrow eosinophilia. Significant marrow extension was seen in one patient who developed clinical stigmata of endomyocardial fibrosis 2 years later as eosinophilia persisted. After 4 months of known eosinophilia, the second patient had a normal ^{52}Fe scan; the eosinophilia disappeared 18 months later and has not recurred in the last 2 years. The third patient, also with a normal scan, had an exfoliative dermatitis as the cause of the eosinophilia and is well now 16 months after diagnosis.

We feel that a radioiron scan is useful in evaluating eosinophilia of unknown cause; if marrow extension is seen, leukemia should be suspected as the diagnosis.

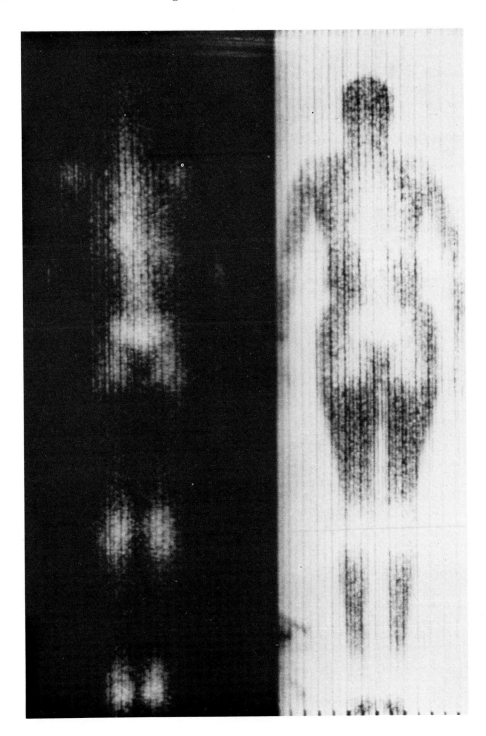

FIG. 9. Whole-body scan (anterior view) shows radio-iron localization in right paravertebral mass diagnostic of extramedullary erythropoiesis.

Detection of Extramedullary Erythropoietic Tissue

Case 8. A 31-year-old woman with severe thalassemia, postsplenectomy and postcholecystectomy, was found to have a right low thoracic paravertebral mass on a routine chest radiograph. It enlarged in a period of 6 months and did not visualize with a ^{99m}Tc–sulfur colloid scan. When referred to us, her hemoglobin was 9.5 g/100 ml, reticulocytes were 17.7 percent and bilirubin was 8.6 mg/100 ml. A ^{52}Fe scan (Fig. 9) showed, in addition to the peripheral marrow extension, brisk pick-up by the chest nodule, diagnostic for extramedullary erythropoiesis. Thoracotomy was avoided.

Paravertebral extramedullary erythropoiesis in patients with severe chronic hemolytic anemia may present a diagnostic problem. The tissue may not contain enough reticuloendothelial cells to significantly concentrate a radiocolloid. Radioiron is the scanning agent of choice. This is of more than academic interest. As discussed by Sorsdahl et al[7] (a) this tissue is quite vascular, and alarming hemorrhage can occur from diagnostic biopsy or surgical excision; (b) epidural extension can cause serious spinal cord damage; and (c) the tissue is radiosensitive.

We therefore advise that radioiron scans be done on these patients, if possible. The diagnosis should be considered established, and potentially dangerous diagnostic procedures—needle or open biopsies—should be avoided. When therapy is indicated (e.g., for epidural involvement), it may be instituted with radiotherapy in preference to surgery, even without a tissue diagnosis.

A similar problem rarely occurs in patients with MMM, who can develop symptomatic myeloid metaplasia in such extrahematic sites as the dura, retroperitoneum, and mesentery. In such cases, radioiron scans of the areas of concern may be useful.

Staging Tumors Involving Bone Marrow

Case 9. This 27-year-old woman had thigh pain due to a biopsy-proven reticulum cell sarcoma of the midfemur. There was no evidence of extraosseous disease. Hip disarticulation was being considered, until a ^{52}Fe scan (Fig. 10) showed ablation of normal marrow activity not only in the femur but in the adjacent hemipelvis. The latter was normal on radiography. Surgery was avoided and a more appropriate radiotherapy field was plotted.

No other radioisotopic scan was performed in this case. It is possible that the use of a radiocolloid might have provided the same information. A "bone-seeking" radioisotope (^{18}F or ^{99m}Tc) may not have been useful; these agents detect primarily abnormalities in bone calcification and/or blood flow and do not completely image the extent of some marrow tumors. We have noted, for example, normal bone scans in some patients with extensive myelomatosis of the marrow.

Evaluation of Marrow Reserve

Case 10. This 72-year-old man had a follicular lymphoma, stage III-A. As initial treatment, he received 4000 rads of radiation to the inguinal, iliac, and retroperitoneal areas. Upper-body radiation was then refused by the patient, and he was referred to us. Over the next 3 years, he developed some abnormal peripheral lymphocytes (cleft-cell type), increasing cervical and axillary adenopathy, and then a mild pancytopenia. Aspiration of the posterior iliac marrow space yielded fat. A ^{52}Fe scan was then done (Fig. 11) and showed no activity of the central pelvis and lower spine due to his previous radiotherapy. A sternal marrow aspiration produced normally cellular marrow containing 45 percent immature lymphocytes.

The hepatic iron pick-up is due to storage and indicates compromised erythropoiesis, in this case due to a combination of radiotherapeutic ablation and tumor infiltration of the remaining active marrow space. We feel that significant hepatic radioiron storage in nonhemochromatotic patients with hematologic disease is an early and sensitive indicator of decreased marrow reserve; if such patients require myelotoxic chemotherapy, the dosage may have to be lowered.

Detection of Unusual Pathology

Case 11. At age 50 this man was found to have a marked erythrocytosis. Even though he lacked splenomegaly and thromboleukocytosis, polycythemia vera was diagnosed, and he was treated with ^{32}P and phlebotomies. He was referred to us at age 60, still with an active erythrocytosis. Urinary erythropoietin was undetectable. A ^{52}Fe scan showed slight peripheral bone marrow extension. In addition, an abnormal focus of erythropoiesis was present in the left lower femur, as indicated in composite scintiphotos (Fig. 12). Aspiration of that site yielded very hypercellular marrow with marked erythroid hyperplasia (M/E ratio 1:5) with an increased number of immature erythroblasts. Chromosomal analysis showed increased polyploidy, "sticky" metaphases, and many anaphase bridges. A concurrent iliac marrow aspirate was only slightly hypercellular due to a mild erythroid hyperplasia (M/E ratio 1:1.5) with no increase in immature erythroblasts and a normal chromosomal pattern. The femoral lesion, containing 5 percent of the total counts, was irradiated, but this had no effect on the polycythemia, which required continued chemotherapy and phlebotomies. At age 65, a repeat ^{52}Fe scan showed normal femora, but slight splenic uptake was present.

An occasional patient with polycythemia vera presents with an isolated erythrocytosis rather than the more typical panmyelopathy and requires thorough evaluation for nonhematic disease as a possible etiology. We have been impressed by the usefulness of erythropoietin assays in these patients. In our experience, polycythemia vera is the only disease causing erythrocytosis associated with depression of erythropoietin production. An abnormal ^{52}Fe scan may also be diagnostically useful in this situation.

Though this is the only patient we have seen with an apparent "erythroblastoma," this observation provides additional evidence that polycythemia vera is a malignancy, albeit chronic. The lesion was irradiated in an unsuccessful attempt at treatment of the polycythemia; it is possible that some benefit did occur from ablation of a more malignant clone of erythroid cells.

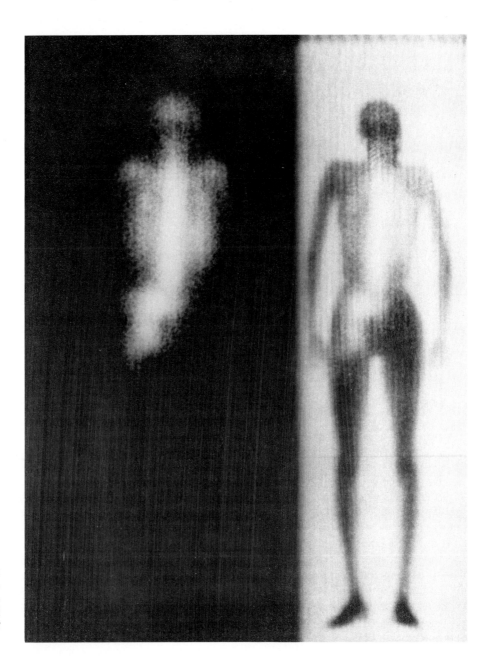

FIG. 10. Whole-body scan (posterior view) shows ablation of normal marrow activity in the right femur and hemipelvis due to tumor involvement.

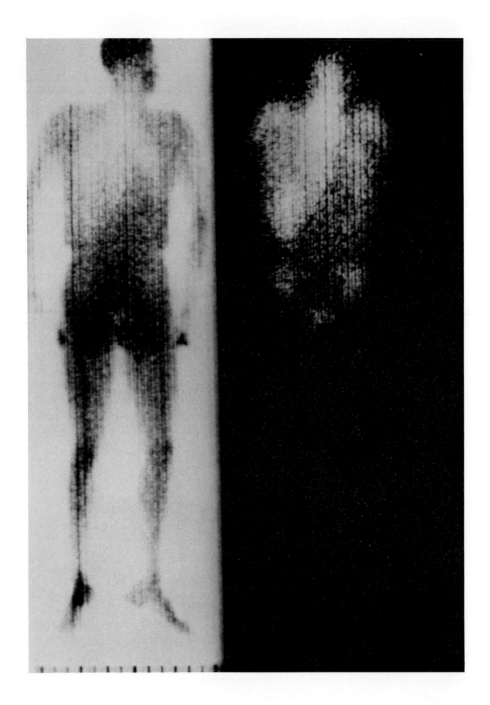

FIG. 11. Whole-body scan (anterior view) shows lack of activity in irridiated central pelvis and lower spine. Hepatic iron pick-up is due to storage and indicates compromised erythropoiesis.

Comparison of Radioiron, Radioindium, and Radiocolloid

The preceding cases make it clear that for marrow imaging to be maximally useful, one needs to obtain information regarding activity of all available hematopoietic sites. Because radioiron nuclides present difficulties in availability (^{52}Fe is cyclotron-produced and has an 8-hour half-life) and external detection (because of the high energies of ^{52}Fe and ^{59}Fe), other radionuclides with better physical characteristics have been used, particularly ^{99m}Tc–sulfur colloid and ^{111}In chloride.[8,9]

Case 12. The scans (Fig. 13) of a 34-year-old man with recently diagnosed polycythemia vera show good correlation of radioindium and radioiron in outlining the bone marrow. However, the striking visualization of the liver and spleen with radioindium obviously labels a nonerythroid component.

This phenomenon severely limits the clinical usefulness of radioindium as a total marrow-imaging agent. The same objection is raised for ^{99m}Tc–sulfur colloid, which labels hepatic–splenic reticuloendothelial cells. In addition, these cells can still be present in bone marrow that is otherwise aplastic, explaining the discrepancies noted by Van Dyke in

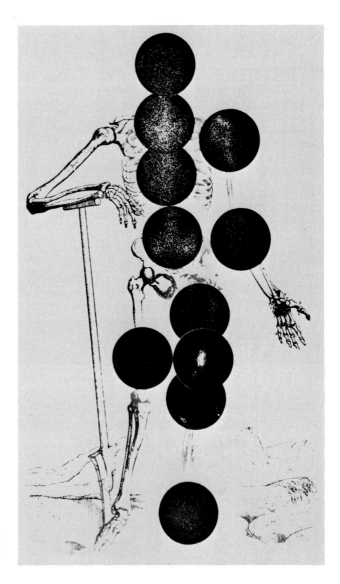

FIG. 12. Composite scintiphoto (anterior view) of a patient with polycythemia vera. Note abnormal focus of erythropoiesis in the left lower femur.

bone marrow imaging comparing radioiron with radiocolloid, particularly in patients with pure red cell aplasia.[10] We have seen occasional patients with significant discordance of bone marrow imaging when comparing radioiron and radioindium.

Case 13. This woman developed polycythemia vera at age 58, with a hemoglobin of 20 g/100 ml. Several phlebotomies were performed. Five years later, the hemoglobin was 16.7 g/100 ml; the white cell count was 21,500/cu mm. At this time, she developed splenomegaly with infarction and weight loss. A marrow biopsy revealed myelofibrosis. The question of splenectomy was raised and scans were obtained (Fig. 14). The radioiron scan showed practically all of her erythropoiesis was in the enlarged spleen, with no bone marrow activity. Splenectomy was not done. The radioindium scan showed an expanded bone marrow compartment down her long bones—this does not represent erythroid tissue. The radioindium scan could have been erroneously utilized in the clinical decision-making process for this patient.

Additional Observations

We have several patients with thrombocythemia vera (essential thrombocytosis, primary hemorrhagic thrombocythemia, idiopathic thrombocythemia) who had slight hepatic and/or splenic uptake of radioiron, signifying myeloid metaplasia of those organs, in the absence of myelofibrosis or a leukoerythroblastic blood picture. The subsequent course of these patients, as regards response to myelosuppressive therapy and longevity, has not appeared different from that of patients with normal scans. The presence of an abnormal scan might be useful

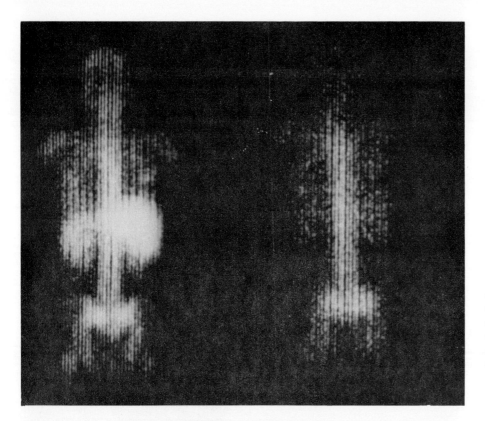

FIG. 13. Indium-111 and ^{52}Fe whole-body scans (posterior view) of a patient with polycythemia vera. There is good correlation between radioindium (left) and radioiron (right) in outlining the bone marrow. However, ^{111}In also labels nonerythroid components in the liver and spleen.

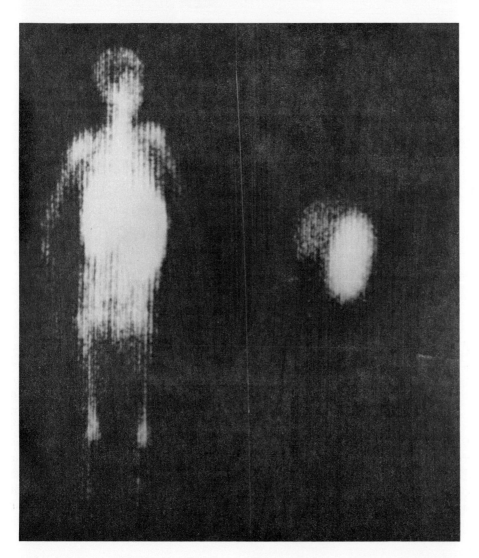

FIG. 14. Indium-111 and ^{52}Fe body scans (posterior view) of a patient with myelofibrosis. Note discordance of bone marrow imaging between radioindium (left) and radioiron (right).

diagnostically for the rare patient in whom one can not otherwise exclude a reactive thrombocytosis.

Peripheral extension of marrow activity is seen not only in chronic severe hemolytic anemias and in myeloproliferative disorders, but also can occur to some degree after extensive truncal radiotherapy. A marked example of peripheral extension was observed by us in a patient with hereditary hemorrhagic telangiectasia who has chronic nasal blood loss of over 500 ml a day. Maintained with 30 ml of intravenous iron–dextran weekly at a hemoglobin of 10 g/100 ml with a 30 percent reticulocytosis, he developed after several years secondary gout and painful leg bones. At this time, a scan showed marrow activity down to his hands and feet.

Comments

In the past, external imaging of radioiron has been limited by lack of adequate instrumentation for the high-energy gamma emitters ^{52}Fe and ^{59}Fe. This problem has been solved, we feel, by the use of the Anger whole-body scanner Mark II. The difficulty in obtaining ^{52}Fe may be obviated by the increasing number of commercial cyclotrons available. Iron-59 has a 46-day half-life and can be produced in a nuclear reactor; it has an advantage over ^{52}Fe in that a follow-up scan in 1 or 2 weeks may be done to detect the presence of splenic hemolysis of labeled red cells; this additional information may be useful in evaluation for splenectomy in occasional patients with myeloid metaplasia. One may also perform ferrokinetic plasma and red cell studies with ^{59}Fe in conjunction with the imaging as shown by Ronai et al.[11] A disadvantage as compared with ^{52}Fe is the inability to repeat the study for a number of months. In addition, the high gamma energy of ^{59}Fe does not permit resolution of small focal areas of uptake.

A radioisotopic scan is of greatest interest when it allows correlation of function and structure. Radioiron imaging of the marrow does this; it can provide unique and occasionally critical information for diagnosis, prognosis, and therapy in such hematologic disorders as myeloproliferative syndromes, anemias, and marrow malignancies.

References

1. Anger HO, Van Dyke DC: Human bone marrow distribution in vivo by iron-52 and the positron scintillation camera. Science 144:1587, 1964
2. Knuspe WH, Rayudu VMS, Cardello M, et al: Bone marrow scanning with ^{52}Iron (^{52}Fe). Cancer 37:1432, 1976
3. Anger HO: Whole body scanner mark II. J Nucl Med 7:331, 1966
4. Pollycove M, Winchell HS, Lawrence JH: Classification and evolution of patterns of erythropoeisis in polycythemia vera as studied by iron kinetics. Blood 28:807, 1966
5. Brown DW, Kirch DL, Trow RS, et al: Quantitation of the radionuclide image. Semin Nucl Med 3:311, 1973
6. Finch CA, Harbr LA, Cook JD: Kinetics of formed elements of human blood. Blood 50:699, 1977
7. Sorsdahl OS, Taylor PE, Noyes WD. Extramedullary hematopoeisis, mediastinal masses, and spinal cord compression. JAMA 189:343, 1964
8. Kniseley RM: Marrow studies with radiocolloids. Sem Nucl Med 2:71, 1972
9. Lilien DL, Berger HG, Anderson DP, et al: ^{111}In-chloride: a new agent for bone marrow imaging. J Nucl Med 14:184, 1973
10. Van Dyke DC, Shkerkin D, Price D, et al: Differences in distribution of erythropoeitic and reticuloendothelial marrow in hematologic disease. Blood 30:364, 1967
11. Ronai P, Winchell HS, Lawrence JH: Whole body scanning of ^{59}Fe for evaluating body distribution erythropoeitic marrow, splenic sequestration of red cells, and hepatic deposition of iron. J Nucl Med 10:469, 1969

The Pediatric Spine as a Clue to Other Diseases

Walter E. Berdon

In the "routine" x-ray examination of the spine of children, important, though not always appreciated findings, are present that can explain a variety of clinical problems.

Genitourinary Disease Explained by Spine Films

Children with incontinence, recurrent urinary tract infections, or constipation may have plain film findings that can lead to a diagnosis of neurogenic problems. This includes the partial absence of the sacrum, so easily missed on frontal films (Fig. 1). Such patients can develop progressive hydronephrosis with renal damage.

The routine use of lateral spine films on initial evaluation (Fig. 2) may therefore lead to early diagnosis and so prevent severe renal damage from developing (Fig. 3).

Less common is the bony defect associated with the anterior sacral meningocele–lipoma complex. Such patients (children or adults) may have similar genitourinary–gastrointestinal problems. CAT scanning is an ideal supplement to plain film study (Fig. 4). Myelography is, of course, definitive in detailing the coexistent problems of the anterior sacral meningocele and the tethered cord that go with this finding.

Finally, the lateral spine film as part of the IVP will discover patients with "occult" neurogenic problems, such as tethered cords, whose sacral canals are markedly widened. Some of these patients will show "soft" neurologic findings and cystometrograms will confirm the neurogenic aspects of their genitourinary problems (Fig. 5).

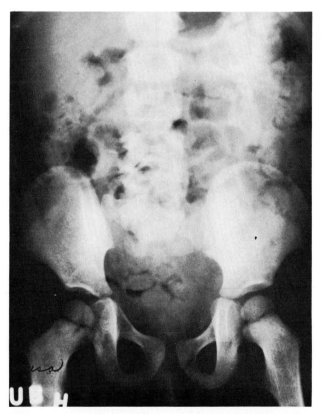

FIG. 1. This constipated child had many urinary tract infections. The missing S3–4–5 segments should be obvious on this film though it had been missed on many earlier studies.

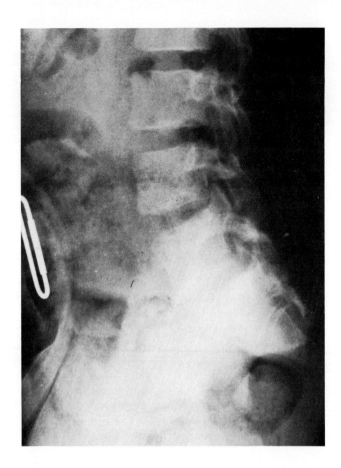

FIG. 2. Another patient had extensive GU surgery for reflux. A lateral sacral view that was subsequently obtained showed the missing sacral segments.

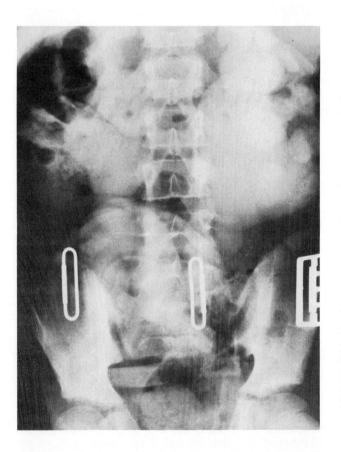
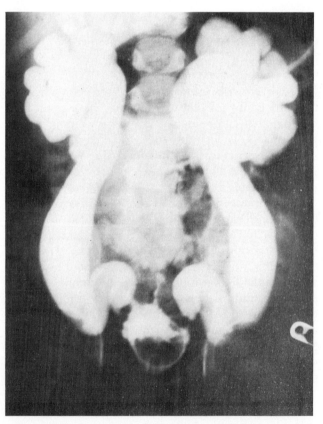

FIG. 3. Left: This is the same patient as in Fig. 2. Failure to appreciate the sacral problem led to years of genitourinary disease and a final ileal diversion. Unfortunately, severe irreversible renal damage had already occurred. Right: This is the same patient as in Fig. 1. This cystogram shows massive reflux from the contracted bladder. Note how easy it is to miss the sacral changes.

The Pediatric Spine

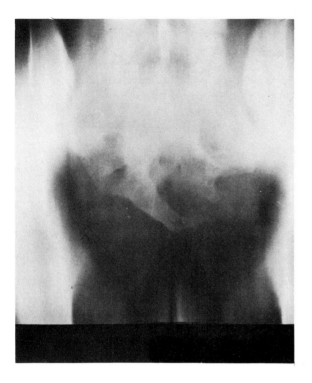

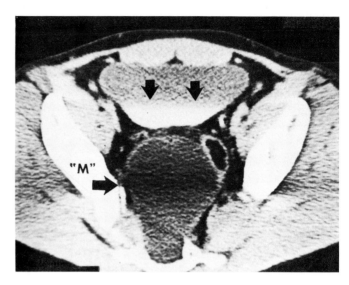

FIG. 4. Left: The "scimitar" defect in the sacrum is found in this teenager with incontinence. A rectal examination disclosed a presacral cystic mass. Right: CAT scan with contrast shows the sacral defect with the huge anterior sacral meningocele (M) in direct continuity with it. A layer of contrast is seen in the bladder.

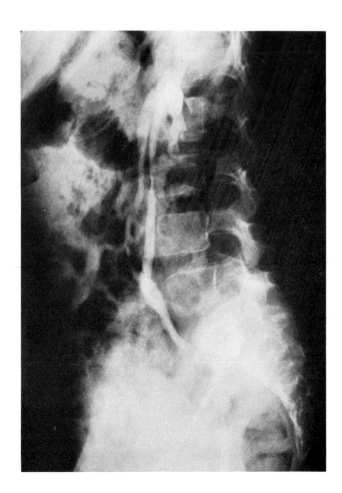

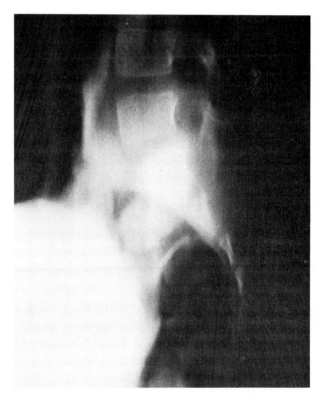

FIG. 5. A. Lateral IVP film in a boy with recurring urinary tract infections shows a widened sacral canal. B. Tomography confirms the finding. Myelography showed a tethered cord. The patient had "soft" neurologic findings. Cystometrogram showed a neurogenic bladder.

Clues to Multisystemic Disease

Clues to multisystemic disease are also provided by the curved spine in association with occult spinal cord tumor, pain from osteoid osteoma, and with spinal involvement by eosinophilic granuloma or infection, all emphasizing the need for multiple views and a high index of suspicion.

Role of Myelography

George F. Ascherl, Jr.

Plain film analysis is always an important part in the interpretation of radiographic contrast studies, particularly in evaluating disorders of the spine. Myelography, which involves the introduction of contrast agents into the spinal subarachnoid space is, to a great extent, performed to verify and provide therapeutically relevant and specific details about a diagnosis that is often already apparent from the plain spine radiographs. Myelography can be performed with both negative (air) and positive contrast agents. Pantopaque, an inert oily medium, is the positive myelographic contrast agent presently used in the United States. A new class of water-soluble contrast agents (e.g., Metrizamide) has been developed that are absorbed and therefore do not require removal from the subarachnoid space. It is anticipated that their use in this country will be legalized in the near future, thereby facilitating the myelographic procedure.

Lesions involving the spinal axis are for descriptive and differential diagnostic purposes placed in three categories: (1) *Extradural* lesions are those that arise extrinsic to the dura and project into the epidural space; the majority of these lesions arise from the intervertebral disks and osseous and ligamentous structures of the spine. (2) *Intradural extramedullary* lesions arise inside the dura and project into the subarachnoid space. (3) *Intramedullary* lesions arise within the spinal cord. While all three categories of lesions may produce osseous changes that herald their presence, if one includes degenerative processes of the spine (spondylosis), the extradural lesions are by far the most common and tend to be the most readily detectable by plain film examination.

Narrow Spinal Canal

Schlesinger and Taveras were the first to describe the spectrum of osseous spine changes resulting in a narrow spinal canal associated with a varying clinical presentation consequent to compression of the nerve roots of the cauda equina (Schlesinger–Taveras syndrome). The ventral border of the spinal canal is formed by the dorsal surface of the vertebral body, the lateral borders by the pedicles, and the dorsolateral borders by the articular facets and laminae, which unite in midline to form the posterior spinous process.

The narrow spinal canal (Schlesinger–Taveras) syndrome appears to occur randomly with no familial pattern, involving both sexes, but more frequently males. While it may be generalized and involve the cervical region, it is usually segmental, most commonly involving the caudal two or three lumbar segments. The narrow spinal canal results from a combination of shorter thicker pedicles, thickened articular facets, and thickened and more vertically oriented laminae. While there may be a decrease in the interpedicular distance, the dorsoventral or sagittal anteroposterior diameter of the spinal canal is most prominently diminished. Consequent to the above, there also tends to be narrowing of the intervertebral foramina.

On the plain lateral spine radiograph (Fig. 1), a measurement from a line connecting the posterior superior and posterior inferior corners of the vertebral body to the point of union of the laminae corresponds to the anteroposterior diameter of the spinal canal. A measurement of 16 mm or less,

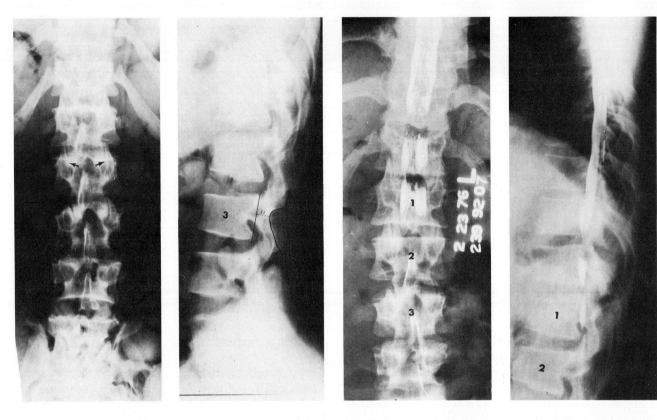

FIG. 1. Narrow canal syndrome in an achondroplastic dwarf. A. Anteroposterior Projection. Posterior apophyseal joints (arrows) characteristically more sagittally oriented in Schlesinger–Taveras syndrome are well seen. B. Lateral Projection. Sagittal anteroposterior diameter of spinal canal is a crucial measurement. C and D. Anteroposterior and lateral myelogram of same patient with contrast instilled from above, in cervical region, demonstrating marked encroachment on subarachnoid space with ultimate complete block to flow of pantopaque at the L1 level.

allowing for magnification (correction factor approximately 0.8), is considered a narrow canal. In accord with the more vertical orientation of the laminae, the plane of the articular facets (posterior apophyseal joints) tends to be more nearly parallel to the sagittal plane and therefore is usually well seen in the frontal projection (Fig. 1).

In the presence of a narrow spinal canal, there is only marginal room to accommodate the neural structures. Therefore, any superimposed degenerative process (hypertrophic osseous or ligamentous change or disk herniation) is poorly tolerated and manifests early because of cord or nerve root compression. Accordingly, the plain radiographs may demonstrate only minimal if any degenerative changes despite marked clinical symptoms. Myelography is necessary to demonstrate the anatomic significance of the narrow spinal canal and the need for widespread laminectomy.

A related disorder, achondroplasia, is probably nature's most exaggerated example of the narrow canal syndrome. The defective enchondral and increased periosteal bone growth seen in achondroplasia results in shortening and thickening of the osseous bounderies of the spinal canal (Fig. 1).

Spondylosis

Hypertrophic osseous and ligamentous changes may result in extradural encroachments on the spinal subarachnoid space or on the exiting roots or nerves. Hypertrophic osseous changes encroaching on intervertebral foramina or on the spinal canal (an anteroposterior diameter of 12 mm or less on a cervical region lateral radiograph signals the possibility of cord encroachment) are evident on plain radiographs, particularly tomograms. However, soft-tissue changes due to uncalcified ligamentous hypertrophy or herniated intervertebral disk can only be well demonstrated with myelography. The indirect sign of disk space narrowing signaling the presence of a herniated disk is frequently marginal or absent. Furthermore, the clinical symptoms may be due to disk herniation at a level other than that of the narrowed disk space. The patterns of neurologic deficit produced by a herniated disk at a given level depends on whether the herniation is central or lateral (Fig. 2). Herniated disks at different levels may potentially produce similar clinical findings. While the vast majority of herniated disks occur at L4–5 or L5–S1 (Fig. 3), myelography is necessary to verify the level of encroachment and thereby provide a roadmap for the neurosurgeon.

Scoliosis

Patients with scoliosis require myelographic evaluation whenever there is an associated neurologic deficit suggesting an intraspinal lesion and prior to surgical correction in any patient even though asymptomatic who in addition to scoliosis has an associated congenital spinal anomaly. These congenital anomalies may involve the vertebral bodies (fusion or segmentation anomalies) or the posterior elements (intersegmental lamina fusion) and are commonly associated with diastematomyelia as well as congenital midline lesions such as dermoid or neuroenteric cysts. A diastematomyelia (Fig. 4) is a congenital fibrocartilaginous or osseous septum usually found below the middorsal level, which divides the spinal canal in a dorsoventral (sagittal) plane transfixing the spinal cord or filum terminale. As a result, the cord may be tethered so that its tip (conus) lies more caudally than the normal (L1–2 level in adult). Recognition of a diastematomyelia is important, since surgical correction of scoliosis with corrective casts or intraoperative placement of Harrington rods may accentuate traction on the cord by the septum with potentially catastrophic neurologic sequelae. Also, it appears that in patients with diastematomyelia, if untreated, there is an increased tendency to develop scoliosis with time. In most cases, myelography is required to verify the presence of diastematomyelia, since the septum is not or is poorly calcified.

Neurofibromatosis

Lateral meningoceles (Fig. 5) (most commonly thoracic) and neurofibromas (neurolemmomas) (Fig. 6) often occur as isolated lesions. However, lateral meningoceles and neurolemmomas, particularly when multiple, are more commonly associated with generalized neurofibromatosis (von Recklinghausen's disease). While both may be associated with adjacent bone abnormalities (widening of intervertebral foramina, erosion of the posterior aspects of adjacent vertebral bodies or ribs) myelography is required to evaluate the extent of the intraspinal component of neurofibromas and indeed at times to distinguish dumbbell neurofibromas from lateral meningoceles.

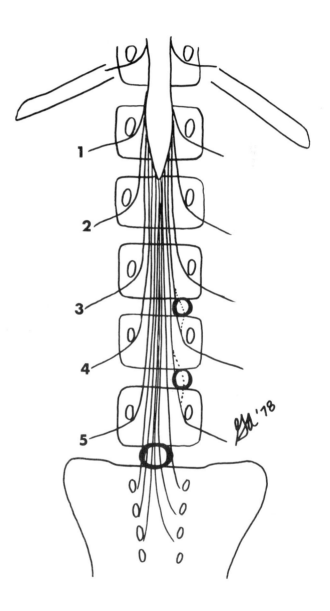

FIG. 2. Simplified scheme of relationship of nerve roots to pedicles and disk spaces. In lumbar region, spinal nerves exit below pedicle of their respective vertebral body. Note that a lateral disk herniation may affect an isolated nerve root, whereas ventral central herniations tend to affect multiple roots.

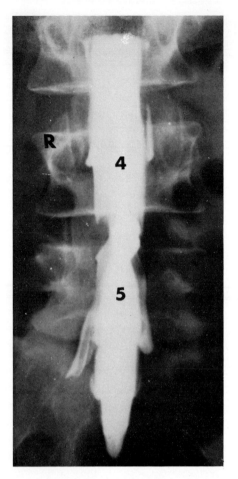 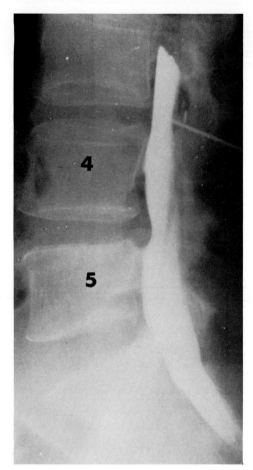

FIG. 3. Typical ventral extradural defect at L4–5 lateralizing to right side, consistent with herniated nucleus pulposus.

Bone Tumors of the Spine

Both primary and metastatic bone lesions may encroach upon the spinal canal and produce neurologic deficit (Fig. 7). Again, myelography is necessary to determine the extent of encroachment and thereby the need for immediate therapeutic measures, usually surgical decompression and/or radiotherapy.

Other Tumors of the Spinal Canal Contents

Intradural, extramedullary, and intramedullary tumors, over a period of time may manifest by erosion of the adjacent osseous structures. Such changes may be subtle or marked (Fig. 8).

Summary

The role of myelography is not primarily to exclude an abnormality, but rather to better define a lesion that is already suspected from plain films and clinical evaluation so that the proper therapeutic approach may be undertaken.

Bibliography

Duvoisin RC, Yahr MD: Compressive spinal cord and root syndromes in achondroplastic dwarfs. Neurology 12:202, 1962

Epstein BA, et al: Cervical spinal stenosis. Radiol Clin North Am 15:215, 1977

Epstein BS, et al: Lumbar spinal stenosis. Radiol Clin North Am 15:227, 1977

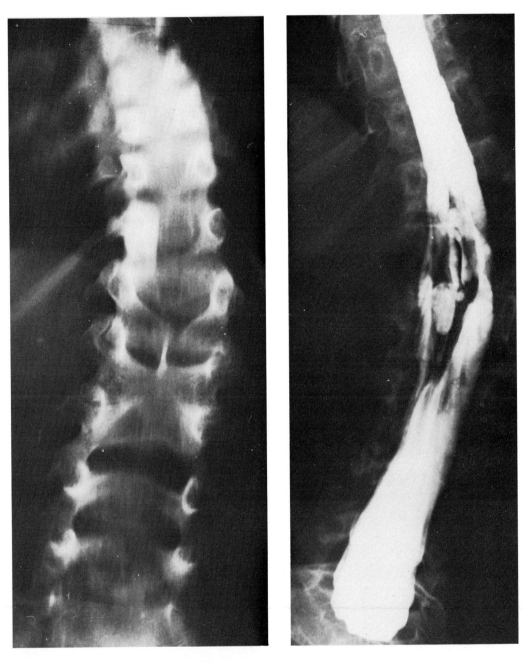

FIG. 4. Diastematomyelia with calcified sagittal septum transfixing spinal cord. Note cord widening and low-lying position of conus.

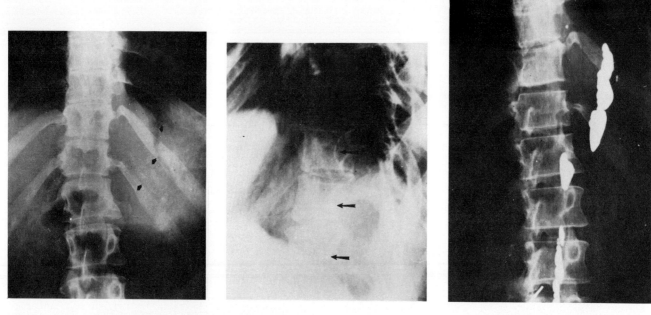

FIG. 5. Lateral meningocele: Note paraspinal "mass" (short arrows) associated with marked pedicle erosion and posterior vertebral scalloping (long arrows). "Mass" fills with pantopaque, confirming herniation of arachnoid and dura (meningocele) outside spinal canal.

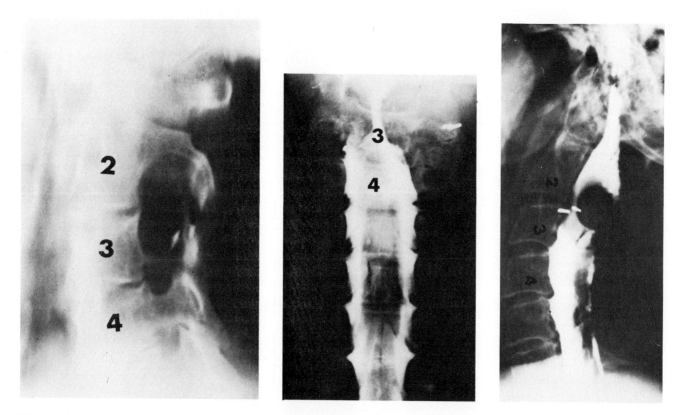

FIG. 6. Cervical "dumbbell" neurofibroma. Note posterior vertebral scalloping and marked enlargement of intervertebral foramen at C2–3 level. Myelogram confirms extension within spinal canal having both intradural and extradural characteristics.

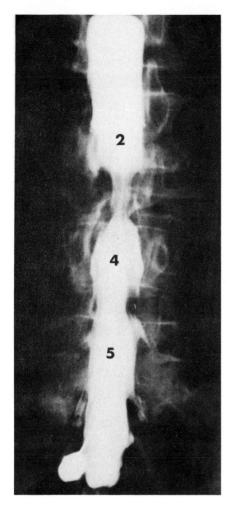 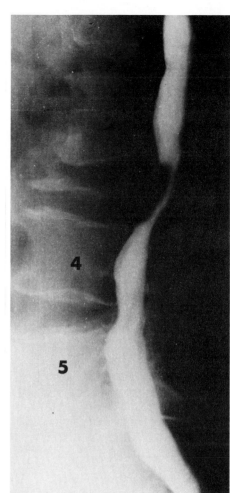

FIG. 7. Plasmacytoma producing complete destruction of L3 vertebral body with marked extradural encroachment on spinal subarachnoid space.

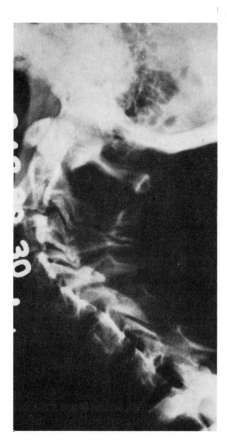 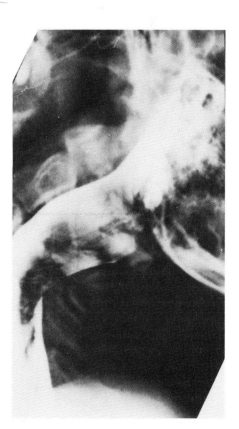

FIG. 8. Seventeen-year-old female with cervicomedullary astrocytoma with consequent pressure erosion and widening of upper cervical canal. Myelogram outlines large intramedullary lesion involving upper cervical cord and medulla.

Hilal SK, et al: Diastematomyelia in children. Radiology, 112:609, 1974

Keim HA, Greene AF: Diastematomyelia and scoliosis. J Bone Joint Surg 55A:1425, 1973

Roberson GH, et al: The narrow lumbar spinal canal syndrome. Radiology 107:89, 1973

Schlesinger EB, Taveras JM: Factors in the production of "cauda equina" syndromes in lumbar discs. Trans Am Neurol Assoc 78:263, 1953

Seaman WB, Schwartz HG: Diastematomyelia in adults. Radiology 70:692, 1958

Shapiro R: Myelography, 3rd ed. Chicago, Yearbook Medical, 1975

Taveras JM, Wood EH: Diagnostic Neuroradiology, 2nd edition, Baltimore, Williams & Wilkins, 1976, Vol I, p 1100

Computed Tomography of the Skull and Spine

Sadek K. Hilal

Computed tomography (CT) contributed greatly to the imaging of soft tissues and to the demonstration of small changes in their radiation absorption characteristics caused by a great variety of lesions. With the improved resolution of the recent scanners, imaging of bone with CT has acquired great diagnostic importance.

The examination of the skull with CT is of great interest in a variety of conditions requiring simultaneous visualization of the soft tissues and the bones. Also, of particular importance are lesions involving several adjacent cavities of the skull, such as the orbit, the paranasal sinuses, and the intracranial cavity. In at least one of the most recent models of scanners (AS&E) 2-mm-thick sections are possible, and visualization of bone detail in the paranasal sinuses and petrous bone is quite good. The sensitivity of CT to the demonstration of small amounts of bone is a very important fundamental aspect of this imaging technique. It permits the visualization of fine septa within large areas of soft-tissue density. Such discrimination is often impossible with conventional tomography, leaving great uncertainty as to whether or not a given thin-bone septum is actually destroyed or surrounded by a large quantity of exudate or soft-tissue density. Appreciation of bone destruction and deformity in these pathologic processes is of great importance. The ability of CT to discriminate soft tissues of different densities is also a very important contribution to the diagnosis of diseases of the paranasal sinuses and orbits. When a sinus is opacified, it may not be possible to differentiate by conventional radiography between a liquid exudate and a soft-tissue mass filling the sinus. CT of the skull and spine offers another important advantage, namely, the display of the anatomy in the transverse axial plane. This view is not obtainable with conventional tomography with the same degree and image quality or with the same degree of soft-tissue discrimination.

It is beyond the scope of this presentation to discuss all pathologic entities involving the head and spine. Nevertheless, a few comments will be made about a variety of conditions that are of interest to the radiologists looking at the nonneurological lesions of the skull and spine.

Lesions of the Cranium

Congenital Anomalies

The abnormal shape of the skull due to hemiatrophy or cranial stenosis can sometimes be better appreciated on CT than on conventional radiographs, particularly as regards the extent of the asymmetry of the two halves of the cranium. Of great interest, of course, is the falx cerebri, which outlines the anatomic midline of the intracranial structures and can be used as an extremely convenient landmark for the comparison of the two halves of the cranium. Of great additional interest is the diagnosis of basal invagination and platybasia that may be made by CT. The severity of these conditions can often be better appreciated on CT. Of importance is the evaluation of hypotelorism in the transverse axial plane because of the recent advent of successful corrective surgery.

Skull Trauma

CT is of great value in the diagnosis of brain injury. In skull fractures, although not as essential as in the case of brain lesions, it does contribute important information sometimes not achievable otherwise. In depressed skull fractures, CT is extremely valuable in evaluating the extent of the depression of the calvarial fragment and any associated brain contusion. Hematomas in the brain or in the surrounding space can be readily recognized. Trauma to the soft tissues outside of the skull can be detected readily, and the delineation of the most affected layer of the scalp can be readily evaluated. Coronal sections are very helpful in outlining fractures of the base of the skull and any entrapment of soft tissues within them.

Inflammatory Processes

Inflammations of the calvarium and petrous bone can be evaluated by CT as to the extent of bone destruction and as to the extent of the involvement of adjacent brain tissue. The determination as to whether or not a given flammatory process has crossed the dural barrier can readily be made by CT.

Neoplasms of the Cranium

CT facilitates the evaluation of neoplasms of the calvarium by demonstrating the extent of involvement of each of the skull tables and the diploic bone. This evaluation is of great diagnostic importance, particularly in establishing whether a process started inside or outside of the skull or within the diploic space.

CT can also evaluate the extent of neoplastic soft tissue associated with a given tumor of the cranium. The violation of the dura and the permeation of the brain can be readily demonstrated. The bone changes due to meningiomas can be readily detected, and their relationships to the soft-tissue components can be evaluated.

Direct extension of a neoplasm from the cranium to adjacent cavities of the skull such as the orbits, sinuses, and subtemporal space are readily demonstrated by CT, particularly if coronal and axial views are used.

The Sella Turcica

The study of the sella turcica with high-resolution CT provides a great advantage by noninvasively demonstrating the following conditions:

1. The extent of a suprasellar neoplasm.
2. The detection of small amounts of calcium in the lesion that cannot be readily appreciated by conventional radiographic techniques.
3. The demonstration of cysts, necrosis, and the collection of low-density material in the tumor.
4. The demonstration of hemorrhage in the tumor.
5. With the advent of high-resolution CT, particularly using the 2-mm-slices technique, the demonstration of an "empty sella" is becoming more readily feasible. Sporadic examples have been shown with small tumors measuring a few millimeters in diameter within the sella.
6. The spread of a neoplasm of the optic nerves into the suprasellar region can be very accurately evaluated with the new high-resolution machines.

Orbits

The CT examination of the orbit has become an indispensable test for the study of diseases of the orbit. The orbit normally contains tissues of high radiographic contrast, namely, the presence of the retrobulbar fat and structures outlined by retrobulbar fat. This area was extremely difficult to study before the advent of CT. The sensitivity of CT in detecting small changes in the retrobulbar fat is more than 98 percent in cases of neoplasms. In cases of enlargement of the retrobulbar muscles due to Graves' disease, CT is positive in all patients with class 3 or higher of thyrotoxic exophthalmus.

CT has been of great value in outlining the extent of neoplastic processes involving the paranasal sinuses and the orbit. It is unsurpassed in evaluating lesions in the orbital apex. The surgical exploration of the orbits for the diagnosis of lesions in the orbital apex can now be carefully planned. Negative diagnostic surgical explorations of the orbit became almost nonexistent after the advent of CT.

The diagnosis and evaluation of the extent of optic nerve gliomas and meningiomas by CT has made a major contribution to the surgical planning and management of these contitions. The recognition of multicentric optic nerve gliomas and evaluation of the extent of intracranial spread is of great help to the treatment.

The use of coronal views of the orbit is very useful in the study of orbital fractures. Blowout fractures of the orbital floor and the entrapment of the extraocular muscles can be readily seen by CT and specifically diagnosed.

Paranasal Sinuses, Facial Bones, and Associated Soft Tissues

The contribution of CT in the diagnosis of lesions of the paranasal sinuses and facial bones is primarily due to its extreme sensitivity in detecting small amounts of bone tissue and in the differentiation between soft tissues of different compositions and between soft tissues and the accumulation of fluids. This distinction of the different types of soft tissues and a fluid-filled sinus cannot be achieved by conventional radiographic techniques, particularly if contrast injection is used. Also, the detection of a thin layer of bone in a mass of pathologic soft tissue may not be achievable with conventional techniques, and a false interpretation of bone destruction may be made. The importance of these findings in differentiating between chronic inflammatory processes of the sinuses and neoplasms is evident.

CT of the facial bones offers another advantage, namely, the display of the anatomy in the transverse axial plane, which may exhibit anatomic relationships difficult to detect by conventional tomography. The differentiation between tissue types of various neoplasms and the tissues of graulomatus processes is not possible by CT. This aspect is another manifestation of the sensitivity of CT in detecting pathologic changes and its limitation in not being sufficiently specific to establish a histologic diagnosis.

Computed Tomography of the Spine

The great advantage that CT offers in the study of the spine and the spinal canal is the display of the anatomy in the transverse axial plane. Conventional tomographic machines, which also display this plane of anatomy, produce images of lesser quality than are attainable by CT. The integrity or deformity of the spinal canal is best appreciated in the transverse axial plane. The extent of paraspinal masses around the spinal column is greatly helped by CT, particularly when there is infiltration of the paraspinal muscles and the body cavity as well. With the new high-resolution scanners, it is possible to see the outline of the spinal cord, dura, and epidural fat without the administration of intrathecal contrast. Encroachment on the spinal canal cavity by a paraspinal neoplasm can be readily appreciated by transverse axial tomography.

The detection of neoplasms involving the spinal cord has been demonstrated by several workers. The main contribution of CT is the differentiation between a solid spinal cord tumor and a cystic lesion of the cord. This is possible in some cases. Neoplasms tend to enhance after the administration of intravenous contrast. The infiltration of the spinal cord with fat can usually be detected with ease (Fig. 1). CT is of particular value in patients where myelography is not possible because of arachnoid adhesions from previous surgery, radiation therapy, or multiple previous myelograms. In these conditions, it remains the only method with which to evaluate the spinal cord. It must be stated, however, that myelography is much more sensitive than CT for the detection of cord enlargement.

The usefulness of CT for the diagnosis of spine disease is outlined in the following sections.

Congenital Anomalies of the Spine

The use of computed tomography for the diagnosis of congenital anomalies of the spine is of great importance. Of particular interest are the following conditions:

1. *Dysraphism.* In dysraphism, CT shows the defect of the bone, whether anterior or posterior; it will show the associated lipoma and its extent into the surrounding soft tissues; and it will demonstrate the meningocele and its contents, whether it is cerebrospinal fluid or fat. Dorsal or ventral meningoceles can be readily appreciated (Fig. 2).
2. *Diastematomyelia.* In diastematomyelia, CT will show the bony spur and any associated defects of the neural cord.
3. *Anomalies of segmentation.* Anomalies of fusion and segmentation can be appreciated by CT but are better seen by conventional radiography.
4. *Arnold–Chiari malformation.* Transverse CT scans at the level of C1 and the occiput, as well as coronal CT scans of this area, may show the low position of the tonsils and the associated small fourth ventricle. The technique is less satisfactory than myelography, which shows a better anatomic detail of this area. Nevertheless, it has the advantage of being noninvasive.
5. *Hydromyelia and lipomas of the spinal cord.* CT is very valuable in evaluating hydromyelia and lipomas of the spinal cord. The cord is lucent in these two conditions, being more lucent with fatty infiltrates. Many cystic hydromyelic cords may not collapse, and conventional air myelography and CT may be the only methods of establishing this diagnosis. The demonstration of a large spinal cord by conventional radiopaque myelography and the subsequent demonstration of the lucent cord by

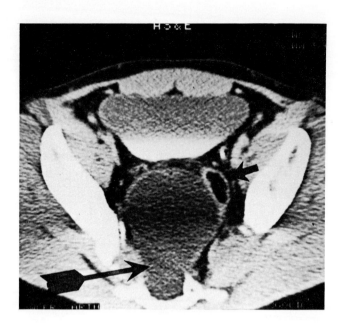

FIG. 1. Lipoma of spinal cord. Note spina bifida of the sacrum (small arrow) and large lucency in the middle of the spinal canal representing lucent fatty tissue in a low cord (large arrow).

CT are strongly suggestive of hydromyelia. The cord involved by lipoma can be recognized by the associated dysraphism, the history, and the more characteristic myelographic appearance (Fig. 1).

6. *Spinal stenosis.* The use of CT for the diagnosis of spinal stenosis has become almost indispensable. The appreciation of the true size and shape of the spinal canal in the transverse axial plane is vital to the evaluation of spinal stenosis. The prominence of the articulating facets, the encroachment on the lateral recesses of the dural sac, and the actual anteroposterior diameter of the canal in the midsagittal as well as adjacent planes can all be readily evaluated and measured, and an accurate statement about spinal stenosis can therefore be readily made.

Trauma

Trauma of the spine has a few important complications. The most important of these is the detection of a bony spicule within the spinal canal. Also of importance is the evaluation of the extent of the fracture, particularly of the neural arch posteriorly. The number of fragments and the shape may be detected with difficulty on conventional tomography. The CT occasionally permits the appreciation of subluxation or dislocation. Its usefulness for this

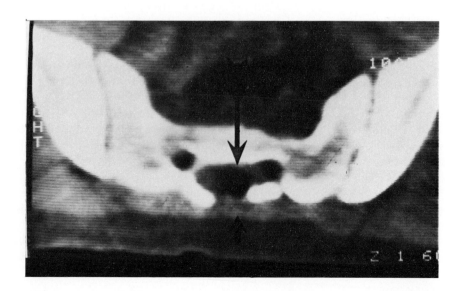

FIG. 2. Anterior sacral meningocele. Note anterior defect (large arrow) and the large soft tissue mass in front of the sacrum filling the pelvis and displacing the colon laterally (arrow) and the urinary bladder anteriorly. Note layering of contrast within the bladder in the supine position.

purpose is more limited than conventional radiography. Hemorrhage around the spine and within the spinal canal can also be easily detected by CT, and in some cases, this technique may be the only way of evaluating epidural, paraspinal, and even intermedullary hematomas.

Degenerative Diseases of the Spine

CT may be used for the demonstration of bony spurs projecting into the spinal canal and encroaching on the dural space. This finding can also be readily demonstrated by conventional radiography. The visualization of these lesions in the transverse axial plane, however, is very helpful. If an intrathecal water-soluble contrast agent is used, one can appreciate the course of the nerve roots to better advantage, and CT may be the only method for demonstrating the encroachment on the nerve root sleeve, which may not be demonstrated by conventional myelography with either an oil or water-soluble contrast agent. The demonstration of the nerve root outside of the dural sac is not readily feasible with conventional myelography but can be achieved on many occasions by CT. CT may be the only method for demonstrating a laterally herniating nucleus pulposus.

Neoplasms

The use of CT for the diagnosis of neoplasms of the spine and paraspinal tissues is helpful in detecting bone erosions and involvement of the spine that are not readily detectable by conventional radiographs. Polytomography or probably nuclear scanning of the bone may be equally effective. The visualization of the anatomy in the transverse axial plane, however, may provide unique information as to the exact extent of the process. Of particular interest are dumbbell lesions that extend both within and outside the spinal canal. The exact size of this lesion is almost impossible to evaluate by other techniques. Tumors of the cord tend to take up contrast media, and CT may be needed to evaluate whether or not they contain any cystic components.

Juvenile Rheumatoid Arthritis

Barbara N. Weissman

In 1896, George Frederic Still described and attempted to classify chronic polyarthritis in children.[1] Similar efforts have continued to the present, and in 1976 a subcommittee of the American Rheumatism Association redefined juvenile rheumatoid arthritis (JRA) as persistent arthritis of one or more joints of at least 6 weeks duration if certain exclusions are eliminated.[2] Unfortunately, the disorders to be excluded are numerous (Table 1). Several subtypes of JRA were also described based on clinical manifestations within the first 6 months of disease.[2] Onset is described as *systemic* (approximately 20%)[3] if JRA is accompanied by intermittent fever with or without other systemic manifestations, *pauciarticular* (40%)[3] if four or fewer joints are involved and systemic onset is excluded, and *polyarticular* (40%)[3] if five or more joints are involved and systemic onset is excluded. Joints are counted individually, except that the cervical spine, carpals, and tarsals are each counted as one joint.

Differential diagnosis may be difficult clinically, since, in contrast to adult rheumatoid arthritis, the latex test may be negative, joint symptoms may be minimal, and systemic signs may predominate.[3-6] Radiographic examination can aid in differential diagnosis, but is of most value in documenting and following articular damage and assessing growth abnormalities. Radiographic features to be reviewed include soft-tissue findings, periosteal reaction, articular abnormalities, and growth disturbances.

The appearance of articular soft-tissue swelling is, in most areas, identical to that seen in adults. Subcutaneous nodules are said to be less common in JRA than in adult rheumatoid arthritis,[5] but when present they are often associated with polyarticular disease.[3] They appear as uncalcified soft-tissue masses in areas of excessive pressure or friction, such as over the olecranon processes or on the backs of the heels.[6] The presence of subcutaneous nodules is thought to herald a poor prognosis.

Soft-tissue calcification is an unusual manifestation of JRA and occurred in 4 of 80 patients studied by Martel.[7] It is usually periarticular (Fig. 1), and if extensive, a diagnosis of dermatomyositis should be considered clinically.

Periosteal reaction may be demonstrated on radiographs taken as early as 2 weeks following the onset of symptoms[8] (Fig. 2). This occurs most frequently along the shafts of the proximal phalanges, metacarpals, and metatarsals near affected joints. Later, this periosteal new bone merges with cortical bone and, although a thickened cortex may result, more often there is thinning of cortex from the endosteal side, leading to a rectangular configuration of metacarpals and phalanges.[7]

Cartilage space narrowing is a late finding in JRA, particularly in young children. For example, children under 7 years of age with severe disease showed cartilage space narrowing only after 1 or 2 years had elapsed since the onset of symptoms.[9] This abnormality is even less frequent in patients with monoarticular JRA.

Occasionally, improvement in cartilage spaces and adjacent subchondral bone may be demonstrated radiographically both in the small joints of the hands and wrists[9] and in large joints such as the hips[10] (Fig. 3). The histologic basis for the improved radiological appearances is not known.

Articular erosion is another late finding in JRA, especially in the large joints of young children.[9,11] Patients with IgM rheumatoid factor may show earlier erosive changes in a distribution more like that

TABLE 1 EXCLUSIONS

A. Other Rheumatic Diseases
   1. Rheumatic fever
   2. Systemic lupus erythematosus
   3. Ankylosing spondylitis
   4. Polymyositis and dermatomyositis
   5. Vasculitis
   6. Scleroderma
   7. Psoriatic arthritis
   8. Reiter's syndrome
   9. Sjögren's syndrome
  10. Mixed connective tissue disease
  11. Behcet's syndrome
B. Infectious Arthritis
C. Inflammatory Bowel Disease
D. Neoplastic Diseases Including Leukemia
E. Nonrheumatic Conditions of Bones and Joints
   1. Osteochondritis
   2. Toxic synovitis of the hip
   3. Slipped capital femoral epiphysis
   4. Trauma
   5. Chondromalacia patellae
   6. Congenital anomalies and genetically determined abnormalities of musculoskeletal system (including inborn errors of metabolism)
F. Hematologic Diseases
   1. Sickle cell anemia
   2. Hemophilia
G. Psychogenic Arthralgia
H. Miscellaneous
   1. Immunologic abnormalities
   2. Sarcoidosis
   3. Hypertrophic osteoarthropathy
   4. Villonodular synovitis
   5. Chronic active hepatitis
   6. Familial Mediterranean fever

From: Brewer EJ, Bass J, Baum J, et al: Arth Rheum 20, 2 suppl., 195–99, 1977

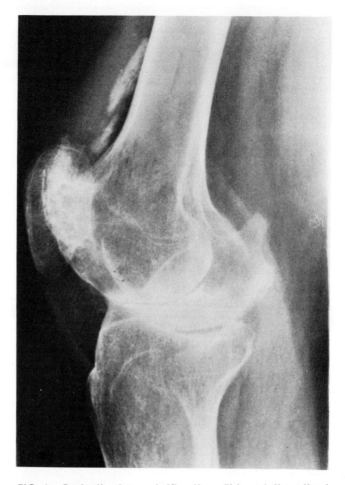

FIG. 1. Periarticular calcification. This adult patient had severe JRA. There is marked cartilage loss and some marginal hypertrophic lipping. Calcification is present in the suprapatellar area.

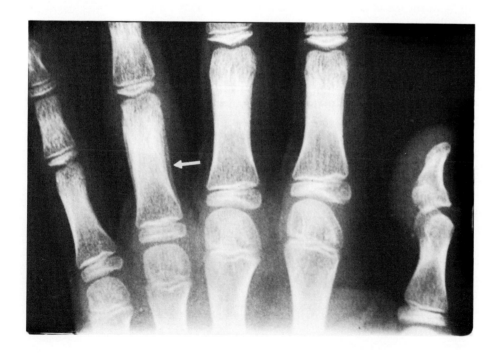

FIG. 2. Periosteal reaction. There is soft-tissue swelling around the ring finger PIP and linear periosteal reaction along the proximal phalanx (arrow). Although periosteal reaction is often an early sign, this patient already has mild narrowing of the PIP joint cartilage space.

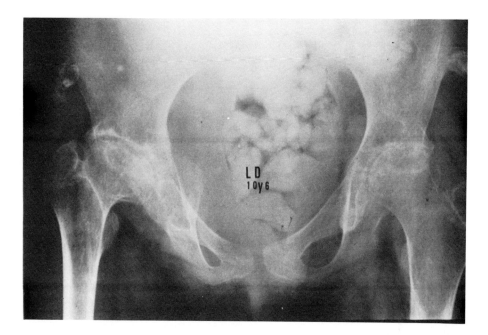

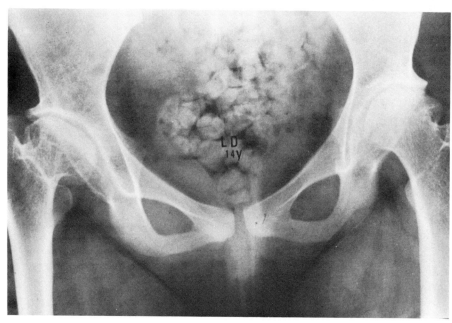

FIG. 3. Healing. Top: An anteroposterior view of the pelvis shows marked cartilage space narrowing in each hip and deformity of the acetabula and femoral heads. There is shortening of the right femoral neck. The femoral epiphyses are closed. Bottom: Widening of cartilage spaces and improvement in subchondral bone is noted.

of adult rheumatoid arthritis.[12] Erosion may occur within the cartilaginous portion of the epiphysis, and if this is the case, it will not become visible until ossification occurs. Therefore, the appearance of erosion in young children may not correlate with current disease activity. Healing of erosion was seen in 3.7 percent of a large series of patients studied by Laaksonen.[9] With healing, a thin sclerotic margin will cover the area of previously absent cortex.

Complete cartilage loss may be followed by bony ankylosis.[7,9,13] This occurs most frequently in the wrists, especially at the carpometacarpal articulations of the index and middle fingers and at the intercarpal articulations. Usually one of the carpal joints remains open allowing some motion[7] (Fig. 4).

Ankylosis of the apophyseal joints in the cervical spine may produce a characteristic radiographic appearance. Usually segments of two, three, or four vertebrae are involved, although occasionally, most or all levels are fused, producing a "rat-tail" appearance (Fig. 5). If apophyseal joint ankylosis occurs early in life, growth abnormalities may be prominent, resulting in a decrease in both the anteroposterior and vertical dimensions of the affected vertebrae. The

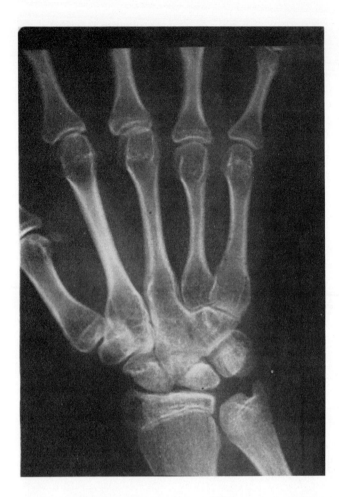
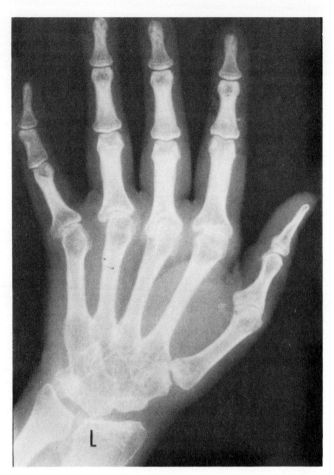

FIG. 4. Carpal ankylosis. Left: Ankylosis of index and middle finger carpometacarpal articulations. A short fourth metacarpal and cartilage space narrowing in the wrist are also present. Right: Ankylosis of carpometacarpal and intercarpal articulation. Note the sparing of the radiocarpal and thumb carpometacarpal joints.

underdeveloped vertebral bodies may flare at the junction with unfused segments (Fig. 6). Disk spaces are diminished in size either due to decreased growth or to subsequent narrowing. Horizontal calcification may occur within the narrowed disk spaces (Fig. 7).

Cervical spondylitis was noted on radiographic examination in 35% of patients with polyarticular disease, 20% of those with systemic onset of disease, and 10% of patients with monoarticular disease.[14] In contrast to patients with ankylosing spondylitis, cervical spine involvement in JRA is usually not accompanied by sacroiliac joint changes, calcification of the annulus fibrosis, or extensive involvement of the lumbar and dorsal levels. Growth abnormalities are uncommon in ankylosing spondylitis.

Atlantoaxial subluxation has been described in detail in patients with rheumatoid arthritis.[15] Normally, the C1–C2 articulations provide rotatory motion, but very little flexion–extension motion. This stability in flexion is maintained by the transverse ligament of the atlas, which runs behind the odontoid, suspended between the lateral masses of the atlas. Lateral views in flexion, neutral, or extension will show an almost constant distance between the posterior inferior aspect of the anterior arch of the atlas and the odontoid. This interval measures less than 2.5 mm in adults,[15] but in children an apparent separation of up to 5 mm may be normal.[16]

In rheumatoid arthritis, JRA, or ankylosing spondylitis, inflammatory changes may occur in a bursa adjacent to the transverse ligament of the atlas. The ligamentous laxity that may result will allow separation of the odontoid from the atlas in flexion, and this will be demonstrated as a widening of the previously noted C1–C2 interval. Such C1–C2 subluxation is more frequent in seropositive JRA patients. It was not seen in patients with monoarticular JRA studied by Cassidy and Martel.[14]

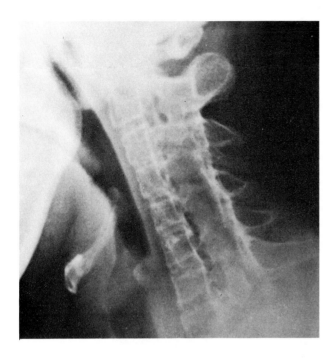

FIG. 5. "Rat-tail" deformity produced by extensive apophyseal ankylosis. The involved vertebrae are small in all dimensions.

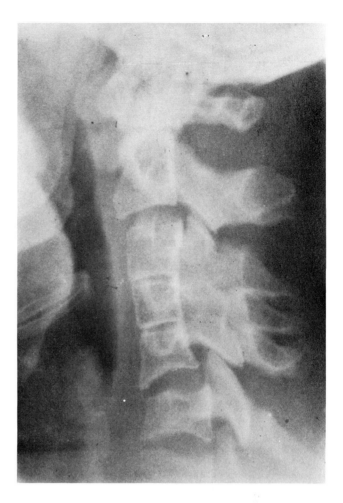

FIG. 6. Apophyseal joint ankylosis. There is deformity of the fused vertebrae and narrowing of the intervening disks.

Alterations in growth, maturation, and epiphyseal appearance are prominent features of JRA. Growth abnormalities include advanced skeletal maturation, enlargement of the secondary centers of ossification, accelerated longitudinal growth, shortening of small tubular bones, early epiphyseal fusion, decrease in the size of the carpal bones, and generalized growth retardation.

Accelerated maturation is most striking in the carpals, particularly when involvement is unilateral (Fig. 8). Eventually, the carpals may be decreased in size in comparison to the contralateral side.

Enlargement of the secondary centers of ossification is often associated with osteopenia and a coarse trabecular pattern. In large joints, enlargement of the secondary centers of ossification is more prominent at the metaphyseal end than at the articular end, resulting in a characteristic tapered appearance (Fig. 9). "Tapering" may be particularly prominent if there is associated marginal erosion. Similar epiphyseal changes may be seen in tuberculosis or hemophilia. A square patella is said to be more common, however, in JRA than in hemophilia.[17] Along with enlargement of the secondary centers of ossification, there may be an increase in longitudinal growth. Eventually, however, some of these epiphyses undergo premature fusion with subsequent limb shortening in comparison to the normal side.[12] In some patients, diffuse atrophy of an entire limb is noted.

Early epiphyseal closure has been reported in 15 to 30% of patients.[9,13] It is most frequent in the metacarpals and metatarsals of older children (Fig. 10). Inflammation at the temporomandibular joints may result in decreased mandibular growth. The muscular imbalance that ensues may produce deformity of the mandible characterized by an upcurving of its inferior border just anterior to the angle of the mandible (antegonial notching).[18] Generalized growth retardation is noted especially in children with long standing active JRA or those with systemic onset JRA.[19] The small size of skeletal structures is of importance in designing total joint prostheses.

An irregular, angular appearance to the secondary centers of ossification, particularly in the carpals, may be an early finding (Fig. 11). When this angular

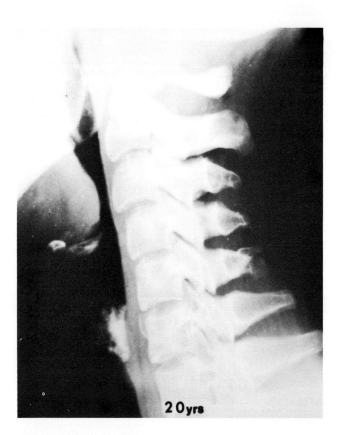 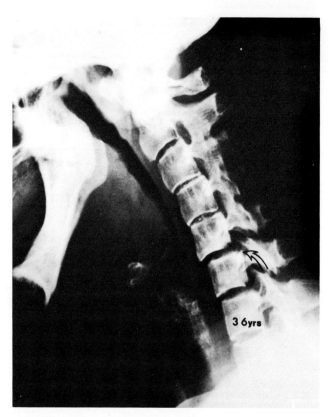

FIG. 7. Cervical spine ankylosis in JRA. Left: Lateral view of the cervical spine shows some narrowing of the C3–C4 apophyseal joints. Right: Repeat view 16 years later shows ankylosis of C2–C5 apophyseal joints, development of disk narrowing and calcification at these levels, C1–C2 subluxation, abnormal motion below the fused segments, and early vertebral erosion. The enlargement of the soft tissues of the base of the tongue is due to amyloid deposition.

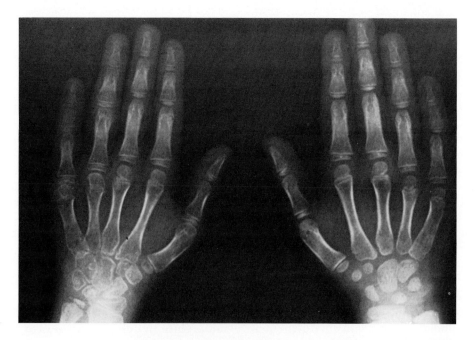

FIG. 8. Accelerated maturation. The left wrist, several metacarpals, and the left ring finger PIP demonstrate advanced maturation in comparison to the normal right side.

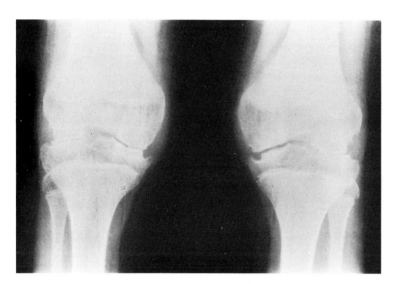
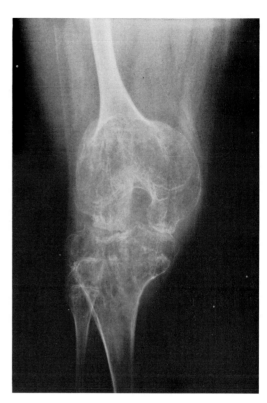

FIG. 9. Epiphyseal enlargement. Left: Mild enlargement of epiphyses with small articular surfaces. Marginal erosion contributes to the small articular surface. Right: Severe changes.

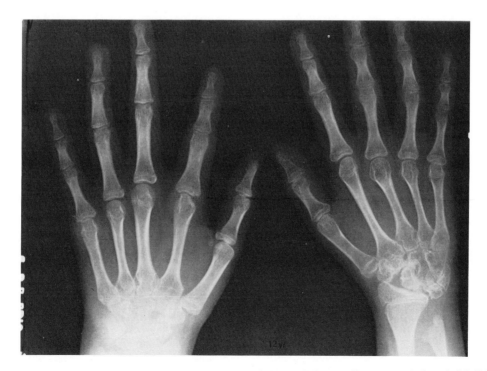

FIG. 10. Early epiphyseal closure. There is shortening of the left index finger and the right third and fourth metacarpals and right ulnar styloid. Note the presence of cartilage narrowing in the right wrist, carpal subluxation, and erosion of the left index finger metacarpal.

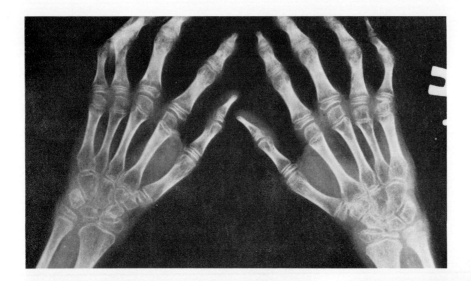
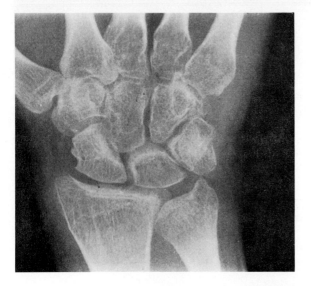
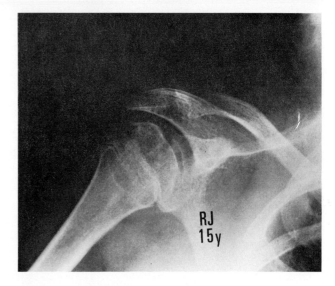

FIG. 11. Angular epiphyses. Opposite: There is an angular appearance to the carpal bones and flattening and irregularity of the metacarpals. Bottom left: Angularity of the carpals and radius and deformity of the ulnar styloid is seen in this adult with JRA. Bottom right: There is irregularity of the shoulder epiphysis in this 15-year-old boy.

appearance persists into adulthood, a diagnosis of JRA can be suggested, even in the absence of other findings.

In summary, the radiographic findings of JRA reflect the articular changes associated with an inflammatory synovitis and the effect of this inflammation on skeletal growth and maturation.

References

1. Still GF: A form of chronic joint disease in children. Brit Med J 2:1446, 1896
2. Brewer EJ, Bass J, Baum J, et al: Current proposed revision of JRA criteria. Arthritis Rheum 20(suppl 2): 195, 1977
3. Stillman JS, Barry PE: Juvenile rheumatoid arthritis: series 2. Arthritis Rheum 20(suppl 2):171, 1977
4. Boone JE, Baldwin J, Levine C: Juvenile rheumatoid arthritis. Pediatr Clin North Am 21:885, 1974
5. Calabro JJ, Holgerson WB, Sonpal GM, Khoury MI: Juvenile rheumatoid arthritis: a general review and report of 100 patients observed for 15 years. Semin Arthritis Rheum 5:257, 1976
6. Calabro JJ: Other extra-articular manifestations of juvenile rheumatoid arthritis. Arthritis Rheum 20(suppl. 2):237, 1977
7. Martel W, Holt JF, Cassidy JT: Roentgenologic manifestations of juvenile rheumatoid arthritis. Am J Roentgenol Radium Ther Nucl Med 88:400, 1962
8. Kapusta MA, Sedlezky I: Periostitis: an early diagnostic sign of juvenile rheumatoid arthritis. J Can Assoc Radiol 18:268, 1967
9. Laaksonen AL: A prognostic study of juvenile rheumatoid arthritis. Acta Paediatr Scand Suppl 166:1, 1966
10. Berstein B, Forrester D, Singsen B, et al: Hip joint restoration in juvenile rheumatoid arthritis. Arthritis Rheum 20:1099, 1977

11. Cassidy JT, Brody GL, Martel W: Monoarticular juvenile rheumatoid arthritis. J Pediatr 70:867, 1967
12. Ansell BM, Kent PA: Radiological changes in juvenile chronic polyarthritis. Skel Radiol 1:129, 1977
13. Sairanen E: On rheumatoid arthritis in children. A clinico-roentgenological study. Acta Rheum Scand Suppl 2:1, 1958
14. Cassidy JT, Martel W: Juvenile rheumatoid arthritis: clinico-radiologic correlations. Arthritis Rheum 20 (suppl 2):207, 1977
15. Martel W: The occipito-atlanto-axial joints in rheumatoid arthritis and ankylosing spondylitis. Am J Roentgenol Radium Ther Nucl Med 86:223, 1961
16. Locke GR, Gardner JI, Van Epps EF: Atlas-dens interval (ADI) in children. Am J Roentgenol Radium Ther Nucl Med 97:135, 1966
17. Chlosta EM, Kuhns LR, Holt JF: The "patellar ratio" in hemophilia and juvenile rheumatoid arthritis. Radiology 116:137, 1975
18. Becker MH, Coccaro PJ, Converse JM: Antegonial notching of the mandible: an often overlooked mandibular deformity in congenital and acquired disorders. Radiology 121:149, 1976
19. Bernstein BH, Stobie D, Singsen BH, et al: Growth retardation in juvenile rheumatoid arthritis. Arthritis Rheum 20(suppl. 2):212, 1977

Arthritides: Complications and Unusual Features

Jack Edeiken

The many manifestations of the arthritides are well documented in the literature, and a consideration of all the arthritides would be impossible in a brief review. However, there are certain complications and unusual features in which the radiologist plays a prime role in the diagnosis.

Rheumatoid Arthritis

The features and complications of rheumatoid arthritis that will be considered are

1. Early diagnosis (involvement of metacarpophalangeal joints).
2. Involvement of the wrist joint
    a. Atlantoaxial dislocation
    b. Giant rheumatoid cysts.

Articular cortical erosions always indicate joint cartilage destruction in rheumatoid arthritis. They may appear to be surface erosions or pseudocysts entirely within the bone substance. The site of the surface erosion is characteristic in each joint and depends upon the extension of the pannus from the synovial reflections onto the bone. At the metacarpophalangeal joints, initial resorptions of the articular cortical bone occur most often on the radial aspect of the distal end of the second and third metacarpals. These changes occur early in the course of the disease and are of great diagnostic value. At the same time, erosion of the proximal end of the first phalanx of the finger occurs in over 85 percent of the cases. These early changes may also occur in the metatarsals and are frequently asymptomatic. The changes may occur as early as 2 to 3 weeks after the onset of symptoms, and the patients may be sera-negative for the rheumatoid factor. They occur in no other arthritis this early in the course and are practically pathognomonic.

Roentgenographic changes of early rheumatoid arthritis are frequently present in the wrist. The changes are usually on the ulnar side of the wrist joint and consist of soft-tissue swelling and early erosion of the ulnar styloid. There may be subluxation of the radioulnar joint. This change may occur in other arthritides such as psoriasis and Reiter's disease, but in no condition will they occur as early in the clinical course as in rheumatoid arthritis. The slightest irregularity of the ulnar styloid or subluxation of the radioulnar joint may occur before the full-blown characteristics of rheumatoid arthritis are present.

Atlantoaxial dislocations occur in a high percentage of patients with severe rheumatoid arthritis. The changes produce very few symptoms and only pain on turning the head may be present. Since these patients usually have severe rheumatoid arthritis, the mild pain in the region of the neck is minor compared to the pain in the extremities, and the dislocation is frequently overlooked. Since it occurs in a high percentage of patients, this area must be examined in all patients, for the complication may lead to sudden death. Many of these atlantoaxial dislocations are stable and are probably of little danger to the patient, but those that are unstable may lead to serious complications, and the instability must be documented early in the course. This is most easily accomplished by performing flexion and extension views in the lateral projection. If the dislocation is unstable, then there will be motion between the first and second cervical vertebrae. With this type of complication, the patient must be protected during the active hours by a neck collar. The patient also should be examined at least once a month for neurologic changes, and if they ensue, fusion is indicated.

Giant rheumatoid "cysts" may develop in patients with long-standing rheumatoid arthritis. When they occur in the region of the gastrocnemius, they are often accompanied by pain and swelling. When a mass is evident, the misdiagnosis of fibrosarcoma is made and the patient is admitted for biopsy of the tumor. If there is no evident mass, then the pain is misinterpreted as thrombophlebitis, and unnecessary treatment for this condition may be given. These giant cysts most often occur in the region of the knee, and they can be evaluated by arthrography of the knee. The opaque and/or air will collect in the cyst in the region of the gastrocnemius.

Large cysts may occur in the hip and the shoulder, and smaller cysts may occur in any synovial joint, such as the temporomandibular joint, the finger joints, and the wrist joint. Any large swelling with bone erosion near a joint in a patient with long-standing rheumatoid arthritis probably has a giant rheumatoid cyst.

Gout

The radiologic changes of gout occur 6 to 8 years after the clinical onset, and the diagnosis is usually made long before there are radiologic features present. An infrequent but important complication is gouty tophaceous material collecting within the spinal canal. These may produce symptoms and signs suggesting spinal cord tumor. It is important for the radiologist to realize that in any patient with long-standing gout with destructive lesions in the spine or cord symptoms, the most likely diagnosis is gouty tophi. The spinal cord involvement may occur without a previous history of gout when the synovitis has not been severe or the synovium not involved at all. This is especially prone to occur in childhood gout, and large erosions of the spinal canal may occur before the onset of any other clinical symptoms. In most of these young patients, both mother and father have gout.

Occasionally, peripheral bone erosions may occur without previous history of gout. These are due to tophi, and since the symptoms of gout depend on involvement of synovia, if they are uninvolved or mildly affected, then symptoms will be absent before the tophaceous stage.

Ankylosing Spondylitis

Stress fracture of a vertebra is the most commonly overlooked complication of ankylosing spondylitis. Most patients with ankylosing spondylitis do not have severe pain that would prevent them from carrying out daily activities. Certainly, the deformity may be severe and the demineralization of the bone especially in the spine may be marked. With long-standing ankylosing spondylitis with fusion and deformity, patients are relatively comfortable. When a stress fracture occurs, there is onset of acute localized and quite severe pain. Since these patients are most difficult to examine, the complication is frequently overlooked. In a patient with ankylosing spondylitis who is complaining of localized pain in the region of the spine, the widening of an intervertebral space with or without sclerotic changes is almost pathognomonic of a stress fracture. When there is total fusion of the spine without fracture, a wide disk space should not occur. Since the deformity and osteoporosis are great, this abnormality is usually overlooked. Tomography will often show the widening of the disk space to extend through the posterior elements, indicating the nature of the disease process. The stress fracture of ankylosing spondylitis usually occurs in the cervical spine, but it can occur anywhere and its next most frequent sites are lower dorsal and upper lumbar spine.

Osteophytes and Syndesmophytes

Spur formation of the spine is usually considered secondary to degenerative changes. It is important to note that there are two different types of spurs—osteophytes and syndesmophytes.

Osteophytes are formed at the level of the disk space or of a joint space. They protrude transversely and then may swing upward and join an osteophyte of the adjacent vertebra. They are due to degenerative changes of the joint or disk space, with stress on the attached soft-tissue structures such as capsules of joints or the annulus fibrosis of the disk space. It is not necessary that the joint space or disk space be narrowed for the degenerative change to occur.

Syndesmophytes originate below the level of the joint or the disk space. They usually originate from the center of the body of the vertebra and project immediately upward to lie parallel to the long axis of the spine or the joint. Syndesmophytes are secondary to inflammatory changes. When they are symmetrical and over a large extent of the spine, they are usually due to ankylosing spondylitis, and the presence of bilateral symmetrical sacroiliac disease confirms this diagnosis. Coarse syndesmophytes that are asymmetrical and have a large knob at the insertion of one end are most frequently due to either Reiter's disease or psoriasis. However, any inflammatory conditions, such as an infection or trauma with hemorrhage, can produce a syndesmophyte.

Systemic Lupus Erythematosus, Dermatomyositis, Progressive Systemic Sclerosis, and Mixed Connective Tissue Disease

Barbara N. Weissman

Systemic Lupus Erythematosus

Systemic lupus erythematosus (SLE) is a systemic disorder with prominent articular symptoms. Diagnosis is based on the presence of four or more of the following clinical or laboratory findings[1]: butterfly rash, discoid lupus, alopecia, photosensitivity, or oral nasopharyngeal ulceration, Raynaud's phenomenon, lupus erythematosus cells, false positive serologic test for syphilis, profuse proteinuria, cellular casts, pleuritis and/or pericarditis, psychosis and/or convulsions, hemolytic anemia and/or leukopenia and/or thrombocytopenia, and arthritis wthout deformity. Articular symptoms occur in about 90 percent of patients.[2] There is often disparity between the marked pain and functional limitation present and the paucity of joint findings on physical or radiologic examination.[3]

Radiographic abnormalities include nonspecific soft-tissue swelling and juxtaarticular demineralization, alignment abnormalities, acral sclerosis, soft-tissue calcification, resorption of terminal tufts, and avascular necrosis.[4-9] Erosion and cartilage loss have been reported, but are uncommon findings if examples of mixed connective tissue disease and patients with clinical and laboratory features of rheumatoid arthritis are excluded.

Deformity without erosion is one of the characteristic radiographic findings in the hands of patients with SLE (Fig. 1). This is an uncommon finding, however, occurring in only 6 of 59 patients in our series.[9] In contrast to rheumatoid arthritis, the deformity in SLE is often reducible early in the disease and is probably related to muscular weakness and flexion contractures rather than to articular erosion and destruction by pannus.[6] Similar findings can be seen in patients with rheumatoid arthritis or Jaccoud's arthropathy.

Sclerosis of the terminal phalanges is a nonspecific finding that has also been described in association with scleroderma, dermatomyositis, rheumatoid arthritis, and sarcoidosis[10,11] and in normal individuals. This finding is of little diagnostic value. Its occurrence is not limited to patients with Raynaud's phenomenon.

Resorption of the tufts of the distal phalanges is most often seen in patients with scleroderma but is occasionally present in patients with SLE. Bone resorption is not as severe as in some patients with scleroderma. Terminal tuft resorption was seen in 3 of 59 patients with SLE, and each affected patient had Raynaud's phenomenon.[9]

Soft-tissue calcification is usually seen in patients with scleroderma or dermatomyositis. However, periarticular subcutaneous, muscular, vascular, and cutaneous calcification has been identified in patients with SLE.[12,13]

Well-defined lytic lesions in the phalanges and carpals were seen in two patients with SLE without other soft-tissue or bony abnormalities. The nature of these lesions and their relationship to SLE is conjectural. Avascular necrosis occurs in both large and small joints.[14]

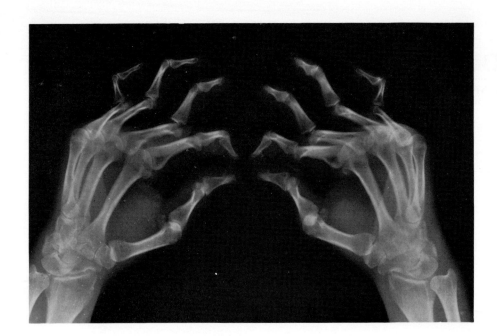

Fig. 1 Systemic lupus erythematosus. Multiple swan neck deformities are present without cartilage space narrowing or erosion. There is calcification in the distal right index fingertip.

Dermatomyositis

Polymyositis is a connective tissue disease characterized clinically by muscle weakness and pain. When characteristic skin lesions are also present, the term *dermatomyositis* is used. Bohan and Peter reviewed these disorders and suggested five criteria useful in diagnosis[15]:

1. Symmetrical weakness of limb girdle musculature and anterior neck flexors. This progresses over weeks or months and may be accompanied by dysphagia or involvement of respiratory muscles.
2. Characteristic findings on muscle biopsy, including necrosis of type I and II fibers, phagocytosis, regeneration of muscle fibers, atrophy in a perifascicular distribution, variation in fiber size, and an inflammatory exudate.
3. Elevation of serum skeletal muscle enzymes, especially creatine phosphokinase (CPK).
4. Characteristic electromyographic findings.
5. Dermatologic abnormalities, including violaceous discoloration of the eyelids (heliotrope) with periorbital edema, and a scaly, erythematous rash involving the dorsum of the hands, especially over the metacarpophalangeal and proximal interphalangeal joints (Grotton's sign) and over the knees, elbows, medial malleoli, face, neck, and upper chest.

A definite diagnosis of dermatomyositis can be made in a patient with a characteristic rash when three or four criteria are present. A definite diagnosis of polymyositis requires four criteria and no rash.

Skeletal and soft-tissue radiographic findings include linear calcifications along muscle planes, nodular subcutaneous calcification, soft-tissue atrophy, osteoporosis, and flexion contractures[16] (Fig. 2). Atrophy of the terminal phalanges, such as is seen in scleroderma, is not seen in patients with dermatomyositis.

Childhood dermatomyositis has been classified separately by Bohan and Peter because of the widespread necrotizing vasculitis that may be present. Early in the course of this disease, radiographic examination may show edema of subcutaneous tissues and muscles, producing increased muscle size and density and loss of definition of soft-tissue planes.[17] Later, contractures, atrophy, and calcification are seen.

Calcification is two to three times more frequent in children than adults.[18] It occurs in fascial planes, skin, subcutaneous tissues, and fat, particularly in the proximal portions of the extremities, the neck, pelvic girdle, and trunk (Fig. 3). Calcification may be noted 6 months to 7 years after the onset of disease and is said to be associated with healing of necrotic tissue. Patients with calcinosis tend to have a better prognosis than patients without calcification, since they have survived the initial phase of the disease and areas of muscle necrosis have healed.[18]

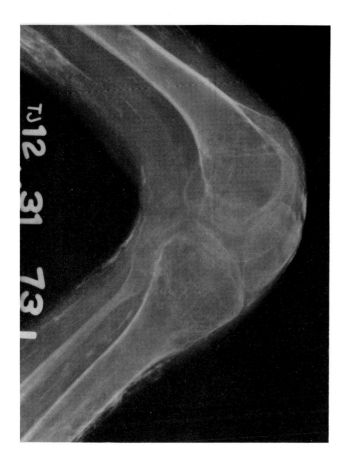

Fig.2 Dermatomyositis. The classical findings of dermatomyositis are present, including osteoporosis, flexion contracture, and plaquelike calcification.

Scleroderma—Progressive Systemic Sclerosis

Progressive systemic sclerosis (PSS) is a disease characterized by edema, induration, and atrophy of connective tissue and vascular lesions affecting the skin, synovial lining, gastrointestinal tract, heart, lungs, and kidneys. In 1964, Winterbauer described a variant of scleroderma termed the *CRST syndrome* based on the presence of calcinosis, Raynaud's phenomenon, sclerodactyly, and telangiectasias.[19] Patients with this variant tend to have a more prolonged course and a less serious prognosis than do patients with PSS. Calcinosis may be much more extensive than in PSS, and skin changes are less widespread. Esophageal dysfunction has since been found in many of these cases, so that the term *CREST syndrome* has been suggested.[20]

Radiographic changes in the hands consist of tapering and volume loss of the distal soft tissues and terminal phalanges, fingertip calcification, articular erosion, and cartilage space narrowing. Straightening of the soft tissue contours of the fingers and loss of the knuckle creases can be seen on oblique views. Distal soft-tissue atrophy accompanies these findings and is said to be present if the vertical thickness of the soft tissues of the fingertip measures 20 percent or less of the width of the base of the distal phalanx as measured on the posterioanterior view.[21] Distal soft-tissue atrophy occurs most frequently in patients with PSS and Raynaud's phenomenon.

Resorption of terminal phalangeal tufts is one of the most frequent and characteristic abnormalities in patients with PSS[22] (Fig. 4). Resorption usually begins distally and progresses proximally. The hands are affected more frequently and more severely than the feet. Occasionally, interphalangeal joints and the radius and ulna[23] may be involved (Fig. 5). Histologic examination shows replacement of bone by granulation tissue.[22]

Soft-tissue calcification is most frequent in the hands but occurs in periarticular subcutaneous tissue as well, particularly around the elbows, knees, and buttocks[24] (Fig. 6). Two patients with intraarticular calcification have been reported.[25] The combination of bone resorption and soft-tissue calcification in the fingers is essentially diagnostic of PSS or the CREST syndrome. These findings may be seen within 6 months following the onset of symptoms.

Rabinowtiz et al. found 7 of 24 patients with PSS to have radiographic features similar to those of rheumatoid arthritis.[26] Findings included marginal erosion of carpals, metacarpals, and phalanges; alignment abnormalities; and ulnar styloid erosion. Cartilage space narrowing occurred in 11 patients, and carpal ankylosis in 1 patient. Articular changes were most frequent in the wrists. Erosion of the superior aspects of the posterior ribs similar to that occurring in rheumatoid arthritis, polio, and hyperparathyroidism is also seen in PSS.[27]

Mixed Connective Tissue Disease

Mixed connective tissue disease (MCTD) is a rheumatologic syndrome that was first described by Sharp et al. in 1972.[28] Patients with this syndrome have overlapping clinical features of SLE, scleroderma, and polymyositis, as well as antibody to the ribonucleoprotein (RNP) component of extractable nuclear antigen (ENA) and high titers of speckled

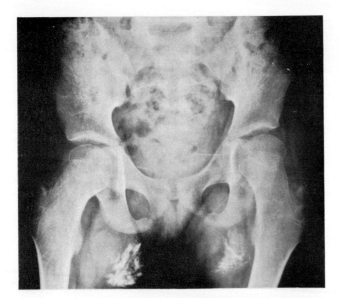
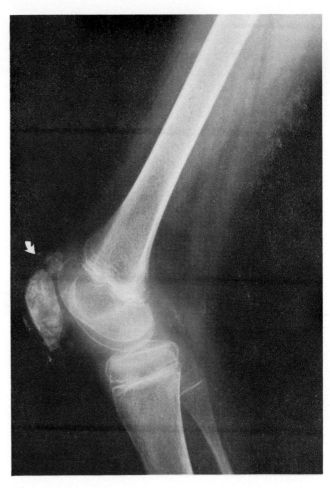

Fig.3 Dermatomyositis. Soft-tissue calcification is present near the ischial tuberosities (above) and in the subcutaneous tissues of the thigh and knee (right). Drainage was present from the superficial collection just above the patella (arrow).

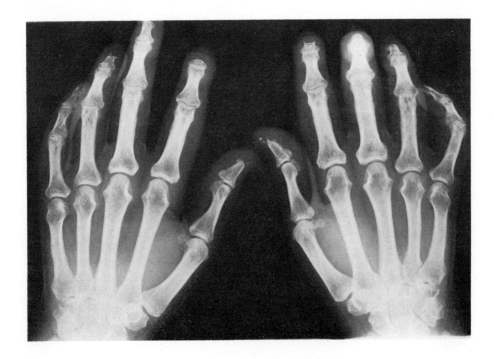

Fig. 4 Scleroderma. There is a marked resorption of terminal tufts and calcification in several fingertips. The fingers are held in flexion.

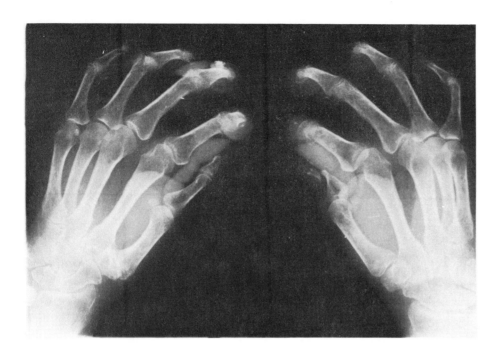

Fig. 5 Scleroderma. There is resorption of the tufts of the distal phalanges and the right thumb IP joint. Distal soft-tissue calcification is present.

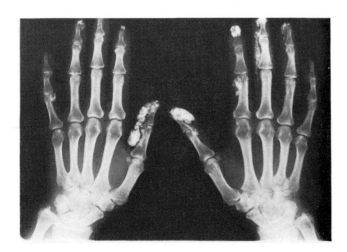

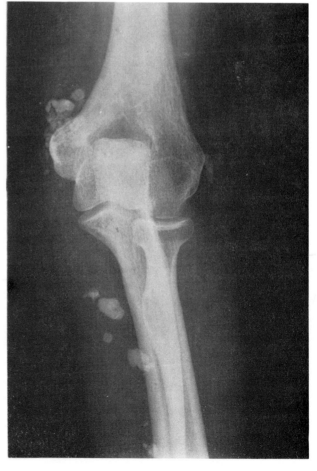

Fig.6 Distal (above) and periarticular (right) calcification in the CRST syndrome.

fluorescent antinuclear antibody (ANA).[28-30] These patients are said to have a favorable clinical course and a lower incidence of nephritis than do patient with SLE.[29-31] Not all patients with antibodies to RNP have clinical features of MCTD, and conversely, patients with overlapping clinical and radiologic findings may not have the serologic findings that characterize MCTD.[31,33]

Clinical features of patients with antibody to RNP include arthritis or arthralgias, Raynaud's phenomenon, swollen hands, serositis, skin changes of scleroderma, abnormal esophageal motility, myositis, and fever.[29,32]

Radiologic findings[33-35] in patients with MCTD reflect the complex clinical manifestations of this syndrome. Periarticular calcification, fingertip calcification, atrophy of distal soft tissues, and terminal tuft resorption (findings usually associated with scleroderma or SLE) may be seen in patients with MCTD. In contrast to most patients with SLE or scleroderma, erosive changes are frequent in the small bones of the hands and feet. Cartilage space narrowing may occur at interphalangeal joints, metacarpophalangeal joints, and wrists.

Visceral abnormalities, including abnormal esophageal motility, esophageal dilatation, colonic pseudodiverticula, pleural and pericardial effusions, interstitial pulmonary fibrosis, and pulmonary artery hypertension may be noted.

Overlapping radiographic features of SLE and scleroderma and the presence of small erosions should suggest a diagnosis of MCTD for clinical and laboratory confirmation. This identification is important because of the more favorable prognosis associated with this syndrome.

References

1. Cohen AS, Reynolds WE, Franklin EC, et al: Preliminary criteria for the classification of systemic lupus erythematosus. Bull Rheum Dis 21:643, 1971
2. Dubois EL: Lupus erythematosus. Los Angeles, Univ. Southern California Press, 1974, pp 259, 357
3. Nesgovorova LI: Lupus polyarthritis: clinical and morphological investigations. Rheumatism 22:99, 1966
4. Gould DM, Daves ML: Roentgenologic findings in systemic lupus erythematosus: an analysis of 100 cases. J Chronic Dis 2:136, 1955
5. Labowitz R, Schumacher HR Jr: Articular manifestations of systemic lupus erythematosus. Ann Intern Med 74:911, 1971
6. Noonan CD, Odone DT, Engleman EP, et al: Roentgenographic manifestations of joint disease in systemic lupus erythematosus. Radiology 80:837, 1963
7. Russell AS, Percy JS, Rigal WM, et al: Deforming arthropathy in systemic lupus erythematosus. Ann Rheum Dis 33:204, 1974
8. Silver M, Steinbrocker O: The musculoskeletal manifestations of systemic lupus erythematosus. JAMA 176:1001, 1961
9. Weissman BNW, Rappoport AS, Sosman JL, Schur PH: Radiographic findings in the hands in patients with systemic lupus erythematosus. Radiology 126:313, 1978
10. McBrine CS, Fisher MS: Acrosclerosis in sarcoidosis. Radiology 115:279, 1975
11. Forrester DM, Nesson JW: The radiology of joint diseases. Philadelphia, Saunders, 1973, p 63
12. Budin JA, Feldman F: Soft tissue calcifications in systemic lupus erythematosus. Am J Roentgenol 124:358, 1975
13. Quismorio FP, Dubois EL, Chandor SB: Soft-tissue calcification in systemic lupus erythematosus. Arch Dermatol 111:352, 1975
14. Green N, Osmer JC: Small bone changes secondary to systemic lupus erythematosus. Radiology 90:118, 1968
15. Bohan A, Peter JB: Polymyositis and dermatomyositis. New Engl J Med 292:344, 403, 1975
16. Nice CM Jr: Clinical roentgenology of collagen diseases. Springfield, Ill, Thomas, 1966, p 124
17. Ozonoff MB, Flynn FJ Jr: Roentgenologic features of dermatomyositis of childhood. Am J Roentgenol 118:206, 1973
18. Steiner RM, Glassman L, Schwartz MW, Yanace P: The radiological findings in dermatomyositis of childhood. Radiology 111:385, 1974
19. Winterbauer RH: Multiple telangiectasia, Raynaud's phenomenon, sclerodactyly and subcutaneous calcinosis: a syndrome mimicking hereditary hemorrhagic telangiectasia. Bull Johns Hopkins Hosp 114:361, 1964
20. Salerni R, Rodnan GP, Leon DF, Shaver JA: Pulmonary hypertension in the CREST syndrome variant of progressive systemic sclerosis (scleroderma). Ann Intern Med 86:394, 1977
21. Yune HY, Vix VA, Klatte EC: Early fingertip changes in scleroderma. JAMA 215:1113, 1971
22. Haverbush TJ, Wilde AH, Hawk WA Jr, Scherbel AL: Osteolysis of the ribs and cervical spine in progressive systemic sclerosis (scleroderma). J Bone Joint Surg 56A:637, 1974
23. Gondos B: Roentgen manifestations in progressive systemic sclerosis (diffuse scleroderma). Am J Roentgenol 84:235, 1960
24. Harper RAK, Jackson DC: Progressive systemic sclerosis. Brit J Radiol 38:825, 1965
25. Resnick D, Scavulli JF, Goergen TG, et al: Intraarticular calcification in scleroderma. Radiology 121:685, 1977
26. Rabinowitz JG, Twersky J, Guttadauria M: Similar bone manifestations of scleroderma and rheumatoid arthritis. Am J Roentgenol 121:35, 1974
27. Keats TE: Rib erosions in scleroderma. Am J Roentgenol 100:530, 1967
28. Sharp GC, Irvin WS, Tan EM, et al: Mixed connective

disease. An apparently distinct rheumatic disease syndrome associated with a specific antibody to extractable nuclear antigen (ENA). Am J Med 52:148, 1972
29. Farber SJ, Bole GG: Clinical features of patients with antibody (Ab) to extractable nuclear antigen (ENA). Arthritis Rheum 18:397, 1975
30. Hench PK, Edgington TS, Tan EM: The evolving clinical spectrum of mixed connective tissue disease (MCTD). Arthritis Rheum 18:404, 1975
31. Sharp GC, Irvin WS, May CM, et al: Association of antibodies to ribonucleoprotein and Sm antigens with mixed connective tissue disease, systemic lupus erythematosus and other rheumatic diseases. New Engl J Med 295:1149, 1976
32. Sharp GC: Mixed connective tissue disease. Bull Rheum Dis 25:828, 1974–1975
33. Silver TM, Farber SJ, Bole GG, Martel W: Radiological features of mixed connective tissue disease and scleroderma—systemic lupus erythematosus overlap. Radiology 120:269, 1976
34. O'Connell DJ, Bennett RM: Mixed connective tissue disease—clinical and radiological aspects of 20 cases. Brit J Radiol 50:620, 1977
35. Udoff EJ, Genant HK, Kozin F, Ginsberg M: Mixed connective tissue disease: the spectrum of radiographic manifestations. Radiology 124:613, 1977

HLA-B27–Associated Arthritis

Barbara N. Weissman

Transplantation or *histocompatibility* antigens are genetically determined antigens on the surfaces of leukocytes and cells of transplanted tissues that are responsible for transplant rejection.[1] Many such antigens have been identified, the better known being designated by HLA (human leukocyte antigen) numbers; and the less well recognized given W (workshop) numbers. Inheritance of these antigens is determined by genes at several loci (i.e., ABCD) on chromosome 6.[1,2] A letter designating the appropriate locus may be inserted before the numerical designation.

In 1973, the high incidence of HLA-B27 was reported to occur in patients with ankylosing spondylitis and their families.[3–5] Since that time, several other disorders have also been noted to be associated with the presence of this antigen, including Reiter's syndrome,[6–8] the spondylitis accompanying inflammatory bowel disease,[9] psoriatic spondylitis,[10] peripheral psoriatic arthritis,[10] *Salmonella* arthritis,[11] and *yersinia enterocoliticia* arthritis.[12]

Both environmental and genetic factors seem to be involved in the etiology of these diseases. The relationship of the HLA system to the presence of these disorders is complex. The HLA gene may directly produce increased susceptibility to disease or may be linked to another gene responsible for increased disease susceptibility.[2] In any event, the presence of this hypothetical gene for disease susceptibility is not, in itself, sufficient to produce disease.[2]

Radiographic evaluation may help to distinguish among patients with HLA-B27 and spondylitis. The radiographic appearances of the spine and peripheral joints in ankylosing spondylitis, Reiter's syndrome, psoriatic arthritis, and inflammatory bowel disease will be reviewed and differential features noted.

Ankylosing Spondylitis

Ankylosing spondylitis is an inflammatory disorder that primarily involves the axial skeleton. Onset is usually between the ages of 16 and 40, and there is a striking male predilection (4:1 to 8:1).[13] HLA-B27 is found in about 88 percent of patients with this disorder.[2]

Pathologic features are similar to those of rheumatoid arthritis involving the peripheral joints. Early changes in synovial joints consist of an accumulation of lymphocytes and plasma cells around small subsynovial blood vessels. Later, there is a proliferation of the synovial lining and gradual outgrowth of granulation tissue, leading to erosion and destruction of articular cartilage and adjacent bone. Eventually, fibrous synostosis or bony ankylosis results.[13]

Sacroiliac Joint Changes

The earliest radiographic findings involve the sacroiliac joints and reflect these pathologic processes. Involvement is always bilateral and usually symmetric.[14,15] Initial radiographic abnormalities consist of loss of definition of the subchondral cortex along the joint, particularly in its mid and lower portions (stage I of Forestier.)[16,17] The iliac side of the joint is involved earliest, since the articular cartilage on that side is thinner than that on the sacral side.[18] Increasing erosion along the joint margins may produce apparent widening of the joint space, termed *pseudowidening*.[16] Progressive erosion is accompanied by areas of sclerosis (stage II of Forestier) that may be so prominent that osteitis condensans ilii is suggested. However, in osteitis condensans ilii, sharply marginated sclerosis occurs in the ilia along the lower portions of the sacroiliac joints without cartilage space narrowing or erosion. In contrast to ankylosing spondylitis, involvement may be unilateral, and women are most often affected (Fig. 1).

Late changes of ankylosing spondylitis consist of gradual decrease in sclerosis and development of bony ankylosis across the sacroiliac joints (stage III of Forestier). Since involvement is not uniform, areas of retained articular cartilage occur, and in those areas, the subchondral cortices will be visualized. Ankylosis involves both the lower two-thirds of the sacroiliac joints (the synovial portion of the joint) and the proximal third of the joint (the site of the interosseous sacroiliac ligaments). In contrast to ankylosing spondylitis, conditions involving synovial joints primarily, such as septic arthritis, involve only the lower portion of these joints.[16]

The sacroiliac joint changes may be identical to those seen in association with ulcerative colitis, Reiter's syndrome, or psoriasis. Unilateral involvement, however, makes a diagnosis of ankylosing spondylitis extremely unlikely.

Spinal Abnormalities

Ankylosing spondylitis involves both the apophyseal joints and the amphiarthrodial intervertebral articulations. Apophyseal joint involvement is thought by some authors to be an essential component of this disease. However, although apophyseal joint erosion may occur early in ankylosing spondylitis, the difficulty in examining these joints radiographically makes their evaluation of little value in early diagnosis.[19]

Squaring

Of considerable practical diagnostic value, however, is the appearance of "squaring" of vertebral bodies.[20] This configuration is thought to be due to erosion of the anterior and anterior–lateral margins of vertebral bodies.[16] Such erosion occurs earliest at the lower dorsal and upper lumbar levels and at S1. Sclerosis of adjacent bone accompanies this erosion producing a characteristic "shining corner" appearance.[16] According to McEwen et al., vertebral squaring occurs in the spondylitis accompanying inflammatory bowel disease, psoriasis, and Reiter's syndrome, as well as in ankylosing spondylitis, although it is found to be more common in patients with ankylosing spondylitis and ulcerative colitis.[15]

Syndesmophytes

The term *syndesmophyte* was first used by Forestier to describe the intervertebral bony bridging in ankylosing spondylitis (Figs. 2 and 3).[17] Syndesmophytes are the result of ossification of the annulus fibrosis of the intervertebral disk and the adjacent connective tissue. Therefore, the classical

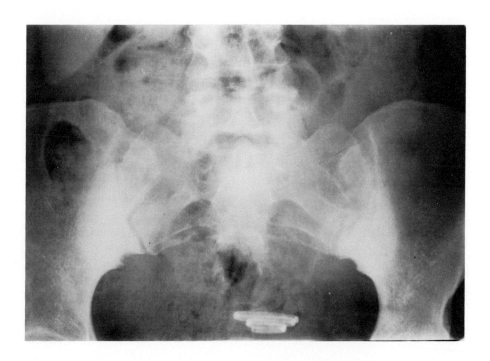

FIG. 1. Osteitis condensans Ilii. There is prominent sclerosis of the iliac sides of the sacroiliac joints. Erosion or cartilage narrowing is not a feature of osteitis condensans ilii.

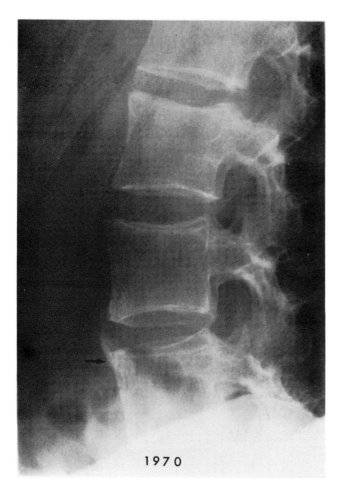 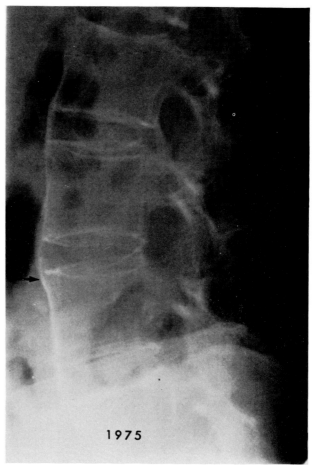

FIG. 2. Progression of findings of ankylosing spondylitis. Left: Lateral view in 1970 shows osteitis of vertebral corners (arrows) and beginning syndesmophyte formation. Right: Repeat examination in 1975 shows extensive bony bridging and disappearance of the "shining corners" previously noted (arrow).

syndesmophyte extends from the margin of one vertebral body to the margin of the next vertebral body.[15] Occasionally, ossification also occurs in the deep portions of the overlying anterior longitudinal ligament.[19] Syndesmophytes are seen earliest in the dorsolumbar junction, and involvement thereafter proceeds in ascending fashion. Syndesmophytes involving only the cervical spine would be most unusual in ankylosing spondylitis. In contrast to syndesmophytes, osteophytes characteristically extend at right angles to the vertebral body, do not usually completely bridge vertebrae, and consist of cortical and cancellous bone, the latter being continuous with the cancellous bone of the vertebral body.

McEwen et al. analyzed the bony bridging in patients with ankylosing spondylitis and the spondylitis accompanying inflammatory bowel disease, psoriasis, and Reiter's disease[15] and found that marginal syndesmophytes occur most frequently in ankylosing spondylitis and ulcerative colitis. Although similar marginal syndesmophytes predominated in patients with spondylitis accompanying psoriasis or Reiter's syndrome, nonmarginal bony bridging, comma-shaped excrescences, and bridging extending between the midportions of vertebral bodies and separated from the vertebrae by a clear space (Bywaters–Dickson syndesmophytes) were commonly seen in these disorders. However, approximately 25 percent of the patients studied with spondylitis accompanying psoriasis or Reiter's syndrome had only the marginal syndesmophytes classically described in ankylosing spondylitis.[15]

Destructive Lesions

Destructive vertebral lesions occur occasionally in ankylosing spondylitis and may involve localized areas or most of a vertebral body end plate.[21–26] The diffuse lesions are of particular importance, since

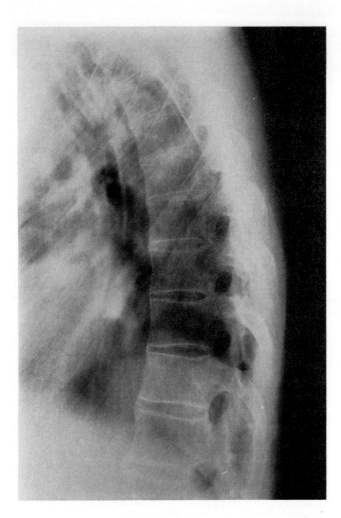

FIG. 3. Ankylosing spondylitis. Syndesmophytes extend from the margin of one vertebral body to the margin of the next.

the atlas (at its inferior border) and the odontoid. Normally, this space is essentially constant in flexion, neutral, and extension and measures less than 2.5 mm.[27] C1–C2 subluxation occurs frequently in rheumatoid arthritis but is seen in only 15 percent of patients with ankylosing spondylitis.[27]

The presence of spinal cord compression is not directly correlated with the presence or degree of C1–C2 subluxation, and most patients with this deformity have no long tract signs.[28]

Additional cervical spine abnormalities include ankylosis of the apophyseal joints, erosion of spinous processes, and sclerosis of the odontoid producing a "shining odontoid."[27]

Costovertebral Joints

The costotransverse and costovertebral joints are affected early in ankylosing spondylitis. However, because of the difficulty in demonstrating these joints radiographically, they are of limited value in early diagnosis.[19] Erosion, sclerosis, and ankylosis may be seen.

their appearance may closely mimic that of infection. These lesions occur most frequently in the dorsolumbar area in patients with long standing disease and ankylosed spines. Irregular destruction and sclerosis of adjacent vertebral body end plates is noted.

The hypothesis that these lesions represent fractures through areas of prior ankylosis with formation of a pseudoarthrosis is supported by the prior history of trauma that is often elicited and the presence of fractures of the posterior elements that may accompany the vertebral body lesions.

Whiskering

Erosion, sclerosis, and eventual new bone production may occur at tendinous insertions, particularly the ischial tuberosities, iliac crests, greater trochanters, and calcanei. The new bone produced extends in multiple linear strands from the bony margins into the soft tissues, an appearance termed *whiskering* (Fig. 4).

Ligamentous Ossification

Frontal radiographs of the spine in patients with severe disease may show three vertical bands of ossification. The central band is due mainly to ossification in the area of the interspinal ligaments. Apophyseal joint ankylosis or ossification of capsular ligaments and connective tissues between the apophyseal joints produce the lateral densities (Fig. 5).

Inflammation apparently begins in the areolar tissue adjacent to avascular ligaments. Subsequent

C1–C2 Subluxation

Laxity or disruption of the transverse ligament of the atlas may result in anterior subluxation of C1 on C2 when the neck is flexed. Ligamentous disruption is a consequence of inflammation of adjacent bursae and is recognized on lateral radiographs as a widening of the distance between the anterior arch of

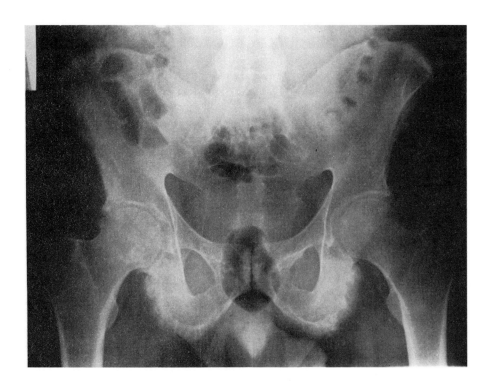

FIG. 4. Whiskering. Irregular bone extends from the margins of the ischia and ilia. Note the syndesmophytes bridging lower lumbar vertebrae, sacroiliac ankylosis, hypertrophic lipping of the right hip, and narrowing and sclerosis of the pubic symphysis.

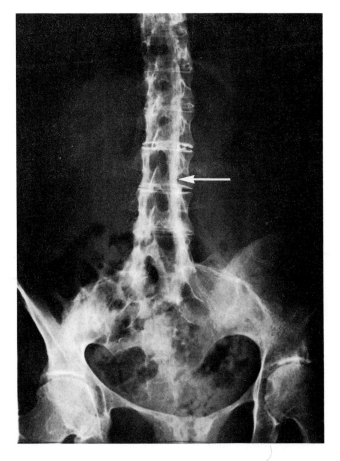

FIG. 5. Ligamentous ossification. A lateral linear band of density bridges the fused apophyseal joints. There is ankylosis of the sacroiliac joints, cartilage space narrowing in the hips, and narrowing, erosion, and sclerosis of the pubic symphysis.

ossification may spread from these areas into the ligaments themselves.[16] Ligamentous ossification is also seen in the spondylitis accompanying ulcerative colitis, Reiter's syndrome, or psoriasis. In ankylosing spondylitis and ulcerative colitis, ossification is primarily seen in the dorsal and lumbar areas, but in psoriasis or Reiter's syndrome, the cervical area is frequently involved.[15]

Peripheral Joint Involvement

Ankylosing spondylitis involves the spine, sacroiliac and costovertebral articulations. Peripheral joint symptoms may occur, however, in 30 to 50 percent of patients,[29,30] and in 20 percent of patients, initial symptoms involve joints distal to the shoulders and hips.[29] Radiographic findings are similar to those of rheumatoid arthritis.

According to Resnick,[29] radiographic findings that favor a diagnosis of ankylosing spondylitis rather than rheumatoid arthritis include markedly asymmetrical joint involvement, bony ankylosis, and periostitis. In the hands, erosions are smaller than in rheumatoid arthritis, and although the metacarpals are frequently involved, distal joint involvement may also occur. Residual deformity and subluxation are more frequent sequelae of rheumatoid arthritis.

Reiter's Syndrome

In 1916, Reiter described patients with nongonococcal urethritis, conjunctivitis, and arthritis.[31] Mucocutaneous lesions (keratoderma blennorrhagica) have been found to accompany this syndrome in 80 percent of patients.[32]

The etiology of Reiter's syndrome is unknown, but theories suggest an infectious etiology acting in a genetically susceptible individual. The most common antecedent events are sexual contact and/or diarrhea. The syndrome may occur following *Shigella flexner* dysentery. HLA-B27 is found in about 75 percent of patients with Reiter's syndrome as compared to 88 percent of patients with ankylosing spondylitis.[2] This antigen is especially frequent in patients with uveitis, a chronic or relapsing course, or severe constitutional symptoms.[33]

Arthritis is the major clinical manifestation of this syndrome. It is usually polyarticular and affects the large weight-bearing joints, the sacroiliac joints, and the spine.[32] Although complete recovery usually follows initial attacks, chronic articular involvement is eventually present in about 50 percent of patients.[32]

Radiologic features early in the course of this syndrome are minimal and reversible and consist of soft-tissue swelling and juxtaarticular osteopenia. With repeated attacks, marginal erosion, cartilage space narrowing, and periosteal reaction may be seen. The sacroiliac joints, heels, and toes are most frequently involved.[34] When the hands are affected, involvement may be asymmetrical and erosion and articular damage is mild in comparison to patients with rheumatoid arthritis.

Sacroiliac Joints

Unlike ankylosing spondylitis, sacroiliac involvement (Fig. 6) in Reiter's syndrome is likely to be unilateral or asymmetrical and bony ankylosis is uncommon.[34] Twenty of 48 patients (42 percent) with Reiter's syndrome studied by Scholkoff had evidence of sacroiliitis.[34] Abnormalities were unilateral in 9 patients, bilateral but markedly asymmetrical in 7, and bilateral and symmetrical in 4. Patients with unilateral involvement usually have the disease less than 1 year. Sacroiliitis is much more common in HLA-B27 positive than in HLA-B27 negative patients.

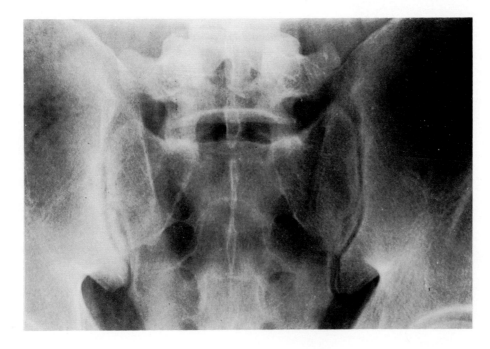

FIG. 6. Reiter's syndrome. There is sclerosis and erosion of the right sacroiliac joint. The left side is normal.

Spinal Involvement

Although sacroiliac findings are common in patients with Reiter's syndrome, spondylitis is unusual. One quarter of the patients with spondylitis accompanying Reiter's syndrome studied by McEwen et al. had a marginal syndesmophyte identical to those in patients with ankylosing spondylitis.[15] However, nonmarginal bony bridging was frequently seen. According to McEwen et al., the spondylitis associated with Reiter's syndrome has many features similar to the spondylitis accompanying psoriasis, and the presence of nonmarginal bony bridging should suggest one of these entities. Nonmarginal bony bridging was not seen to occur in patients with ankylosing spondylitis.

Peripheral Joint Changes

In Reiter's syndrome the feet are involved more frequently and more severely than the hands. Soft-tissue swelling and erosion of the great toe interphalangeal joint and the tuft of the distal phalanx are the most common abnormalities.[34] This joint is also involved in psoriatic arthritis, but the lack of distal joint involvement and the more minimal findings in the hands may suggest the diagnosis of Reiter's syndrome. Periosteal reaction may occur particularly along the shafts of the metatarsals and phalanges[36] (Fig. 7).

Calcaneal involvement was seen in 59 percent of patients in one series[34] and consists of plantar and posterior calcaneal spurs, Achilles tendonitis, sub-Achilles bursitis, posterior calcaneal erosion, and periostitis. Two types of plantar calcaneal spurs have been described. The degenerative plantar spur is smooth, well defined, and curves superiorly.[37] In contrast, the spur classically described in Reiter's syndrome is fluffy and irregular and is thought to be the result of localized periostitis. The calcaneus may be dense adjacent to the spur. These irregular spurs were seen in 20 percent of patients with Reiter's syndrome, 4 percent of patients with rheumatoid arthritis, and 5 percent of patients with ankylosing spondylitis.[35]

Periosteal reaction may be prominent in Reiter's syndrome, but it is also seen in patients with psoriatic arthritis and juvenile rheumatoid arthritis.

Psoriatic Arthritis

HLA-B27 is present with peripheral psoriatic arthropathy alone (18 percent) and in patients with psoriasis and sacroiliitis or spondylitis (50 percent) in significantly higher frequency than in control groups.[2] It has been suggested, therefore, that being HLA-B27 positive increases the risk of a psoriatic patient developing either peripheral arthropathy or spondylitis.[2,39] Comparisons between patients with

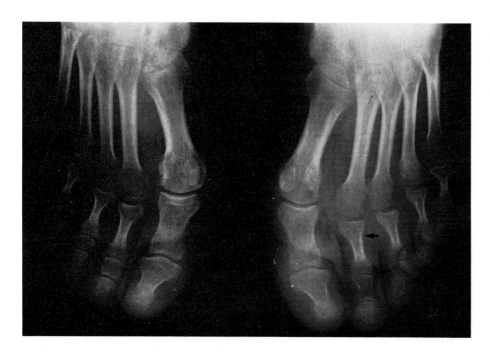

FIG. 7. Reiter's syndrome. There is marked asymmetrical involvement characterized by severe osteopenia, soft-tissue swelling, and periosteal reaction (arrow).

radiologic evidence of sacroiliitis and patients with sacroiliitis and spondylitis show no significant difference in the percentages of B27 positive individuals.[2]

Psoriatic arthritis should be considered when psoriasis and inflammatory polyarthritis occur together without subcutaneous nodules or serum rheumatoid factor.[40] In 87 percent of patients studied by Baker et al., psoriasis preceded or coincided with the onset of arthritis.[40] In 13 percent, arthritis preceded skin changes by 6 to 25 years.

Several radiographic features may help differentiate psoriatic arthritis from rheumatoid arthritis. In contrast to the fusiform soft-tissue swelling around joints seen in rheumatoid arthritis, the swelling in psoriatic arthritis may involve the entire digit ("sausage digit"). This is attributed to simultaneous involvement of the interphalangeal joints, tendon sheaths, and surrounding tissues[40] (Fig. 8).

Another classical finding in psoriatic arthritis is its distal distribution. However, predominant distal interphalangeal (DIP) joint involvement was seen in only 8 of 23 patients with psoriatic arthritis and peripheral erosive changes.[40] There is no clear relationship between DIP joint changes and overlying nail changes. When DIP involvement occurs in rheumatoid arthritis, there are usually associated severe changes in the proximal joints of the hand and wrist.

In addition to the distal distribution, the type of involvement of joints may differ from rheumatoid arthritis in that bony ankylosis of interphalangeal joints or marked bone destruction on either side of the joint occurs more frequently in psoriatic arthritis (Fig. 9). Periosteal reaction also appears to be more frequent in psoriatic arthritis than in rheumatoid arthritis.

As noted in Reiter's syndrome, destruction of the interphalangeal joint of the great toe occurs frequently in psoriatic arthritis, sometimes associated with bony proliferation at the base of the distal phalanx.

Analysis of the radiographs of 155 patients with psoriasis and arthritis by Avila et al. showed that 31 percent of patients had findings typical of psoriatic arthritis.[41] In an additional 31 percent of patients, some of the findings of psoriatic arthritis were present, accompanied by findings typical of rheumatoid arthritis. In 38 percent of patients, the findings present were those of rheumatoid arthritis alone.

Sacroiliitis

Sacroiliitis is noted in one-quarter to one-third of patients with psoriatic arthritis.[38,40,42] Additional patients may have early involvement detectable by bone scan.[42] Damage is usually bilateral and symmetrical but, in contrast to ankylosing spondylitis, may be asymmetrical or unilateral.[15,43]

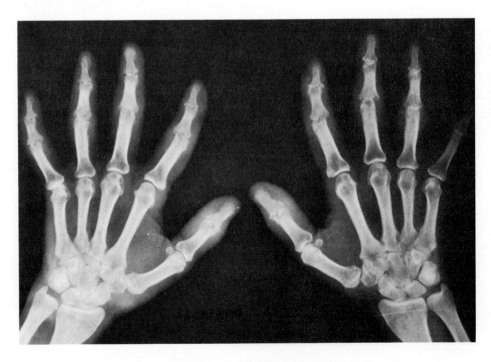

FIG. 8. Psoriatic arthritis. There is diffuse swelling of most of the fingers ("sausage digits"). Erosive changes are present at most distal and proximal interphalangeal joints. The metacarpals and wrists are much less involved.

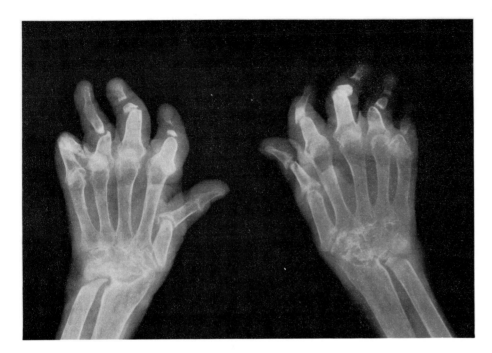

FIG. 9. Severe psoriatic arthritis. There is marked bone loss at several interphalangeal joints and ankylosis of some distal interphalangeal joints. Severe destruction changes are also present in the wrists. The fusion of the left thumb interphalangeal joint was surgical.

Spondylitis

The spinal manifestations of psoriasis have been reviewed by McEwen et al.[15] Squaring of vertebral bodies occurs in psoriatic arthritis less frequently than in ankylosing spondylitis. Bony bridging between vertebrae may originate at the margins of the vertebral bodies as in ankylosing spondylitis or may arise at other sites (other than marginal) along the vertebral surface. About one-quarter of patients with psoriatic spondylitis or the spondylitis of Reiter's syndrome show only marginal syndesmophytes, similar to patients with ankylosing spondylitis.[15] Most patients with psoriatic spondylitis have both marginal and other than marginal bony bridging. Bridging is rarely only other than marginal (Fig. 10). One type of other than marginal paravertebral bridging seen in patients with psoriasis was described by Bywaters and Dixon.[44] In this type, bone extends from the midportion of one vertebra laterally to the midportion of the next vertebral body and is separated from the spine by a clear zone. This has also been described in patients with Reiter's syndrome.[38]

Inflammatory Bowel Disease

Peripheral arthritis, sacroiliitis, and spondylitis may occur in association with inflammatory bowel disease. The reported incidence of these findings varies widely. Fernandez-Herlihy studied 555 cases of chronic ulcerative colitis and found articular manifestations in 17 percent.[45] In this series, 4.2 percent of patients had arthralgias, 3.2 percent had rheumatoid arthritis, and 5 percent had ankylosing spondylitis. In 10 additional patients, arthritis involved large joints and was distinctive in that in 9 cases it developed with or after the onset of colitis and flared with exacerbations of bowel symptoms. Colectomy cured all 10 patients. The exacerbation of peripheral joint symptoms with flares of colonic disease has since been confirmed.

Peripheral Joint Changes

The ankles and knees are the most frequently involved peripheral joints. Elbows, proximal interphalangeal joints of the fingers, wrists, metacarpals, and shoulders are affected less frequently. Permanent peripheral joint damage is unusual. In 52 patients studied by McEwen et al., permanent changes were detectable in 13 patients on clinical examination.[46] Radiographic findings are identical to those of rheumatoid arthritis.

Sacroiliac and Spine Changes

The sacroiliac joints are the most frequent areas to be involved in patients with inflammatory bowel disease. Moderate to severe sacroiliac joint changes

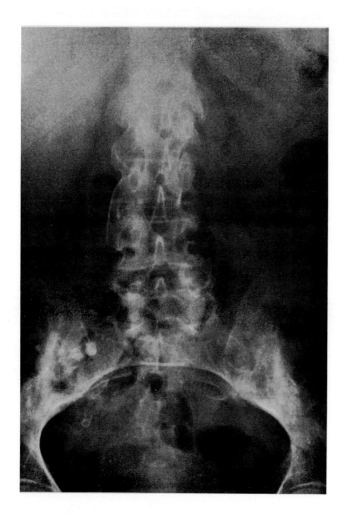

FIG. 10. Psoriatic spondylitis. There is asymmetrical, nonmarginal bony bridging and bilateral sacroiliitis.

were noted in 18 percent of patients with chronic ulcerative colitis.[47] Spondylitis is less frequent. In contrast to peripheral joint involvement, damage to the sacroiliac joints and spine is progressive and less affected by the course of bowel disease. Radiographic features are identical to those in ankylosing spondylitis.

References

1. Brewerton DA, James DCO: The histocompatibility antigen (HL-A27) and disease. Semin Arthritis Rheum 4:191, 1975
2. Woodrow JC: Histocompatibility antigens and rheumatic diseases. Semin Arthritis Rheum 6:257, 1977
3. Caffrey MFP, James DCO: Human lymphocyte antigen association in ankylosing spondylitis. Nature 242:121, 1973
4. Brewerton DA, Caffrey M, Hart FD, et al: Ankylosing spondylitis and HL-A27. Lancet 1:904, 1973
5. Schlosstein L, Terasaki PI, Bluestone R, Pearson CM: High association of an HL-A antigen, W27, with ankylosing spondylitis. New Engl J Med 288:704, 1973
6. Brewerton DA, Caffrey M, Nicholls A, et al: Reiter's disease and HL-A27. Lancet 2:996, 1973
7. McClusky OE, Lordon RE, Arnett FC, Jr: HL-A27 in Reiter's syndrome and psoriatic arthritis: a genetic factor in disease susceptibility and expression. J Rheumatol 1:263, 1974
8. Morris RI, Metzger AL, Bluestone R, et al: HL-A W27—a clue to the diagnosis and pathogenesis of Reiter's syndrome. N Engl J Med 290:554, 1974
9. Morris RI, Metzger AL, Bluestone R, Terasaki PI: HL-A W27—a useful discriminator in the arthropathies of inflammatory bowel disease. N Engl J Med 290:1117, 1974
10. Brewerton DA, Caffrey M, Nicholls A, et al: HL-A27 and arthropathies associated with ulcerative colitis and psoriasis. Lancet 1:956, 1974
11. Aho K, Ahvonen P, Alkio P, et al: HL-A 27 in reactive arthritis following infection. Ann Rheum Dis 34(Suppl 1):29, 1975
12. Aho K, Ahvonen P, Lassus, et al: HL-A 27 in reactive arthritis; a study of *Yersinia* arthritis and Reiter's disease. Arthritis Rheum 17:521, 1974
13. Boland EW: Ankylosing spondylitis in arthritis and allied conditions. Hollander JL, ed. Philadelphia, Lea and Febiger, 1966, p 633
14. Edeiken J, Hodes PJ: Roentgen Diagnosis of Diseases of Bone. Baltimore, Williams and Wilkins, 1973, p 733
15. McEwen C, DiTata D, Lingg C, et al: Ankylosing spondylitis and spondylitis accompanying ulcerative colitis, regional enteritis, psoriasis and Reiter's disease. Arthritis Rheum 14:291, 1971
16. Romanus R, Ydén S: Pelvo-Spondylitis Ossificans. Chicago, Year Book, 1955
17. Forestier J: The importance of sacroiliac changes in the early diagnosis of ankylosing spondyloarthritis. Radiology 33:389, 1939
18. Goss CM: Gray's Anatomy, Philadelphia, Lea and Febiger, 1973, pp. 312–13
19. Berens DL: Roentgen features of ankylosing spondylitis. Clin Orthop 74:20, 1971
20. Boland EW, Shebesta EM: Rheumatoid spondylitis: correlation of clinical and roentgenographic features. Radiology 47:551, 1946
21. Cawley MID, Chalmers TM, Kellgren JH, Ball J: Destructive lesions of vertebral bodies in ankylosing spondylitis. Ann Rheum Dis 31:345, 1972
22. Lorber A, Pearson CM, Rene RM: Osteolytic vertebral lesions as a manifestation of rheumatoid arthritis and related disorders. Arthritis Rheum 4:514, 1961
23. Wholey MH, Pugh DG, Bickel WH: Localized destructive lesions in rheumatoid spondylitis. Radiology 74:54, 1960
24. Kanefield DG, Mullins BP, Freehafer AA, et al:

Destructive lesions of the spine in rheumatoid ankylosing spondylitis. J Bone Joint Surg 51A:1369, 1969
25. Rivelis M, Freiberger RH: Vertebral destruction at unfused segments in late ankylosing spondylitis. Radiology 93:251, 1969
26. Sutherland RIL, Matheson D: Inflammatory involvement of vertebrae in ankylosing spondylitis. J Rheumatol 2:296, 1975
27. Martel W: The occipito-atlanto-axial joints in rheumatoid arthritis and ankylosing spondylitis. Am J Roentgenol 86:223, 1961
28. Sharp J, Purser DW: Spontaneous atlanto-axial dislocation in ankylosing spondylitis and rheumatoid arthritis. Ann Rheum Dis 20:47, 1961
29. Resnick D: Patterns of peripheral joint disease in ankylosing spondylitis. Radiology 110:523, 1974
30. Dilsen N, McEwen C, Poppel M, et al: A comparative roentgenologic study of rheumatoid arthritis and rheumatoid (ankylosing) spondylitis. Arthritis Rheum 5:341, 1962
31. Reiter H: Über ein bisher unerkannte spirochaelinfektion (spirochaetosis arthritica). Dtsch Med Wochenschr 42:1535, 1962.
32. Weinberger HJ: Reiter's syndrome re-evaluated. Arthritis Rheum 5:202, 1962
33. McClusky OE, Lordon RE, Arnett FC, Jr: HL-A27 in Reiter's syndrome and psoriatic arthritis: a genetic factor in disease susceptibility and expression. J Rheumatol 1:263, 1974
34. Sholkoff SD, Glickman MG, Steinbach HL: Roentgenology of Reiter's syndrome. Radiology 97:497, 1970
35. Mason RM, Murray RS, Oates JK, Young AC: A comparative radiological study of Reiter's disease, rheumatoid arthritis and ankylosing spondylitis. J Bone Joint Surg 41B:137, 1959
36. Weldon WV, Scalettar R: Roentgen changes in Reiter's syndrome. Am J Roentgenol 86:344, 1961
37. Berens DL, Lin R: Roentgen Diagnosis of Rheumatoid Arthritis. Springfield, Ill, Thomas, 1969
38. Peterson CC Jr, Silbiger ML: Reiter's syndrome and psoriatic arthritis. Their roentgen spectra and some interesting similarities. Am J Roentgenol 101:860, 1967
39. Eastmond CJ, Woodrow JC: The HLA system and the arthropathies associated with psoriasis. Ann Rheum Dis 36:112, 1977
40. Baker H, Golding DN, Thompson M: Psoriasis and arthritis. Ann Intern Med 58:909, 1963
41. Avila R, Pugh DG, Slocumb CH, Winkelmann RK: Psoriatic arthritis: a roentgenologic study. Radiology 75:691, 1960
42. Barraclough RAS, Percy JS: Psoriatic spondylitis: a clinical, radiological and scintiscan survey. J Rheumatol 4:282, 1977
43. Jajic I: Radiological changes in the sacroiliac joints and spine of patients with psoriatic arthritis and psoriasis. Ann Rheum Dis 27:1, 1968
44. Bywaters EGL, Dixon A St J: Paravertebral ossification in psoriatic arthritis. Ann Rheum Dis 24:313, 1965
45. Fernandez-Herlihy L: The articular manifestations of chronic ulcerative colitis. An analysis of 555 cases. New Engl J Med 261:259, 1959
46. McEwen C, Lingg C, Kirsner JB, Spencer JA: Arthritis accompanying ulcerative colitis. Am J Med 33:923, 1962
47. Wright V, Watkinson G: Sacroiliitis and ulcerative colitis. Brit Med J 2:675, 1965

Osteoarthrosis (Osteoarthritis, Degenerative Joint Disease)

Jack Edeiken

Osteoarthritis may be defined as a local disorder of individual diarthrodial joints independent of infection and systemic disease, brought out by abnormal mechanical conditions in conjunction with senescence, trauma, or other degrading change in the articular hyaline cartilage.

Cartilage is composed of a ground substance, chondroitin sulfate, in which collagen fibrils are embedded. The collagen fibrils are laid down in several layers, and chondroitin sulfate takes part in the building of the fibers. The collagen fibrils determine the elasticity of cartilage. The content of chondroitin sulfate decreases with increasing age. It is believed that the initial lesion of osteoarthritis is a decrease in the chondroitin sulfate, which leads to a disintegration of the unsupported collagen fibrils.

Pathologically, osteoarthritis begins in the articular cartilage as fibrillation, the splitting of cartilage along its fibrillatory planes. As it progresses, surface continuity to bone is lost and the cartilage is lifted from the bone. The cartilage damage provides a stimulus for blood vessel growth in the subchondral bone, which becomes eburnated and sclerotic, after which the spurs and shells of new osteophytic bone are formed. Blood vessels also invade the cartilage and initiate ossification. Eventually, the cartilage is completely destroyed, but by this time, the subchondral bone is well fortified by sclerosis and eburnation.

Osteoarthritis is often explained as the physiologic wear and tear of aging, but this concept cannot be completely accepted, since young individuals are sometimes affected. Also, damage to articular cartilage will not cause osteoarthritis unless a joint is used. It seems more apt to describe the osteoarthritis as a destructive change in the cartilage due to multiple causes, some of them obscure and associated with the mechanical trauma of misuse. Therefore, osteoarthritis may be divided into primary and secondary osteoarthritis.

Primary generalized osteoarthritis occurs as a regressive alteration in articular cartilage of the body. Originally, this was thought to be due to vascular deficiencies or ischemia, but recent evidence suggests the opposite. Probably there is a local vascular proliferation and dilatation with growth of the new blood vessels into the subchondral areas and into the cartilage itself. Whenever new blood vessels invade these areas, new bone is formed, which explains the production of the osteophytes as well as the weakening in the weight-bearing areas of the cartilage.

Mechanical factors are fundamental in secondary osteoarthritis, especially in young individuals whose joints are subjected to unusual stress from congenital or acquired orthopedic deformities. A single episode of infection, trauma, or vascular anomaly of underlying bone may be the initiating factor. Endocrine imbalance may play a part in the weakening of cartilage, since the pituitary growth hormone is necessary for its formation and maintenance.

Primary osteoarthritis must be a congenital diathesis affecting the articular cartilages, manifested by a regressive affectation of the joint. The cause may depend on cartilage metabolism, which determines the resistance to stress. Secondary osteoarthritis depends on a previous insult to the cartilage, such as infection, trauma, vascular abnormality, or perhaps even hormone imbalance, with degeneration of specific joints. It is not a generalized disease.

The roentgenographic appearance of osteoarthritis is well known, since it is the most common abnormality encountered in the skeleton. At first, there is narrowing of the joint space without destruction of the margins of the articular cortex or synovial thickening. This is followed by the formation of osteophytes at the margin of the articular surface. These are osseous outgrowths consisting of cortical and cancellous bone that blend with the normal bone beneath. When the articular cortex is badly eroded, the subchondral bone becomes irregular and eburnated, and cystlike lesions may form near the joint. Sometimes they are conspicuous and give the false impression of a destructive inflammatory process. Synovial cysts may form before or after joint space narrowing. They produce large bone erosions with sclerotic margins. Loose osteocartilaginous bodies form within the joint. Because osteoarthritis varies in different joints, individual variations occur.

Fingers

Osteoarthritis of the fingers (Heberden's nodes) is of two varieties—traumatic and idiopathic. Although the lesions look alike, the patient's history will distinguish them. Traumatic Heberden's nodes arise immediately after significant trauma to one or more fingers. The injury is usually acute and vividly recalled. After a period of time, the Heberden's nodes change from a hard, painful swelling to a softer, nontender area.

Idiopathic Heberden's nodes arise spontaneously with no known trauma. They may be tender and red initially, but within 2 years they reach a resting state; they become soft and nontender and do not change in size. Successive involvement eventually affects most of the fingers. The idiopathic type of Heberden's nodes affects women ten times more frequently than men.

The radiographic changes are best seen in the lateral views as large spurs arising from the dorsal aspects of the proximal ends of the distal phalanges and the distal ends of the middle phalanges. Sometimes, the prominence of these spurs resembles the symmetrical soft-tissue enlargement of rheumatoid arthritis. When the changes become extreme the joint spaces are lost completely and the articular surfaces become irregular. When marked destruction occurs, it may be mistaken for infection; however, there is no loss of articular cortical margins as occurs in infections. Sometimes the proximal interphalangeal joints show osteoarthritic changes, although the literature principally mentions this condition in the distal interphalangeal joints.

Erosive osteoarthritis is most dramatic in the hands and will be discussed below.

Ankylosis of the interphalangeal joints may result from osteoarthritis. The joints become bridged by bony trabeculae best demonstrated in the lateral projection. Clinical evaluation should distinguish the condition from rheumatoid arthritis and other causes of ankylosis.

Hips

Osteoarthritis of the hip is frequent. It may develop early in adult life. The earliest roentgenographic changes are best seen at the margins of the head of the femur, where osteophytes form. Since there is often considerable irregularity of the superior margin of the acetabulum, this is not a reliable area in which to look for osteoarthritis. The hip is usually not painful until there is loss of motion and bony impingement by spurs. Late in the disease, pain is troublesome, but as the hip becomes more fixed, pain virtually disappears. Osteoarthritis of the hip may be due to underlying dysplasia; marked flattening of the head has been described. At times, the femoral head undergoes severe deformation, with absorption adding to its poor fit into the head of the acetabulum and increasing the destructive osteoarthritic changes.

The two aspects of the hip joints where narrowing occurs, the medial and the superolateral, have features that implicate different conditions. Unilateral hip osteoarthritis usually remains so, and the joint space narrowing is the superolateral area and may be related to leg length disparity. Bilateral osteoarthritis usually features narrowing of the medial joint space. A superomedial form may be an intermediate type but is less well defined. Gofton has recorded the incidence and his classified hip osteoarthritis is shown in Tables 1 and 2.

Murray believes that minor anatomic abnormalities are responsible for incongruity of the hip articulation and cause premature degenerative changes in other joints. He found that only 14 percent of men with anatomically normal hips had osteoarthritis and suggested that these may be secondary to subclinical inflammatory arthritis. Anatomically normal hips are unlikely to develop spontaneous osteoarthritic change.

Knees

The cause of osteoarthritis of the knees is usually obscure. The first evidence is frequently crepitus without pain, which is secondary to cartilage destruc-

Table 1 Incidence of Types of Osteoarthritis of Hip in 67 Patients*

Idiopathic		57
With subluxation		9
Secondary to trauma		1
		67

Type	Unilateral	Bilateral
Superolateral		
Idiopathic	36	1
With subluxation	5	0
Supermedial		
Idiopathic	4	3†
With subluxation	1	2
Medial		
Idiopathic	2	7
Unclassified		
Idiopathic	4	0
With subluxation	1	0
Secondary to trauma	1	0
Total	54 (80%)	13 (20%)

Modified from Gofton, JP, Can Med Assoc J 104:679, 1971
*Patients with ankylosing spondylitis or rheumatoid arthritis were excluded.
†One patient with bilateral disease had supermedial osteoarthritis in one hip and superolateral in the other.

tion and fibrillation. When the knee is painful, crepitus is almost always produced on motion. Few early roentgenographic changes are found. Eventually, the joint space becomes thin due to destruction of the joint cartilage; it may be confined to one compartment of the knee, such as the lateral or medial compartment, and affect only the lateral medial condyle of the femur. When one compartment of the knee is involved, it is very suggestive of osteoarthritis rather than other forms such as rheumatoid arthritis. Spurs form, first at the posterior surface of the patella and later on the lateral and medial borders of the tibia and fibula.

Spine

Osteoarthritis of the spine should refer to changes in the intervertebral apophyseal joints rather than in the intervertebral disks, which are not true joints. Early manifestations in the apophyseal joints are difficult to demonstrate radiographically. The term *osteoarthritis* should not be used to indicate spur formation of the bodies of the vertebrae; this is not a

Table 2 Classification of Osteoarthritis of Hip*

Types	Criterion	Other Characteristics
Superolateral	Narrowing of the joint space occurs first in superolateral segment	Medial, lateral, and inferior osteophytes are common. A buttress on the femoral neck is common. Femoral head tends to flatten as the disease progresses. Sclerosis, pseudocysts, and femoral head collapse are seen in advanced disease. This form is usually unilateral.
Medial	Joint space narrowing occurs medially	The femoral head usually becomes deeply seated, and protrusioacetabuli of varying degrees may develop. A buttress is occasionally seen. A medial osteophyte is not seen, but small peripheral osteophytes may be present. This form is usually bilateral.
Superomedial	Joint space narrowing begins medial to the lateral margin	Peripheral osteophytes, a buttress and femoral head distortion may occur. This form may be unilateral or bilateral.
Unclassified	Includes many cases where advanced disease does not permit classification into one of the above categories. Most of these will have been superolateral at onset, and very few will have been medial. Occasionally, the hip will deteriorate in a pattern resembling that seen in rheumatoid arthritis or ankylosing spondylitis. There is widespread narrowing of articular cartilage, with circumferential erosions at no specific site.	

*Modified from Gofton JP, Can Med Assoc J, 104:679, 1971

true arthritis and is called *arthrosis* or *osteophytosis* of the spine. Spurs confined to the centra of the vertebrae are usually associated with diskogenic disease. Laborers and athletes develop large spur formations (presumably in areas of disk compression), that are most evident on the anterior surface of the centra and in the lateral margins, especially in the lumbar and dorsal areas. Osteoarthritis is not associated with deossification, and the changes are usually painless.

Erosive Osteoarthritis (Degenerative Joint Disease)

Jack Edeiken

Erosive osteoarthritis is a form of degenerative joint disease with inflammation. It is usually most severe in the hands, but many of the joints may be involved, particularly the knees. It is usually the hands, however, that show the earliest and most dramatic changes. Since there is an inflammatory component, it may be easily mistaken for rheumatoid arthritis. The roentgenographic features are so characteristic that this error in diagnosis should not occur, in spite of the parrallel clinical symptoms.

Roentgenographic Features

The distal joints of the fingers are most severely involved, showing destructive changes, predominantly in the distal interphalangeal joints and in some proximal interphalangeal joints. Small spurs form, and eventually, fusion may occur. If the metacarpophalangeal joints are involved, there is usually narrowing without erosive changes. The first carpometacarpal joint frequently shows degenerative joint disease. There is evidence of sclerosis and joint space narrowing. The mineralization of the hand remains excellent, even though severe erosive disease is present. The ulnar styloid is rarely if ever affected. It may be distinguished from rheumatoid arthritis by the lack of osteoporosis, the involvement of the distal interphalangeal joints as compared to the metacarpophalangeal joints in rheumatoid arthritis, and the absence of ulnar styloid irregularity, which is very common in rheumatoid arthritis.

Neurotrophic Arthropathy

Jack Edeiken

Neurotrophic arthropathy may occur in any joint if two conditions are present—a change in the sensory nerves and trauma. The trauma is produced by continued minor stresses on an unstable joint that has been deprived of the accurate weight distribution of the muscles and of the stabilizing protection of an intact reflex nervous mechanism.

Neurotrophic joint was formerly most frequently thought to be secondary to syphilis of the central nervous system. However, today, diabetes is probably the most common cause in the United States. Syringomyelia, peripheral nerve injury, spinal cord tumors, spina bifida, and damage to the spinal cord by infection or trauma may also be etiologic factors. Approximately 25 percent of syringomyelia patients and 10 percent of tabetics develop neuropathies. The early clinical signs are joint swelling, deformity, and instability. Pain, the early absence of which often masks the disease, may be a troublesome late complaint.

The apparent paradox of pain in an insensitive joint can be explained by the fact that sensation is lost only on the joint surfaces and periosteum, while the overlying soft tissues and muscles remain sensitive. Obviously, then, the distention of soft tissues by fluid or trauma to ligaments and muscles, and not the joint damage, causes the pain.

The deep reflexes are usually absent in patients with neurotrophic joints. Analgesia and hypalgesia are commonly present and often so are ataxia and Argyll–Robertson pupils. The blood and spinal serologic tests are positive in only half the patients with tabes dorsalis. The joints most commonly affected by tabes are the knees, hips, tarsals, and ankles. Neurotrophic joints in the upper extremities are most often caused by syringomyelia, with the shoulder most commonly affected.

Pathologic Features

Potts listed the pathologic changes in this order: loss of protective joint sensibility, relaxation of the lateral ligaments with consequent minor marginal joint fractures, destruction of the articular cartilage and the intraarticular ligaments, sclerosis of the bone ends denuded of cartilage, periarticular and parosteal bone production, continued erosion and fracture of the articulating ends of bone, and, finally, atrophy, when the patient is bedridden and the bones no longer articulate. The process may cease at any phase, or any of the above characteristics may predominate, depending on the joint involved and the extent of nerve injury.

Relaxation from tearing of the lateral ligaments is the first gross change in the knee joint, resulting in increased lateral mobility. This instability and continued everyday usage injures the joint. Contusions, loosening of osteophytes, and fractures with effusions follow. Small detached fragments of bone remain in the joint of the capsule where they grow and form joint mice. Bony spurs at the joint margins are indistinguishable from the spurs of osteoarthritis, but as the disease progresses, these spurs grow larger and more irregular and extend to the bone shaft. Parosteal new bone formation may extend for 10 cm along the bone.

Roentgenographic Features

Tabetic neurotrophic arthropathy occurs most often in the lower extremities and spine, whereas syringomyelic arthropathy is most frequent in the upper extremities. More than one joint may be affected, especially in tabetic patients.

The earliest roentgenographic sign is a persistent joint effusion. The joint capsule may be greatly distended due to the effusion and periarticular ligament relaxation.

Narrowing of the joint space occurs after the cartilage has been destroyed. Minor hypertrophic spurs then form on the margins of the joint. As the condition progresses, destruction of the articular cortex becomes evident, followed by increased density of the subchondral bone. Fragmentation and bone proliferation around the joint continue. Synovial membrane calcifications form and separate in the joint to form loose bodies. Fragmentation of the eburnated subchondral bone begins, and irregular masses of bone form at the margin of the articular surfaces. Joint subluxations result from the laxity of the periarticular soft-tissue structures. Eventually, a distinctive appearance evolves made up of marked disorganization of the bone, with multiple fragments floating in the joint space, and severe eburnation of bone with subluxation. The hip and the shoulder may present a slightly different appearance, since there is progressive and rapid resorption of the head and neck with very little dense bone or loose body formation. At the margin of the destructive area, there is a clear-cut definition between the destroyed bone and the host bone—almost a "guillotine effect." The acetabulum and glenoid may be partially destroyed. In the spine, there is usually increased density of multiple vertebral segments, with large spur formations and narrowing of the disk spaces. The spine is usually malaligned.

Infections of Bones and Joints

Jack Edeiken

The vascular supply of the long bone dictates the changes due to osteomyelitis and septic arthritis. When the long bone is infected, thrombosis of the vascular systems dictates the extent of the ischemic necrosis and the roentgenographic changes.

A long bone is supplied by the nutrient artery, the periosteal vessels, and the epiphyseal metaphyseal arteries. The nutrient artery is the main vessel supplying the blood to the shaft of the bone in both the young and in the adult. It is responsible for the supply of the whole marrow and of the inner two-thirds of the cortex. The periosteal vessels supply the outer part of the cortex. The metaphyseal vessels alone are incapable of keeping the bone marrow and the deep part of the cortex alive, but after their union with the epiphyseal vascular network, following the fusion of the growth cartilage, enough blood is provided by the nutrient artery through the distal branches to secure the life of the marrow and bone. Simple division or thrombosis of the nutrient artery does not cause a significant effect on the shaft. In the young, the compensatory circulation is secured by the periosteal vessels; in the adult by the metaphyseal epiphyseal vascular network.

Infection in the bones and joints is different in infants, children, and adults. In the infant, many vessels penetrate the epiphyseal plate into the epiphysis and vice versa; therefore, infections in the metaphyseal area are easily spread to the epiphysis; since the epiphysis is intraarticular, spread into the soft tissues of the joint is facilitated. As the child matures, fewer vessels penetrate the epiphyseal plate, and there is less chance for epiphyseal infection and joint involvement from metaphyseal infection. Infants have a loose periosteum, and infections easily strip the periosteum and can extend to the articular end of bone.

In the child, the metaphyseal vessels loop backward into sinusoids, which are a fertile ground for the implantation of infection. When the epiphyseal plate closes, there is a direct connection with epiphyseal vessels, and infections easily extend to the joint. Joint involvement is a less common complication of osteomyelitis in children than in adults and infants because of the lack of vascular communication across the epiphyseal plate.

Pathologic Features

Osseous infection produces the same pathologic changes as infection elsewhere, i.e., hyperemia, edema, white cell infiltration, necrosis, and pus. The infection usually spreads throughout the bone marrow, but it may remain localized, depending on the virulence of the organism in relation to the resistance of the host. The infection may reach the subperiosteal space by breaking through the cortex, thus elevating the periosteum and allowing the infection to spread along the shaft. Eventually, the cortical bone becomes necrotic, and the periosteum is stimulated to form successive layers of new bone. The infection may even break through the periosteum and extend into the soft tissues.

Acute pyogenic arthritis is a frequent complication of osteomyelitis in infants and adults, and occurs by direct extension of the infection into the distal end of the bone and then out into the joint space. Pyogenic arthritis also may occur after surgical intervention for osteomyelitis. The infection rarely crosses the epiphyseal plate during the growth period, but it does tend to break out of the bone near the epiphysis. If this occurs in the intercapsular portions of the metaphysis, pyogenic arthritis will occur. When the metaphysis is not intracapsular, infection may reach the joint by extending along the shaft or by breaking through the bone at the articular end, and when osteomyelitis is distant from the joint, pyogenic arthritis may result from hematogenous spread.

Reparative changes of osteomyelitis begin roughly 10 ten days after onset. Hodges et al. grouped these changes into three divisions: (1) changes in the necrotic bone, (2) the formation of new bone, and (3) changes in the old living bone.

Changes in Necrotic Bone

Dead bone is absorbed by the action of granulation tissue, which develops about its surface. Absorption takes place earliest and most rapidly at the junction of dead and living bone. If the dead bone is small in amount, it is entirely destroyed by granulation tissue, leaving behind a cavity. The necrotic cancellous bone in localized osteomyelitis is usually all absorbed, even though extensive in amount. Dead cortex in an appreciable amount is gradually detached from living bone to form a sequestrum. Lines of sequestration are usually jagged and irregular. The organic elements present throughout dead bone are largely broken down by the action of proteolytic ferments of the exudate. Because of loss of blood supply, the dead bone is whiter than living bone. The time required for separation is variable according to the density and thickness of the involved bone.

Spongy bone is absorbed rapidly and may be completely sequestrated or destroyed in from 2 to 3 weeks, but the necrotic cortex may require from 2 weeks to 6 months for separation, depending on its size and thickness and on the general condition of the patient. After complete sequestration, the dead bone is less readily attacked by granulation tissue and more slowly absorbed. The rate of absorption is greater when dead bone is surrounded by an involucrum. When free in a pocket or located along a discharging sinus, there is practically no absorption, since healthy granulation tissue does not come into contact with it. When periosteum is extensively killed or no involucrum forms over the dead bone, its surface may remain uneroded for long periods. The same may be true of the endosteal surface when the shaft is destroyed in the greater part or all of the circumference. Healthy granulation cannot come into contact with it, and even the cancellous bone may be found years after the occurrence of necrosis.

New Bone Formation

New bone forms from the surviving portions of periosteum, endosteum, and cortex in the region of the infection. It may be laid down along both periosteal and endosteal surfaces. The surviving periosteum about the dead bone forms an involucrum, which gradually increases in density and thickness to form part or all of the new shaft. This is continuous with new bone formed on the end of the cortex that is not killed with the infection. Recurrence of infection may result in the formation of superimposed layers of involucrum, and cloacae are present at the point where periosteum has been killed. The new bone increases in amount and density for weeks or months, according to the size of the bone and extent and duration of the infection. The endosteal new bone may obstruct the medually canal at the limits of the infection. After discharge or removal of sequestra, the remaining cavity may be filled with new bone. This is especially apt to occur in children. However, in adults, particularly when the walls are dense, the cavities persist. In some instances, the space may become filled with fibrous tissue or it may be lined with fibrous tissue and communicate with the outside by means of a discharging sinus.

Changes in Old Living Bone

The surviving bone in a field of osteomyelitis usually becomes osteoporotic during the active period of infection. This is the result of both atrophy or disuse and of the variable degree of inflammatory reaction that permeates it. After subsidence of the infection and resumption of function of the part, its density increases again and it may undergo extensive transformation to meet the lines of stress and strain resulting from loss of substance of the shaft. Eventually, it may be difficult to distinguish between the old living bone and the newly formed bone. Areas of chronic infection frequently persist in the old living bone years after healing has occurred. In other cases, especially in children, the infection is completely overcome and the bone so transformed that years afterward all traces of the disease have disappeared.

The pathogenesis of osteomyelitis described above accounts for the majority of bone infections between the ages of 2 and 16. There are basic differences among infant, childhood, and adult types:

1. The soft-tissue component of infant osteomyelitis is more striking than in adults and children.
2. Subperiosteal abscess extensions and involucrum formations are more frequent in infants than in

children or adults. The periosteal new bone formed in adults is delicate and not as extensive as that seen in the two younger groups.
3. Sequestrations are more frequent in childhood osteomyelitis than in infant and adult types.
4. Joint involvement is a less common complication of osteomyelitis in children than in adults and infants.

Several factors that may contribute to these differences:
1. Streptococcal osteomyelitis is common in infants after the newborn period; staphylococcus infection is the more common bacterial type in childhood and adult infections.
2. The structure of infants' bones is less rigid than that of adults and children, having more soft-tissue components and larger vascular spaces. The cortex is particularly thin in infant metaphyseal bone regions, through which communication to the subperiosteal space is facilitated. The periosteum is easily perforated and also more loosely attached, so that purulent material easily extends along the subperiosteal space. This material may progress readily from the primary focus in the bone through the metaphyseal cortex to the subperiosteal space and along the shaft of the bone and may even rupture through the periosteum. These factors allow for an orderly decompression of the purulent material and may account for the rapid healing and lack of sequestrations in infants.
3. Healing is rapid in infants and necrotic material is resorbed much more quickly than in the older age groups.
4. The epiphyseal plate is an effective barrier to infection in infants.
5. Trueta believes that the changing vascular arrangements at each age level may explain the diverse clinical features of osteomyelitis. In children, the vessels in the metaphyseal region immediately adjacent to the growth plate loop to the plate and then return to ramify in the metaphyseal area. These capillaries are the last ramifications of the nutrient artery terminal vessels communicating with the large sinusoids, which spread throughout the entire metaphyseal area. In sinusoids, the blood flow is sluggish, an ideal environment for the propagation of pathogenic bacteria, which may explain the frequent metaphyseal distribution of childhood osteomyelitis. Trueta found that the vascular barrier at the growth plate became evident at the age of 8 months and was fully formed by the eighteenth month. Until the age of 8 months, the growth plate forms no effective barrier to the spread of infection into the epiphyses and the joint.

Localized Bone Abscess

Localized bone abscess rather than diffuse osteomyelitis may occur, with a limited type of osteomyelitis resulting from either low virulence of the organism or high resistance in the host. Only a small focus of cancellous or cortical bone is destroyed and becomes limited by a surrounding sclerotic bone. Initially, the small destroyed area contains purulent exudate, granulation tissue invades the adjacent bone, and eventually the destroyed area may be completely replaced by fibrous tissue. If the area is small, final replacement by bone can occur, leaving no evidence of infection.

Roentgenographic Features

As noted in the clinical and pathologic features, the infant, child, and adult show fundamental differences in pyogenic infections of bone. The primary roentgenographic focus of osteomyelitis in children is most frequently the metaphysis, occasionally the cortex of the bone, and on rare occasions the epiphysis. The initial osseous manifestation will not occur until 5 to 14 days after the onset of symptoms; 10 days is usually required. Adjacent soft-tissue swelling commonly appears much earlier than the bone destruction, but since this swelling may be due to primary cellulitis, it is only presumptive evidence of osteomyelitis.

In the metaphysis, small single or multiple osteolytic areas are the first osseous change, in most instances, and represent resorption of necrotic bone. If the organism is relatively virulent, the areas of destruction are larger. Subperiosteal spread of the infection elevates the periosteum, and layers of new bone are laid down parallel to the shaft. This change occurs sometimes after the destruction of bone is manifest. This periosteal new bone usually requires 3 to 6 weeks to appear and is the first evidence of involucrum formation. A periosteal layer or lamellation may form, and then a mound of bone may be deposited on the external surface. Occasionally it is not layered but appears as a relatively thick band of new bone formation. These are good indications of a neoplastic process.

The extent of the periosteal reaction along the shaft is proportional to the extent of the subperiosteal abscess and therefore may be limited to a small area or may encompass the entire shaft. The periosteal new bone continues to be laid down, and eventually a large area of new bone (involucrum) appears around the cortex. The contours of the involucrum become irregular and wavy, until the affected segment of bone is enveloped in a thick bony sleeve. Interval defects in the involucrum (cloaca) are the result of focal periosteal necrosis. These channels allow the drainage of pus and small sequestra. The involucrum blends imperceptibly with viable bone but is separated from the dead bone, so that sequestrations may be identified before the changes in density between viable and dead bone appear. Bone sequestra become evident 3 to 4 weeks after the onset of the infection. These portions of nonviable bone actually maintain their original density, while the surrounding viable bone is deossified.

Apparently there are differences among sequestra. Most common are those lying in an involucrum housing infected granulation and pus, which remain sharp and well demarcated in their beds. Less common are sequestra not bathed in pockets of pus but in intimate contact with overlying bone. In the former, infection is still active and a major factor; in the latter, infection is less obvious and indeed may be quiescent. Such a sequestrum is less distinct and not as easily separated from its parent bone.

As healing takes place, the living bone assumes its normal density. Cavities filled with fibrous tissue appear as areas of rarefaction in thickened and sclerosed bone. For a long time, the site of the old infection will show dense bone sclerosis with considerable irregularity and thickening. If low-grade infection persists, bone eburnation may be marked and contain cavities and fistulas. In such instances, a careful search for a sequestrum should be made.

In infants and children, the bone may acquire an almost normal appearance after years have passed, but in adolescents and adults, sclerosis and irregularity of the shaft usually persist. In children, bones affected by osteomyelitis may be either elongated or shortened. If there is actual destruction of the epiphyseal cartilage, shortening is inevitable, usually with marked deformity caused by a change in the inclination of the articular surface. Where the epiphyseal cartilage is destroyed, growth may be stimulated by chronic hyperemia, which leads to accelerated growth and maturation of adjacent epiphyses.

The localized form of osteomyelitis appears most frequently at the end of the bone but may occur in the midshaft. The focus of necrosis is indicated by a well-circumscribed region of rarefaction surrounded by a sclerotic zone. Occasionally, a metaphyseal serpiginous channel with a sclerotic border marks the tract of infection. The tract abuts on the epiphyseal plate, and smaller tracts may extend into the ossification center. This is characteristic of nontuberculous infection and may be due to staphylococcus or streptococcus. Frequently, the lesion is near the cortex, and lamellated periosteal thickening may occur.

Clinical Applications of Magnification in Skeletal Radiography

Harry K. Genant

Magnification techniques in skeletal radiography have received increased attention in recent years.[1-19] This expanded use has resulted from three factors: (1) advances in x-ray technology, (2) optimization of physical parameters and exposing factors, and (3) delineation of the meaningful areas for clinical application. High resolution magnification can be achieved by two different techniques.[4,9] The first is optical magnification of fine-grain films, and the second is direct radiographic magnification.

The optical magnification technique consists of contact exposures using conventional X-ray equipment and fine grain industrial films such as Kodak type M. Clinical studies with this technique are not new. Fletcher[5,6] in the early 1950s reported his experience with the use of fine-grain film and photographic enlargement for studying peripheral arthritis. In 1969, Berens and Lin[1] published a monograph on their findings in rheumatoid arthritis using industrial film and optical magnification. More recently, Meema,[15-16] Weiss,[19] Genant,[7-10] and Mall,[13-14] have reported extensive experience with this technique in a variety of metabolic and arthritic skeletal disorders. Thus, the clinical importance of the optical magnification technique for assessing selected skeletal disorders appears established.

Direct radiographic magnification for skeletal radiography has received far less attention. It is only with the recent development of X-ray tubes having very small focal spots (100–150 μm in size) and adequate output for clinical examinations that this technique has become available. Limited clinical experience with direct radiographic magnification has been reported by Takahashi,[18] Ishigaki,[12] Doi,[4] and Genant.[9-11] The initial results have been promising, and applications for both thin and thick body parts are being established.

Radiographic Techniques

The standard technique[9] for optical magnification uses nonscreen industrial type M or AA film, which is exposed with approximately 50 kvp, 500 ma, and 0.2–0.6 second at 100 cm focus-to-film distance. A conventional x-ray tube with a 1.2-mm focal spot is used for these contact exposures, and inherent magnification for thin parts is very low (approximately 1.01 to 1.04x magnification).[3] Thus, exposure times are kept relatively short, and geometric unsharpness is minimal due to the low degree of inherent magnification. The industrial film must be developed manually or by means of an industrial processor. For routine purposes, we use the Kodak industrial B processor, which uses standard chemicals and has a variable speed drive to accommodate industrial films requiring 8-minute processing. Alternatively, a fine-grain rapid-process (90-second) industrial film such as RPM can be substituted for type M film with a resultant slight loss of image quality. It should be pointed out that nonscreen industrial films should be exposed fairly "dark" to an optical density of approximately 2 in order to achieve high contrast. If industrial film is underexposed or "light," the contrast is very poor.[9] The finished industrial radiographs are surveyed without magnification initially and then viewed with optical magnification using a hand lens or a projector.[9,13] For individual viewing, a loupe or hand lens is most convenient, whereas for demonstration purposes or for group viewing, a special projector is most convenient.

The radiographic magnification technique[4,9] consists of direct geometric magnification of 2 to 3x using a microfocus tube having a nominal focal spot size of 100 μm (RSI MAG 2 or GE grid-biased X-ray tubes).

The images are recorded onto relatively high-resolution rare earth recording systems such as single-screen/single-emulsion systems for the thin body parts and double-screen/double-emulsion systems for the thicker body parts. Because of the relatively long exposure times (0.3 to 3.0 seconds), motion unsharpness can be introduced unless measures for immobilization are carefully undertaken. The radiographs are processed in a standard X-omat processor and are viewed with the unaided eye.

Clinical Applications for Optical Magnification

The relative usefulness of the optical magnification technique can be appreciated from the following assessment.[10] Two hundred consecutive patients having hand or foot x-rays were examined simultaneously with the nonscreen type M technique and nonscreen cardboard RP film technique. In Table 1, the distribution of clinical disorders can be seen. From the clinical comparative study, it is apparent that the cardboard medical film technique was adequate for correct interpretation and delineation of findings in the majority of examinations. This was particularly true for trauma and advanced arthritis, where gross changes are obvious. In the remaining cases, the optical magnification technique provided additional or, in some cases, essential information for correct interpretation. The areas in which type M technique was particularly helpful were in the delineation of (1) subtle surface erosions related to early rheumatoid arthritis, hyperparathyroidism, or infection; (2) early periosteal or juxarticular reactive bone formation related to psoriatic arthritis, Reiter's disease, plumonary osteoarthropathy, or infection; and (3) intracortical bone resorption related to high bone turnover states, such as thyrotoxicosis, renal osteodystrophy, Sudek's atrophy, or primary hyperparathyroidism (Figs. 1, 2, and 3). It should be emphasized that the difference between the two radiographic techniques in assessing clinical cases became apparent and critical only when optical magnification was applied to the type M film. If the findings were readily detected with the unaided eye, i.e., without magnification, the cardboard technique was generally adequate. It is apparent that the appropriate examination can be selected on the basis of clinical data provided to the radiologist and that the optical magnification technique can be applied selectively in the circumstances delineated. In the remaining cases, which constitute a majority, the cardboard medical film or fine-screen–film system will be adequate. The use of a par-speed or faster screen–film systems even for routine hand and foot examination is inadequate and not recommended.

Clinical Applications of Direct Magnification

The direct radiographic magnification technique, unlike the optical magnification technique, can be readily applied to either thin or thick body parts. Insight into the relative value of the direct radiographic magnification compared with conventional technique was provided by reviewing a clinical assessment of 200 patients.[11] For thin parts, the comparison consisted of (1) direct radiographic magnification onto a high-resolution screen–film system such as the Dupont Lo-Dose system and (2) conventional contact exposure using cardboard technique. For thick parts, the comparison consisted of (1) direct magnification using a rare earth screen–film system such as Trimax Alpha 4 and (2) contact exposure using a par-speed screen–film system with Buckey technique. Again, similar categories of diseases were defined and the relative usefulness of magnification was assessed (Table 2).

It can be seen that for arthritis, metabolic disorders, and infection, magnification was particularly helpful is assessing subtle resorptive changes at cortical surfaces and early periosteal reaction (Fig. 4). Most of these cases involved the thin body parts such as hands, feet, and long bones. For trauma, magnification was generally not helpful, although it was occasionally instrumental in delineating subtle insufficiency fractures in the extremities or small avulsion fractures in the spine. For the evaluation of neoplasms, which constituted the single largest group, magnification was considered helpful in over 60 percent. These examinations were largely for the

Table 1 Clinical Value of Type M Technique over RP Cardboard Technique

Disorder	No. of cases	Not helpful (%)	Helpful (%)
Arthritis	75	49	51
Metabolic	44	18	82
Infection	11	18	82
Trauma	44	91	9
Other	26	88	12
Total	200	55	45

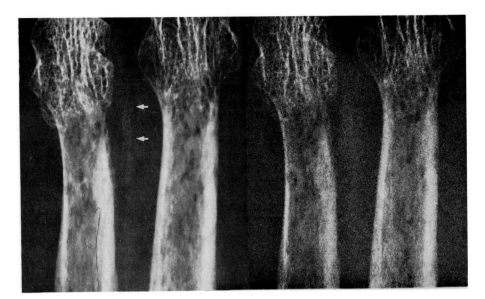

FIG. 1. Aggressive bone resorption in a diabetic patient with rapidly evolving disuse osteoporosis. Intracortical and endosteal bone resorption are readily detected with the type M technique on the left. These changes, as well as vascular calcification (arrows), are not well delineated with cardboard technique on the right.

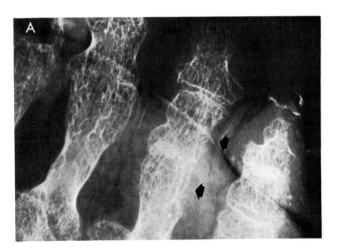
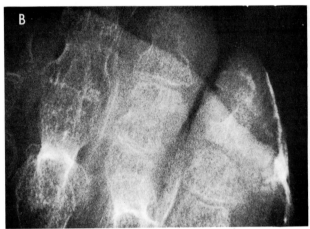

FIG. 2. A. Distal foot of diabetic patient with suspected osteomyelitis. Cortical destruction of the lateral aspects of the middle and proximal phalanges of the fourth digit are readily detected with type M technique (arrows). B. The cortical outlines are inadequately delineated for accurate interpretation with the cardboard technique.

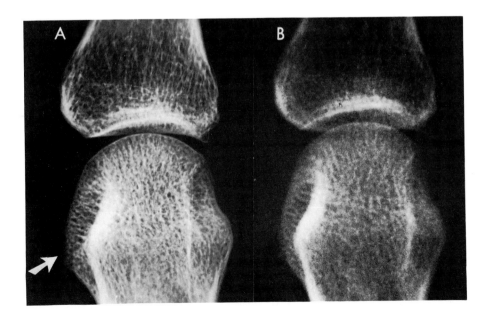

FIG. 3. The "dot–dash" appearance of subtle surface erosion (arrow) is clearly identified with fine-detail radiography (A) but not with conventional radiography (B).

FIG. 4. Proximal tibia of a patient with metastatic transitional cell carcinoma of the bladder. The direct radiographic magnification film on the left demonstrates irregular lysis and sclerosis permeating the medullary space and cortex associated with subtle periosteal reaction and bone formation in the adjacent soft tissues. These important features are difficult to delineate using the conventional technique on the right, consisting of contact exposure and a midspeed screen–film system.

thick body parts, such as pelvis, hips, and spine (Figs. 5 and 6). Here, the conventional radiographs were frequently normal or equivocal, and magnification films were instrumental in delineating permeative lytic destruction. The direct magnification study in these cases frequently followed positive bone scans with normal conventional x-ray films.

Table 2 Clinical Value of Direct Magnification Technique over Conventional Techniques

Disorder	No. of cases	Not helpful (%)	Helpful (%)
Arthritis	39	23	77
Metabolic	32	25	75
Infection	31	7	93
Trauma	17	54	46
Neoplasm	71	36	64
Other	10	40	60
Total	200	30	70

Summary and Conclusion

It is apparent from the above discussion that magnification, both in our own experience and that of others, can be helpful or sometimes essential in reaching an appropriate interpretation or correct diagnosis. The magnification techniques, however, result in moderately increased radiation exposure to the patient compared with conventional techniques and should therefore be used selectively. It should be applied to those cases in which diagnostic uncertainty frequently exists, such as arthritis, osteomyelitis, and metabolic bone disease, and to those cases in which

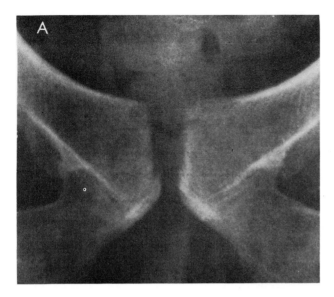 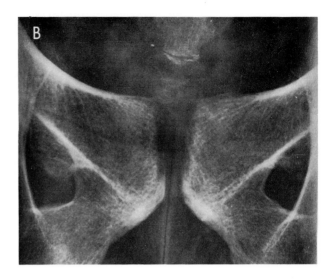

FIG. 5. The conventional radiograph (A) and the magnification view for comparison (B) demonstrate widening of symphysis pubis. The magnification study, in addition, shows irregular destruction of the subchondral cortical line, producing a ragged lacelike appearance. These features indicate an aggressive, evolving process, and support the diagnosis of infectious osteitis pubis.

symptoms, clinical findings, or a positive bone scan strongly suggest a skeletal abnormality ill-defined or undetected on conventional films. For the thin parts, one can choose between optical magnification of industrial film or direct radiographic magnification, whereas for thick parts, only the latter technique can be used. With optical magnification, special processing is required; however, conventional x-ray equipment can otherwise be used. With direct magnification, a special microfocus x-ray tube is necessary; however, conventional processing and standard viewing procedures can be used. These latter advantages, along with the lower radiation exposure, may favor this technique over optical magnification.

It appears that with further optimization of technical parameters and with greater clinical experience, these newer high-resolution radiologic modalities may become diagnostic procedures of major importance.

References

1. Berens DL, Lin RK: Roentgen Diagnosis of Rheumatoid Arthritis. Springfield, Thomas, 1969
2. Calenoff L, Norfray J: Magnification digital roentgenography: a method for evaluating renal osteodystrophy in hemodialyzed patients. Am J Roentgenol 118:282, 1973
3. Doi K, Genant HK, Rossmann K: Effect of film graininess and geometric unsharpness on image quality in fine-detail skeletal radiography. Invest Radiol 10:35, 1975
4. Doi K, Genant HK, Rossmann K: Comparison of image quality in optical and radiographic magnification techniques for fine-detail skeletal radiography: effect of object thickness. Radiology, 00:000, 1976
5. Fletcher DE, Rowley KA: Radiographic enlargements in diagnostic radiology. Br J Radiol 24:598, 1950
6. Fletcher DE, Rowley KA: The radiological features of rheumatoid arthritis. Br J Radiol 25:282, 1952
7. Genant HK, Heck LL, Lanzl LH, et al: Primary hyperparathyroidis. A comprehensive study of clinical, biochemical and radiographic manifestations. Radiology 109:503, 1973
8. Genant HK, Kozin F, Bekerman C, McCarty DJ, Sims J: The reflex sympathetic dystrophy syndrome. Radiology 117:21, 1975
9. Genant HK, Doi K, Mall JC: Optical versus radiographic magnification for fine-detail skeletal radiography. Invest Radiol 10:160, 1975
10. Genant HK, Doi K, Mall JC: Comparison of non-screen techniques (medical versus industrial film) for fine-detail skeletal radiography. Invest Radiol 11:486, 1976
11. Genant HK, Doi K, Mall JC, Sickles EA: Direct radiographic magnification for skeletal radiology: an assessment of image quality and clinical application. Radiology 123:47, 1977
12. Ishigaki T: First metatarsal–phalangeal joint of gout—macroroentgenographic examination in 6 times magnification. Nippon Acta Radiol 48:839, 1973
13. Mall JC, Genant HK, Rossmann K: Improved optical

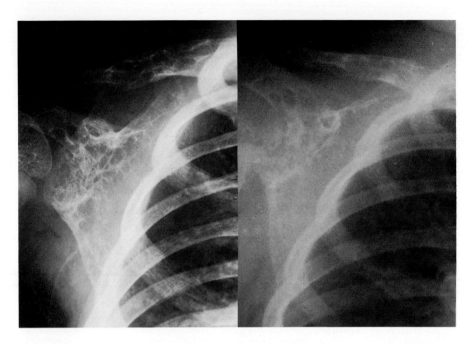

FIG. 6. Multiple well-defined cystic and lytic destructive lesions in the scapula and clavicle in a patient with hemangiomatosis of bone. Although the larger lesions can be seen with both techniques, evaluation of the extent of involvement is far superior on the magnification film.

magnification for fine-detail radiography. Radiology 108:707, 1973
14. Mall JC, Genant HK, Silcox DC, et al: The efficacy of fine-detail radiography in the evaluation of patients with rheumatoid arthritis. Radiology 112:37, 1974
15. Meema HE, Schatz DL: Simple radiologic demonstration of cortical bone loss in thyrotoxicosis. Radiology 97:9, 1970
16. Meema HE, Meema S: Comparison of Macroradioscopic and morphometric findings in the hand bones with densitometric findings in the proximal radius in thyrotoxicosis and in renal osteodystrophy. Invest Radiol 7:88, 1972
17. Rossman K: Image quality. Radio Clin North Am 7:3, 1969
18. Takahashi S, Sakuma S, Ayakawa Y, Maekoshi H, Ohara K: Radiation levels of macroradiography. Radiology 112:709, 1974
19. Weiss A: A technique for demonstrating fine-detail in bones of the hands. Clin Radiol 23:185, 1972

Paget's Disease of Bone: Clinical and Therapeutic Considerations

Robert E. Canfield and Ethel S. Siris

One hundred years ago Sir James Paget described the first case of *osteitis deformans,* a disease now recognized to be a common entity in the aging population and one which now bears his name. Estimates of incidence of this disease, based upon post mortem studies, range from 1 to 3 percent between the ages of 40 and 50 and approximately 10 percent after the age of 80. Males appear to be more commonly affected than females. Although the etiology of the disease is unknown, some researchers have felt that the disease reflects a benign neoplasm of bone[1] and others have recently demonstrated virus-like inclusions in Pagetic osteoclasts, and have speculated about a possible viral cause[2] leading to osteoclast transformation.

The disease appears to be most prevalent in Western Europe and the United States and is reported to be rare in the Orient. For many years it had been accepted that Paget's was not frequently encountered in Africa, but experience in American clinics suggests that blacks of African descent may be affected.[3]

Although the disease generally presents in a sporadic fashion, it is often found to cluster in families. McKusick has proposed an automsomal dominant mode of transmission as a possible pattern of inheritance,[4] but there have been inadequate numbers of patients studied to verify this hypothesis.

Research related to Paget's disease has been supported by NIH Grant AM-09579 and RR-00645.

Pathophysiology

It is generally agreed that the primary abnormality in Paget's disease emanates from the osteoclast. Histological sections taken from Pagetic bone demonstrate osteoclasts which are abundant in number and individually enlarged, bearing increased numbers of nuclei. These characteristics appear to lead to a markedly accelerated rate of bone resorption. In response to the increased resorption, osteoblastic new bone formation is rapidly increased, with the development of a mosaic of irregular islands of newly formed, incompletely calcified bone in place of normal lamellar bone. This chaotic architectural pattern eventually results in structurally weakened bone which is susceptible to alterations in shape secondary to weight bearing (e.g., bowing) and has a predilection to fracture. Pain ensues most probably from several factors, including stress due to altered anatomic positioning of bone, joint destruction, increased blood flow through the bone, and multiple microscopic fractures.

The increased metabolic activity in Pagetic bone and its response to various forms of therapy are reflected by two clinically useful chemical markers. As a result of increased osteoclastic resorption there is an enhanced degradation of bone collagen and a concomitant increase in the urinary excretion of hydroxyproline. Elevations in urinary hydroxyproline are closely correlated with elevations in a second

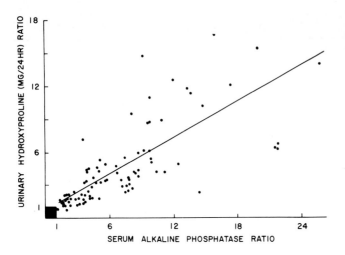

FIG. 1 Values of serum alkaline phosphatase and urinary hydroxyproline obtained in 90 untreated patients with Paget's disease. The values are plotted as the ratio of the patients' values to the upper limits of normal, i.e. normal values would fall within the black square at the lower left.

marker, serum alkaline phosphatase, an enzyme elaborated by osteoblasts, and produced in great quantities as accelerated new bone formation attempts to compensate for the increased rate of bone resorption (Fig. 1).

Clinical Presentation

Paget's disease can affect virtually any and all bones in the skeleton. Commonly it is monostotic, and appears most often in one of the long bones of the lower extremity, a portion of the pelvic bones, or the skull. Polyostotic disease can involve all of these sites as well as the vertebral bodies, bones of the upper extremities (though rarely the hands), the ribs, scapulae, and clavicles. Often it is asymmetric, with extensive disease located in a single tibia or hemipelvis and with completely normal bone on the opposite side. The disease may end abruptly one-half or two-thirds of the way down the shaft of a femur or humerus, with normal bone adjacent to a line of demarcation.

Patients may present with Paget's disease in a number of ways. Most commonly the disease is discovered incidentally, after an unrelated x-ray is taken which discloses silent involvement of adjacent bone, or following a routine blood screen in which an elevated serum alkaline phosphatase is revealed. Bone pain, particularly in the back, hips, or skull, may be a distressing presenting symptom, and occasionally, fracture of what may be a painless but deformed extremity may bring the patient to medical attention. Finally, extensive bone involvement, which has been present for many years and has led to deformity of the skeleton, may result in physical limitation and mechanical disability. Such individuals may be in chronic, severe pain and may be either wheelchair bound or completely bedridden.

While many patients are only mildly symptomatic, others, because of the extent of their disease or the anatomic location of it, can develop serious medical complications. Polyostotic disease, which is highly metabolically active, is associated with enhanced skeletal vascularity and an increased cardiac output. In the older population, in whom the disease is most prevalent, underlying coronary artery or hypertensive cardiovascular disease is common and the additional burden of increased blood flow in the vicinity of Pagetic lesions may accentuate problems with cardiac output. We have recently treated two such patients with underlying severe cardiac disease whose symptoms of angina and congestive heart failure were improved when their very severe Paget's disease (alkaline phosphatases of 5,000 IU/liter and 2,000 IU/liter respectively) was metabolically quieted after mithramycin and calcitonin therapy.

Neurologic complications of Paget's disease can be the cause of severe morbidity and of mortality. Nerve root compression by Pagetic bone may be a source of considerable discomfort. More seriously, spinal cord compression can produce progressive paraplegia. Surgical decompression of the spinal cord is wrought with great hazard in patients with highly vascular bone, and the surgery is often complicated by massive bleeding.

Paget's disease of the skull can be complicated by cranial nerve palsies, commonly involving the 8th nerve and providing one of the several mechanisms by which Paget's can result in deafness, but also involving the 3rd, 6th, or 7th cranial nerves. Moreover, basilar invagination of the skull can cause acute or chronic brain stem compression, with direct bony extension into cerebellar or brain stem tissue or with obstruction of cerebral spinal fluid flow and concomitant internal hydrocephalus (Fig. 2).

Neoplastic degeneration arising in Pagetic bone is a rare but extremely serious complication affecting less than 1 in 100 patients. The lesions which can develop include osteogenic sarcomas, histopathologically similar to the same tumors arising in normal bone, and less commonly fibrosarcomas and chondrosarcomas. It is unclear which factors, if any, predispose patients with Paget's disease to the

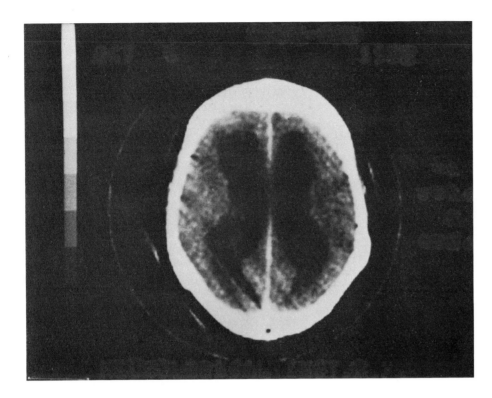

FIG. 2 Computerized tomography (CAT) scan of the brain of a patient with Paget's disease and severe basilar invagination, who developed the acute onset of a unilateral dilated pupil and coma. The dilated lateral ventricles shown here reflected obstruction of cerebral spinal fluid flow due to aqueductal blockage by overgrown Pagetic bone. The patient underwent a ventriculo-peritoneal shunt with successful resolution of her acute symptoms.

development of malignant changes. However, relatively acute development of severe new pain in a bone known to be affected with Paget's disease should be a clear warning to look for radiographic evidence of sarcomatous degeneration.

Another type of neoplastic abnormality which is believed to occur less frequently than sarcomatous degeneration is giant cell tumor of bone. This lesion is often thought to represent sarcoma when it presents clinically or radiographically, but is actually a benign tumor which poses clinical problems through expansion in proximity to nearby vital tissues, such as brain and spinal cord. In our experience,[5] these tumors are exquisitely sensitive to high dose decadron; and in those situations where the tumor is accessible to surgical curettage, it can often be effectively removed after steroid-induced shrinkage.

Treatment

In 1969, when Barry's classic treatise on Paget's disease was published, the portion of the book describing therapy was brief and pessimistic.[6] For many years trials of fluoride, aspirin, corticosteroids, radiation therapy, and on occasion, amputation were the available modes of treatment. Although salicylates and other analgesics continue to be helpful adjuncts for pain relief, none of the above-mentioned drugs provided significant arrest of progression of the disease or of its symptoms. However, since the early 1970s three new and relatively effective forms of therapy have become available and merit discussion.

Calcitonin

Calcitonin is a polypeptide hormone secreted by the parafollicular cells of the thyroid in mammals, including man, and by the ultimobranchial gland of birds and fish (Fig. 3). Its action upon bone results in a decrease in the rate of bone resorption; this fact has led to the use of porcine, human, and salmon calcitonin in patients with Paget's disease. The preparation of calcitonin which is currently being utilized clinically in the United States is a synthetic polypeptide of 32 amino acids whose primary structure is identical to that of salmon calcitonin. This medication is injected either daily or several times weekly, intramuscularly or subcutaneously, at a dose of 50 to 100 Medical Research Council (MRC) units. Reductions in serum alkaline phosphatase and urinary hydroxyproline excretion usually result, occasionally to normal levels but more commonly by 40 to 60 percent, and bone pain is often improved.[7] While calcitonin has a low incidence of relatively mild gastrointestinal side effects (primarily nausea), the problems with the drug include its expense (a year's supply can cost up to $1,500 at the time of this writing) and the need for parenteral administration.

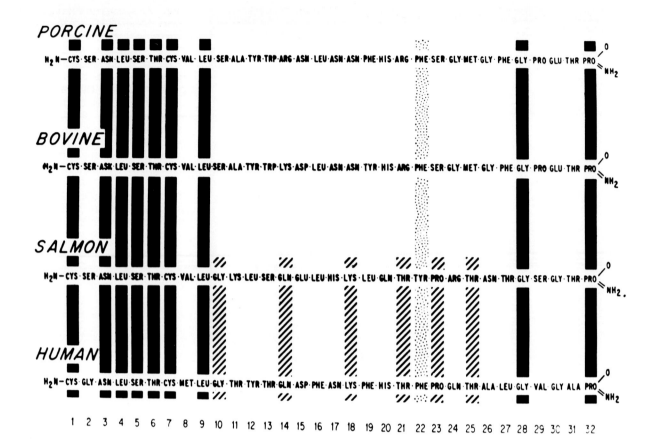

FIG. 3 Primary structure of porcine, bovine, salmon, and human calcitonin. The solid bars highlight the amino acids which are identical in all four species, while the hatched bars represent identical areas between the closely related human and salmon molecules.

When porcine calcitonin first became available several years ago, its use was occasionally limited by a loss of effectiveness with time, presumably due to antibody formation. This type of "escape phenomenon" appears to be less of a problem with the synthetic salmon preparation, whose primary structure more closely resembles that of the human hormone. When calcitonin therapy is discontinued, persistent clinical and chemical benefit may remain for up to one year, but most patients eventually relapse and prolonged therapy is often indicated.[8]

Diphosphonates

Ethane-1-hydroxy-1, 1-diphosphonic acid (EHDP), a diphosphonate preparation, is a pyrophosphate analogue which appears to bind selectively to bone mineral (Fig. 4). Current thinking is that osteoclasts "ingest" hydroxyapatite-bound EHDP during bone resorption and undergo metabolic derangement with loss of function. Clinical trials with EHDP at doses of 5 mg/kg/day for 6 month periods have shown it to be effective in lowering levels of alkaline phosphatase and urinary hydroxyproline (Fig. 5) and in providing symptomatic relief in a majority of patients.[9] When 20 mg/kg/day of EHDP was used for 6 to 12 months during the same clinical trials, some patients experienced a worsening of bone pain and a few individuals actually suffered fractures. Histologic examination of bone biopsy specimens obtained from these patients revealed evidence of widened osteoid seams. Therefore, most patients are placed on a regimen of 5mg/kg/day for 6 month cycles of therapy; at this dose, the agent appears to be similar to calcitonin with respect to efficacy. Since experience with EHDP is relatively limited, long-term studies will be required to evaluate

Paget's Disease of Bone

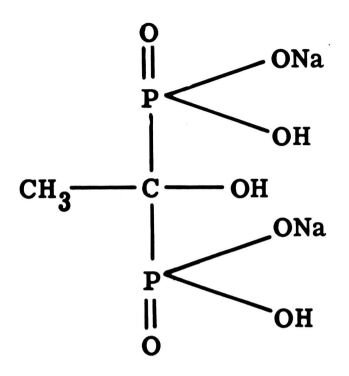

FIG. 4 Structure of ethane-1-hydroxy-1, 1-diphosphonic acid (EHDP).

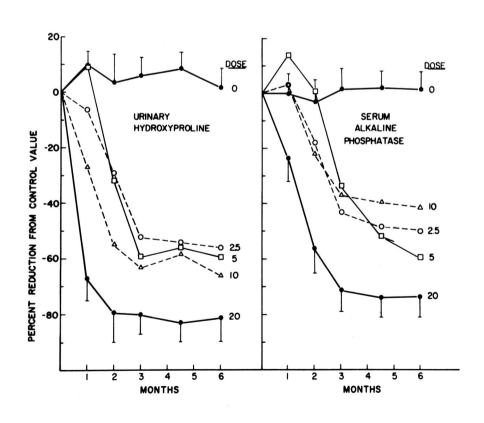

FIG. 5 Serum alkaline phosphatase and urinary hydroxyproline values in patients treated either with placebo (0 dose) or EHDP at doses of 2.5, 5, 10 or 20 mg/kg/day.

Osteolytic Paget's Disease

W. B. Seaman

The radiologic appearance of Paget's disease is usually quite typical and the knowledgeable radiologist has no difficulty in recognizing this mysterious disease. The osteolytic type is less familar and correct diagnosis may be difficult, especially when young patients are affected. The osteolytic type does not include the relatively radiolucent cystic areas found in advanced Paget's disease, but represents instead the early, initial phase of bone resorption.

The characteristic histologic picture of Paget's disease consists of irregular pieces of lamellar bone separated by deeply staining cement lines. These bony fragments are distributed in a random, disorganized fashion forming a pattern usually described as mosaic. This mosaic architecture is the result of an intricate cycle of bone resorption and deposition. The cement lines are the result of incomplete resorption surrounded by newly formed bone. These processes of bone resorption and repair go on almost simultaneously. Possibly in some areas, resorption is so rapid that bone formation apparently has not had time to occur. In this instance, the moasic pattern is absent and only replacement of bone by very vascular connective tissue is visible microscopically. In rare instances, the latter is the only process visible and this is the pathologic basis for osteolytic Paget's disease. It occurs most commonly in the cranium, a bone not subject to the stress which may be necessary to stimulate the bone deposition so characteristic of the reparative process. This theory would not hold true for the tibia, which is the second most common site of osteolytic Paget's disease.

The gross pathologic appearance of this osteolytic phase is one of striking vascularity, labelled by Schmorl as a "hemorrhagic infarct." It is certainly not an infarct since circulation through the diseased bone can be so great as to triple the cardiac output and lead to high output congestive failure. This hemodynamic phenomenon was previously ascribed to the presence of arteriovenous shunts, and this concept was supported by the demonstration of increased oxygen content of the venous blood draining the involved areas. The concept was definitively disproven when Rhodes et al injected radioactive labelled protein particles of 15 to 30μ in diameter intraarterially and demonstrated that they remained trapped in the capillary network of the Paget's bone.

M Sosman, in 1927, was the first to suggest that the circumscribed areas of osteoporosis in the skull, described and named by Schuller as osteoporosis circumscripta, actually represented Paget's disease. In 1926, he speculated about "the peculiar apparent decalcification of the bones descriptively named by Schuller as osteoporosis. It may well be that this is simply the absorptive or destructive phase of Paget's disease at work." Six years later, Kasabach and Dyke reported 12 patients with osteoporosis circumscripta of the skull, six of whom showed characteristic Paget's disease elsewhere in the body. Four patients were followed long enough to demonstrate progression of the lesions with typical Paget's disease developing in and around the osteoporotic areas.

In the skull, the osteolytic changes begin either in the frontal or occipital areas and extend towards the vertex, involving chiefly the diploe and outer table. Occasionally, almost the entire skull becomes osteolytic before the reparative or osteoblastic phase begins so that it may resemble the osteopenic skull of hyperparathyroidism. The margins of the osteolytic areas are quite sharp and readily cross suture lines. The time interval between the osteolytic and osteoblastic phases is usually of matter of years.

Osteolytic Paget's disease is even rarer in the long bones, and the tibia is most frequently affected. This

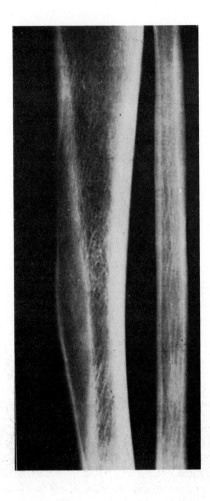 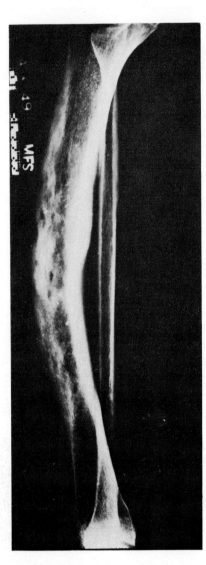

FIG. 1. Left: Lateral view of tibia shows a radiolucent expansion of the anterior cortex with sharply defined borders. Note the characteristic V-shaped upper and lower limits of the lesion. Right: Thirteen years later the tibia exhibits the roentgen characteristics of classical Paget's disease. From Seaman WB: Roentgen appearance of early paget's disease. From Am J Roentgenol 66:587, 1951

was first reported in 1944 by Brailsford who described a lucent ("cystic") lanceolated lesion in the tibia of a 27-year-old man. Sixteen months later, this "cyst" exhibited features characteristic of Paget's disease and 14 years later, the disease appeared in other bones. Because of the typical mosaic pattern was not found in the original biopsy, the diagnosis was in doubt until the characteristic roentgenographic pattern was observed later.

The following case illustrates the progression of osteolytic Paget's disease of the tibia into the sclerotic appearance of classical Paget's disease.

Case 1. A 27-year-old man complained of a mild, dull, aching pain in the right lower leg of two months' duration. X-rays showed a radiolucent area limited to the anterior tibial cortex with clearly defined margins and ending, both superiorly and inferiorly, in a sharp, V-shaped configuration (Fig. 1).

A biopsy showed evidence of increased osteoclastic and osteoblastic activity with many multinucleated giant cells. Pink staining matrix, with an occasional purple fringe, was also described with vascular connective tissue. No diagnosis was offered. These slides were reviewed by Dr. Fuller Albright who thought it was consistent with Paget's disease. Both Albright and Brailsford thought the x-rays unquestionably represented Paget's disease. Thirteen years later, the tibia showed all of the classical roentgen features of Paget's disease with anterior bowing, sclerosis, and coarse, thick trabeculae.

The roentgenographic features that are helpful in the recognition of osteolytic Paget's disease are the following:

Osteolytic Paget's Disease

1. Usually affects a long bone and the tibia is by far the most common site, although it has been reported in the vertebra and ilium.
2. In the tibia, it commonly begins on the anterior aspect of the mid-shaft (Fig. 2).
3. The proximal and distal margins of the lesion have a characteristic V, inverted V, flame-shaped, or lanceolated configuration (Figs. 3, 4).
4. The affected segment of bone is enlarged.

Clinically, it is important to emphasize that the initial osteolytic phase may appear at an early age, as young as 18 years. One of Paget's original patients was only 28 at the time of onset of his disease.

Vertebral involvement by osteolytic Paget's may be accompanied by compression fracture and associated injury to the spinal cord, which occurred in 30 percent of the reported cases of cervical spine involvement with Paget's. The compression tends to mask the underlying disease process so that it is usually diagnosed as metastatic disease.

Differential diagnosis of osteolytic Paget's disease includes fibrous dysplasia, metastatic disease, hyperparathyroidism, and simple cysts.

Case 2. Numbness and weakness of both upper and lower extremities of 6 months' duration was present in a 42-year-old man.

Skull x-rays showed osteoporosis circumscripta. An osteolytic process also involved the bodies, pedicles, and neural arches of the fifth and sixth cervical vertebrae, with an associated compression fracture that forced bony fragments into the spinal canal.

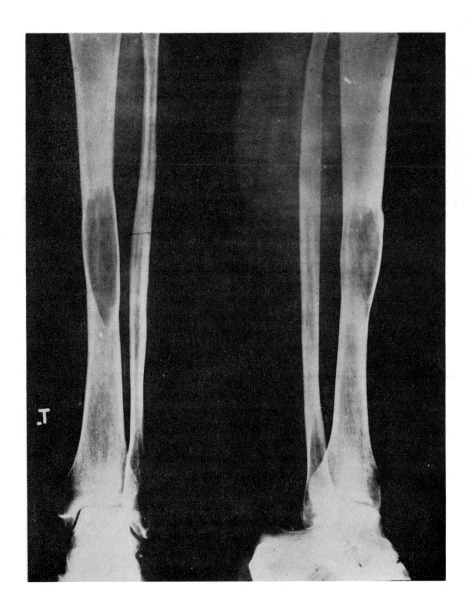

FIG. 2. Left: Oval radiolucent lesion of the tibial diaphysis with V-shaped lower margin on the frontal projection. Right: On the lateral view the appearance of the lesion is less characteristic.

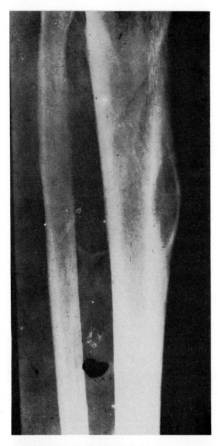 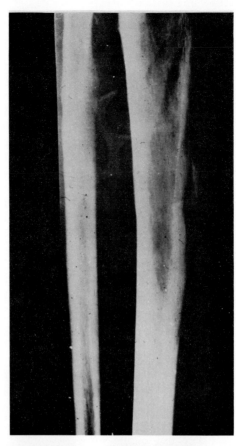

FIG. 3. Left: 43-year-old man with a similar lanceolate lucent area in the anterior cortex of the tibia. Right: One year later there was partial filling in of the lucent area with coarse trabeculation. (Courtesy of Fuller Albright)

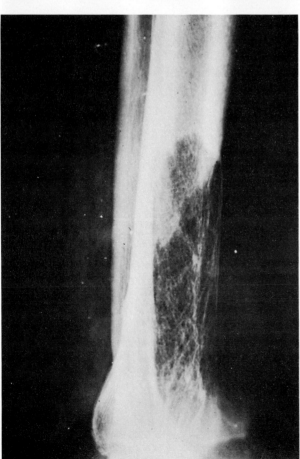 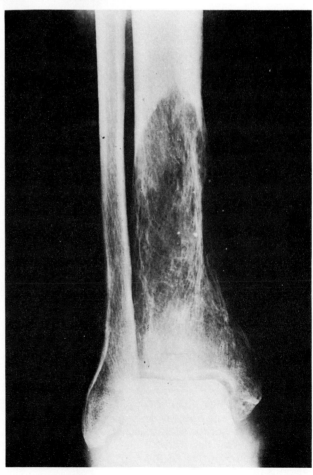

FIG. 4. Left: Paget's disease of the distal tibia extending to the articular surface. Right: Note V-shaped upper margins of the lesion and coarse trabeculation.

Although the appearance of the cervical spine was thought to be more compatible with metastatic carcinoma, a diagnosis of Paget's disease was considered in view of skull findings. Biopsy verified the diagnosis of Paget's disease.

Bibliography

Dickson DD, Camp JD, Ghormely RK: Osteitis deformans: Paget's disease of the bone. Radiol 44:449, 1945

Feldman F, Seaman WB: The neurologic complications of Paget's disease of the cervical spine. Am J Roentgenol 105:375, 1969

Kasabach HH, Dyke CG: Osteoporosis circumscripta of the skull as a form of osteitis deformans. Am J Roentgenol 28:192, 1932

Kasabach HH, Gutman AB: Osteoporosis circumscripta of the skull and Paget's Disease; Fifteen new cases and review of the literature. Am J Roentgenol 37:577, 1937

Rhodes BA, Greyson ND, Hamilton CR, et al: Absence of anatomic arteriovenous shunts in Paget's disease of bone. N Engl J Med 287:686, 1972

Seaman WB: The roentgen appearance of early Paget's disease. Am J Roentgenol 66:587, 1951

Osteoporosis, Osteomalacia, and Hyperparathyroidism

Stanley S. Siegelman

The orderly incorporation of calcium salts into the organic matrix of bone is a vital facet of skeletal physiology. The mechanical strength and the radiographic density of bone are chiefly attributable to calcium content. Bones may be deprived or depleted of minerals by a variety of mechanisms. Mineral-deficient bone appears less dense on radiographs, becomes deformed by mechanical stress, and fractures more readily than normal bone. Osteoporosis, osteomalacia, and hyperparathyroidism are a trio of conditions which account for a generalized decrease in skeletal density. It is a common experience for radiologists to encounter metabolic bone disease as an incidental finding. As an example, compression fractures of the thoracic spine may be detected on a routine lateral chest examination. Without knowledge of the patient's serum calcium, phosphate, urea nitrogen, or alkaline phosphatase; and without information as to the status of the remainder of the skeleton, it may be impossible to diagnose the nature of the bone disease. In such situations many prefer to employ the term osteopenia. Osteopenia is not a specific disorder; it implies that a more precise term is available if more detailed information were at hand. With further study the patient with osteopenia will be found to have osteoporosis, osteomalacia, hyperparathyroidism, or some combination of these specific disorders. In this presentation we shall attempt to outline the key features of osteoporosis, osteomalacia, and hyperparathyroidism. Table 1 is a rough guide which may serve as an overview.

Osteoporosis*

Osteoporosis is a common skeletal condition characterized *dynamically* by a significant period in which bone resorption exceeds bone production;[1] *physically* by a reduction of calcified bone mass per unit volume of bone;[2] *histologically* by an increased porosity of bone due to a diminution in the size and number of trabeculae;[3] *chemically* by normal mineral structure and a normal degree of mineralization;[4,5] *mechanically* by decreased tensile and compression strengths of cancellous bone;[6] and *clinically* by an increased incidence of fractures of spongy bone. A list of causes of osteoporosis is provided in Table 2. The radiologic manifestations of osteoporosis reflect these essential features of the disorder.

Clinical Localization of Osteoporosis

The clinical manifestations of osteoporosis are a consequence of alterations in the cancellous bone of the weight-bearing segments of the skeleton: the lower dorsal spine, the lumbar spine, the proximal humerus, and the neck of the femur. A predilection

---

* Some of the text material on osteoporosis has been reproduced from a prior publication on this subject by the author: Siegelman SS: The Radiology of Osteoporosis. In Brazel, US (Ed) Osteoporosis. Chapter 5, pp 68-79. New York, Grune & Stratton, 1970.

Table 1 Metabolic Bone Disease

Feature	Osteoporosis	Osteomalacia	Hyperparathyroidism
General term	Brittle bone	Malleable bone	Osteitis fibrosa cystica
Frequency	Most common	Less common	Less common
Specificity	Least specific Encountered in a range of diverse disorders	Intermediate Found in a limited number of disorders which affect metabolism of calcium, phosphorus, or Vitamin D	Most specific Bone disease always associated with increase in serum parathormone
Symptoms	Episodic pain associated with fractures	Muscle weakness, bone pain	Renal stones, weakness, anorexia
Pathology	Inadequate osteoid production. Decreased bone per unit volume of bone. Bone chemically normal	Inadequate matrix mineralization, excess osteoid (nonmineralized osseous tissue)	Escessive bone resorption associated with increased osteoclastic activity
Radiologic features	Fractures of spine and hips, cortical thinning	Bone softening fractures Pseudofractures	Subperiosteal resorption
Serum calcium	Normal	Low or normal ⎱ Product	Elevated (low or normal in 2°)
Serum phosphorus	Normal	Low or normal ⎰ decreased	Decreased (elevated in 2°)
Serum alkaline phosphatase	Normal	Elevated	Elevated

for spongy bone is a result of the imbalance between the rates of bone formation and bone resorption. Since the production and removal of bone are processes which occur on the endosteal surface of preexisting bone, trabecular bone, with its higher surface-to-bone volume ratio, is lost faster than cortical bone.[7] The compromise of cancellous bone is also linked to an inherent lesser capacity of spongy bone to provide support. The normal human vertebra can sustain a compressive force of 900 pounds per square inch, but osteoporotic centra will fracture at pressures of 300 pounds per square inch.[8] Compact bone, which has much greater strength, has breaking stresses approaching those of case iron.[9]

Radiologic Findings in Osteoporosis of the Vertebra

Ordinary roentgenograms of the spine contain considerable data which can provide a reasonable evaluation of the degree of osteoporosis. Normal vertebral bodies have a dense network of interlacing osseous trabeculae which provide a striking radiopacity when contrasted to the adjacent disc cartilages. The superior and inferior endplates are parallel, straight surfaces which merge imperceptibly with the contiguous cancellous bone of the centra.

Alterations without Deformity

An interesting and important feature of osteoporosis is the maintenance of the external dimensions of the skeleton.[10] A bone involved with osteoporotic atrophy will maintain its normal volume until it is fractured. Severe osteoporosis ultimately

Table 2 Causes of Osteoporosis

I. Congenital disorders: Osteogenesis imperfecta
II. Primary osteoporosis = idiopathic osteoporosis
   A. Juvenile osteoporosis Below age 20
   B. Adult osteoporosis Age 20 to 40
   C. Postmenopausal Age 40 to 60
   D. Senile osteoporosis Over age 60
III. Nutritional Disturbances: Scurvy, malnutrition, chronic liver disease, alcoholism
IV. Endocrinopathy: Cushing's disease, hypogonadism, thyrotoxicosis, acromegaly
V. Immbolization
VI. Protein catabolism: Rheumatoid arthritis, myeloma, lymphoma, leukemia
VII. Drugs
    Heparin, methotrexate
VIII. Marrow hypertrophy
    Hemolytic anemia
IX. Localized osteoporosis: Sudeck's atrophy, transient osteoporosis of the hip, migratory osteolysis of lower extremities.

produces compression deformities of the spine, but there is a phase of the disorder characterized by alterations without deformity.

DIMINISHED RADIOGRAPHIC DENSITY Skeletal radiopacity is largely attributable to the mineral content of bone.[11] In the presence of considerable cortical bone, roentgenograms are not a sensitive indicator of losses of underlying cancellous bone.[12] The vertebral bodies, however, have very little cortical bone[6] and reductions in mineral content can be appreciated on ordinary roentgenograms. Caldwell demonstrated an excellent correlation between radiographic density and calcium content of lumbar vertebral bodies removed at autopsy.[2,3]

Unfortunately, the density of vertebrae in clinical radiographs is influenced by a number of variables including positioning of the patient, soft tissue variations, uniformity of the x-ray beam, and film development.[13] The estimation of radiographic density is useful for gross evaluation of skeletal mineral content, but it is not a reliable indicator of minor variations in the degree of bone rarefaction.

VERTICAL STRIATIONS Examination of sectioned osteoporotic vertebrae reveals a definite pattern to the bone atrophy: the horizontal trabeculae are strikingly deficient but there is a tendency for preservation of the vertically oriented bony framework.[14] Retention of vertical trabeculae has been attributed to the added mechanical stimulus to bone formation produced by the stress of weight bearing.[8] Radiographically, the pattern of atrophy is manifested by an appearance of vertical striations. In some cases, the retained trabeculae may actually be thicker and coarser than normal.[5,10] In a study by Caldwell and Collins in which the radiographic texture of vertebrae was correlated with histologic appearance and mineral content, the loss of transverse markings and accentuation of vertical markings were found to be an extremely reliable sign of osteoporosis.[3] Hurxthal et al, carefully examining spinal radiographs with the use of a magnifying glass, occasionally visualized vertical striations in otherwise normal vertebrae.[15] Some patients with osteoporosis never exhibit this sign. In others, the vertical striations are present initially but disappear as further demineralization occurs. In our experience readily detectable vertical situations indicate a significant loss of bone mineral.

PROMINENCE OF ENDPLATES Prominence of the vertebral endplates is an additional sign of osteoporosis. Much of the increased contrast is due to the resorption of cancellous bone from the centrum. There are several studies which also suggest an absolute increase in the mineral content of cortical bone with aging.[16]

ABSENCE OF OSTEOPHYTES Another feature of osteoporosis is a notable lack of osteophytes and an absence of ossification of the anterior and lateral spinal ligaments.[10]

Compression Deformities

An analysis of the mechanical factors involved in the weight-support function reveals that the capacity of vertebrae to resist compression is related in an exponential fashion to loss of bone mineral.[6] Thus, moderate atrophy of cancellous bone results in exaggerated losses of vertebral strength. Advanced osteoporosis eventually produces deformities of the vertebral bodies.

CONCAVITIES As the osteoporotic spine endures the burden of chronic weight bearing, there is an interaction between vertebral bodies and adjoining disc cartilages. The disc cartilages expand and the vertebrae become biconcave. This results in a measurable deformity of vertebral endplates. The ratio between the height of the midportion of the vertebral body and the height of the anterior portion is a useful index of the degree of biconcavity. When this ratio is less than 80 percent, osteoporosis is usually present.[4,17] In a study by Caldwell, there was a fairly good correlation between the presence of concavity and the existence of osteoporosis as determined by chemical analysis of vertebrae at autopsy.[2] There were many exceptions, however; some vertebrae of normal density and normal calcium content had increased concavity, and not all vertebrae with moderate osteoporosis exhibited increased concavity. One must take care to obtain properly centered radiographs in evaluating this parameter; variations in projection may produce pseudo-concavities.[18]

ALTERATIONS IN VERTICAL DIMENSIONS Comparison of the vertical dimensions of centra and disc spaces may make it possible to diagnose osteoporosis even when increased concavity of vertabral endplates is not present. The height of the disc spaces in the upper lumbar area should not exceed 35 percent of the height of adjacent vertebral bodies.[18] Osteoporosis is generally present when the vertical disc measurement exceeds 40 percent of the vertical measurement of the mid centrum (Fig. 1). Another feature of the osteoporotic spine is a tendency toward progressive flattening of vertebrae. Thus, on lateral roentgenograms of the spine we note a decreasing height/width ratio of vertebral bodies. It is interesting to observe that the converse of these relationships is also true.[19] Vertebral bodies are taller, disc spaces are shorter, and vertebral endplates may be convex rather than concave in spines which are not subjected to the usual stress of weight bearing. The compressive force of

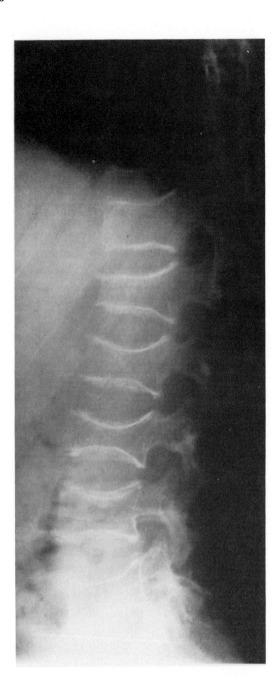

FIG. 1 Osteoporosis. A patient with homocystinuria. Note generalized demineralization of the spine, the flattening of the vertebral bodies so that the disc spaces have a greater vertical dimension, and the prominence of the vertebral endplates.

weight bearing has an inhibitory effect on longitudinal growth of vertebral bodies. The expansion of disc cartilages and the concavities of the endplates are thus interpreted as gradually induced reshaping of structures as a result of mutual stress and interaction. These changes do not occur acutely and do not represent fractures.

COMPRESSION FRACTURES Osteoporosis is the most common condition responsible for compression fractures of the spine.[20] The fractures involve the lower dorsal and upper lumbar vertebral bodies. As the severity of osteoporosis increases, there is an increasing likelihood of fracture. Fractures generally occur when there is a sudden attempt at extension of the flexed spine, such as occurs in lifting an object from a bed or table.[21]

Osteoporosis is defined as a diminution in bony mass per unit volume of bone. Ironically, compression fractures improve the osteoporotic state of the affected vertebra by reducing volume without reducing the amount of bone present. It is not unusual for a patient who fractures one or two vertebrae to be free of additional fractures for many years. Should another compression fracture occur, a previously nonfractured vertebra is more likely to be involved. This mechanism, whereby fractured vertebrae are promoted to a less osteoporotic state, may be beneficial to the patient. Although somewhat shorter in stature, the patient with mild-to-moderate osteoporosis will usually live on with a surprising lack of disability.

Reliability of Radiologic Signs

In the preceding paragraphs, radiologic criteria for assessment of the severity of osteoporosis of the spine have been discussed. This approach is founded on the premise that other causes of skeletal demineralization are not present. Osteomalacia, hyperparathyroidism, myeloma, and metastatic bone disease must be excluded on the basis of distinctive clinical and radiologic features.[5,22-24] When this has been accomplished, a reasonable estimate of the extent of spinal osteoporosis can be obtained by consideration of all of the radiologic parameters. Quantitative evaluations of spinal mineral content based on roentgenograms have been successfully employed by several investigators.[25,26]

Cushing's Syndrome

Osteoporosis is not a specific disease, but rather a state of the skeleton encountered in a heterogeneous group of disorders which include nutritional deficiencies, endocrine disturbances, and senile involution.[1,5,8,10,22] Distinctive radiographic features may make it possible to identify the precise condition which is responsible for the skeletal alterations.

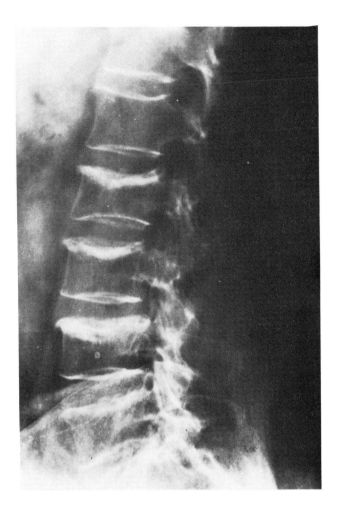

FIG. 2 Cushing's disease due to adrenal tumor. There is a dense, white band underlying each fractured vertebral endplate due to the presence of pseudocallus.

Cushing's syndrome, for example, is associated with characteristic radiologic findings. Osteoporosis of the skull was present in 41 of 43 patients with Cushing's syndrome studied by Howland et al.[27] Patchy demineralization or diffuse decalcification of a moderate to severe degree were the usual findings. The striking radiologic appearance of the skull mimics the findings in multiple myeloma or metastatic carcinoma. Impressive demineralization of the skull is not recognized in other conditions associated with osteoporosis. Irregular zones of rarefaction, however, may be seen elsewhere in the skeleton, especially in the metaphyseal region in any condition associated with osteoporosis of acute origin. The skull may remineralize following the removal of an adrenal cortical adenoma. In the series of Howland et al, 6 patients showed improvement and 3 patients had recalcification among 11 patients with post-treatment roentgenograms of the skull.[27]

Pseudocallus is another special radiographic feature of Cushing's syndrome.[28] In the spine, horizontal bands of increased density are seen adjacent to each fractured vertebral endplate (Fig. 2).

Fracture sites in the ribs and pelvis also contain luxurious masses of callus. Actually, these dense areas represent an abortive attempt at bone restoration. In Cushing's disease, the process of fracture repair progresses normally up to the stage of early cartilage bridging of the fracture site. Subsequently the cartilage calcifies but bone formation is delayed because of a marked inhibition of osteoblastic function. The densities about the fractures, therefore, represent calcified cartilage, not true callus.[28]

The pseudocallus formation in Cushing's syndrome may be related to the bone remodeling rate. It has been suggested that osteoporosis associated with a diminished bone remodeling rate is characterized by frequent fractures and an excess of defective callus.[10] The rates of bone formation and bone resorption (which usually are coupled) are both decreased in Cushing's syndrome. The radiographic occurrence of pseudocallus is also encountered in osteogenesis imperfecta, another condition with a decreased bone remodeling rate. In senile osteoporosis, where the bone remodeling rate is probably elevated, fractures heal with minimal callus.[29]

Osteoporosis and Cortical Bone

Osteoporosis is now recognized as a generalized disorder which affects the entire osseous system. Decreasing skeletal mass is also associated with a reduction in the width of compact cortical bone in the appendicular skeleton. Careful study of serial roentgenograms has yielded two important facts: accurate, reliable, reproducible quantitative measurements of cortical bone may be easily obtained; and the changes in cortical bone are closely related to parallel alterations of cancellous bone.

Progressive loss of compact bone in later adult life is a phenomenon which has been documented in many diverse populations.[30] Reduction in the width of cortical bone is produced by progressive endosteal resorption.[31] Bone loss is most pronounced in the cortex of the metaphyseal region, but the diaphyseal cortex is also involved.[22] Skeletal wasting begins earlier and is more pronounced in the female. Thus Garn et al, employing metacarpal measurements, report a 3 percent loss of cortical bone per decade in the male as contrasted to a loss of 8 percent per decade in the female.[30] A pattern of progressive cortical thinning is well established before the menopause, but it may be accelerated by early castration.

The following radiologic measurements of cortical bone have been employed.

SECOND METACARPAL A standard roentgenogram of the hand is required. This can readily be obtained; suitable films have been gathered using portable equipment in the field.[30] Positioning is not critical. Barnett and Nordin proposed a measurement of the midshaft of the second metacarpal which they termed the "hand score." The sum of the thickness of the medial and lateral cortices of the shaft are divided by total shaft diameter and the ratio is multiplied by 100.[4,17] Although it is notoriously difficult to provide precise endpoints in the evaluation of bone atrophy, a hand score of less than 44 strongly suggests osteoporosis.[17] Using the same measurements, it is also possible to record total cortical thickness or to calculate percent cortical area[30] or cross-sectional area of the cortex.[26] Any of these parameters can be used over a period of years to follow cortical bone losses in the individual patient. A single determination offers an excellent guide to the general status of skeletal mineral content, and correlates with the degree of spinal osteoporosis.[25,26]

PROXIMAL RADIUS This valuable measurement was devised by Meema.[33,34] A lateral roentgenogram of the supinated elbow is employed to determine the combined cortical thickness (CCT) of the proximal radius. The site of measurement is the radial shaft, 1 to 2 cm distal to the tuberosity, where the endosteal and periosteal cortical outlines are parallel.[33,34] There is a gradual thinning of the radial cortex with age. There is also a relationship between the amount of cortical bone in the radius and the incidence of compression fractures of the spine. No patient with a CCT of 5 or more had compression fractures of the spine. Forty-three percent of those with a CCT of 2 and 31 percent of patients with a CCT of 3 had multiple spinal compression fractures. A CCT of 4 is a borderline value associated with a 5 percent incidence of spine fractures.[34]

CLAVICLE This measurement, which was recently proposed by Anton,[35] requires a standard posteroanterior chest radiograph. The thickness of the upper cortex of the clavicle is measured at its midpoint. There is a significant correlation between clavicular cortical thinning and the other radiologic and clinical manifestations of osteoporosis. A reading of 1.5 mm or less is definitely indicative of osteoporosis; 58 percent of patients with this value had evidence of spinal compression fractures or increased spinal bioconcavity.

MID-FEMUR A "femoral score" derived from the thickest portion of the femoral cortex on an anteroposterior roentgenogram of the thigh, was one of the objective measurements of cortical bone introduced by Barnett and Nordin.[17] The sum of the thicknesses of the medial and lateral cortices of the femoral shaft is divided by the total shaft diameter at the point of maximal cortical thickness. As with the hand score, the fraction is multiplied by 100. A value below 46 is consistent with cortical bone atrophy.[17] With advancing age the cortical width decreases from endosteal resorption while the total shaft diameter increases due to periosteal apposition of new bone.[36] The femoral score has been less reliable than the metacarpal or proximal radius measurements as an index of systemic osteoporosis. Atkinson et al have demonstrated that endosteal resorption is more active in the juxtametaphyseal area of the femur.[32] Measurements obtained from the mid-femur are, therefore, less sensitive indications of cortical bone losses.

OTHER METHODS An alternative radiographic means of quantitative assessment of skeletal mass is the "photometric approach." A standard aluminum-alloy calibration wedge is included in the field of the part which is radiographed (e.g. calcaneus, wrist, finger). A comparison of the photographic densities of the bone and the wedge derived from the roentgenograms is used to obtain an objective indication of skeletal mineral content.[37-39] Accuracy with this method requires careful control of the x-ray beam and film development.[16]

Sorenson and Cameron have developed an excellent, reliable method for evaluating bone mineral content which employs a monoenergetic radioactive source, a collimated photon detector, and an electronic computer.[40] Photon-beam transmission measurements will probably become the method of choice for in vivo serial estimation of skeletal mass.

Osteomalacia

Osteomalacia is the second metabolic bone disorder associated with decreased skeletal density. In osteomalacia, there is insufficient mineralization of bone matrix due to an inadequate concentration of calcium and phosphate in tissue fluids. Bone biopsy reveals that osteoid seams (nonmineralized osseous tissue) cover bone trabeculae and line Haversian canals. Rickets is the childhood counterpart of osteomalacia and the inadequately mineralized osteoid is present at the growth plates.

Vitamin D Metabolism

The late 1960s and the 1970s were a period of renewed interest in the metabolism and function of

Osteoporosis, Osteomalacia, and Hyperparathyroidism

Vitamin D. A brief description of our new understanding of the Vitamin D endocrine system is in order at this point. Several excellent reviews are available for the reader who is interested in more detailed information.[41-44] It was known for some time that following administration of an appropriate dose of Vitamin D to experimental animals, a disturbing lag of at least 10 hours occurs before physiologic response is observable. It was this puzzling lag period which led DeLuca to raise the question of whether Vitamin D had to be metabolically processed before it could function.[41] In 1966, radioactive synthesized Vitamin D became available. Experiments with radioactive Vitamin D established that the molecule was converted to polar metabolites with very high Vitamin D-like activity. We now know that Vitamin D_3 is converted by the liver to 25 hydroxy D_3. The reaction takes place in the microsomes or endoplasmic reticulum of hepatic cells. Subsequently, 25 hydroxy D_3 is converted in the kidney to 1,25 dihydroxy D_3. This latter reaction occurs in the mitochondria of renal epithelial cells. The final product 1,25 dihydroxy Vitamin D_3 is the metabolically active form of Vitamin D. One of the key actions of 1,25 dihydroxy D_3 is to serve as a moderator of ribonucleic acid metabolism in the synthesis of a calcium binding protein in intestinal epithelium.[45] Vitamin D and 25 hydroxy D are precursors of the active hormone. 25 hydroxy D has no physiologic activity in animals without kidneys.

Vitamin D Promoted to Hormone 1,25 Dihydroxy Vitamin D

The upshot of our newly acquired knowledge of calcium and phosphate metabolism is that we can now consider that a vitamin (Vitamin D_3) is converted by the liver and the kidney to a hormone (1,25 dihydroxy D_3). Although no precise definitions exist, the term *vitamin* is generally applied to a chemical compound necessary for normal metabolism which must be supplied preformed in very small amounts in the diet. The term *hormone* generally refers to a substance with biologic activity derived from endogenous synthesis by the organism. The biologic activity is directed toward a target organ or organs and the production of the hormone is regulated by a feedback mechanism. The functions of the hormone 1,25 dihydroxy D_3 are presented in Table 3. Rickets or osteomalacia is a consequence of a failure in the process whereby hydroxyapatite mineral is deposited in newly synthesized organic matrix of bone. All of the activities of our new hormone are consistent with maintaining an adequate concentration of calcium

Table 3 Vitamin D Functions to Raise Serum Calcium and Phosphate

*	(1)	Absorption of calcium from bowel
PTH	(2)	Mobilization of calcium and phosphate from previously formed bone
	(3)	Absorption phosphate from bowel
PTH	(4)	Absorption calcium from renal tubule
	(5)	Absorption phosphate from renal tubule

and phosphate in the serum. The chief function is to promote absorption of calcium from the bowel. By a separate process absorption of phosphate from the intestine is also stimulated. Calcium absorption appears to occur chiefly in the proximal bowel, while phosphate absorption is mainly from the distal small intestine.

Parathyroid hormone works in synergy with 1,25 dihydroxy D_3. Both hormones contribute to absorption of calcium from the renal tubule and to the mobilization of calcium and phosphate from previously formed bone. Parathormone is necessary for the final conversion of 25 hydroxy D_3 to 1,25 dihydroxy D_3 which serves as an example of a specific interhormonal feedback mechanism. The production of 1,25 dihydroxy D_3 is also stimulated by low serum phosphate. When serum calcium levels are adequate, serum parathormone falls and less 1,25 dihydroxy D_3 is produced by the kidney. In this situation (adequate serum calcium) the undesirable mobilization of calcium and phosphate from previously formed bone is discontinued. It is believed that an alternate metabolite, 24,25 dihydroxy D_3, is produced when serum parathormone is low and serum phosphate levels are adequate. The 24,25 dihydroxy D_3 is about half as effective as 1,25 dihydroxy D_3 in promoting intestinal transport of calcium and it is much less active in mobilizing calcium from bone and in promoting phosphate absorption from the intestine.[41]

Harrison and Harrison have proposed a new classification of rickets and osteomalacia with a distinction between conditions directly attributable to abnormalities in Vitamin D metabolism (Group I) and disorders in which the problem is a defective target organ response (Group II).[46] In this latter group (Fanconi's syndrome, primary hypophosphatemia, and oncogenous rickets), osteomalacia exists despite the presence of physiologic quantities of 1,25 dihydroxy D_3. A slightly modified version of the Harrison and Harrison classification is presented in Table 4. Several practical considerations may be derived from an appreciation of the dichotomy in the

Table 4

I. Abnormality in Vitamin D metabolism
  A. Vitamin D deficiency
    1. Dietary lack of Vitamin D
    2. Malabsorption of Vitamin D
      a. abnormalities of pancreas and biliary tract
      b. postgastrectomy
      c. inflammatory bowel disease
  B. Defective conversion of Vitamin D to 25 hydroxycholecalciferol in the liver
    1. Liver disease
    2. Anticonvulsant drugs
  C. Defective conversion of 25 hydroxycholecalciferal to 1,25 dihydroxycholecalciferol in the kidney
    1. Renal failure
    2. Vitamin D dependent rickets
II. Osteomalacia despite normal Vitamin D metabolism
  A. Intestinal malabsorption of calcium or phosphate
    1. Consumption of substances which form chelates with calcium
    2. Ingestion of aluminum salts which form insoluble complexes with phosphate
    3. Severe malabsorption states (also may cause Type I osteomalacia)
  B. Target cell abnormalities
    1. Fanconi syndrome
    2. Primary hypophosphatemia (so-called Vitamin D resistant rickets)
    3. Oncogenous rickets

etiology of osteomalacia. Group I conditions, in which the essential abnormality is deficiency of 1,25 dihydroxy D_3, will all manifest a period of hypocalcemia due to impaired absorption of calcium from the gastrointestinal tract. The hypocalcemia stimulates the parathyroid glands and hence, Group I osteomalacia is associated with reactive hyperparathyroidism.[47] In Group II conditions, the serum calcium remains normal, and significant secondary hyperparathyroidism is unusual. All of the conditions in Group I can be adequately treated with therapeutic doses of Vitamin D. In Group II, therapy with various metabolites of Vitamin D is no longer considered appropriate. The current approach to treatment is oral administration of large doses of inorganic phosphate.[46]

VITAMIN D-DEPENDENT RICKETS (GROUP I) This is an autosomal recessive hereditary disorder characterized by a presentation in childhood with growth failure, very low serum calcium (accompanied by tetany or convulsions) and hypoplasia of dental enamel.[48] The condition can be treated with large doses of Vitamin D_3 or 25 hydroxy D_3. Since patients respond to physiologic doses of 1,25 dihydroxy D_3, (one μg per day), the disorder is attributable to a defect in renal hydroxylation of 25 hydroxy D_3.[47]

Osteomalacia due to Target Cell Abnormality

There is a group of genetic and acquired disorders in which the primary abnormality is a functional disturbance of renal tubular cells resulting in hypophosphatemia. Normally appropriate inorganic phosphate concentrations in the extracellular fluid are maintained by a balance between glomerular filtration rate and tubular reabsorption of phosphate which is modulated by Vitamin D and parathyroid hormone. The target cell abnormality of Harrison and Harrison refers to a diverse group of renal tubular disorders in which there is defective retrieval of phosphate by the renal tubule which results in hypophosphatemia. Chronic hypophosphatemia results in osteomalacia despite the presence of normal amounts of 1,25 dihydroxy D_3. Type II osteomalacia can be one of the manifestations of a diverse group of genetic, metabolic, toxic, or acquired disorders in which the enzyme systems in the proximal renal tubular cells are damaged or defective.

Oncogenous Rickets

Oncogenous or tumor-induced osteomalacia is an example of Type II osteomalacia. It is an unusual but fascinating entity in which a tumor of bone or soft tissue appears to release a humoral substance which inhibits tubular resorption of phosphate.[49,50,51] The hypophosphatemia results in osteomalacia despite normal Vitamin D metabolism. It is a disease of adults and older children who present with fatigue, weakness, back pain, and multiple fractures accompanied by radiologic evidence of osteomalacia. The tumors are usually vascular with histologic evidence of some combination of osteoid formation and fibrosis. Specific diagnoses have included sclerosing hemangioma, hemangiopericytoma, ossifying mesenchymal tumor, and non-ossifying fibroma.[50] Among the first 13 reported cases, 6 of the tumors involved bone and 7 involved soft tissue. The smallest tumor was 1 cm in diameter. Two of the lesions appeared to be malignant. Many of the patients recovered (average time following tumor resection 16 weeks) following total removal of the offending neoplasm.

Osteomalacia in the Adult

Osteomalacia in the adult is an underdiagnosed disease. The etiology is apt to be a disorder related to Vitamin D metabolism (Table 4, Group I) such as malabsorption of Vitamin D due to prior gastrec-

Table 5 Senile Osteoporosis Versus Osteomalacia

Feature	Senile osteoporosis	Osteomalacia
(1) Bone pain		
A. Description	Episodic: Related to fractures. Pain-free intervals.	Continuous: Poorly defined initially. Later pain localized to specific sites in bone.
B. Location	Axial skeleton. Thoracic and lumbar spine at fracture sites.	Shoulders, pelvis, forearms, thighs. Pseudofractures are tender.
(2) Fractures		
A. Appearance	Compression fracture spine, complete fracture extremities.	Compression fracture spine, complete fracture extremities + stress fractures, greenstick fractures.
B. Predilected sites	Spine	Pelvis, thorax
C. Healing	Usually complete	Markedly delayed
(3) Radiographic appearance of nonfractured bone		
A. Cortices	Thinned, sharply defined (more sharply defined than normal).	Thinned, fuzzy, and indistinct outline
B. Cancellous bone	Trabeculae decreased in size and number but sharply defined. Trabeculae clearly distributed along lines of stress.	Trabeculae decreased in size and number. Irregularly coarsened and less well defined.
C. Deformities	Only those attributable to prior fractures of hips, shoulders, and vertebra.	Bowing and bending of bones. Hourglass thorax, buckled pelvis.
D. Pseudofractures	Absent.	Present.
E. Secondary hyperparathyroidism	Absent.	May be present.

tomy. The radiologic manifestations are easily confused with senile osteoporosis since both conditions may be manifested by osteopenia with vertebral compression fractures. It is important to make a distinction between these two entities since, unlike senile osteoporosis, osteomalacia is a readily treatable disorder. Table 1 outlines some basic differences between osteoporosis and osteomalacia and Table 5 provides additional distinctions between senile osteoporosis and osteomalacia in the adult. Whenever osteomalacia is suspected because of an appropriate history (e.g. prolonged dilantin therapy without adequate Vitamin D supplementation) or suggestive x-ray findings, it is important to determine the serum calcium, phosphorus, and alkaline phosphatase. In osteomalacia, the serum phosphorus will generally be decreased and the serum alkaline phosphatase will be elevated.

X-ray Findings—Adult Osteomalacia

Osteomalacia of the adult skeleton has distinctive features. The experienced radiologist should be able to suggest the diagnosis even in the absence of pseudofractures. Bone cortices are thinned as in senile osteoporosis but instead of being sharply defined the endosteal surface is fuzzy and indistinct. Trabeculae of cancellous bone are reduced in number but (as opposed to senile osteoporosis) there is a lack of sharp definition of individual trabeculae (Fig. 3). The trabeculae appear coarse, irregular, and frayed. I have frequently noted that the roentgenograms in osteomalacia bear a superficial resemblence to Paget's disease of bone. Both conditions give coarsening of trabeculae and bone softening with deformities. The deformities of osteomalacia are attributable to a reduced capacity for mechanical stress by the mineral-depleted bone. The function of mineralized matrix is to provide resistance to mechanical compression.[52] Due to the repeated mechanical stresses of breathing, walking, and weight bearing, the thorax acquires an hourglass configuration, the long bones become bowed, and the pelvis is buckled and compressed. There is an increased incidence of complete fractures of the extremities but stress fractures and greenstick fractures also occur.[52] The presence of a stress fracture or a greenstick fracture or markedly delayed union of a complete fracture in an adult should lead to investigation for osteomalacia.

Pseudofractures are the hallmark of osteo-

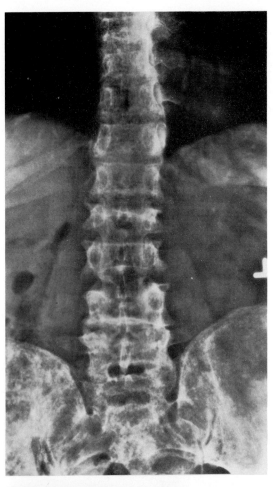
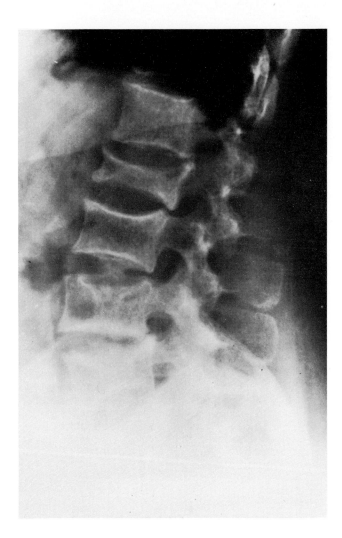
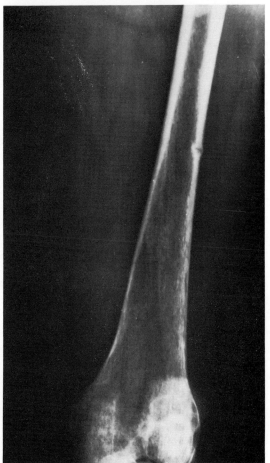

FIG. 3 Osteomalacia in a 60-year-old female due to hypophosphatemia of renal tubular origin. Top: Spine. Several compression fractures are present. The trabeculae appear coarse, irregular, and frayed. There is a superficial resemblance to Paget's disease of bone. Bottom: View of the femur shows a single, tiny pseudofracture of the lateral aspect of the femoral cortex.

malacia.[53] My own view is that pseudofractures represent stress fractures with nonunion. Pseudofractures occur where repeated stresses from mechanical compression and muscle pull are produced during routine physical activities such as breathing and walking. The thorax and the pelvis are prediclected and the lesions tend to be bilateral and symmetrical. Specific sites are the femoral necks, the ischiopubic junctions, the scapula, and the ribs (Figs. 4–6). Pseudofractures are also seen in other conditions in which mechanically defective bone is subjected to repeated mechanical stress: osteogenesis imperfecta, neurofibromatosis, Paget's disease, fibrous dysplasia, and hypophosphatasia.

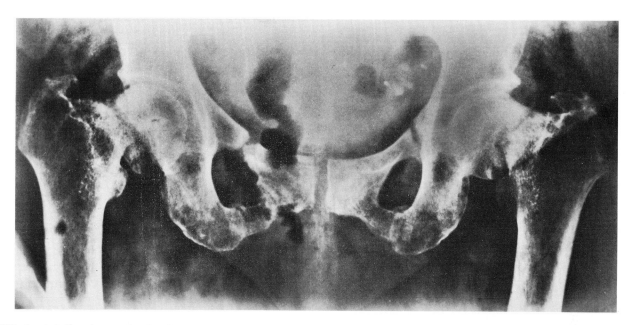

FIG. 4 Adult osteomalacia. Symmetrical pseudofractures are present at the ischiopubic junction superiorly and inferiorly. Bilateral pseudofractures of the femoral necks are also present.

Hyperparathyroidism

Hyperparathyroidism, our final generalized metabolic bone disease, is another potential cause of diffuse osteopenia. Overt skeletal disease as seen in Figure 7 is not a surprising consequence of an unattended parathyroid adenoma. In the 1930s and 1940s, the majority of patients who came to surgery for parathyroid adenomas had objective evidence of skeletal disease. Today with earlier diagnosis due to more sophisticated medical practice, radiologic detection of skeletal disease accompanying parathyroid adenomas is much less common.

The primary mission of the parathyroid gland is to maintain the level of ionized calcium in the blood. All of the activities of parathormone are related to this key function, and hypocalcemia is the most potent parathyroid stimulant. Parathormone has two prime target organs: the skeleton and the kidney. Parathormone releases calcium and phosphate by osteocytic bone resorption and by increased osteoclastic activity. In recent years, it has been established that calciolysis and osteolysis mediated by the osteocyte are more important than osteoclastic activity in maintaining serum calcium.[54] In the renal tubules, parathormone enhances absorption of calcium and inhibits resorption of phosphate. 1,25 Vitamin D_3 acts in concert with parathormone by augmenting release of minerals from bone and by enhancing renal tubule renal resorption of calcium. It is helpful to remember that the effective action of parathormone is to raise serum calcium and lower serum phosphate, whereas 1,25 Vitamin D_3 produces increase in both calcium and phosphate.

In bones made osteopenic under the influence of parathormone there is evidence of diffuse bone resorption highlighted by increased osteoclastic activity. With advanced disease, cortices are riddled by cutting cores of osteoclasts, and cancellous bone exhibits networks of patchy resorption with a dissecting osteitis mediated by waves of osteoclasts. Other sequelae of aggressive osteolysis include lysis of the collagenous matrix of bone with replacement of osseous tissue and marrow by a fibrous stroma. Since multiple areas of fibrous replacement contain hemorrhage and cystic degeneration, the term "osteitis fibrosa cystica" is appropriate. Brown tumors, which appear as lytic lesions on x-ray,[55] are focally expanded areas which represent the sequelae of osteolysis and have been likened to a "battlefield of destruction."[53] Brown tumors contain nests of osteoclasts (giant cells), a network of fibrous tissue, and scattered zones of hemorrhage, necrosis, and bone fragments.

Types of Hyperparathyroidism

Overproduction and increased secretion of parathormone (PTH) may be found in a variety of

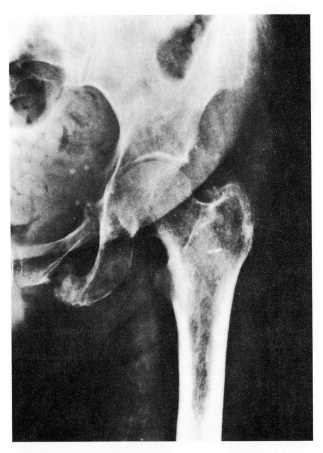 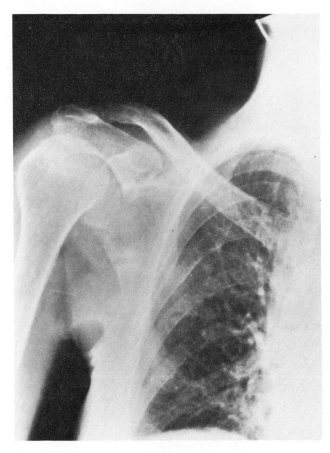

FIG. 5 Adult osteomalacia. Pseudofractures are present at the ischiopubic junction superiorly and inferiorly.

FIG. 6 Adult osteomalacia due to longstanding biliary cirrhosis. A pseudofracture is present in the lateral aspect of the scapula.

situations. In primary hyperparathyroidism, parathyroid hyperactivity exists despite the absence of any detectable provocative stimulus. The commonest cause of primary hyperparathyroidism is a solitary adenoma (80 percent), but diffuse hyperplasia (10 percent), multiple adenomas, and (rarely) carcinoma may be responsible. Secondary hyperparathyroidism is a reactive hyperplasia of all of the parathyroid glands in response to a systemic disorder which results in hypocalcemia. Renal failure is currently the most common cause of secondary hyperparathyroidism but Vitamin D deficiency and gastrointestinal disorders with hypocalcemia due to severe malabsorption may also cause prolonged parathyroid stimulation. In secondary hyperparathyroidism, the increased production of PTH is generally appropriate because it compensates for hypocalcemia. The activity of the hypertrophied parathyroids is mediated by the severity and the duration of the hypocalcemic stimulus. If hypocalcemia is corrected by other measures PTH production should decrease accordingly. Occasionally, with severe prolonged parathyroid stimulation (especially with renal failure), the parathyroids apparently acquire autonomy, i.e., parathyroid gland activity is not suppressed by measures which raise serum calcium and lower serum phosphate. When parathyroid autonomy appears in the setting of secondary hyperparathyroidism, *tertiary hyperparathyroidism* is said to exist. The term was first used by St. Goar in reference to a patient with chronic renal failure who developed a functioning parathyroid adenoma.[56] Generally, however, severe reactive hyperparathyroidism is due to hyperplasia. Hypersecretion of PTH will gradually respond to successful restoration of serum calcium and phosphate,[57] but many groups will opt for subtotal parathyroidectomy in the face of tertiary hyperparathyroidism.

The term *pseudohyperparathyroidism* was introduced by Fry with reference to the combination of hypercalcemia and hypophosphatemia secondary to neoplasms without skeletal metastases.[58] Lafferty col-

Osteoporosis, Osteomalacia, and Hyperparathyroidism

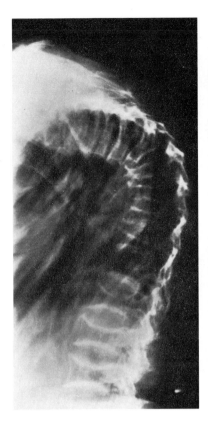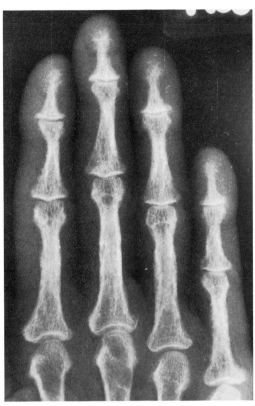

FIG. 7 Left: A patient with primary hyperparathyroidism due to a parathyroid adenoma. The spine shows multiple vertebral fractures, kyphosis, and a marked loss of mineral content. Right: Hand of the same patient shows impressive subperiosteal resorption of bone.

lected 50 cases of pseudohyperparathyroidism from the literature; 60 percent were related to bronchogenic carcinomas and renal carcinomas.[59] Some neoplasms synthesize parathormone-like polypeptides to account for the syndrome. Significant reduction of calcium follows adequate resection of the neoplasm. With recurrence or metastases hypercalcemia reappears. Schotz et al reported a case of pseudohyperparathyroidism in which skeletal changes detectable by x-ray were present.[60]

Hyperparathyroidism—X-ray Findings

Significant abnormalities are detectable by radiologic examination in 20 to 30 percent of patients with hyperparathyroidism.

Subperiosteal resorption of bone is the single most important radiographic finding. It is detectable in many locations:

Hands: The radial sides of the middle phalanages of the middle fingers plus the tufts of the terminal phalanges (Fig. 7,8).
Outer ends of clavicles
Sites of bone modeling: These include the medial aspects of the upper tibial shafts, the inferior aspect of the femoral necks, and the medial aspect of the proximal humerus.
The pelvis. Cortical bone absorption may be present adjacent to the sacroiliac joint, the symphysis pubis, and the ischial tuberosity.

Chondrocalcinosis can be a clue to the existence of a parathyroid adenoma.[61,62] Dodds and Steinbach reported that 18 percent (16 of 91) of patients with primary hyperparathyroidism had evidence of cartilage calcification in the wrists, knees, shoulders, hips, or elbows.[63] In half of the patients (8 of 16) there was no evidence of subperiosteal resorption of bone.

Brown tumors are aggressive, expansile, cortically-oriented lesions which may occur at any site in the skeleton (Fig. 9). Since patients present with weakness and hypercalcemia, brown tumors are frequently tentatively diagnosed as metastatic carcinoma or myeloma. After removal of a parathyroid adenoma, a true brown tumor will heal with some degree of sclerosis.

Absorption of the lamina dura. Although x-rays of teeth will frequently demonstrate absorption of the lamina dura, the sign is of limited usefulness since it is nonspecific. Absorption of the lamina dura may also be seen with periodontal disease.

Generalized demineralization. As previously men-

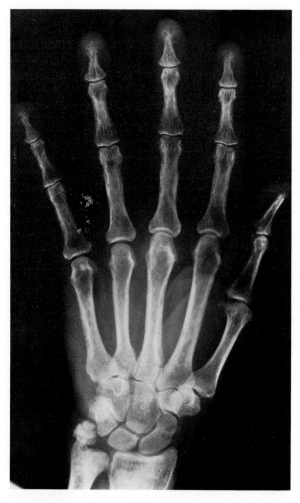

FIG. 8 Secondary hyperparathyroidism due to renal failure. There is subperiosteal resorption of bone along the radial aspects of the middle phalanges of the middle fingers and the tufts of the terminal phalanges. In addition there is vascular calcification noted in the index finger and periarticular globular calcification noted distal to the ulna styloid.

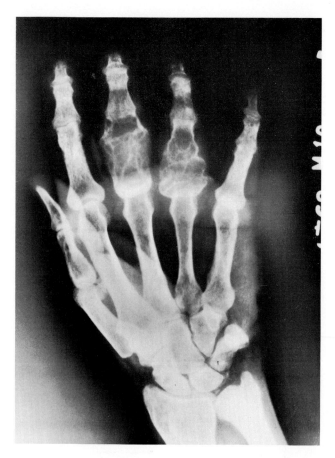

FIG. 9 Brown tumors. The proximal phalanges of the third and fourth fingers have aggressive-looking expansile lesions due to brown tumors.

tioned, hyperparathyroidism may also be responsible for generalized osteopenia with compression fractures of the spin.

Renal Osteodystrophy

Renal osteodystrophy is a term applied to the bone disease which accompanies chronic renal failure. It is a complicated disorder because of an interplay of multiple diverse factors involved in the pathogenesis of the bone disease. The factors which contribute to the complexity are:

1. Disturbance of Vitamin D metabolism leading to a deficiency in endogenous production of 1,25 Vitamin D_3. Inadequate absorption of calcium from the intestine is an immediate consequence. This is followed by hypocalcemia and rickets (renal rickets) or osteomalacia (Fig. 10).
2. Secondary hyperparathyroidism is produced because of the hypocalcemia.
3. Reduction in glomerular filtration rate eventually results in retention of phosphate. Hyperphosphatemia is an important element which further impairs Vitamin D metabolism, affects the action of PTH on bone, and predisposes to soft tissue calcification.[64,65]

Renal osteodystrophy varies from case to case due to variations in individual patients. Generally the bone disease consists of a variable mixture of osteomalacia and secondary hyperparathyroidism. Table 6 outlines the distinctions between primary

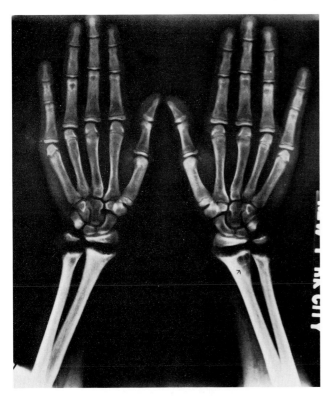

FIG. 10 Renal rickets. The diagnosis of rickets can be made when the radiolucent growth plate measures more than 1 mm in width. The growth plates are markedly widened in the radius and ulna bilaterally.

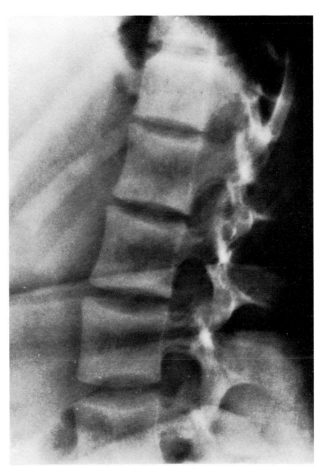

FIG. 11 Renal osteodystrophy (rugger-jersey spine). Patient with chronic renal failure showing the increased density of the spine with prominent dense bands of bone adjacent to the vertebral endplates.

hyperparathyroidism and renal osteodystrophy. The appearance of subperiosteal resorption of bone is identical in both conditions (Fig. 8). Following a prolonged period of renal failure, osteosclerosis appears as an additional element to complicate the picture. Osteosclerosis is particularly prominent in the spine (Fig. 11) where the vertebral endplates acquire sclerotic bands (rugger-jersey spine). There is no current adequate explanation for the genesis of the osteosclerosis.

Table 6 Distinctions between Primary Hyperparathyroidism and Renal Osteodystrophy

Factor	Primary hyperparathyroidism	Renal osteodystrophy
Subperiosteal resorption of bone	Present	Present
Vascular calcification	Generally absent	Generally present
Chondrocalcinosis	Frequent (20-30%)	Unusual
Brown tumors	More common	Less common
Periarticular calcification	Rare	More common
Osteosclerosis	Rare	Common
Rickets	Absent	Common
Serum phosphate	Depressed	Elevated

References

1. Harris WH, Heany RP: Skeletal renewal and metabolic bone disease. New Engl J Med 280:193-202, 253-259, 303-311, 1969
2. Caldwell RA: Observations on the incidence, etiology, and pathology of senile osteoporosis. J Clin Pathol 15:421-431, 1962
3. Caldwell RA, Collins DH: Assessment of vertebral osteoporosis by radiographic and chemical methods, post-mortem. J Bone Joint Surg 43B:346-361, 1961
4. Barnett E, Nordin BEC: The clinical and radiological problem of thin bones. Brit J Radiol 34:683-692, 1961
5. Casuccio C: Concerning osteoporosis. J Bone Joint Surg 44B:453-463, 1962
6. Bell GH, Dunbar O, Beck JS, Gibb A: Variations in strength of vertebra with age and their relation to osteoporosis. Calcif Tissue Res 1:75-86, 1967
7. Frost HM: Bone dynamics in metabolic bone disease. J Bone Joint Surg 48A:1192-1203, 1966
8. Cooke AM: Osteoporosis. Lancet 1:929-937, 1955
9. Bell GH, Cuthbertson DP, Orr J: Strength and size of bone in relation to calcium intake. J Physiol 100:299-317, 1961
10. Steinbach HL: The roentgen appearance of osteoporosis. Radiol Clin North Am 2:191-207, 1964
11. Lachman E: Osteoporosis: The potentialities and limitations of its roentgenologic diagnosis. Am J Roentgenol 74:712-715, 1955
12. Lodwick GS: Reactive response to local injury in bone. Radiol Clin North Am 2:209-219, 2:#2, Aug 1964
13. Doyle FH, Gutteridge DH, Joplin GF, Fraser R: An assessment of radiological criteria used in the study of spinal osteoporosis. Brit J Radiol 40:241-250, 1967
14. Vost A: Osteoporosis: A necropsy study of vertebrae and iliac crests. Am J Pathol 43:143-151, 1963
15. Hurxthal LM, Vose GP, Dotter WE: Densitometric and visual observations of spinal radiographs. Geriatrics 24:93-106, 1969
16. Pridie RB: The diagnosis of senile osteoporosis using a new bone density index. Brit J Radiol 40:251-255, 1967
17. Barnett E, Nordin BEC: The radiological diagnosis of osteoporosis: A new approach. Clin Radiol 11:166-174, 1960
18. Hurxthal LM: Measurement of anterior vertebral compressions and biconcave vertebrae. Am J Roentgenol 103:635-644, 1968
19. Houston CS, Zeleski WA: The shape of vertebral bodies and femoral necks in relation to activity. Radiology 89:59-66, 1967
20. Nicholas JA, Wilson PD: Diagnosis and treatment of osteoporosis. JAMA 171:2279-2284, 1959
21. Burrows HJ, Graham G: Spinal osteoporosis of unknown origin. Q J Med 14:147-170, 1945
22. Dent CE, Watson L: Osteoporosis. Postgrad Med J [Suppl] 42:583-608, 1966
23. Nickholas JA, Savaille PD, Bronner F: Osteoporosis, osteomalacia, and the skeletal system. J Bone Joint Surg 45A:391-405, 1963
24. Rose GA: The radiological diagnosis of osteoporosis, osteomalacia, and hyperparathyroidism. Clin Radiol 15:75-83, 1964
25. Saville PD: A quantitative approach to simple radiographic diagnosis of osteoporosis: Its application to the osteoporosis of rheumatoid arthritis. Arthritis Rheum 10:416-422, 1967
26. Smith RW Jr, Frame B: Concurrent axial and appendicular osteoporosis. New Engl J Med 273:73-78, 1965
27. Howland WJ, Pugh DG, Sprague RG: Roentgenologic changes of the skeletal system in Cushing's syndrome. Radiology 71:69-78, 1958
28. Murray RO: Radiological bone changes in Cushing's syndrome and steroid therapy. Brit J Radiol 33:1-19, 1960
29. Dent CE: Idiopathic osteoporosis. Proc R Soc Med 48:574-578, 1955
30. Garn SM, Rohmann CG, Wagner B: Bone loss as a general phenomenon in man. Fed Proc 26:1729-1736, 1967
31. Atkinson PJ: Changes in resorption spaces in femoral cortical bone with age. J Pathol Bacteriol 89:173-178, 1965
32. Atkinson PJ, Weatherell JA, Weidmann SM: Changes in density of the human femoral cortex with age. J Bone Joint Surg 44B:496-502, 1962
33. Meema HE: The occurrence of cortical bone atrophy in old age and in osteoporosis. J Can Assn Radiol 13:27-32, 1962
34. Meema HE: Cortical bone atrophy and osteoporosis as a manifestation of aging. Am J Roentgenol 89:1287-1295, 1963
35. Anton HC: Width of clavicular cortex in osteoporosis. Brit Med J 1:409-411, 1969
36. Smith RW, Walker RR: Femoral expansion in aging women. Implications for osteoporosis and fractures. Science 145:156-157, 1964
37. Doyle FH: Ulnar bone mineral concentration in metabolic bone disease. Brit J Radiol 34:698-712, 1961
38. Mack PB, LaChance PA, Voxe GP, Vogt FB: Bone demineralization of foot and hand of Gemini-Titan IV, V and VII astronauts during orbital flight. Am J Roentgenol 100:503-511, 1967
39. Mayo KM: Quantitative measurement of bone mineral in normal adult bone. Brit J Radiol 34:693-698, 1961
40. Sorenson JA, Cameron JR: A reliable in vivo measurement of bone-mineral content. J Bone Joint Surg 49A:481-497, 1967
41. DeLuca HF: Vitamin D endocrinology. Ann Int Med 85:367-377, 1976
42. DeLuca HF: Recurrent advances in our understanding of the Vitamin D endocrine system. J Lab Clin Med 87:7-26, 1976
43. Haussler MR, McCain TA: Basic and clinical concepts related to Vitamin D metabolism and action. N Engl J Med 297:974-983, 1041-1050, 1977
44. Pett MJ, Haussler MR: Vitamin D: Biochemistry and clinical applications. Skeletal Radiol 1:191-208, 1977
45. Fullmer CS, Wasserman RH: Bovine intestinal calcium

bending proteins (Ca^{B}P): Purification and some properties. Fed Proc 31:693, 1972
46. Harrison HE, Harrison HC: Rickets then and now. Pediatrics 87:1144-1151, 1975
47. Scriver CR: Rickets and the pathogenesis of impaired tubular transport of phosphate and other solutes. Am J Med 57:43-49, 1974
48. Arnaud C, Maijer R, Reade T, Scriver CR, Whelan DT: Vitamin D dependency. An inherited postnatal syndrome with secondary hyperparathyroidism. Pediatrics 46:871-880, 1970
49. Renton P, Shaw DG: Hypophosphatemic osteomalacia secondary to vascular tumors of bone and soft tissue. Skeletal Radiol 1:21-24, 1976
50. Linovitz RJ, et al: Tumor-induced osteomalacia and rickets: A surgically curable syndrome. J Bone Joint Surg 58A:419-423, 1976
51. Harrison HE: Oncogenous rickets: Possible elaboration by a tumor of a humoral substance. Inhibiting tubular reabsorption of phosphate. Pediatrics 52:432-433, 1973
52. Chalmers J, Conacher WDH, Gardner DL, Scott PJ: Osteomalacia—A common disease in elderly women. J Bone Joint Surg 49B:403-423, 1967
53. Mankin HJ: Rickets, osteomalacia and renal osteodystrophy. J Bone Joint Surg 56A:101-128, 352-386, 1974
54. Vitalli PH: Osteocyte activity. Clin Orthop 56:213-226, 1968
55. Case Records of the Massachusetts General Hosp. N Engl J Med, Case 4-1978, 298:266-274, 1978
56. St. Goar WT: Discussion. Case records of the Massachusetts General Hospital. (Case 29-1963). New Engl J Med 268:949-953, 1963
57. Vosik WM, et al: Successful medical management of osteitis fibrosa due to "tertiary" hyperparathyroidism. Mayo Clin Proc 47:110-113, 1972
58. Fry I: Pseudohyperparathyroidism with carcinoma of bronchus. Brit Med J 1:301, 1962
59. Lafferty FW: Pseudohyperparathyroidism. Medicine 45:247, 1966
60. Scholz DA, Riggs BL, Purnell DC, Goldsmith RS, Arnaud CD: Ectopic Hyperparathyroidism with renal calculi and subperiosteal bone resorption. Mayo Clin Proc 48:124-126, 1973
61. Wang C, Miller LM, Weber AL, Krane JM: Pseudogout. A diagnostic clue to hyperparathyroidism. Am J Surg 117:558-565, 1969
62. Grahame R, Sutor DJ, Mitchener MB: Crystal deposition in hyperparathyroidism. Ann Rheum Dis 30:597-604, 1971
63. Dodds WJ, Steinbach HL: Primary hyperparathyroidism and articular cartilage calcification. Am J Roentgenol 104:884-892, 1968
64. Naidich DP, Siegelman SS: Paraarticular soft tissue changes in systemic diseases. Semin Roentgenol 8:101-116, 1973
65. Parfitt AM: Soft-tissue calcification in uremia. Arch Int Med 124:543-556, 1969

Radioimmunoassay—
Parathyroid Hormone and Calcitonin

Robert E. Canfield and Ethel S. Siris

Introduction

Radioimmunoassay (RIA) is a laboratory technique that permits a precise, sensitive, and specific measurement of a variety of biologic substances in body fluid and tissues. Since the initial description of the principle of RIA by Berson et al in 1956,[1] its applications in both research and clinical settings have grown enormously. In particular, the contribution of RIA as a means of measuring physiologic concentrations of hormones has greatly expanded the scope of clinical endocrinology.

The principle of RIA is both simple and elegant. The concentration of the substance (e.g., hormone) to be measured can be determined based upon its ability to compete for binding sites on a specific antibody with a trace amount of the same (or structurally very similar) hormonal material that has been radiolabeled. In this system, both labeled and unlabeled hormone will attempt to bind to a finite number of binding sites on the antibody. Increasing amounts of unlabeled hormone will displace proportionately more radiolabeled hormone from antibody binding sites, with the result that fewer radioactive counts will remain bound in the hormone–antibody complex. By measuring the radioactivity that is bound to the hormone–antibody complex at each of several incremental doses of unlabeled hormone, a dose–response relationship or standard curve can be established; samples of patients' blood can then be tested for their ability to displace labeled hormone from the antibody, and by use of the standard curve as a reference, the quantity of hormone in the patient sample can be estimated.

Obviously, the antibody for which labeled and unlabeled hormone compete is the cornerstone of the RIA. The antibody confers both sensitivity and specificity upon the system. Sensitivity permits the measurement of levels of hormone that are in the range at which the hormone acts physiologically. Specificity provides that the concentration of hormone measured actually represents levels of that hormone and not a combination of the hormone plus other substances that may resemble it structurally.

The successful generation of sensitive and specific antibodies for RIA depends in large part on the degree of purity of the hormonal material that has been used to stimulate their production. In some systems, the generation of technically excellent antibodies yielding highly reliable and precise measurements has been relatively easy to accomplish; in others, including the RIAs of parathyroid hormone and calcitonin, optimal antibodies have proven to be more elusive, so that the development and refinement of RIAs for these two substances continues to evolve.

Parathyroid Hormone

Parathyroid hormone (PTH) is a single polypeptide containing 84 amino acids (Fig. 1), (personal communication from the authors) that is secreted by the parathyroid glands. Its major physiologic role is to regulate the concentration of ionized calcium in extracellular fluid within the limits required for normal calcium-related metabolic and neuroregulatory functions. Specifically, a fall in ionized calcium results in an enhanced secretion of PTH and an increase in PTH action at several sites. These include

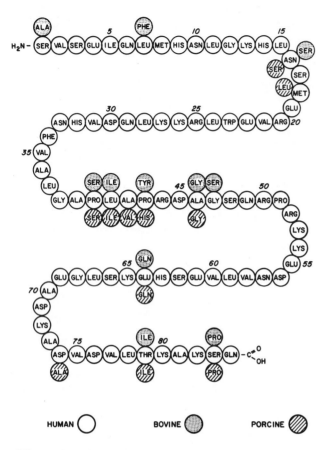

FIG. 1. Structure of parathyroid hormone.

major form of immunoreactive PTH that is detected in serum by the RIA is a large fragment termed the C-fragment, which comprises the middle and carboxy terminal portion of the hormone molecule. Carboxy terminal fragments have relatively long half-lives in serum but are biologically inactive. Conversely the biologically active forms of PTH include both the intact hormone and the amino terminal (or N-terminal) fragments of the intact hormone, both of which have short half-lives compared with C-terminal fragments and both of which contribute much less of the immunoreactive PTH in serum.

The RIA of PTH may be clinically useful in a number of situations, including the differential diagnosis of both hypocalcemic and hypercalcemic states, in the differentiation between primary and secondary hyperparathyroidism, and in preoperative localization of abnormal parathyroid glands by selective venous catheterization of the thyroid vessels. Whenever PTH measurements are made the values obtained must be interpreted in the context of a concurrent serum calcium measurement.

RIA of PTH in patients with primary hypoparathyroidism (i.e., idiopathic or surgical) should reveal either no detectable PTH or extremely low levels (i.e., background in the assay), which are inconsistent with low levels of serum calcium. Hypocalcemia associated with elevated levels of PTH is compatible either with the diagnosis of pseudohypoparathyroidism or with secondary hyperparathyroidism resulting from such conditions as osteomalacia, malabsorption, or chronic renal failure.

The diagnosis of primary hyperparathyroidism as a cause of hypercalcemia can usually be confirmed by the PTH RIA. In the presence of hypercalcemia, absolutely elevated or inappropriately high levels of PTH for the level of serum calcium are virtually diagnostic for primary parathyroid hyperfunction or, less commonly, the ectopic secretion of PTH by a malignant tumor. Conversely, hypercalcemia associated with low or absent levels of PTH is almost certainly not due to parathyroid overactivity, and the patient must be carefully evaluated for either underlying malignant disease with or without bony metastases, sarcoidosis, hyperthyroidism, vitamin D intoxication, or other causes not related directly to parathyroid function.

Finally, in situations where primary hyperparathyroidism is strongly suspected but where previous neck surgery has been unsuccessful or where preoperative localization is imperative for other reasons, selective venous catheterization of the vessels of the neck may be helpful.[3] By analyzing samples of peripheral venous blood and multiple venous samples taken from sites in the neck as shown

(a) effects on bone to increase osteoclastic resorption and release calcium into the extracellular fluid from skeletal stores, (b) effects on the kidneys to increase tubular reabsorption of calcium, limiting its excretion, and (c) effects on the gastrointestinal tract via PTH-induced activation of 1,25-dihydroxy vitamin D to increase calcium absorption. The resulting rise in serum calcium will in turn decrease PTH secretion through a negative feedback mechanism. The action of PTH is mediated by cyclic 3′5′-adenosine monophosphate (3′5′-AMP), which is produced through activation of adenylate cyclase by PTH in both bone and kidney.[2] In addition to its effects on calcium, PTH also causes an increase in the urinary excretion of phosphate through a direct action on the kidneys.

PTH circulates in the blood in a number of forms, including intact native hormone and various fragments of the molecule that may or may not have biologic activity. Antibodies to PTH that have been developed for use in the PTH RIA may be directed at any of several different pieces of the hormone. The

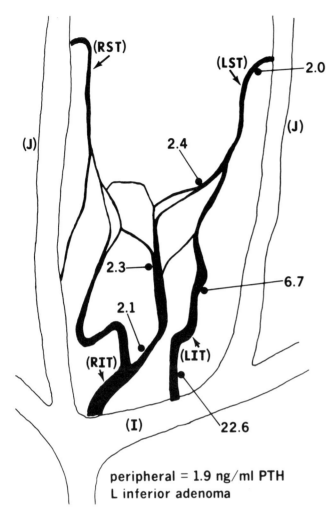

FIG. 2. Selective venous catheterization study from a patient with a left inferior parathyroid adenoma showing parathyroid hormone concentrations in venous samples taken at multiple sites. The numbers represent PTH concentrations expressed in nanograms per milliliter. The peripheral venous concentration was 1.9 ng/ml. Abbreviations: J, jugular vein; I, innominate vein; RST, right superior thyroid vein; LST, left superior thyroid vein; RIT, right inferior thyroid vein; LIT, left inferior thyroid vein. The step-up in PTH on the left side localized the adenoma. (From Bilezikian JP, et al: Am J Med 55:505, 1973.)

in Fig. 2, an area of step-up of PTH secretion may be identified, predicting the side of the neck containing an adenoma.

Calcitonin

Calcitonin is a polypeptide hormone consisting of 32 amino acids that is secreted by the parafollicular cells of the mammalian thyroid. The structure of human calcitonin in known (see Fig. 3 in chapter on Paget's disease of bone) and it has been synthesized in the laboratory. Control of the rate of the secretion of calcitonin is a direct function of the concentration of ionized calcium in the circulation; the hypocalcemic effect of this hormone is achieved through an inhibition of bone resorption—at least in part via an inhibition of PTH-induced osteoclastic activity—and is mediated through an adenylate–cyclic AMP system.[4]

In normal man, the physiologic role of calcitonin remains unclear. Despite the known calcium-lowering effects of calcitonin, hypercalcemia rarely develops following ablation of calcitonin production by thryoidectomy, and hypocalcemia is rarely a feature of medullary thyroid carcinoma, the disease associated with calcitonin hypersecretion. Although pharmacologic amounts of parenterally administered calcitonin are effective in slowing the accelerated bone resorption that characterizes Paget's disease of bone, there is, to date, no defined role of this hormone in the pathogenesis of metabolic bone disease.

In general, it has not been possible to study extensively the secretion of calcitonin in normal human subjects because of the very low concentration of hormone as measured by RIA in most laboratories. Many assays have been unable to measure quantities of calcitonin of less than 100 pg/ml, with the result that levels in normal subjects are generally undetectable. However, the recent development of a more sensitive RIA by Parthemore and Deftos has permitted the quantitation of as little as 10 pg/ml of calcitonin in plasma, opening the way to defining its physiologic range.[5] Parthemore and Deftos have described the mean basal level of calcitonin in normal human subjects to be 24 pg/ml (range <10 to 75 pg/ml), without significant age or sex differences. With improvements in assay methods, it is hoped that valuable information will soon begin to emerge regarding the function of this hormone in normal man.

In patients with medullary carcinoma of the thyroid, the RIA of calcitonin has proven to be of great clinical value. This disease is associated with increased production of calcitonin. A demonstration of elevated basal levels of calcitonin or of marked hypersecretion following a provocative stimulus such as an intravenous calcium infusion or pentagastrin administration can indicate the presence of a tumor preoperatively and subsequently be used to follow patients for recurrence after surgery. Since medullary carcinoma of the thyroid may have a familial as well as a sporadic distribution—either alone or as a part of the multiple endocrine neoplasia type II syndrome, with pheochromocytoma and often with parathyroid hyperplasia—its measurement in family members at risk for the disease is a valuable screening mechanism. Since basal or stimulated levels of calcitonin in affected individuals can be elevated for years before evidence of disease appears on thyroid scan or by physical examination, the use of the RIA may lead to early surgery and an excellent prognosis for complete cure of the tumor.

References

1. Berson SA, Yalow RS, Bauman A, Rothschild MA, Newerly K: Insulin-I^{131} metabolism in human subjects; demonstration of insulin binding globulin in the circulation of insulin treated subjects. J Clin Invest 35:170, 1956
2. Aurbach GD, Keutmann HT, Niall HD, Tregear GW, O'Riordan JLH, Marcus R, Marx SJ, Potts JT: Structure, synthesis and mechanism of action of parathyroid hormone. Recent Prog Horm Res 28:353, 1972
3. Bilezikian JP, Doppman JL, Shimkin PM, Powell D, Wells SA, Heath DA, Ketcham AS, Monchik J, Mallette LE, Potts JT, Aurbach GD: Preoperative localization of abnormal parathyroid tissue. Cumulative experience with venous sampling and arteriography. Am J Med 55:505, 1973
4. Aurbach GD: Beeson–McDermott Textbook of Medicine, 14th ed, Philadelphia, Saunders 19   p 1808
5. Parthemore JG, Deftos LJ: Calcitonin secretion in normal human subjects. J Clin Endocrinol Metab 47:184, 1978

Bone Mineral Determinations: Methods and Techniques to Date

Peter M. Joseph

It has become apparent that a method for accurately measuring the mineral content of bones in vivo is of crucial importance for diagnosing and managing many types of metabolic bone diseases. Indeed, the single disease entity known as senile osteoporosis is so widespread that an inexpensive, simple, and accurate method of measuring bone mineral content (BMC) would enable earlier therapy to be given to millions of postmenopausal women. For this reason, considerable effort has been expended during the last two decades in searching for the ideal physical measurement technique. The result has been a collection of techniques, most of which work on a variation of the basic concept of measuring x-ray attenuation in bones and converting the measurement into some kind of "bone mineral density."

Before delving into the physics of the various measurement processes, it is worthwhile to clarify the various meanings which the word "density" could have in this context. The fundamental meaning of density is mass per unit volume, symbolized as $\rho(x,y,z)$, and expressed as grams per cubic centimeter. The notation $\rho(x,y,z)$ is used to remind the reader that this is basically a "point" or "microscopic" density, which will vary from point to point in a volume of bone. If a segment of bone of some particular volume, V, is cut out and found to have mass m (by weighing), then the average density is $\bar{\rho} = m/V$.

The distinction between ρ and $\bar{\rho}$ would be significant in differentiating osteoporosis—where bone mineral together with its protein (osteoid) matrix is diminished, from osteomalacia—where the osteoid matrix is present but poorly calcified with inorganic bone mineral.[1] In the case of osteomalacia, both ρ and $\bar{\rho}$ would be reduced. In the case of osteoporosis (OP), the microscopic ρ of the bone mineral itself would be unchanged. However, averaging over several trabeculae in cancellous bone would indicate decreased $\bar{\rho}$. On the other hand, in OP, cortical bone tends to diminish in *size* rather than *density*, so that both ρ and $\bar{\rho}$ would remain unchanged within the cortical region. Only a technique which was sensitive to cortical *thickness* would indicate any loss of BMC. A long osteoporotic bone would still appear to be less dense on a radiographic film because the thinner cortex would cast a less dense shadow on the film.

Finally, any physical measurement which purports to measure bone mineral density must make certain assumptions about the nature of the task; for example, that fat is absent from the bone, or that connective tissue has the same x-ray attenuation coefficient as water. If these assumptions are untrue, there will inevitably be a certain systematic error, or lack of accuracy. That is, accuracy refers to the relationship between the measurement obtained and the true density. Nevertheless, even "inaccurate" data may be extremely useful if they are precise. *Precision* refers to the variation in errors from repeated measurements on the same bone in a given patient. Many BMC techniques have demonstrated precisions of 1 to 2 percent, while the accuracy may be 5 to 20 percent, depending on many anatomic and physical factors.

Geometric Measurements on X-ray Films

It is possible to see the effects of various osteopenic diseases by carefully measuring the sizes of certain bones on radiographs. For example, the width of the cortex of the radius or ulna, and the heights of the vertebra, will be measurably decreased in osteoporosis.[2] Also, in severe disease, the radiodensity of the vertebra will be visibly diminished.[3] However, most studies have concluded that only relatively advanced disease, representing a loss of at least 30 percent of bone mineral, can be reliably detected by these visual techniques.

Three Concepts of Density

The fundamental concept of density is the $\rho(x,y,z)$ discussed in the Introduction. There are two other closely related concepts, however, which are often used in quantifying osteopenic disease.

Let us define *area density*, T, as the product of ρ and a particular path length, L. That is, if the density is constant along a particular path of length L, then

$$T = \rho L \qquad (1A)$$

More generally, if $\rho(x,y,z)$ varies along the path, then T is defined as an integral along L.

$$T(x,z) = \int_L \rho(x,y,z)\, dy \qquad (1B)$$

This concept is central in discussing the attenuation of x-rays which pass along a line. This is illustrated in Figure 1A, which shows a vertical line of length L passing through a tubular bone.

Note that T has dimensions of grams per cm² of area: in this case the area is perpendicular to the line L. It is important to realize that T does not represent the mass per unit area of axial cross section, since line L is not along the z axis. Further note that $T(x,z)$ will vary dramatically as we scan through the bone, as illustrated in Figure 1B. All linear type bone scanning machines essentially measure $T(x,z)$ as they move along the x direction.

The physical significance of $T(x,z)$ is that, when multiplied by the mass attenuation coefficient of bone, it determines the amount of x-ray attenuation due to bone in the path.

The third concept of density is commonly called "bone mineral content" (BMC) and is the area under the $T(x)$ curve. Mathematically, the definition is

$$BMC(z) = \int T(x,z)\, dx \qquad (2)$$

Physically, BMC has dimensions of grams per cm of length; that is, it is exactly the mass of the mineral content of a piece of bone of length 1 cm along the z axis. Equation (2) emphasizes that BMC will depend on z; that is, on the position at which the bone is scanned.

From the point of view of a mechanical engineer, one would expect that BMC would determine the strength of the bone to withstand mechanical stress. Indeed, this has been verified experimentally.[4] The underlying reason is that BMC is equal to the product of bone density and axial cross-sectional area:

$$BMC(z) = \rho \times Area(z). \qquad (3)$$

In osteoporosis, cortical bones lose strength and

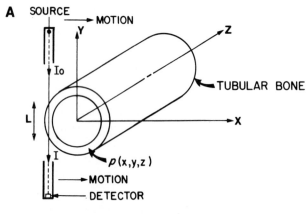
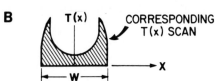

FIG. 1. (A,B) Illustration of the geometric arrangement for scanning a long tubular bone. As the source and detector move along the x direction, the measured curve T(x) is obtained. T(x) is the number of grams/cm² of bone mineral along path length L. The area under the T(x) curve is the BMC at the z position of the scan.

BMC because the cross-section area shrinks, while spongy bones lose BMC primarily because the density is diminished.

In the case of long bones, such as the radius or ulna, the area will depend both on the thickness of the cortex and the overall diameter. That is, a person whose original bone had a large diameter might have a BMC in the normal range even though the cortex was eroded. For this reason, some authors[5] have proposed dividing BMC by overall bone width, W, in an attempt to reduce the normal anatomic variations in BMC. Referring to Figure 1B, and recalling the definition of BMC (equation 2), we see that the quantity BMC/W is just the average value of T(x) seen as the scanner crosses the bone:

$$BMC/W = \overline{T}$$
$$= \text{average } T(x) \quad (4)$$

From this point of view, it is clear that $\overline{T}$, with dimensions of grams per cm², is still likely to be dependent on normal variations in bone size. That is, normal variations in cortical thickness will still be reflected in $\overline{T}$. In fact, most studies have found that $\overline{T}$ is of marginal value.[6]

Single Energy X-ray Transmission Method

With the aforementioned concepts of bone density clarified, we can now discuss the most common methods for measuring bone density. While these methods appear to be very different in terms of apparatus and procedure, they are all based on the same physical premise: that bone density can be determined in vivo by measuring the attenuation of a single spectrum of x-rays as they pass through the bone in question. In general, if we start with an x-ray beam of intensity, I_o, which passes through T grams per cm² of a substance whose *mass* attenuation coefficient is μ_m, then the transmitted intensity, I, is given by the simple equation

$$I = I_o e^{-T \mu_m} \quad (5)$$

That is, transmitted intensity, I, depends directly on area density, T.

If we had an isolated bone we could, starting with measurements of I and I_o, calculate T of the bone by using the mass attenuation coefficient (MAC), μ_m, of bone. Obviously, any in vivo measurement will also involve soft tissue (ST) in the beam, and the MAC of soft tissue differs considerably from that of bone. Since it is impossible to determine both the T of bone and the T of soft tissue from one measurement, it is necessary to somehow eliminate one variable. The method which has proven useful here for extremities (especially the radius, ulna, or calcaneus) is to immerse the arm or foot in water. This fixes the total path length of the x-ray beam. Hence, the loss of bone mineral is presumed to be accompanied by its replacement by "soft tissue" on an equal volume basis. That is, the change in transmitted x-ray intensity depends on the *difference* of linear attenuation coefficients (μ) of bone mineral and soft tissue.

In practice, it is essential that x-ray transmission also be determined for a path just outside of the bone of interest. This serves as a control or reference measurement representing zero bone mineral for each particular patient. The value of T (mineral) for any path can be calculated from the following formula:

$$T(\text{mineral}) = -\rho(\text{mineral}) \frac{\ln(I(\text{bone})/I(\text{reference}))}{\mu(\text{mineral}) - \mu(\text{soft tissue})} \quad (6)$$

If the reference path contains an appreciable portion of fat, then an error will be made in the value of T (mineral) obtained.

All of the following methods basically measure T (mineral), from which BMC can be computed.

Film Densitometry Method

In this method,[7] a radiographic image is used in a manner that allows, in principle, the measurement of T (mineral) at each point in the field. The arm is immersed in a radiolucent box of water, along with an aluminum wedge. The basic principle is that T (mineral) is to be inferred by measuring the photographic optical density (OD) of the film image of the bone. This inference is made possible by the aluminum wedge; the wedge establishes a direct comparison between film density and T (aluminum) on each exposure. Thus the T (mineral) values are necessarily expressed as an equivalent quantity of aluminum. (This is reasonable because the effective atomic number of aluminum is close to that of bone.) Alternatively, the step wedge could be made of calcium hydroxyapatite (CHA), so that T (mineral) would be expressed as grams/cm² equivalent of CHA.

This method suffers from two important disadvantages. The use of a large x-ray field means that scatter will contribute a significant error.[8] Furthermore, the extraction of a quantitative measure from an x-ray film involves the use of an optical densitometer and its associated error. Obviously, any nonuniformity of film processing or x-ray field will lead to errors in determining T (mineral).

The Cameron Scanning Method[9]

This method (Fig. 1), which has been implemented in a commercially available clinical instrument, uses a radioactive source (usually ^{125}I) and a NaI crystal detector. The arm is either immersed in water or is surrounded by a water-bag type of jacket so that total thickness is constant. The transmitted beam is electronically processed to give an instantaneous measure of T (x,z) of bone. The source and detector move together in a simple linear path to scan across the radius or ulna. The BMC is automatically computed by electronically integrating the T (x,z) signal according to equation (2). In addition, a measurement of BMC/W or $\overline{T}$ can be obtained automatically, as in equation (4).

It is generally agreed that when used carefully, the Cameron isotopic scanning method gives greater accuracy than the film densitometry method. One important reason is the highly collimated beam which reduces scattered photons to a negligible level. Another reason is the use of electronic equipment which counts each detected photon; this system is clearly less susceptible to unknown processing errors than is the film system.

Probably the greatest drawback to the Cameron system is the problem of positioning of the arm. Theoretically, a rotation of the arm around an axis parallel to, say, the ulna should not affect the BMC, since BMC represents grams per cm length along the z axis. However, if the scanning direction is not perpendicular to the long axis of the ulna, an erroneous BMC value will result. Since no image is provided in this technique, careful positioning is essential.

The question of the absolute accuracy of the BMC values obtained is complex and difficult. Theoretically, the most important variable which is not included in the analysis is the fat content of either the marrow or the soft tissues adjacent to the bone. In principle, it is only a *variation* in fat content across the bone which will cause an error, the magnitude of which has been estimated to range from 2 to 9 percent.[10] Studies[9] have been conducted comparing measured BMC values on cadavers with direct physical determination of BMC (ash weight) on excised bones. These results generally show inaccuracies ranging from 5 to 8 percent.[12] However, studies[6] of the correlation of BMC in various bones within a single skeleton indicate variations on the order of ±15 percent. Thus the clinical significance of the absolute accuracy remains problematic.[11]

In principle, any such scanning machine should be tested by scanning a bone, with a soft tissue equivalent covering, to establish absolute calibration. A more convenient, nonbiologic calibration phantom is an aqueous solution of K_2HPO_4; this substance is used because its effective atomic number closely approximates that of calcium hydroxyapatite. In practice, any clinical machine should be checked daily with a simple bone simulating phantom, such as an aluminum rod in a plastic base.

Other Variations on the Single Energy Transmission Method

It is, in principle, possible to use a conventional rectilinear scanner of the nuclear medicine type to do Cameron type scans. However, one would still need a water bath to equalize path lengths, and identification of bone edges may be problematic. It is possible to utilize the *imaging* ability of nuclear cameras to obtain an image of, say, the T values of the calcaneus with the heel immersed in water.[12] Recently, more sophisticated detectors using multiwire xenon proportional chambers with computer readout have been proposed.[13,14] The motivating idea here is to obtain an image in which quantitative transmission data are also available. The physician can then clearly indicate the particular path through the bone along which the BMC is to be computed.

All of these imaging techniques will still suffer from the problem of scatter from a broad area beam.

Effect of Photon Energy on BMC Measurements

In general, the linear attenuation coefficient, μ, of any substance will depend on the photon energy being used. However, substances with high atomic number, Z, will show a more pronounced dependence on energy than will low Z substances. The reason for this behavior is the fact that the photoelectric absorption coefficient, which is much more energy dependent than Compton scattering, is proportional to (approximately) Z cubed. Consult any text[15] on radiologic physics for further discussion of this basic physics.

Thus, the choice of the photon energy will strongly affect the accuracy of BMC measurements. In Figure 2 we see that bone, because of its calcium content, shows much more energy dependence in its μ_m than does soft tissue or fat. What this means is that, in principle, it is possible to separately measure two different tissue components by using photons of two different energies. It does not follow, however, that by using N photon energies we can quantify N unknowns. The versatility of multienergy technique is

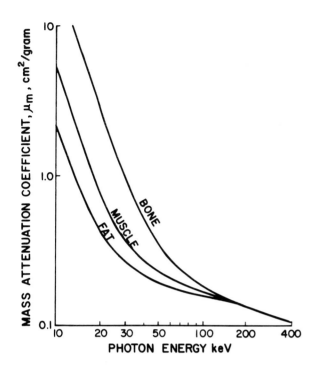

FIG. 2. Mass attenuation coefficients for fat, muscle, and bone as they depend on photon energy. The bone is assumed to contain 14 percent calcium.

limited by the fact that there are only two important attenuation processes: Comptom scattering and photoelectric absorption. In principle, a three-variable measurement is possible if coherent scattering is included. However, in practice, any attempt to determine three densities using three energies would give very poor precision. Hence, without the use of a water bath it is not possible to simultaneously measure the mineral, soft tissue, and fat components of T.

Dual Energy Transmission Methods

There are two possible points of view regarding exactly what advantage should be gained by measuring transmission using two photon energies. The most popular goal is to simultaneously measure both T (mineral) and T (soft tissue) for each beam path through the patient.[16] This then allows us to dispense with the water jacket needed with the single photon technique. It then becomes theoretically possible to scan any part (or the whole) of the body for BMC. Considerable effort has been devoted to quantifying vertebral BMC, as well as whole body calcium, with this technique.

A second possible goal would be to keep the water jacket, with its implied anatomical limitations, and instead use the second photon energy to simultaneously measure both T (mineral) and T (fat).[17] In this way, the error in BMC induced by varying fat content along the scan can be eliminated.

One obvious technique for dual energy scanning is to use two different radioactive isotope sources (with different photon energies) in a linear scanning machine. The NaI detector crystal is then connected to two different pulse height discriminators, with each discriminator set to respond to one of the two photon energies. The two intensity measurements are then processed[18] using analogue computing circuits to provide T (mineral) for each path.

With any two such energy methods, it is imperative that the photon energies used be as different as possible. If the photon energies are too similar, then there will be only a slight difference between the mass attenuation coefficients of bone and muscle (Fig. 2), and the value of T (bone) obtained will be extremely sensitive to small errors of measurement (due to photon statistics, for example). As a result, most dual energy systems will usually give improved accuracy at the cost of decreased precision.

One ingenious variant of the dual photon method is called "x-ray spectrophotometry" by Jacobson.[19] This system uses an x-ray tube which is electronically switched between two different kilovoltages. The machine works by automatically adjusting the thickness of a "soft tissue" wedge and a "bone mineral" wedge in the beam to maintain constant intensity. That is, the wedge thicknesses provide a direct readout of T (mineral) and T (soft tissue).

Scattered Photon Techniques

These methods approach the problem of density measurement from an entirely different physical basis than x-ray transmission. They work because the number of photons scattered by a given volume of tissue is proportional to the density of the tissue. That is, they extract information primarily from the scattered beam, not the transmitted beam. In each case, one must use a highly collimated gamma ray source and a highly collimated detector (Fig. 3); it is the overlap in space of these two collimated "beams" that determines the volume element being measured.

Compton Scattering

In this technique, relatively high photon energies, on the order of 100 to 500 keV, are used.[20,21] This

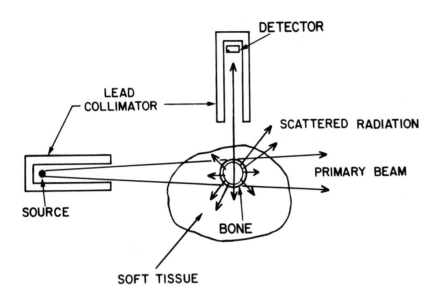

FIG. 3. Geometric arrangement for measuring radiation scattered from a bone. The region measured is determined by the overlap of the source and detector collimators.

means that photoelectric absorption will be small, and that the detected scatter will be proportional to electron density, independent of the atomic number, Z. For this reason, the method is about one-half as sensitive to calcium as are the photon absorption techniques.[22] Major problems in achieving accuracy are photon attenuation in tissues outside of, and multiple scattering of photons within, the region of interest. The use of a high photon energy will mitigate these problems, but at the expense of increased radiation dose. Possible correction schemes include the use of radiographs[21] to determine the amount of tissue absorption, and pulse height analysis of the scattered photon[23] to correct for multiple photon scatterings. To date, this method has not yet demonstrated sufficient accuracy for clinically useful scans of vertebrae.

Coherent Scattering

An ingenious extension of the scattering method has been proposed[24] which may overcome some of the disadvantages of the Compton scatter technique. The idea is to measure both Compton and coherent scattering of photons in, say, cancellous bone. This is possible using energy discrimination on the scattered photons since there is a small energy loss in Compton scattering and none in coherent scattering. Furthermore, coherent scattering, like photoelectric absorption, depends on the cube of Z. Therefore, the ratio of coherent to Compton scattered photons is a measure of effective atomic number, independent of density. The ratio should, therefore, be decreased in any osteopenic disease of cancellous bone. Furthermore, the ratio is almost unaffected by extraneous scatter, so elaborate corrections for this are not necessary.

Computed Tomography Scanning

Since computed tomography (CT) scanning provides, in theory, a direct measure of the linear attenuation coefficient, μ, at each point in a given transverse slice, it can in principle be used to measure the mineral content at each point within any bone. This is true because μ is related to density as

$$\mu = \mu_m \rho \qquad (7)$$

Remember that the mass attenuation coefficient (MAC), denoted as μ_m, depends only on the chemical composition and not on the density of the tissue. This leads us to a paradox, however, since biologically the chemical composition of bone is indirectly related to density; that is, osteopenia will lower both ρ and μ_m.

This Gordian knot is most easily cut by simply accepting μ itself as a valid measure of mineralization.[25] That is, rather than attempt to measure CHA density, one can study μ of bone in normal and osteoporotic subjects. The problem with this approach is that μ is strongly dependent on photon energy (Fig. 2). Furthermore, the effective energy of a *bremsstrahlung* beam will be different for thick and thin body parts, even if KVP is kept constant. For this reason, it is better to use monoenergetic photons from a radioactive source. Such a special purpose scanning machine has been built for measuring μ

values of extremities in vivo.[26] Preliminary results[26,27] indicate that precisions on the order of 1 to 2 percent are easily achieved with this technique. Furthermore, animal experiments[25] indicate that CT may be twice as sensitive, when used to measure μ of trabecular bone, as other methods as an index of osteoporosis.

If absolute accuracy is desired then the problems become more subtle. The basic problem is that each pixel represents an unknown mixture of bone mineral and soft tissue (ST), where ST represents an unknown mixture of water, protein, and fat. If the differences among the chemical compositions of the components of ST can be neglected, then it is possible to obtain a measurement of the mineral density, ρ (mineral), independent of ST density. The technique used[28] involves scanning the patient with two different photon energy spectra. This has the drawback of sacrificing precision to the level of about 10 percent.

It should be noted, however, that fat will effect CT ρ(mineral) values for a completely different reason than it affects BMC values of the Cameron scanning technique. In CT, only the fat content at the point in question will matter, while for the Cameron scanning, it is a *variation* in fat content along the path that matters.[10]

CT still has not developed to the point where the absolute accuracy is reliable. This is especially true of whole body scanners which do not use a water bath. In addition to the engineering problems involved in making sufficiently accurate detectors, hardening of the photon beam in tissue is a fundamental problem whose practical solution has just begun to attract serious effort.[29]

Appendix: Significance of the Correlation Coefficient

The purpose of the correlation coefficient, r, is to tell us quantitatively just how correlated two variables, Y and X, are. For example, X could be the BMC of the radial bone and Y could be the BMC of a vertebra in the same patient. (It is not necessary that X and Y even refer to the same kind of measurements.) It is common in the field of BMC measurements to find experimenters quoting "high" correlation values of, say, 0.80 or 0.90. Here we wish to clarify, in as simple a way as possible, just what this r value tells us about the accuracy of such measurements.

Given a large population of individual measurements of X and Y, there will be a certain variation in the X and Y values obtained. Let $\sigma(X)$ and $\sigma(Y)$ be the statistical standard deviations of the X and Y populations taken separately. If X and Y were perfectly correlated, then $\sigma(Y)$ would be determined directly by $\sigma(X)$. If, however, there were an additional cause for variation in the Y variable which was not present in the X variable (or vice versa), then one σ will be larger than the other.

By oversimplifying somewhat, we can represent this additional variation as an extra error, ϵ, in Y which was not present in X.

$$Y = X + \epsilon \qquad (A-1)$$

ϵ is often called the "standard error estimate," or SEE.

Clearly it is ϵ, not r, which directly tells us the extent to which we err when inferring the value of Y from a measurement of X. ϵ is given by the formula

$$\epsilon = \sqrt{1 - r^2}\ \sigma(Y) \qquad (A-2)$$

Clearly, if r = 1 or -1 (perfect correlation) then ϵ = 0. In general, however, ϵ will be surprisingly large if r is much less than 1. For example, r = 0.95 does *not* mean that the SEE is 5 percent. Furthermore, equation (A-2) says that knowing r alone does not tell us how small ϵ will be; it depends also on the amount of variance within the sample population, $\sigma(Y)$.

An excellent practical discussion is found in a paper by Christiansen and Rödbro.[6] For further explanation of correlation analysis, consult any text on statistics.[30]

Acknowledgement

The author is grateful to Dr. J. Silver for a critical and constructive reading of the manuscript.

References

1. Nordin BEC: Osteoporosis and Calcium Deficiency. Rodahl K, Nicholson JT, Brown EM (eds): Bone as a Tissue. New York, McGraw-Hill, 1960, pp 46-48
2. Meema HE: Cortical Bone Atrophy and Osteoporosis as a Manifestation of Aging. Am J Roentgenol 89:1287, 1963
3. Urist MR: Observations on the Problem of Osteoporosis. Rodahl K, Nicholson JT, Brown EM. Op Cit pp 34-38
4. Wilson CR: Prediction of Femoral Neck and Spine Bone Mineral Content from the BMC of the Radius and the Relationship between Bone Strength and BMC. Proc International Conf on Bone Mineral Measurement, Chicago, Ill, 1973, Mazess RB (ed.) pp 51-59
5. Johnston CC, Smith DM, Yu PL, Deiss WP: *In vivo* Measurement of Bone Mass in the Radius. Metab Clin Exp 17:1140, 1968
6. Christiansen C, Rödro P: Estimation of Total

Body Calcium from the Bone Mineral Content of the Forearm. Scand J Clin Lab Invest 35:425, 1975
7. Keane BE, Spiegeler G, Davis R: Quantitative Evaluation of Bone Mineral by a Radiographic Method. Brit J Radiol 32:162, 1959
8. Mayer EH, Trostle HG, Ackerman E, Schraer H, Sittler OD: A Scintillation Counter Technique for the X-ray Determination of Bone Mineral Content. Radiol Res 13:156, 1960
9. Cameron JR, Mazess RB, Sorenson JA: Precision and Accuracy of Bone Mineral Determination by Direct Photon Absorptiometry. Invest Radiol 3:141, 1968
10. Wooten WW, Judy PF, Greenfield MA: Analysis of the Effects of Adipose Tissue on the Absorptiometric Measurement of Bone Mineral Mass. Invest Radiol 8:84, 1973
11. Dalen N, Jacobson B: Bone Mineral Assay: Choice of Measuring Sites. Invest Radiol 9:174, 1974
12. De Puey EG, Thompson WL, Alagarsamy V, Burdine JA: Bone Mineral Content Determined by Functional Imaging. J Nuc Med 16:891, 1975
13. Horsman A, Reading DH, Connolly J, et al: Bone Mass Measurement Using a Xenon-filled Multiwire Proportional Counter as a Detector. Phys Med Biol 22:1059, 1977
14. Zimmerman RE, Lanza RC, Tanaka T, et al: A New Detector for Absorptiometric Measurement. Am J Roentgenol 126:1272, 1976 (Abs)
15. Meredith WJ, Massey JB: Fundamental Physics of Radiology, 2nd ed. Baltimore, Williams and Wilkins, 1972, pp 59-81
16. Wilson CR, Madsen M: Dichromatic Absorptiometry of Vertebral Bone Mineral Content. Invest Radiol 12:180, 1977
17. Rassow J: A Two-energy Densitometry Method for Measuring Bone Mineral Concentrations and Bone Densities *in-vitro* and *in-vivo*. Am J Roentgenol 126:1268, 1976
18. Kan WC, Wilson CR, Witt RM, Mazess RB: Direct Readout of Bone Mineral Content with Dichromatic Absorptiometry. Proc Internat Conf Bone Mineral Meas, Chicago, Ill, 1973. Mazess RB (ed), pp 66-72
19. Gustafsson L, Jacobson B, Kosoffsky L: X-ray Spectrophotometry for Bone Mineral Determinations. Med Biol Eng 12:113, 1974
20. Webber CE, Kennett TJ: Bone Density Measured by Photon Scattering. I. A System for Clinical Use. Phys Med Biol 21:760, 1976
21. Hazan G, Leichter I, Loewinger E, Weinreb A: The Early Detection of Osteoporosis by Comptom Gamma Ray Spectroscopy. Phys Med Biol 22:1073, 1977
22. Olkkonen H, Karjalainen P: A ^{170}Tm Gamma Scattering Technique for the Determination of Absolute Bone Density. Brit J Radiol 48:594, 1975
23. Kennett TJ, Webber CE: Bone Density Measured by Photon Scattering. II. Inherent Sources of Error. Phys Med Biol 21:770, 1976
24. Puumalainen P, Uimarihuhta A, Alhava E, OlkkonenH: A New Photon Scattering Method for Bone Mineral Density Measurements. Radiology 120:723, 1976
25. Pullan BR, Roberts TE: Bone Mineral Measurements Using an EMI Scanner and Standard Methods: A Comparative Study. Brit J Radiol 51:24, 1978
26. Ruegsegger P, Elsasser U, Anliker M, et al: Quantification of Bone Mineralization Using Computed Tomography. Radiology 121:93, 1976
27. Posner I, Griffiths H: Comparison of CT Scanning with Photon Absorptiometric Measurement of Bone Mineral Content in the Appendicular Skeleton. Invest Radiol 12:542, 1977
28. Genant HK, Boyd D: Quantitative Bone Mineral Analysis Using Dual Energy Computed Tomography. Invest Radiol 12:545, 1977
29. Joseph PM, Spital RB: A Method for Correcting Bone Induced Artifacts in Computed Tomography Scanners. J Comp Assist Tomo 2:100, 1978
30. Larson HJ: Statistics: An Introduction. New York, Wiley, 1975, pp 307-315

Quantitative Bone Mineral Analysis Using Computed Tomography

Harry K. Genant

Currently available in vivo techniques for quantifying skeletal mass are generally restricted to measurements of the peripheral tubular bones and reflect primarily cortex. A technique capable of separately quantifying cancellous bone may be a more sensitive determinant of metabolic bone disease. To this end, we have assessed the potential of computed tomography (CT) for bone mineral determination.

Methods

For preliminary evaluation, the EMI head unit was used, and phantoms were constructed to simulate mineral, soft tissue, and fat, either separately or in composite. Dipotassium hydrogen phosphate (K_2HPO_4) solution was chosen, since it has absorption properties[1] similar to those of calcium hydroxyapatite ($Ca_{10}(PO_4)_6(OH)_2$).[2] Water was used to simulate lean soft tissue and ethyl alcohol (C_2H_5OH) to simulate fat.[3-5] Single-chamber cylinders (16-mm diameter) were filled with combinations of these solutions. Coaxial dual-chamber cylinders were constructed with the outer circumferential chamber (42- or 33-mm diameters) containing high concentrations of phosphate solution and the inner chamber (31- or 22-mm diameters) containing lower concentrations to simulate cancellous bone. Variable amounts of alcohol were added to the central chamber to simulate marrow fat. Monitor displays of single and coaxial phantoms and the corresponding digital printouts were used for quantitation.

In vivo determination of cancellous bone, however, measures mineral in the presence of soft tissue with variable amounts of fat, which, if uncorrected, can introduce substantial errors.[6] The dual energy beam technique may correct these inaccuracies. Thus, the mean linear attenuation coefficient was determined at two divergent energies—80 and 140 kvp.

Results

In Figure 1, the measured integral CT number is plotted against the mineral content calculated on the basis of the known concentration and size of phantoms. The correlation coefficient is high and the scatter of points about the regression line low, giving an accuracy of 1.9 percent.

Figure 2 shows a plot of the mean CT numbers versus the fraction or percent potassium phosphate solutions for the coaxial phantom simulating cortical and cancellous bone. The high correlation previously shown for integral bone is likewise shown here for cancellous determination and for cortical determination done separately. This separate determination cannot be done with the Norland-Cameron photon absorption technique.

Figure 3 demonstrates graphically the basis for the dual determination. It is made possible by the large change in linear attenuation coefficient for the mineral component as a function of kvp. The fat (ethanol) and soft tissue (water) components show a smaller, predictable, and paralleling change in linear attenuation coefficient. Thus, if we know the change in CT number for an unknown mixture, the mineral fraction can be derived independent of fat and water.

The accuracy and precision of the dual- and single-energy techniques are shown in Figure 4. Here we have used a series of phantoms with a constant 5 g%

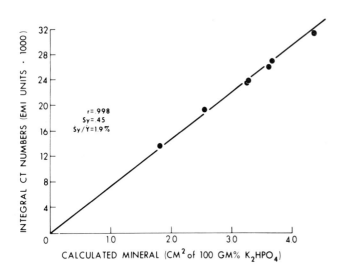

FIG. 1. Summation of integral of CT number (140 kvp) for a slice of solid cylinder plotted against the known standards of dipotassium hydrogen phosphate expressed as square centimeters of 100 g% K_2HPO_4 equivalent.

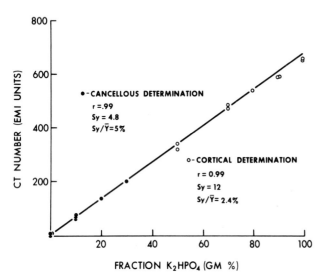

FIG. 2. Mean CT numbers versus the fraction for coaxial phantoms simulating cortical and cancellous bone.

potassium phosphate concentration but with the various mixtures of alcohol ranging from 0 to 40 percent by volume and plotted on the axis. The measured value for mineral content should be 5 percent for all determinations. With the single-energy techniques, however, the measured value becomes progressively lower with increasing alcohol content such that, at 40 percent added alcohol, one obtains a result of approximately 2 g% for the 80- to 140-kvp techniques. The error is greater than 50 percent. The precision, however, is very high, as indicated by the very small standard deviation bars, approximately 1 to 2 percent.

With the dual energy technique, the error is greatly reduced; we obtain values close to the correct 5 g% value. In other words, the accuracy of this determination relative to the single-energy mode is considerably improved, although its magnitude was not determined by multiple measurements. The precision, however, is poorer, as evidenced by the standard deviation of approximately 10 percent. Thus, the dual-energy technique reduces the inaccuracy resulting from unknown amounts of fat at the sacrifice of precision.

It appears that for serial determinations in a given patient, the single energy technique will be most useful, since precision is high. For a diagnostic determination, however, to separate a patient from a normal population, accuracy is required, and thus the dual-energy technique must be used.

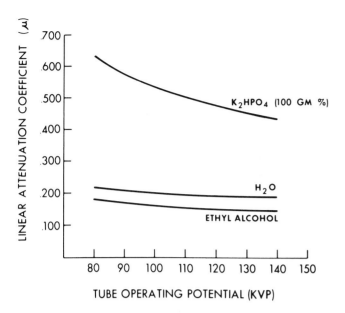

FIG. 3. Linear attenuation coefficient versus kvp for the tissue equivalent substances. The large change in linear attenuation coefficient for the mineral component is demonstrated. Water and ethyl alcohol show a smaller and paralleling change with tube operating potential.

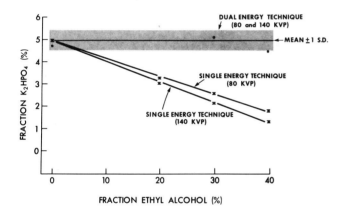

FIG. 4. Measured dipotassium hydrogen phosphate fraction determined from fixed 5 g% solutions with various admixtures of alcohol for dual- and single-energy techniques. The higher precision but poorer accuracy for single-energy techniques is shown.

Discussion

Potential advantages of the computed tomographic technique[7-11] over other modalities for bone mineral determination include (1) transaxial display of data, which permits identification of anatomy and separate determination of cortical, cancellous, or integral bone; (2) capability of determining linear absorption coefficient for a readily defined volume of bone; and (3) in the dual-energy mode, the ability to determine mineral content in the presence of variable fat and soft tissue. Other procedures available for determining bone mineral content in the peripheral skeleton include the following:

First, there is the simple measurement of cortical thickness, which is easy to perform, is reproducible, and is backed by a large body of normative data.[12-15] This technique may lack sensitivity in assessing many metabolic bone disorders, since only endosteal bone resorption is reflected by this measurement. Intracortical resorption (cortical porosity) and trabecular bone resorption, which are important determinants of high bone turnover states, are not measured.[16-22]

Second, photodensitometry[20,23-25] is a technique using an X-ray source, radiographic film, and a known standard wedge that has proved reproducible in experienced hands and is possibly more sensitive than simple cortical measurement. Complicated technical requirements have limited the clinical scope of this technique, however.

Third, a more widely used, easier, and more

precise technique is photon absorptiometry using an iodine-125 source interfaced with a sodium iodide scintillation detector.[21,26-28] Many such devices have been used, but the most widely accepted is the Norland-Cameron[27] device, which measures the radial shaft (although other tubular bones can be examined). Considerable normative data are available, and many clinical studies have supported its usefulness.[21,29] A reproducibility of 2 percent and an accuracy of approximately 6 percent have been demonstrated.[27] The measurement is primarily an integral of cortical bone, since the diaphysis, where measurements are generally made, contains little cancellous bone and the metaphysis, which contains proportionally more trabecular bone (up to 25 to 40 percent of total integral bone),[30] is more difficult to measure due to repositioning errors and consequently precision may be poor. The impetus for measuring cancellous bone is that this bone tissue has a greater surface-to-volume ratio than cortical bone and shows alteration earlier and more dramatically in many metabolic disorders.[12,19,22,29,31] It is for this reason that the quantified cross-sectional display of CT may prove particularly valuable. Additionally, a technique such as CT, capable of measuring various sites in the skeleton, may be advantageous.[12,13,32]

A study recently reported by Reich[10] utilized a Delta CT scanner and correlated the results with those obtained using the Norland–Cameron densitometer, both techniques measuring an integral of bone in the radial shaft. A fairly high correlation was obtained, $r = 0.72$. Additionally, accuracy was determined by measuring tubular bones devoid of soft tissue and placed in a water bath. The CT results correlated well with subsequent calcium determination in specimens ($r = 0.97$), indicating a high accuracy achievable in vitro. Precision was not determined, and there was no attempt to separately determine the mineral content of cortical and cancellous bone or the mineral in the presence of variable soft tissue components.

A specially devised CT technique for bone mineral analysis recently reported by Ruegsegger[11] used an iodine-125 source and computer-assisted reconstruction. It showed excellent capability for separate determination of cortical and cancellous bone in the radius and provided approximately 2 percent precision. Accuracy was not determined. However, it should be noted that the error due to unknown fat will depend greatly upon the beam energy as well as the relative amount of fat.[6] Thus, with a low-energy source such as iodine-125 (mean energy 27 kev), the attenuation by mineral is accentuated relative to fat and soft tissue due to the Z^3-dependence of the photoelectric attenuation process, which dominates at low photon energy. At this low energy, the error in determining bone mineral content due to variable fat in the marrow space of the radius has been estimated at less than 3 percent.[33] Moreover, beam-hardening error, which can be considerable with a low-energy polychromatic source, is eliminated by the nearly monochromatic iodine-125 photon spectrum. It is apparent, then, that this technique of CT reconstruction using an iodine-125 source may be superior to that achievable with commercially available CT scanners for peripheral small bones.

The current study addressed the problems of separate mineral quantitation of cortical and cancellous bone in the presence of variable fat complicated by beam-hardening of the polychromatic photon beam. The EMI head scanner was used at 80 and 140 kvp to measure coaxial phantoms containing solutions of dipotassium hydrogen phosphate, water, and ethyl alcohol, which simulate variable bone, soft tissue, and fat. The results indicated that the CT technique at a single energy can readily and accurately determine mineral content for cortical, cancellous, or integral bone in a two-component system containing mineral and soft tissue. It can do this in dual-component phantoms with accuracy matching or surpassing the Norland-Cameron technique, which can only measure integral bone. Additionally, in the dual-energy mode,[34] mineral content of a three-component system containing mineral, soft tissue, and fat can be relatively accurately determined at a considerable sacrifice of precision. Furthermore, for the small tubular bones, such as the radius, the hardening of the polychromatic beam at the relatively high energy levels used is rather minimal. Thus, the CT technique discussed here shows considerable promise for quantitating mineral in the peripheral skeleton, although it may not compare favorably with the low-energy, monochromatic CT technique of Ruegsegger.[11]

For analysis of the axial skeleton, however, particularly the spine, where complications of bone loss frequently become manifest,[19,22] the dual-energy CT technique discussed herein may prove particularly valuable. Since higher energies must be used for statistical and dosimetric reasons, the relative error due to marrow fat may be considerable (up to 30 percent).[6,7,33] This indicates the necessity for dual-energy technique for the thicker body parts. The beam-hardening effects for a polychromatic source, however, will necessarily be greater and repositioning errors may be large. Further analysis of computed tomography for spinal mineral quantitation will be necessary to determine the potential of this technique in comparison with other modalities currently under investigation, i.e., single-projection dual-energy photon absorptiometry, X-ray spectrophotometry,[33,35-37] total or partial body neutron activation,[38-40] and Compton scattering.[41]

References

1. Witt RM, Cameron JR: An improved bone standard containing dipotassium hydrogen phosphate solution for the intercomparison of different transmission bone scanning systems. Progress Report AEC Grant No. AT-(11-1)-1422:1970
2. Blumenthal NC, Betts F, Posner AS: Nucleotide stabilization of amorphous calcium phosphate. Mater Res Bull (in press)
3. McCullogh EC: Photon attenuation in computed tomography. Med Phys 2:307, 1975
4. Rao PS, Gregg EC: Attenuation of monoenergetic gamma rays in tissues. Am J Roentgenol Radium Ther Nucl Med 123:631, 1975
5. Ter-Pogossian MM, Phelps ME, Hoffman EJ, et al: The extraction of the yet unused wealth of information in diagnostic radiology. Radiology 113:515, 1974
6. Sorenson JA, Mazess RB: Effects of fat on bone mineral measurements. In JR Cameron (ed): Proc. Bone Measurement Conference. U.S. Atomic Energy Comm. Conf. 700515, 1970, p. 255
7. Boyd DP, Genant HK, Korobkin MT: Improving the accuracy of CT body scanning as applies to bone mineral quantitation (Abstract). Annual Session of the Radiological Society of North America, Chicago, November 1976
8. Bradley JG, Huang HK, Ledley RS: Evaluation of calcium concentration in bones from CT scans (Abstract). Annual Session of the Radiological Society of North America, Chicago, November 1976
9. Genant HK, Boyd D, Korobkin MT, et al: Quantitative bone mineral analysis using dual energy computerized tomographic scanning. Invest Radiol 12:545, 1977
10. Reich NE, Seidelmann FE, Tubbs RR, et al: Determination of bone mineral content using CT scanning. Am J Roentgenol Radium Ther Nucl Med 127:593, 1976
11. Ruegsegger P, Elsass U, Anliker M, et al: Quantification of bone mineralization using computed tomography. Radiology 121:93, 1976
12. Dequeker J: Bone and aging. Ann Rheum Dis 34:100, 1975
13. Garn SM Poznanski AK, Nagy JM: Bone measurement in the differential diagnosis of osteopenia and osteoporosis. Radiology 100:509, 1971
14. Newton-John HF, Morgan DB: The loss of bone with age, osteoporosis, and fractures. Clin Orthop 71:229, 1970
15. Virtama P, Helela T: Radiographic measurements of cortical bone. Acta Radiol (Suppl) (Stockholm) 293:7, 1969
16. Duncan H: Cortical porosis: a morphological evaluation. In Jaworshi Z (ed): Proceedings of the First Workshop on Bone Morphometry. University of Ottawa, Canada, March 1973
17. Genant HK, Heck LL, Lanzl LH, et al: Primary hyperparathyroidism. A comprehensive study of clinical, biochemical and radiographic manifestations. Radiology 109:513, 1973
18. Genant HK, Kozin F, Bekerman C, et al: The reflex sympathetic dystrophy syndrome. A comprehensive analysis using fine-detail radiography, photon absorptiometry, and bone joint scintigraphy. Radiology 117:21, 1975
19. Gordan GS, Vaughan C: Clinical Management of the Osteoporoses. Acton, Mass, Publishing Sciences Group, 1976
20. Memma HE, Meema S: Comparison of microradioscopic and morphometric findings in the hand bones with densitometric findings in the proximal radium in thyrotoxicosis and in renal osteodystrophy. Invest Radiol 7:88, 1972
21. Smith DM, Johnston C Jr, Yu Pao-Lo: In vivo measurement of bone mass. JAMA 219:325, 1972
22. Steinback HL: The roentgen appearance of osteoporosis. Radiol Clin North Am 2:191, 1964
23. Colbert C, Mazess RB, Schmidt PB: Bone mineral determination in vitro by radiographic photodensitometry and direct photon absorptiometry. Invest Radiol 5:336, 1970
24. Doyle FH: Some quantitative radiological observations in primary and secondary hyperparathyroidism. Br J Radiol 39:161, 1966
25. Mack PB: Radiographic Bone Densitometry. Washington, D.C., NASA, March 25, 1965, p 31
26. Aitken JM, Smith CB, Horton PW, et al: The interrelationships between bone mineral at different skeletal sites in male and female cadavers. J Bone Joint Surg 56B:370, 1974
27. Cameron JR, Mazess RB, Sorenson JA: Precision and accuracy of bone mineral determination by direct photon absorptiometry. Invest Radiol 3:11, 1968
28. Lanzl LH, Strandjord N: Radioisotopic device for measuring bone mineral. Proceedings of Symposium on Low-Energy X- and Gamma Sources and Applications. Illinois Institute of Technology Research Institute, Chicago, October 20, 1964
29. Hahn TJ, Boisseau VC, Aviolo LV: Effect of chronic corticosteroid administration on diaphyseal and metaphyseal bone mass. J Clin Endocrinol Metab 39:274, 1974
30. Schlenker RA, von Segen WW: The distribution of cortical and trabecular bone mass along the lengths of the radius and ulna and the implication for in vivo bone mass measurements. Calcif Tissue Res 20:41, 1976
31. Roos B, Rosengren B, Skoldborn H: Determination of bone mineral content in lumbar vertebrae by a double gamma-ray technique. In Cameron JR (ed): Proc. Bone Measurement Conference. U.S. Atomic Energy Comm. Conf. 700515, 1970, p 243
32. Wilson CR: Prediction of femoral neck and spine bone

mineral content from the BMC of the radius or ulna and the relationship between bone strength and BMC (Abstract). Department of Radiology (Medical Physics), University of Wisconsin Hospitals, Madison, Wisc
33. Jacobson B: Bone salt determination by x-ray spectrophotometry. In Cameron JR (ed): Proc. Bone Measurement Conference. U.S. Atomic Energy Comm. Conf. 700515, 1970, p 237
34. Rutherford RA, Pullan BR, Isherwood I: Measurement of effective atomic number and electron density using an EMI scanner. Neuroradiology 11:15, 1976
35. Dalen N, Jacobson B: Bone mineral assay: choice of measuring sites. Invest Radiol 9:174, 1974
36. Krokowski E: Quantitative analysis of calcium in the spine using x-rays of different energy qualities. Symposium on Bone Mineral Determination. Stockholm/Studsvik, May 1974
37. Reiss KH, Killig K, Schuster W: Dual photon x-ray beam applications. In Mazess RB (ed): International Conference on Bone Mineral Measurement DHEW Publ. 75-863, Washington, D.C. U.S. Department of Health, Education and Welfare, 1974, p 80
38. Al-Hiti K, Thomas BJ, Al-Tikrity SA, et al: Spinal calcium: its in vivo measurement in man. Int J Appl Radiat Isot 27:97, 1976
39. Manzke E, Chesnut CH III, Wergedal JE, et al: Relationship between local and total bone mass in osteoporosis. Metabolism 24:605, 1975
40. McNeil KG, Thomas BJ, Sturtridge WC, et al: In vivo neutron activation analysis for calcium in man. J Nucl Med 14:502, 1973
41. Webber CE, Kennett TJ: Bone density measured by Compton photon scattering. I. A system for clinical use. Phys Med Biol 21:760, 1976

Plasma Cell Dyscrasias: Multiple Myeloma and Related Conditions

Elliott F. Osserman

Plasma Cell Dyscrasias

The major criteria that define the plasma cell dyscrasias (PCDs) include (1) proliferation of immunologically competent cells (plasma cells), generally in the absence of an identifiable antigenic stimulus; (2) elaboration of homogeneous M-type gamma globulins and/or comparably homogeneous polypeptide subunits of these proteins; and (3) deficient synthesis of normal immunoglobulins. The term *monoclonal gammopathy* is used essentially interchangeably with PCD.

The conditions included among the plasma cell dyscrasias can be classified in two main groups (Table 1). Group I is comprised of the symptomatic states, i.e., those conditions associated with overt clinical and pathological manifestations, specifically, multiple myeloma, macroglobulinemia, the heavy-chain diseases, amyloidosis of the pattern I and mixed pattern I and II distribution, lichen myxedematosus, and the recently defined "deleted H and L chain disease."[10]

Group II includes the asymptomatic or presymptomatic PCDs, in which the characteristic protein abnormalities are discovered in the absence of any signs or symptoms of myeloma, macroglobulinemia, or the other clinically distinct manifestations. It is known that a certain proportion of these cases will ultimately progress to overt myeloma or another symptomatic form of PCD after periods of a few months to many years. However, it is also well documented that some individuals exhibit monoclonal-type (M-type) protein abnormalities for 20 or more years without progression to a clinical pattern of overt multiple myeloma, macroglobulinemia, amyloidosis, etc. In these latter cases, it is probable that the M-type protein abnormality reflects an excessive reticuloendothelial response to a specific stimulus, particularly a stimulus that is unusually protracted and resistant to elimination by the cellular and humoral reactions that it evokes. As indicated in Table 1, this hypothesis is strongly suggested by the frequency with which these asymptomatic PCDs are associated with a variety of chronic inflammatory and infectious processes.

Finally, in certain cases, PCDs with characteristic monoclonal proteins are transiently associated with hypersensitivity reactions to drugs or with bacterial and viral diseases. Thus, the finding of a monoclonal protein abnormality does not invariably imply neoplasia. Unfortunately, at present it is not possible to predict accurately the eventual course of the process in a particular case.

Multiple Myeloma

Multiple myeloma is the most common form of clinically overt PCD. As implied by the term *myeloma* (i.e., marrow tumor), the presenting and usually predominant clinical manifestations are those related to marrow infiltration and bone destruction by neoplastic (plasma) cells (Fig. 1).

The incidence of myeloma has apparently increased in recent years, and mortality statistics now show incidence figures in the range of 2 to 3 per 100,000. In most large clinics in the United States and Europe, myeloma is now seen with approximately the same frequency as Hodgkin's disease and chronic lymphatic leukemia. Males and females are approximately equally affected. The age of onset of symp-

Table 1. Plasma Cell Dyscrasias

I. *Clinically overt forms* (with distinctive clinical and pathologic features)
    Plasma cell myeloma
        Multiple myeloma, myelomatosis, "solitary" and multiple plasmocytomas, plasma cell leukemia
    Waldenström's (primary) macroglobulinemia
    γ Heavy-chain (Franklin's) disease
    α Heavy-chain (Seligmann's) disease
        Abdominal lymphoma
    μ Heavy-chain disease
    Deleted H and L chain disease
    Amyloidosis
        Pattern I and mixed pattern I and II amyloidosis
    Lichen myxedmeatosus
        Papular mucinosis, lichen amyloidosis

II. *Clinically occult (asymptomatic or presymptomatic) forms* (benign or "essential" monoclonal gammopathy)
    Plasma cell dyscrasia of unknown significance (PCDUS)
        Associated with chronic inflammatory and infectious processes, e.g., osteomyelitis, tuberculosis, chronic biliary tract disease, pyelonephritis, rheumatoid arthritis, chronic pyoderma
        Associated with nonreticular neoplasms, particularly cancers of the bowel, biliary tract, and breast
        Associated with lipodystrophies, particularly Gaucher's disease, familial hypercholesteremia, and xanthomatosis
    Transient plasma cell dyscrasias
        Associated with drug hypersensitivity, sulfonamides
        Associated with (presumed) viral infections
        Associated with cardiac surgery; valve prostheses

toms of myeloma ranges from young adulthood to advanced years, with a peak incidence in most series in the midfifties.

Presymptomatic Myeloma (Premyeloma)

It is now established that the clinically overt, symptomatic stage of myeloma is preceded by a significant asymptomatic or presymptomatic period in most and possibly in all cases. This fact has been established by the chance finding of characteristic monoclonal (M-type) protein abnormalities in asymptomatic individuals who have ultimately developed overt symptoms and signs of multiple myeloma. At present, the duration of the presymptomatic period in myeloma is impossible to define, but it apparently can span 20 years or longer. In many cases, the first suggestion of the presence of an M-type protein abnormality is the finding on routine examination of an increased erythrocyte sedimentation rate or unexplained and persistent proteinuria. In all cases of unexplained proteinuria, the possibility of Bence Jones proteinuria should be considered and an electrophoretic analysis of the urinary proteins carried out. This is of particular importance because of the well-documented risk of precipitating irreversible renal shut-down by intravenous pyelography in patients with Bence Jones proteinuria. It is now well established that urine protein electrophoresis is a far more sensitive and reliable technique for detecting Bence Jones proteins than the classical heat precipitation (at 50 to 60 C) and resolution (at 90 to 95 C) method. Since urine electrophoresis is a relatively simple procedure and the best available method for qualitatively screening undefined urinary proteins, it should be increasingly employed as a routine laboratory procedure.

Skeletal Manifestations

When myeloma becomes symptomatic, skeletal pains are the presenting and predominant manifestations in most cases. Skeletal pains may initially be mild and transient, or the onset may be sudden, with severe back, rib, or extremity pain following an abrupt movement or effort. The abrupt onset of pain frequently indicates a pathologic fracture. As the disease progresses, more and more areas of bone destruction develop, frequently resulting in marked skeletal deformities, particularly of the sternum and rib cage, and in shortening of the spine, causing a decrease of as much as five or more inches in stature.

In most cases, multiple osteolytic (punched-out) lesions are apparent on initial x-ray examination, with an increase in number and size with disease progression. In many cases, however, the initial skeletal x-rays may be negative or may show diffuse osteoporosis without discrete osteolytic lesions.

It should be emphasized that the relentless progression of skeletal lesions, which has previously been considered so characteristic of myeloma, is no longer inevitable. Thus, the significant advances that have been made in the chemotherapy and general management of myeloma in recent years can now, when properly applied, in the majority of cases achieve a marked suppression of the disease process for many months and years, with control of the skeletal as well as the other manifestations of myeloma.

In occasional cases, myeloma presents as an apparently single skeletal lesion. Although these lesions are commonly designated as solitary plasmacytomas, most cases ultimately develop disseminated disease, even if the original lesion is radically

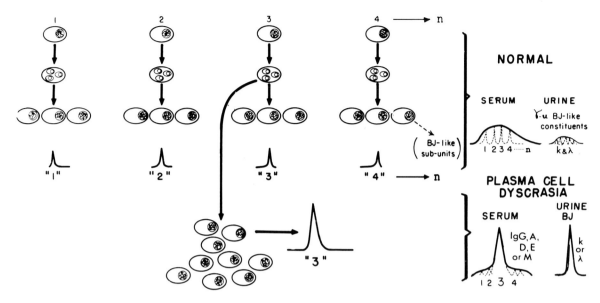

FIG. 1. Schematic representation of the cellular (plasma cells) and biochemical (gamma globulin) aspects of normal immunoglobulin synthesis and the postulated mechanisms underlying the development of myeloma and the other related plasma cell dyscrasias.

excised or irradiated. In cases of apparently solitary skeletal plasmacytomas, M-type protein abnormalities are almost invariably demonstrable. M-type proteins are less frequent in association with plasmacytomas of the soft tissues, e.g., in the respiratory tract. Osteosclerotic lesions in myeloma are rare, and most of the reported cases of myeloma with osteosclerosis have exhibited other clinical features quite atypical for myeloma, including polyneuritis and extramedullary hematopoiesis.

Bone Marrow Documentation

Bone marrow examination either by aspiration or biopsy is required for the documentation of myeloma. Increased numbers of plasma cells and abnormal forms are found in essentially all cases, although more than one attempt may be necessary. In most cases, aspiration of marrow from the iliac crest, sternum, or a vertebral body will disclose an increased number of plasmacytic forms, i.e., in excess of 5 to 10 percent, along with large, immature, and multinucleated forms. When the proportion of plasma cells exceeds 15 to 20 percent, and particularly when clusters or sheets of plasma cells with a large proportion of immature and abnormal forms are found, the diagnosis of myeloma is virtually certain.

In all cases, however, the marrow findings must be interpreted in conjunction with the clinical features and other laboratory data, particularly the associated protein abnormalities.

Hematologic Abnormalities

Virtually all cases of myeloma exhibit anemia either at the time of diagnosis or subsequently with disease progression. The severity of the anemia is variable. In many cases, it remains of moderate proportions (hemoglobin in the range of 7 to 10 g) and is well tolerated, whereas in others it assumes major proportions and necessitates repeated transfusions. The anemia is usually normocytic and normochromic, but may be macrocytic and associated with a megaloblastic bone marrow. It is characteristically refractory to iron, vitamin B_{12}, folic acid, and liver therapy. In occasional cases a hypochromic anemia may be related to blood loss from the gastrointestinal tract associated with a defect in coagulation or to plasmacytic or amyloid infiltrates of the intestinal wall. In these cases, iron administration may be indicated. Several factors participate in the production of anemia in myeloma, including marrow replacement, accelerated erythrocyte destruction, blood loss, renal insufficiency, the effects of radiotherapy and chemotherapy, associated infections, nutritional factors, and so on; it is generally difficult to define the precise role of each of these factors in a particular case.

Leukocyte and platelet counts are usually within normal limits prior to cytotoxic therapy. Occasionally, however, moderate to severe leukopenia and/or thrombocytopenia may be observed prior to treatment, and indeed, with therapy, the white cell and platelet counts may rise toward normal. The differential white cell count frequently reveals a relative lymphocytosis of the order of 40 to 55 percent, with a variable proportion of immature lymphocytic and plasmacytic forms. These have been shown to be predominantly B lymphocytes bearing the same idiotypic markers as the monoclonal protein in the serum or urine.

A clinical pattern consistent with a diagnosis of plasma cell leukemia, with hepatosplenomegaly and white cell counts in excess of 15,000 and over 50 percent plasma cells, is occasionally observed. The symptoms and signs in these cases are generally similar to those exhibited by patients with other types of leukemia, with weakness, anemia, and bleeding manifestations. The cells in the peripheral blood range from typical plasmacytes to immature and atypical forms. Bence Jones proteinuria and abnormal serum globulins occur with the same frequency as in myeloma. The majority of these cases pursue an acute or subacute clinical course. It is of interest that both of the reported cases of IgE myeloma displayed the pattern of plasma cell leukemia, and there are three reported cases with IgD globulins and this clinical pattern.

Enhanced susceptibility to bacterial infections, particularly pneumococcal pneumonias, is exhibited in many cases of myeloma and is at least partially the result of an impaired capacity for antibody formation. This is generally reflected in a decrease in the serum concentration of normal IgG, IgA, and IgM immunoglobulins, irrespective of the type of M protein elaborated. The basis for these deficiencies is presently unknown, but available studies indicate that both diminished production and enhanced catabolism are contributory factors. Herpes zoster and generalized varicella infections occur with increased frequency in myeloma and other plasma cell dyscrasias, but apparently no more so than in Hodgkin's disease and lymphosarcoma. Susceptibility to other virus infections is not apparently increased. Studies of the homograft rejection reaction have demonstrated a significant prolongation of rejection time, indicating that the immunologic deficiencies in myeloma may include cell-mediated immune mechanisms.

Although the antibody-deficient status of most myeloma patients suggests that prophylactic gamma globulin administration might be of value, this has only rarely been found to be indicated. In the majority of cases, the bacterial infections that do develop respond readily to appropriate antibiotic therapy.

Serum and Urinary Protein Abnormalities

The demonstration of characteristic monoclonal (M-type) serum globulins and/or urinary Bence Jones proteins is extremely useful for establishing the diagnosis of myeloma (Fig. 2). When appropriate electrophoretic and immunoelectrophoretic studies of both serum and urine are carried out, M-type proteins can be found in over 99 percent of cases of myeloma. The characteristic that distinguishes M-type proteins is their structural homogeneity. This is best demonstrated by their appearance as sharp, homogeneous peaks (spikes) in the electrophoretic patterns of serum and/or urine. Structural homogeneity is generally further confirmed by appropriate immunoelectrophoretic analyses.

Each monoclonal serum globulin and each Bence Jones protein is structurally unique and remains qualitatively unchanged throughout the patient's course. The structural specificity of these proteins is apparently related to the structural specificity of functional antibodies, and indeed, there is increasing evidence that many, if not all, of these monoclonal M-type proteins, both human and murine, may be antibodies directed against still-to-be-defined antigens. The determination of these specificities should help to elucidate the pathogenesis of these conditions and ultimately be of great diagnostic value.

Relationship of Protein Abnormalities to Clinical Manifestations Particular monoclonal proteins display specific physicochemical properties such as solubility, intrinsic viscosity, thermostability, and reactivity with other proteins, which may be correlated with specific clinical and pathologic manifestations in individual cases, e.g., coagulation defects secondary to the interaction (protein–protein complexing) of specific M-type globulins to specific coagulation factors, including platelets, fibrinogen, factors V and VII, prothrombin, and the antihemophilic globulin; symptoms related to cold-precipitable proteins, (cryoglobulins) such as Raynaud-type phenomenon, circulatory impairment, and gangrene; and circulatory impairment, particularly in the central nervous system and retina, related to monoclonal globulins with a high intrinsic viscosity. Many of these symptoms can also be observed in macroglobulinemia and related to IgM macroglobulins with similar physical properties.

An association between Bence Jones proteinuria and renal functional impairment (so-called myeloma kidney) has long been implied and generally ascribed to the tubular precipitation of these proteins and blockage due to formation of casts. Recent studies, however, make this simplistic mechanism unlikely, since casts per se appear to contain little if any Bence Jones protein. Bence Jones proteins are catabolized by the kidney in addition to being excreted. With progressive renal failure, this catabolic function is also decreased. Specific defects in renal tubular reabsorption mechanisms, including the adult Fanconi syndrome, have been reported in association with Bence Jones proteinuria, indicating that certain Bence Jones proteins interfere with specific tubular transport mechanisms. The fact that a nephrotoxic potential is not common to all Bence Jones proteins is indicated by the maintenance of apparently normal renal function in many cases with protracted and profound Bence Jones proteinuria. Several factors in addition to Bence Jones proteinuria contribute to renal damage in myeloma, particularly hypercalciuria, hyperuricosuria, and dehydration. Since the advent of improved chemotherapy and greater emphasis on the maintenance of hydration and ambulation, the frequency of serious renal dysfunction in myeloma has greatly decreased.

Additional Clinical Manifestations of Myeloma

Neurological manifestations of myeloma develop in a significant percentage of cases of myeloma as the result of direct pressure on the spinal cord, nerve roots, or cranial or peripheral nerves or as a result of a pathologic fracture of a vertebral body or long bone. Spinal cord compression leading to paraparesis and ultimately paraplegia is an extremely serious complication, which fortunately has become considerably less common in recent years with the availability of more effective therapy and increased emphasis on the maintenance of ambulation. The development of symptoms and signs of cord compression requires prompt attention, generally with surgical decompression by laminectomy followed by local radiation.

Infiltration of peripheral nerves and nerve roots by amyloid can cause peripheral neuropathies or root symptoms that are usually symmetrical and associated with other evidence of amyloidosis, such as macroglossia, cardiac manifestations, and the carpal tunnel syndrome. Rarely, polyneuropathies develop in myeloma and the other plasma cell dyscrasias without demonstrable tumor or amyloid infiltration. The pattern of these polyneuropathies is nonspecific and apparently similar to the polyneuropathies of obscure origin occasionally associated with other neoplastic diseases. Multifocal leukoencephalopathy has also been reported in association with myeloma and monoclonal gammopathy. Myopathies involving principally proximal muscle groups are also rarely associated with myeloma and other PCDs, and again, the pathogenic mechanisms are obscure.

Management and Therapy of Multiple Myeloma

Both the comfortable and useful life of the majority of patients with myeloma can be significantly extended for many months or years by proper management. Stated differently, all but a small percentage of cases of myeloma can be benefited both subjectively and objectively by proper general medical management combined with judicious chemotherapy. This clearly represents a significant change from the situation that prevailed up to 20 years ago, when the prognosis for a myeloma patient was essentially relentless progression and death within a few months to a year. In contrast, the present outlook for most myeloma patients, while still far from satisfactory, is better both qualitatively and in duration of survival.

Local radiotherapy can be extremely useful for the control of limited areas of symptomatic involvement, particularly a painful area of the spine or a long bone that may be hindering mobilization or threatening to produce cord compression or pathological fracture. Most plasma cell tumors are quite radiosensitive, and a tumor dose of 1800 to 2400 rads is usually adequate for the control of symptoms. Since radiotherapy is palliative and not curative, both the total dosage and the field of irradiation should be as limited as possible in order to spare the normal marrow reserves.

The complex of hypercalcemia, dehydration, and renal failure represents a major threat, particularly to the immobilized patient. Hypercalcemia is secondary to release of calcium from involved bones in amounts exceeding the excretory capacity of the kidney; dehydration results from the obligatory solute diuresis and tubular dysfunction caused by hypercalciuria plus the anorexia and vomiting, which are frequently associated with hypercalcemia. Hyperuricemia and hyperuricosuria also contribute to the renal dysfunction. This combination starts slowly and insidiously, but builds rapidly into a life-threatening emergency. Therapy of the hypercalcemic–dehy-

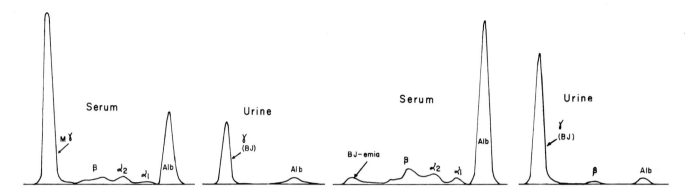

FIG. 2. Serum and urine protein electrophoretic patterns in the plasma cell dyscrasias. Above, left: Case of multiple myeloma exhibiting both a λG serum spike and also a Bence Jones (BJ) urinary peak. Note the small amount of albumin in the urine relative to the larger amount of BJ. Note also that the serum Mγ peak is in the slow γ mobility range, whereas the BJ is in the fast γ range. Above, right: Case of multiple myeloma producing only light chains with a very prominent Bence Jones urinary protein peak (and a minimal albuminuria). A very small peak in the serum was shown by immunoelectrophoresis to be Bence Jones proteinemia. Opposite: Case of macroglobulinemia with a mid-γ serum spike. This could only be identified as a γM globulin by immunoelectrophoresis and/or ultracentrifugation. The urine shows only nonspecific proteinuria with albumin as the predominant constituent.

drated–azotemic myeloma patient consists primarily of rehydration in combination with prednisone and allopurinol (Zyloprim), as necessary. Oral and parenteral fluids should be given in quantities sufficient to establish a urine output of 1500 to 2500 ml per day. Prednisone, 50 to 60 mg/day, should be given in cases that fail to respond promptly to hydration alone. Diuretics and phosphates are not recommended. In most cases, hydration, prednisone, and allopurinol will accomplish a significant reduction in serum calcium and uric acid levels and an improvement in renal function within 24 to 48 hours. Only after this is accomplished can chemotherapy with an alkylating agent be instituted with reasonable safety, since the cytotoxic effects of these drugs will necessarily increase the urate load on kidneys already jeopardized.

Of the chemotherapeutic agents presently available, melphalan (l-phenylalanine mustard, Alkeran) and cyclophosphamide (Cytoxan) are the most effective in the long-term management of myeloma. Both these alkylating agents can be administered by mouth and are generally well tolerated. Both are marrow-suppressive agents, but when properly administered, the abnormal plasma cells and their precursors are apparently inhibited to a greater degree than normal hematopoiesis.

With melphalan, therapy is usually initiated with a loading dose of 8 to 10 mg per day for 7 to 10 days, i.e., an initial course of 56 to 100 mg depending on the patient's size, hematologic status, and general condition. Immediately following this initial loading period, maintenance therapy with a daily dose of 2 mg is instituted and continued indefinitely. Interrupted "pulse" regimens of melphalan and prednisone have also been recommended, but our experience indicates that continuous low-dose administration may be preferable.[3]

Serial quantitation of myeloma serum globulins and/or Bence Jones proteinuria are excellent indices of the efficacy of chemotherapy. Because Bence Jones proteins have a shorter half-life than myeloma serum globulins, a decrease in Bence Jones proteinuria is usually observed sooner after starting chemotherapy than a decrease in the abnormal serum globulin peak.

Symptomatic improvement may be noted within 2 to 3 days of instituting chemotherapy or may not be evident for 2 or 3 weeks. In this period, analgesics and local radiotherapy should be used as necessary to relieve pain and assist ambulation. Blood counts should be checked twice weekly for the first month, weekly for the second month, and at progressively longer intervals once maintenance therapy is well established and stabilized. Repeated bone marrow studies are sometimes useful to appraise the efficacy of therapy, but in most cases an adequate assessment can be made from peripheral blood counts, protein studies, and symptoms.

Because of the complexities of myeloma and its treatment, it is extremely useful and almost necessary to maintain an ongoing graph of each patient's hematologic, protein, and clinical status. The small additional time and effort required to maintain these charts is more than compensated for by the critical perspective they provide to each patient's course.

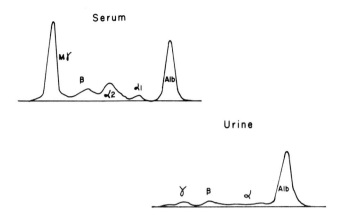

Macroglobulinemia (Primary or Waldenström's Macroglobulinemia)

Macroglobulinemia is the plasma cell dyscrasia involving those cells normally responsible for the synthesis of IgM. The excessive proliferation of these cells results in the elaboration of large quantities of electrophoretically homogeneous IgM and a variable clinical pattern, with anemia, bleeding manifestations, and symptoms related to the serum macroglobulins as the predominant features. Males and females are approximately equally affected. Symptoms generally begin in the fifth or sixth decade. As the disease slowly evolves, lymphadenopathy, splenomegaly, and hepatomegaly develop in a variable percentage of cases, producing a clinical pattern resembling a malignant lymphoma or lymphatic leukemia. Skeletal lesions of the type seen in myeloma are exceptionally rare.

Lymph nodes demonstrate proliferation of lymphocytic–plasmacytic forms often arranged in a pattern of follicular hyperplasia. Lymphocytic–plasmacytic forms are also seen in the peripheral blood and bone marrow.

When macroglobulinemia becomes symptomatic, anemia is the most common presenting manifestation and is frequently profound, with hemoglobin levels in the range of 4 to 6 g/100 ml. Usually, the anemia is due to a combination of factors, including accelerated red cell destruction, blood loss, and decreased erythropoiesis. Coating of erythrocytes with IgM globulin is apparently responsible for the marked rouleaux formation, positive Coombs reactions, and cross-matching difficulties encountered in many cases.

A large percentage of IgM globulins have specific physicochemical properties that are responsible for specific symptom patterns. These properties include cold insolubility (cryoglobulins), high intrinsic viscosity, and the capacity to form complexes with coagulation factors and other plasma proteins. Cryoglobulin-related symptoms include Raynaud's phenomenon, cold sensitivity, cold urticaria, and vascular occlusion, with gangrene following exposure to cold. Viscosity-related manifestations are most evident in the retinal vasculature, in which a pattern of patchy venous bulging and localized narrowing ("sausage effect" or fundus paraproteinemicus) develops, frequently associated with hemorrhages, exudates, and visual impairment. Circulatory impairment in the central nervous system owing to increased plasma viscosity produces changing patterns of neurologic signs and symptoms, e.g., transient paresis, reflex abnormalities, deafness, impairment of consciousness (coma paraproteinemicum), frequently terminating with cerebral vascular hemorrhage. Cardiac decompensation and pulmonary symptoms may also develop secondary to increased viscosity in the systemic and pulmonary vascular beds. Protein–protein interaction with formation of complexes between IgM globulins and coagulation factors (fibrinogen, prothrombin, factors V and VII, etc.) is an important contributing factor to the bleeding diathesis (particularly epistaxes, oral mucosal bleeding, and purpura) exhibited in many cases. Interference with platelet function (platelet agglutination) and capillary damage secondary to increased serum viscosity are additional factors contributing to bleeding manifestations. It must be recognized that all these symptom patterns are also observed in occasional cases of myeloma with M-type IgG and IgA globulins possessing similar physicochemical properties.

Bence Jones proteinuria is present in approximately 10 percent of cases of macroglobulinemia, but renal functional impairment is much less common than in myeloma, presumably because of the absence of the contributing factors of hypercalcemia and hypercalciuria. Amyloidosis has been observed in only a few cases of macroglobulinemia, and in all of these, the liver, spleen, and parenchymal organs have been the major areas of involvement (pattern II) in contrast to the primary, atypical mesenchymal distribution (pattern I) of amyloid usually observed in myeloma.

Peripheral neuropathies (Bing–Neel syndrome) and myelopathy may be progressive and incapacitating. Myopathies and rheumatoid-like arthropathies have also been observed.

In the Columbia College of Physicians and Surgeons series of 57 cases, 15 had clinical or postmortem evidence of an associated nonreticular neoplasm, and an additional 12 had a background of long standing infection, particularly tuberculosis.

The majority of IgM globulins are euglobulins and

give a positive Sia water-dilution reaction, but this is not specific for macroglobulins, because certain IgG globulins are also Sia positive euglobulins. Approximately one-third of IgM globulins are cryoglobulins, yielding a white precipitate or a thick clear gel on cooling. The temperature and duration of cooling needed for precipitation varies with individual cryoglobulins. With some cryoglobulins, precipitation or gel formation occurs almost immediately following venipuncture, and a prewarmed syringe is necessary for blood sampling. Others require several hours at 10 C for precipitation. Similarly, the increased viscosity of serum containing a viscous M-type macroglobulin may be readily apparent when a tube of serum at room temperature is inverted, or viscosimetric determinations at different temperatures may be required.

Hematologic abnormalities, in addition to anemia, include an absolute lymphocytosis with "atypical, immature, and plasmacytic" forms in many cases, occasionally reaching leukemic proportions. Polymorphonuclear leukopenia, thrombopenia, and eosinophilia are also observed. Bone marrow aspirations characteristically reveal an increase in lymphocytic–plasmacytic forms accompanied by eosinophils and mast cells in many cases.

Additional laboratory abnormalities in certain cases of macroglobulinemia include positive flocculation reactions, false positive serologic reactions, and positive rheumatoid factors. The last are occasionally associated with rheumatic symptoms and arthropathy.

Treatment

The principal indications for therapy in macroglobulinemia are anemia, bleeding manifestations, and symptoms related to increased plasma viscosity. When the latter symptoms are severe and threaten central nervous system function and vision, plasmapheresis is indicated as a temporary measure. Repeated plasmapheresis for a period of several weeks may be required until effective chemotherapy can be instituted.

Chlorambucil (Leukeran) is presently regarded as the chemotherapeutic agent of choice in macroglobulinemia. This should be administered continuously at a dosage level of 8 to 10 mg daily. Although this dosage level is significantly higher than the average employed in most cases of lymphatic leukemia or lymphosarcoma, it is usually well tolerated in macroglobulinemia. A large number of patients have been maintained in objective and subjective remission for periods of up to 9 years with continuous chlorambucil therapy. Prednisone may be of some value in the control of capillary bleeding.

Amyloidosis

Amyloidosis (Table 2) is a term applied to a variety of conditions associated with tissue infiltrates comprised of insoluble proteins and/or protein–polysaccharide complexes. With the relatively recent development of methods for the isolation and chemical and immu-

Table 2. Patterns of Amyloidosis

Pattern	Previous designations	Distribution	Associated conditions	Type of amyloid protein
Pattern I	"Primary," atypical, para-amyloid, pericollagen, mesenchymal	Tongue, heart, GI tract, muscles ligaments, skin	Plasma cell dyscrasia, esp. BJ only; overt myeloma	Ig light chain
Pattern II	"Secondary," typical, perireticular	Liver, spleen, kidney, adrenal	Chronic infections, rheumatoid arthritis, familial Mediterranean fever	Protein A
Mixed Pattern	—	Various combinations of I and II sites	Plasma cell dyscrasia with complete IgG, A, D, and M globulins	Not known
Localized	Tumor-forming	Lung, GI tract, bladder, eye	Plasma cell dyscrasia	Not known

nologic analysis of amyloids, it has become evident that there are at least two chemical types of amyloid. In one (Ig-type), immunoglobulin light chains appear to be the principal protein component; in the other, an apparently nonimmunoglobulin constituent (protein A) is the major protein. It must be stressed that other chemical types of amyloid will probably be defined by further studies, but the Ig-type and protein A-type probably comprise the majority of amyloidosis cases encountered clinically. In both types, the tissue infiltrates apparently represent noncovalently bonded polymers of the basic protein subunit, either alone or in combination with other proteins or polysaccharides. The polymers are sufficiently large to yield a characteristic fibrillar structure, which can be visualized by electron microscopy and which display the distinctive green-yellow dichroic birefringence when examined by polarization microscopy after Congo red staining.

As outlined in Table 2, there is some correlation between chemical type of amyloid and the clinical pattern of amyloid distribution. Thus, IgG-type amyloid has been mainly found in cases with either occult plasma cell dyscrasia or overt multiple myeloma. The pattern of distribution of amyloid in these cases is most frequently that previously referred to as *primary type,* i.e., principal involvement of the tongue, heart, gastrointestinal tract, skeletal and smooth muscles, carpal ligaments, nerves, and skin. This distribution pattern referred to as pattern I was manifested in 50 percent of the cases in our series.[10]

Protein A-type amyloid has thus far been mainly found in cases of amyloidosis associated with chronic infections (e.g., tuberculosis), rheumatoid arthritis, Hodgkin's disease, and familial Mediterranean fever. The pattern of distribution of amyloid in these cases most frequently is that previously referred to as *secondary type,* i.e., principal involvement of the liver, spleen, kidneys, and adrenals. This distribution pattern, now designated pattern II, was displayed by 17 percent of our cases.

In a significant proportion of cases (30 percent in our series), amyloid involves both pattern I and II sites to varying degrees, and these cases are now referred to as *mixed pattern I and II.* Monoclonal IgG, A, D, and M globulins, indicative of associated plasma cell dyscrasias, are demonstrable in the majority (82 percent) of mixed pattern I and II cases, but the specific role of these proteins in the amyloid deposits has not been determined. In about 3 percent of cases, amyloid is apparently localized to one tissue or organ. In all our cases of localized amyloidosis, abnormal collections of plasma cells were found in immediate juxtaposition to the amyloid deposits, and monoclonal immunoglobulins were also demonstrated in all three cases.

The diagnosis of amyloidosis depends on biopsy documentation. If possible, an accessible site with apparent involvement should be biopsied (e.g., skin, muscle). Tongue biopsy should be avoided because of pain and the risk of infection. Liver and kidney biopsy are possibly associated with a risk of bleeding, but this has not been a significant problem in most series. Because amyloidosis of both the pattern I and pattern II distribution frequently involves the arterioles of the rectal submucosa, rectal biopsy is a relatively safe and useful diagnostic procedure. All biopsy material should be stained with Congo red and examined by polarization microscopy. This technique not infrequently discloses amyloid deposits not revealed by the usual staining procedures.

In all cases of amyloidosis, detailed examination of the serum and urinary proteins by electrophoresis and immunoelectrophoresis is indicated, along with bone marrow studies, to document a plasma cell dyscrasia. In a case of amyloidosis with proteinuria, the finding of a negative Bence Jones reaction on heat testing does not exclude the presence of an abnormal constituent of the Bence Jones type, because albumin and other serum proteins in the urine may mask a Bence Jones reaction.

Treatment

There are no known methods to reverse the deposition of amyloid infiltrates except the identification and control of associated conditions, particularly osteomyelitis, pulmonary abscesses, and tuberculosis. Several cases have been reported in which effective antimicrobial therapy or surgery or both for chronic suppurative processes has resulted in cessation of amyloid deposition and reabsorption of existing infiltrates. Unfortunately, this is rare; in most cases, the amyloidosis continues despite apparently effective control of associated conditions.

Experience with melphalan therapy in cases of amyloidosis with overt myeloma or occult plasma cell dyscrasia is limited, but results have been generally discouraging. Most patients have continued to pursue a relentless downhill course, succumbing to cardiac decompensation or complications related to tongue and gastrointestinal involvement. Corticosteroids occasionally provide slight symptomatic benefit. In cases of macroglossia, careful attention to oral hygiene is important to avoid irritation and ulceration of the tongue.

Gamma Heavy-Chain Disease

Gamma heavy-chain disease is a relatively rare form of plasma cell dyscrasia characterized by the elaboration of excessive quantities of polypeptide related to the heavy chains and, more specifically, to the Fc fragment of IgG globulin. The clinical pattern resembles a lymphoma, with lymphadenopathy, splenomegaly, and hepatomegaly as the predominant manifestations associated with the nonspecific symptoms of weakness, fever, weight loss, and marked susceptibility to bacterial infections. Several patients have exhibited transient palatal erythema and edema resembling that of infectious mononucleosis, and in some there was transient spontaneous regression of the lymphadenopathy after an initially rapid onset. None have had clinical or roentgenographic evidence of skeletal destruction. Most have shown anemia, leukopenia, and thrombopenia—presumably owing to hypersplenism—and moderate to marked eosinophilia. Bone marrow aspirations and lymph node biopsies demonstrated proliferation of plasmacytic and lymphocytic forms, along with eosinophils and large reticulum or reticuloendothelial cells.

Identification of the abnormal protein as related to the γ heavy chain (Fc fragment) and unrelated to Bence Jones and light-chain polypeptides must be accomplished by immunochemical and physicochemical analyses.

Alpha Heavy-Chain Disease

IgA heavy-chain (α chain) disease is a recently identified plasma cell dyscrasia, previously referred to as *Mediterranean-type abdominal lymphoma*. The predominant features are a diffuse lymphomalike proliferation in the small intestine and mesentery, chronic diarrhea, and malabsorption unresponsive to gluten withdrawal. Most cases have been in either non-Ashkenazi Jews or Israeli Arabs, although it has also been described in a South American (Colombian) male of Spanish and Indian (mestizo) descent.

Chronic diarrhea, malabsorption, and progressive wasting are the major clinical features. Small intestinal biopsies demonstrate a profound infiltration of the lamina propria with abnormal plasma cells. Intestinal absorption studies demonstrate impaired absorption of vitamin B_{12}, glucose, lactose, and fat. Roentgenograms of the small intestines show thickened mucosal folds, segmentation, and dilated intestinal loops. Bone marrow aspiration reveals moderate increases in plasma cells, and skeletal roentgenograms show moderate diffuse osteoporosis but no destructive lesions.

On electrophoretic analysis of serum, the distinctive increase in IgA heavy chains (α chains) has been evidenced by a markedly elevated broad peak traversing the β and α_2 mobility range. The electrophoretic polydispersity has been shown to be due to polymeric heterogeneity of the α chains. This tendency to polymerize also explains the relatively low concentration of α chains in the urine as compared with the excretion of γ chains in the IgG heavy-chain disease. Confirmation of the identity of the abnormal serum protein as free α chains is accomplished by demonstrating the absence of associated light chains by appropriate immunologic and chemical analyses.

The association of this particular form of plasma cell dyscrasia with the small intestine is of particular interest in view of the known preponderance of IgA-producing plasma cells in the intestinal tract. The role of antecedent chronic infection or irritation of the intestines in the pathogenesis of IgA heavy-chain disease cannot be presently defined.

Plasma Cell Dyscrasia of Unknown Significance (PCDUS) and PCD Associated with Chronic Infections, Biliary Disease, and Nonreticular Neoplasms

Use of serum electrophoresis as a routine clinical laboratory procedure has resulted in the detection of M-type protein abnormalities in otherwise asymptomatic persons, some of whom develop signs and symptoms of myeloma, macroglobulinemia, or amyloidosis after many months or years. The term *plasma cell dyscrasia of unknown significance* is considered preferable to *premyeloma* or *essential or benign monoclonal gammopathy*.

In general, asymptomatic M-type protein abnormalities are more commonly observed in older subjects. In many cases, there is a background of tuberculosis, syphilis, or other chronic infection; chronic biliary tract disease; or a nonreticular neoplasm, particularly large-bowel, breast, oropharyngeal, and biliary tract carcinomas. Although a coincidental association cannot be excluded, particularly in older subjects, there is increasing evidence that certain plasma cell dyscrasias in man may be induced by diverse forms of protracted reticuloendothelial stimuli comparable to those described in mice. Because of the frequency of these associations, a careful search for an occult infection or neoplasm should be periodically carried out in all patients with asymptomatic M-type protein abnormalities, with particular attention to the bowel and biliary tracts.

To date, chemotherapy with melphalan or other agents has not been deemed warranted in asymptomatic subjects, despite the knowledge that a certain percentage of them will later develop overt myeloma. Although there are obvious theoretical advantages to earlier chemotherapy, the unpredictability of the course in any one case and the significant toxicities of presently available agents are strong arguments against the institution of chemotherapy before a distinct clinical pattern is evident.

Development of Monocytic Leukemia (Dyscrasia) after Long-Term Chemotherapy of Myeloma

In recent years, since the availability of melphalan and cyclophosphamide, monocytic leukemia with markedly elevated serum and urine lysozyme (muramidase) levels has developed as the terminal event in a small but significant number of cases of myeloma.[14] In our four cases, melphalan and/or cyclophosphamide had been given for 2 to 6 years with excellent remissions in all instances. Because of severe leukopenia and/or thrombocytopenia, melphalan or cyclophosphamide therapy had been discontinued for 2 to 10 months prior to emergence of the monocytic dyscrasia. In all cases, there was only minimal clinical and biochemical evidence of residual myeloma during the terminal monocytic dyscrasia. The onset of the monocytic leukemia was apparently abrupt in three of the four cases, with rapid transition from leukopenia to leukemic peripheral blood and bone marrow patterns.

The pathogenic mechanisms responsible for this complication of myeloma are obscure, but there are several possible contributory factors. Since the monocyte–histiocyte–macrophage system is functionally related to the plasma cell system in the processing of antigens and synthesis of antibodies, it is possible that both systems may be susceptible to the same neoplastic stimuli. Thus, the long-term chemotherapeutic suppression of the plasma cell myeloma in these cases may have simply permitted the expression of a corresponding neoplastic proliferation of the monocytic population. Alternatively, the alkylating therapy per se may have been responsible for the monocytic dyscrasia, since these agents are known to be potentially carcinogenic. Certainly, at present, the potential risk of inducing monocytic leukemia by melphalan or cyclophosphamide therapy of myeloma is far outweighed by the benefits derived from long-term treatment with these drugs.

In monocytic and monomyelocytic leukemia, serum and urinary lysozyme (LZM) levels are markedly elevated. As the disease progresses, increasing quantities of LZM are excreted in the urine, ranging up to 4 g/day. Assay of LZM levels in serum and urine is useful for both diagnosis and serial study of cases of monocytic and monomyelocytic leukemia.

References

1. Azar HA, Potter M: Multiple Myeloma and Related Disorders, vol 1. Hagerstown, Maryland, Harper and Row, 1973
2. Broder S, Humphrey R, Durm M, et al: Impaired synthesis of polyclonal (nonparaprotein) immunoglobulins by circulating lymphocytes from patients with multiple myeloma. New Engl J Med 293:887, 1975
3. Farhangi M, Osserman EF: The treatment of multiple myeloma. Seminars Hemat 10:149, 1973
4. Farhangi M, Osserman EF: Myeloma with xanthoderma due to an IgG$_\lambda$ monoclonal anti-flavin antibody. New Engl J Med, in press
5. Frangione B, Franklin EC: Heavy chain diseases: clinical features and molecular significance of the disordered immunoglobulin structure. Seminars Hemat 10:53, 1973
6. Franklin EC, Zucker-Franklin D: Current concepts of amyloid. Dixon FJ, Kunkel HG (eds): Advances in Immunology, vol 15, New York, Academic Press, 1972, p. 149
7. Glenner GG, Terry WD, Isersky C: Amyloidosis: its nature and pathogenesis. Seminars Hemat 10:65, 1973
8. Isobe T, Osserman EF: Patterns of amyloidosis and their associations with plasma-cell dyscrasia, monoclonal immunoglobulins and Bence Jones proteins. New Engl J Med 290:473, 1974
9. Isobe T, Osserman EF: Pathologic conditions associated with plasma cell dyscrasia: a study of 806 cases. Ann NY Acad Sci 190:507, 1971
10. Isobe T, Osserman EF: Plasma cell dyscrasia associated with the production of incomplete IgG$_\lambda$ molecules, gamma heavy chains, and free lambda chains containing carbohydrate: description of the first case. Blood 43:505, 1974
11. MacKenzie, MR, Fudenberg HH: Macroglobulinemia: an analysis of 40 patients. Blood 39:874, 1972
12. Osserman EF: Multiple myeloma and related immunoglobulin-producing neoplasms. Proc. Myeloma Workshop, Geneva, Switzerland, April 1974. UICC Tech. Report Series, vol 13, 1974
13. Osserman EF, Takatsuki K, Farhangi M, Pick AI, Isobe T: Plasma cell dyscrasias: general considerations,

plasma cell myeloma, primary macroglobulinemia, amyloidosis, heavy chain diseases. Williams WJ et al (eds): Hematology, New York, McGraw-Hill, 1972, p 950
14. Osserman EF, Lawlor DP: Serum and urinary lysozyme (muramidase) in monocytic and mono-myelocytic leukemia. J Exp Med 124:921, 1966
15. Osserman EF, Takatsuki K, Talal N: The pathogenesis of "amyloidosis." Studies on the role of abnormal gamma globulins and gamma globulin fragments of the Bence Jones (λ-polypeptide) type in the pathogenesis of "primary" and "secondary amyloidosis," and the "amyloidosis" associated with plasma cell myeloma. Seminars Hemat 1:3, 1964
16. Seligmann M, Brouet JC: Antibody activity of human myeloma globulins. Seminars Hemat 10:163, 1973
17. Solomon A: Medical Progress. Bence Jones proteins and light chains of immunoglobulins: biochemical and clinical significance. New Engl J Med, in press
18. Waldenström J: Diagnosis and Treatment of Multiple Myeloma. New York, Grune and Stratton, 1970
19. Zawadski ZA, Edwards GA: Dysimmunoglobulinemia associated with hepatobiliary disorders. Am J Med 48:196, 1970

The Radiology of Plasma Cell Dyscrasias

Stanley S. Siegelman

Plasma cells are highly sophisticated members of the B cell lymphocyte defense system. The B cell team of lymphocytes is derived from bone marrow, and its chief function is to produce antibodies in response to invasion of the body by foreign antigens.[1,2] The plasma cell is an extremely active cell. Kinetic studies by Salmon have shown that each plasma cell contains 5 to 10 million immunoglobulin molecules and that an active plasma cell produces 5,000 to 80,000 immunoglobulin molecules per minute.[3] The B cell system achieves its mission when antibody molecules successfully immobilize and inactivate foreign materials which penetrate the mechanical defenses of the body. As may be seen in Figure 1, not every vigorous proliferation of B cells constitutes a plasma cell dyscrasia. Increased proliferative activity by B cells may be a salutary and appropriate phenomenon, particularly in the setting of an inflammatory reaction in association with penetration of the mucosal barriers of the respiratory and gastrointestinal tracts. As applied by Osserman, the term "plasma cell dyscrasia" refers to a group of disorders in which there is an inappropriate, uncontrolled proliferation of plasma cells in the absence of a recognizable antigenic stimulus.[4,5]

Monoclonal Disorder

The plasma cell is a dedicated cell. Each of the vast multitude of immunoglobulin molecules produced by a single plasma cell is identical, with a characteristic pair of heavy chains and light chains. Each of the members of a group of plasma cells produced from a common B cell precursor will also produce an identical molecule. Such a collection of functionally identical cells derived from a single common ancestor is called a clone. In a sense a specific antibody molecule is the signature of the clone of cells. Plasma cell dyscrasias are monoclonal disorders. The existence of a neoplastic clone of B cells (i.e., a plasma cell dyscrasia) can be detected by identification of the homogeneous immunoglobulin secretory products of the clone in the serum or the urine. A large population of cells is required before a distinct secretory product can be isolated and identified. It has been estimated that a population of at least 2×10^{10} cells is required for the production of a detectable immunoglobulin or immunoglobulin product in the serum or urine.[3] The structure of the secretory product of the aberrant clone of cells may provide a clue to the likely nature of the parent cell (Fig. 1). When a monoclonal spike of IgG is present in the serum, a clone of mature plasma cells is generally responsible. A clone of plasmocytoid lymphocytes or more primitive B cells is generally responsible for the production of IgM. In some B cell proliferative disorders, the cells are more primitive lymphocytes and only a fragment of an immunoglobulin molecule is produced. There are many analogies to be drawn among the very specific disorders which comprise the plasma cell dyscrasias. In each instance, a large homogenous population of cells is distributed throughout the marrow, lymph nodes, and reticuloendothelial system. The disability engendered by the specific plasma cell dyscrasia is always a combination of the mechanical burden produced by virtue of a considerable physical bulk of the clone plus functional disorders related to the pathophysiological consequences of the proteinaceous secretory products of the B cells.

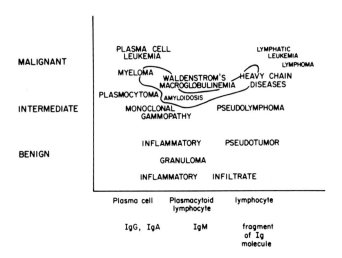

FIG. 1. Interrelationships of B cell proliferative processes.

Multiple Myeloma—The Mechanical Burden

In multiple myeloma, the radiologic manifestations are directly related to the accumulation within the bone marrow of a sizable population of tumor cells. The mass of myeloma cells displaces and erodes the internal bony architecture of the skeleton. Salmon's studies of tumor kinetics (already cited), have revealed that myeloma is rarely diagnosed before the clone of abnormal plasma cells reaches a population of 2×10^{11} cells. The most common early radiologic manifestation of myeloma is diffuse skeletal demineralization which is usually a sign that 1×10^{12} cells (or 1 kg of tumor) are present. Most patients with myeloma eventually manifest diffuse osteoporosis, lytic lesions, and fractures,[6] an indication that the clone of cells has reached a population of 2×10^{12} cells (Fig. 2). With 7.5×10^{12} cells, death generally ensues.[3]

Occasionally (less than 5 percent of the time), a solitary plasmocytoma is produced when the proliferation of plasma cells is confined to a focal area in the skeleton. Such plasmocytomas are generally lytic, slightly expansile lesions (Fig. 3) containing sheets of plasma cells. Although bone marrow aspiration from other sites may show no increase in plasma cells and the mass of cells present may be insufficient to produce a detectable immunoglobulin spike in the serum or urine, all such patients eventually will develop disseminated myeloma.

Myeloma is almost invariably manifest initially in the red-marrow containing areas of the skeleton: the dorsal spine, lumbar spine, pelvis, proximal femur, proximal humerus, and the ribs.[7] As the myeloma cells proliferate and displace the normal marrow, the yellow marrow in the peripheral skeleton becomes converted to red marrow. Myeloma cells gradually extend to the peripheral skeleton as well, and lytic lesions of the tibia or the radius are not unusual when the femur and the humerus have been involved for a prolonged period.

In the spectrum of plasma cell dyscrasias, myeloma appears to be the disorder with the largest cell population and hence, is most often associated with disability from the direct mechanical effects of the mass of tumor cells. Myeloma is by no means unique in this respect, however, and masses of plasmocytoid lymphocytes may create lytic skeletal lesions in Waldenstrom's macroglobulinemia.[8] Malfunction from mechanical interference by huge populations of plasma cells may also occur outside the skeleton, as witnessed by the occasional occurrence of malabsorption due to diffuse intestinal infiltration with plasma cells.[9]

Amyloidosis

The disability produced by plasma cell dyscrasias is also attributable to the pathophysiologic consequences of release into the body fluids of large quantities of the homogeneous, secretory products of plasma cells. Amyloidosis is an excellent example of this concept. Amyloidosis is a complex systemic disorder in which the hallmark is a widespread distribution of a perivascular, extracellular, eosinophilic, fibrillar proteinaceous material. Study of these deposits in primary amyloidosis has disclosed that a series of identical, regularly arranged polypeptide chains is an essential component of amyloid fibrils.[10] In a number of patients, the sequence of regularly arranged amino acids in the polypeptide chains of amyloid protein deposits has been found to correspond exactly to the specific structure of the Bence Jones proteins in the affected patient's urine. In these instances, the signature of the monoclonal plasma cell dyscrasia, the specific immunoglobulin fragment, appears not only in the serum and the urine but in the amyloid deposits as well. It is, therefore, an attractive hypothesis to consider that primary amyloidosis is attributable to a clone of plasma cells producing light chains.[11] Osserman has attempted to verify this concept with clinical data. In a survey of a series of cases of amyloidosis, monoclonal immunoglobulin or Bence Jones proteins were demon-

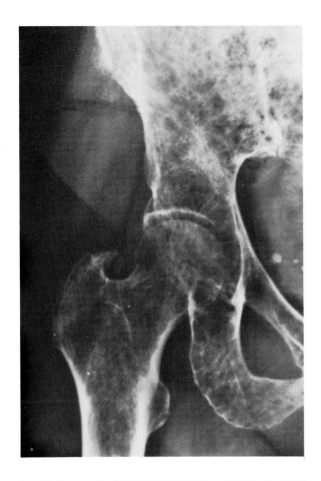
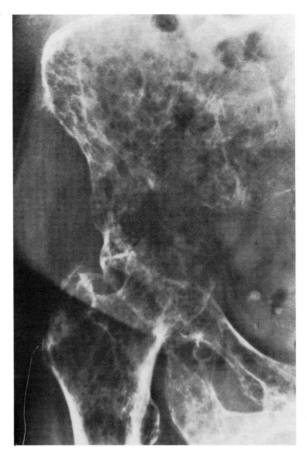
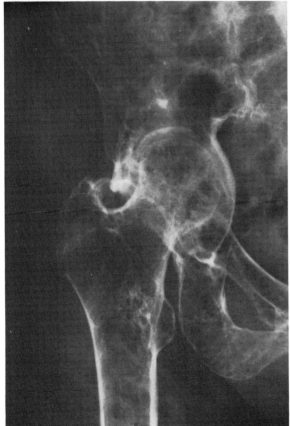

FIG. 2. Progression of plasma cell dyscrasia. The patient has plasma cell dyscrasia with a clinical diagnosis of multiple myeloma catheterized by a monoclonal spike of IgG. Top left: 1/75. There are scattered lytic lesions throughout the right iliac bone and right ischium. Top right: 11/75. The displacement of the bony trabeculae by the proliferating plasma cells has weakened the bone structure and the femoral head has been thrust through the acetabulum. Opposite: 4/77. The patient has been treated with chemotherapy with reduction in the number of plasma cells in the malignant clone. The skeleton has restored itself with a reparative process.

strated in all 50 patients with primary amyloidosis.[12] The deposition of amyloid within mesenchymal and parenchymal structures produces a wide range of abnormalities including cardiac enlargement, hepatosplenomegaly, and submucosal thickening of the small intestine.[13]

The proteinaceous fibrillar material derived from the accumulation of altered light chains may be deposited in and around the joints in synovial membranes, joint capsules, bursae, tendons, and muscles. The amyloid deposition produces joint tenderness and periarticular thickening and, hence, the disability may clinically resemble rheumatoid

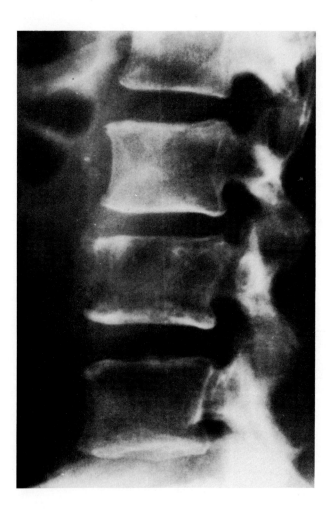

FIG. 3. Plasmocytoma. A middle-aged male presenting with back pain has a solitary lytic lesion on the centrum of L3. There was no evidence of increased plasma cells in the sternal bone marrow. Biopsy showed sheets of plasma cells. This is the initial presentation of a plasma cell dyscrasia. The patient developed disseminated bone lesions after a relatively asymptomatic interval of 6 years.

arthritis.[14] The patients exhibit morning stiffness, easy fatigability, and limitation of articular motion, but generally to a much less marked degree than that characteristic of rheumatoid arthritis.[15]

The thickening of tendons and ligamentous structures predisposes to the development of carpal tunnel syndrome, a condition which is often a feature of rheumatoid arthritis as well. One-third of all patients with amyloid arthropathy will have a concurrent or antecedent bilaterally symmetrical carpal tunnel syndrome.[17] Unlike rheumatoid arthritis, the larger joints (shoulder, hip, and knee) are predilected, but involvement of wrists, metacarpal-phalangeal, and interphalangeal joints can be striking.[15,16] The absence of inflammatory changes on examination of joint fluid is one of the keys to the distinction from rheumatoid arthritis. The sparse, yellow, viscous joint fluid often contains masses of free floating amyloid and a low cell count (500/mm^3 or less) of predominantly mononuclear cells. Electrophoresis of the joint fluid will frequently disclose a monoclonal immunoglobulin fracture, usually a kappa light chain.[15] Biopsy of the synovium will not reveal the inflammatory infiltrate, villous hypertrophy, and pannus typical of rheumatoid arthritis. Instead the articular and periarticular tissues will be found to contain an amyloid infiltration in synovial membranes, joint capsule, and periarticular cartilages. A confusing feature is the very frequent occurrence of subperiosteal amyloid collections which superficially resemble the subcutaneous nodules of rheumatoid arthritis. The most characteristic location is the ulna near the olecranon.[16] Since these amyloid nodules are fixed by a periosteal attachment, they are not as freely movable as rheumatoid nodules. The classic clinical appearance of amyloid arthritis is that of a man in his mid-50s with a prior history of carpal tunnel syndrome who presents with bilaterally symmetrical joint tenderness of the shoulders, hips, wrists, and hands and shows evidence of periarticular swelling, joint effusion, and flexion contractions. The radiologic findings may be limited to effusion and periarticular soft tissue swelling of the affected joints (Fig. 4). Amyloid infiltration of the synovium may produce cystic deposits typically found in the humeral and femoral heads (Fig. 5) or juxtaarticular osteolytic lesions with a punched-out appearance similar to gout. Neuropathic joint disease, mediated by a neuropathy attributable to amyloid infiltration of peripheral nerves, has also been reported.[18]

Amyloidosis may also be manifested radiologically by lytic lesions of the skeleton indistinguishable from myeloma (Fig. 6). The skeletal defects are produced by focal accumulations of amyloid in the marrow space. In such cases, amyloidosis is invariably associated with an overt plasma cell dyscrasia with a diffuse and marked increase of plasma cells in the bone marrow and monoclonal light chains in the urine. The patient frequently has other lytic lesions which are typical of plasmocytomas. It is customary to describe such patients as cases of multiple myeloma complicated by amyloidosis. In another sense, the disorder can be viewed as a primary plasma cell dyscrasia with manifestations of myeloma and amyloidosis. Such patients do not have two "diseases" but rather two common manifestations of a single basic disorder: an uncontrolled proliferation of a clone of plasma cells (Fig. 7).

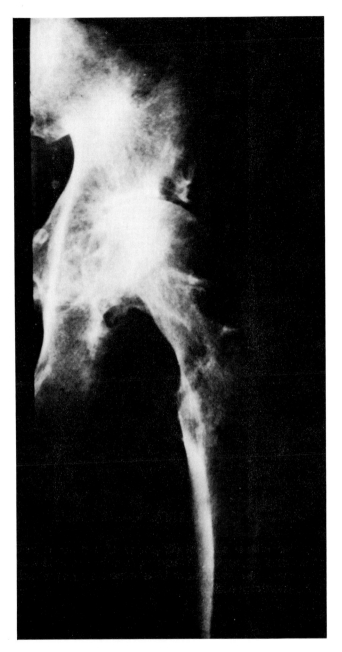

FIG. 4. Amyloidosis involving the hip. The infiltration by amyloid of the cartilage and synovial structures in and around the hip has resulted in juxtaarticular cystic erosions.

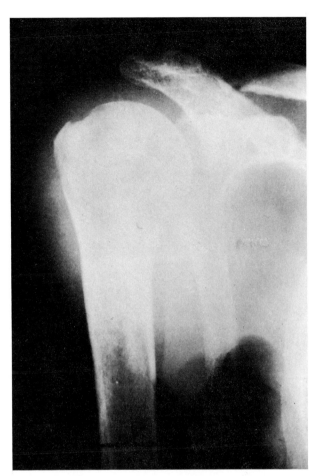

FIG. 5. Amyloid of the shoulder. There is diffuse soft-tissue swelling about the shoulder due to amyloid infiltration of the synovium. A single lytic defect is present in the lateral aspect of the humeral head.

Fanconi's Syndrome

Plasma cell dyscrasias may be responsible for the development of a proximal renal tubular dysfunction state with loss of sugar, phosphate, and amino acids in the urine—the Fanconi syndrome.[18] The entity is not commonly recognized, but it is an important example of the principle that the secretory products of plasma cells may be the prime cause of disability in plasma cell dyscrasias (Table 1). Plasma cells produce light chains and heavy chains which become incorporated into immunoglobulin molecules. Usually synthesis of light chains is excessive. Light chains are rapidly cleared from the serum glomerular filtration. Most of the filtered light chains are resorbed and degraded by the cells of the proximal and renal tubules—an important site of light chain catabolism. The nonabsorbed light chains are excreted in the urine.[19] Excessive light chain catabolism probably is nephrotoxic.[20] The accumulation of light chains may damage enzyme systems in the proximal renal tubular cells resulting in loss of sugar, amino acids, and phosphate in the urine. With chronic hypophosphatemia, osteomalacia develops. Maldonado et al have presented a series of 17 cases of Fanconi's syndrome associated with Bence Jones proteinuria; 7 of the patients had evidence of pseudofractures.[18]

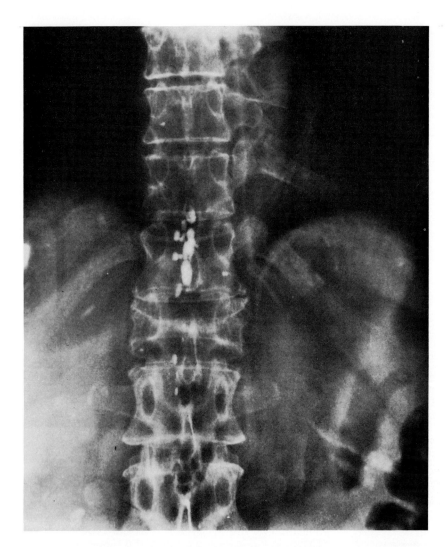 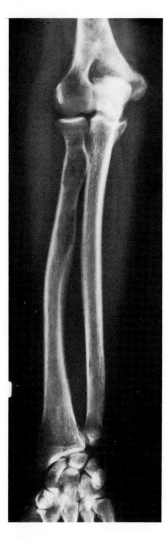

FIG. 6. Amyloidosis with lytic lesions of bone. This is a patient with a clinical diagnosis of a plasma cell dyscrasia manifested by a spike of IgG in the serum and in the urine. A lytic lesion leading to collapse of L1 was present. There is also a lytic lesion in the proximal radius. Biopsy of the lesion from L1 revealed only amyloid. Left: AP spine showing collapse of L1. Right: Lytic lesion in the proximal radius.

Systemic Light Chain Deposition

Light chains are chemically altered during the conversion to amyloid. There is one report, however, of the parenchymal deposition of unaltered light chains which produced hepatosplenomegaly.[19]

Hyperviscosity Syndrome

The viscosity of a protein solution such as blood serum is a function of both the molecular weight and the concentration of the proteins.[21] Since IgM has such a high molecular weight (close to 1 million), increased serum IgM produces increased serum viscosity. Normally the viscosity of serum is 1.5 times the viscosity of water. When the serum viscosity reaches 8 times that of water, symptoms develop. IgG can also be responsible for hyperviscosity syndrome but only when the concentration exceeds 10 g per 100 ml.[22] Hyperviscosity syndrome is manifested by neurological symptoms (vertigo, ataxia, lethargy, and peripheral neuropathy), visual disturbances, bleeding from the mucous membranes, and congestive heart failure. Hyperviscosity syndrome is classically seen as a manifestation of Waldenstrom's macroglobulinemia, a plasma cell dyscrasia with a monoclonal spike of IgM in the serum. It is a chronic lymphoma-like illness with hepatosplenomegaly and lymphadenopathy.[23]

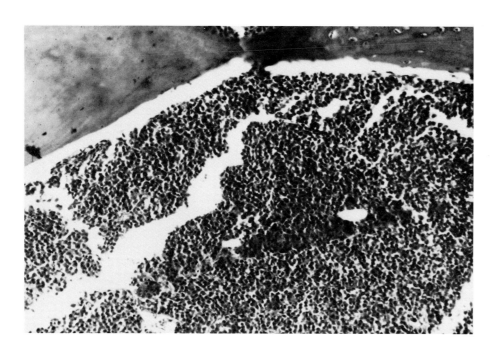

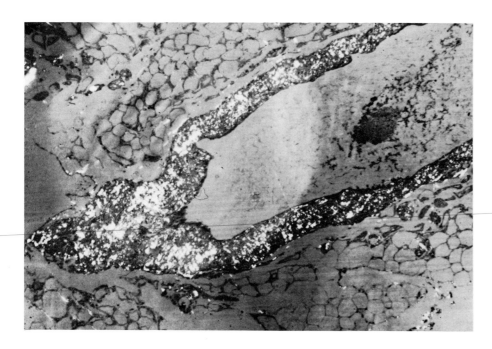

FIG. 7. Interrelationships of the plasma cell dyscrasias. A patient with a five-year history of clinically diagnosed multiple myeloma with a lytic lesion in the cervical spine eventually developed renal failure and chronic distention of the gastrointestinal tract. At autopsy there was evidence of both myeloma and amyloidosis. Top: Bone marrow showed sheets of plasma cells. Bottom: Mesenteric blood vessel exhibits birefringence due to infiltration with amyloid.

Table 1 Products of Plasma Cells. Functional Abnormalities and Radiological Findings

Substance produced	Consequences	Disorder	X-ray findings
I. Light chains MW 22,000	Light chains converted to amyloid. Amyloid deposited in vascular and perivascular structures.	Amyloidosis	(A) Periarticular soft tissue swelling (B) Cystic erosions due to synovial hypertrophy (C) Lytic lesions long bones
	Light chains filtered by glomerulus, enter cells in proximal renal tubules, damage cells	Fanconi syndrome	Osteomalacia
	Retention and tissue deposition of light chains	Systemic light chain deposition	Hepatosplenomegaly
II. Immunoglobulin molecules MW 150,000-900,000	Presence of large numbers of molecules increases viscosity of serum	Hyperviscosity syndrome	Congestive heart failure ischemic injury to bone, lung, bowel
	Large molecules precipitate when temperature falls	Raynaud's syndrome	
III. Osteoclast stimulating factor MW 18,000	Osteoclastic activity stimulated	Hypercalcemia	Focal and diffuse lytic bone lesions
IV. Erythropoiesis inhibitory factor	Plasma cell elaborate a polypeptide which inhibits erythropoiesis	Refractory hypochromic microcytic anemia	Lymphadenopathy

Other Biologic Properties of Plasma Cell Products

We have indicated two other examples in Table 1 of pathologic phenomena directly related to the activity of substances produced by plasma cells. Supernatant fluids from bone marrow cultures of patients with myeloma contain a factor which stimulates osteoclastic bone resorption in organ culture.[24] This osteoclast stimulating factor or osteoclast activating factor appears to be an important mediator of bone resorption and the hypercalcemia which frequently accompanies myeloma. Patients with giant lymph node hyperplasia of the plasma cell type, a disorder characterized by enlarged lymph nodes in the thorax and abdomen, may present with refractory hypochromic microcytic anemia, hypergammaglobulinemia, and a diffuse increase in plasma cells in the bone marrow.[25] The enlarged lymph nodes also contain an abundant population of plasma cells. There is suggestive evidence that such patients produce a serum factor which inhibits erythropoeisis. In summary, therefore, we have seen that the clinical manifestations of plasma cell dyscrasias are attributable to organ infiltration by tumor cells and to direct physicochemical properties of the primary protein product of the clone of plasma cells.

References

1. Siegelman SS: Plasma cell dyscrasias. In Jacobson HG, Murray RO: The Radiology of Skeletal Disorders, 2nd edition. Churchill Livingstone, Edinburgh, London, and New York, 1977, pp 1823-1833
2. Siegelman SS: Radiological features of plasma cell dyscrasias. In Margulis and Gooding: Diagnostic Radiology. Univ of California Press, 1977, pp 455-466
3. Salmon SE: Immunoglobulin syntheses and tumor kinetics of multiple myeloma. Semin Hematol 10:135, 1973
4. Osserman ET: Plasma cell dyscrasias. In Beeson PB, McDermott W (eds): Textbook of Medicine, 13th edition. Philadelphia, Saunders, 1971
5. Isobe T, Osserman EF: Pathologic conditions associated with plasma cell dyscrasias: A study of 806 cases. Ann NY Acad Sci 190:507, 1971
6. Kyle RA: Multiple myeloma. Review of 869 cases. Mayo Clin Proc 50:29, 1975
7. Jacobson HG, Siegelman SS: Some miscellaneous solitary bone lesions. Semin Roentgenol 1:314, 1966
8. Vermess M, Pearson KD, Einstein AB, Fakey JF:

Osseous manifestations of Waldenstrom's macroglobulinemia. Radiology 102:497, 1972
9. Chantar C, et al: Diffuse plasma cell infiltration of the small intestine with malabsorption associated to IgA monoclonal gammopathy. Cancer 34:1620, 1974
10. Glenner GG, Terry WD, Isersky C: Amyloidosis: Its nature and pathogenesis. Semin Hematol 10:65, 1973
11. Glenner GG, Ein D, Terry WD: The immunoglobular origin of amyloid. Am J Med 52:141, 1972
12. Isobe T, Osserman EF: Patterns of amyloidosis and their association with plasma-cell dyscrasia, monoclonal immunoglobulins and Bence-Jones proteins. New Engl J Med 290:473, 1974
13. Pruzanski W, Warren RE, Goldie JH, Katz A: Malabsorption syndrome with infiltration of the intestinal wall by extracellular monoclonal macroglobulinemia. Am J Med 54:811, 1973
14. Gordon DA, Pruzanski W, Ogryzeo MA, Little HA: Amyloid arthritis simulating rheumatoid disease in five patients with multiple myeloma. Am J Med 53:142, 1973
15. Goldberg A, Brodsky I, McCarty D: Multiple myeloma with paramyloidosis presenting as rheumatoid disease. Am J Med 37:653, 1964
16. Wiernik PH: Amyloid joint disease. Medicine 51:465, 1972
17. Scott RB, et al: Neuropathic joint disease (Charcot joints) in Waldenstrom's macroglobulinemia with amyloidosis. Am J Med 54:535, 1973
18. Maldonado JE, et al: Fanconi Syndrome in adults. A manifestation of a latent form of myeloma. Am J Med 58:354, 1975
19. Randall RE, Williamson WC, Mullinax F, Yung MY, Still WJS: Manifestations of systematic light chain deposition. Am J Med 60:293, 1976
20. Clyne DH, et al: Renal effects of intraperitoneal kappa chain injection: Induction of crystals in renal tubular cells. Lab Invest 31:131, 1974
21. Fahey JL, Barth WF, Solomon A: Serum hyperviscosity syndrome. JAMA 192:120, 1965
22. MacKenzie MR, Fudenberg HF, O'Reilly RA: The hyperviscosity syndrome. I. In IgG myeloma. The role of protein concentration and molecular shape. J Clin Invest 49:15, 1970
23. Cohen RJ, Bohannon RA, Wallerstein R: Waldenstrom's macroglobulinemia. Am J Med 41:274, 1966
24. Mundy GR, et al: Evidence for the secretion of an osteoclast stimulating factor in myeloma. N Engl J Med 291:1041, 1974
25. Burgert EO, et al: Intra-abdominal, angiofollicular lymph node hyperplasia (plasma-cell variant) with an antierythropoietic factor. Mayo Clin Proc 50:542, 1975

Round Cell Lesions: Pathology and Newer Classifications

D. C. Dahlin

The three entities of Ewing's sarcoma, malignant lymphoma (reticulum cell sarcoma), and myeloma are considered in the usual discussion of round cell lesions of bone. Ewing's sarcoma[1,2] is a distinctive small round cell sarcoma that is the most lethal of all of the bone tumors (Fig. 1). The literature regarding this disease is controversial because of the somewhat nonspecific histologic characteristics of the tumor, which is composed of solidly packed small cells. Especially in former years, other entities, such as small cell osteosarcoma, most of the "reticulum cell" sarcomas, and even benign conditions such as eosinophilic granuloma, were sometimes classified with the Ewing's tumors. A practical approach is to regard as Ewing's sarcomas those highly anaplastic small round to oval cell sarcomas that have the clinical and radiologic characteristics of a primary osseous lesion. This concept provides for the exclusion of cytologically incompatible lesions such as myeloma, malignant lymphoma, and histiocytosis X. Production of a chondroid or osteoid matrix by the tumor cells, even if in small amounts, excludes the diagnosis of Ewing's sarcoma. In a small biopsy specimen, it may be very difficult or impossible to distinguish metastatic malignant tumors such as neuroblastoma, small cell carcinoma of the lung, or even leukemic infiltrate from a specimen of Ewing's sarcoma. Ancillary considerations such as the total clinical picture, including at times study of urinary catecholamines, clarifies the issue.

Some have proposed that electron microscopic study is necessary for the diagnosis of Ewing's sarcoma and its differentiation from some of these other diseases; such studies are neither necessary nor practical. Speculation regarding the possible origin of the cells that comprise Ewing's tumor has been fruitless, and it seems best to regard them as of undifferentiated mesenchymal origin.

The 299 Ewing's sarcomas in the Mayo Clinic's surgical experience up to January 1, 1976, occurred in a group of somewhat more than 6200 primary benign and malignant lesions of bone. There is a distinct predilection for male patients. More than 75 percent of the patients were less than 20 years of age. Although the bones of the extremities provided the majority of the Ewing's sarcomas, practically any bone of the body may be involved.

The usual symptomatology includes pain and swelling in the region of the tumor. Some patients with Ewing's sarcoma have an elevated temperature and increased sedimentation rate of erythrocytes.

Grossly these tumors are characteristically gray-white, moist, glistening, and somewhat translucent. Occasionally they are so soft as to be almost liquid in consistency. Generally, the gross characteristics mimic those of malignant lymphoma. Although metastases from Ewing's sarcoma are characteristically to the lungs, an unusually large number of them produce metastases to other bones as well as to other sites.

Histologically, even at low magnification, these tumors appear to be remarkably cellular and to have little intercellular stroma except where they are invading other tissue. The nuclei are noteworthy for their regularity in shape and size and are round to oval (Fig. 1). The cytoplasm surrounding these nuclei is slightly granular, and the cell outlines are indistinct. The nuclei contain a rather finely dispersed chromatin that imparts a "ground-glass" appearance. The

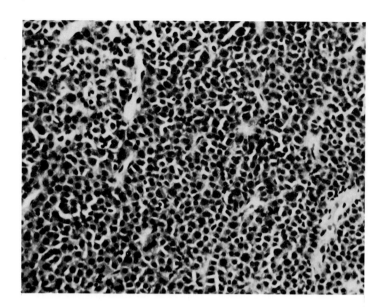

FIG. 3. Myeloma of ilium. Note that the cells contain considerable cytoplasm and have eccentric nuclei. H&E, ×250.

multiple myeloma. Patients with such lesions nearly always develop myeloma, but sometimes after an interval of 5 to 10 years after the diagnosis is made.

So-called plasma cell granuloma is considered by some to be a difficult differential diagnostic consideration. These inflammatory foci that contain a predominance of plasma cells are ordinarily distinguished readily from myeloma. The differentiation is based on the findings of scattered clusters of lymphocytes and polymorphonuclear leukocytes among the plasma cells of inflammatory disease. Additionally, proliferating capillaries and fibroblasts, present in inflammation, are never found in the nodules of neoplastic myeloma.

Grossly, myeloma is similar to Ewing's sarcoma or malignant lymphoma. The symptoms of myeloma are often related to systemic disease, but sometimes the osseous lesion or extensions of it produces local pain and swelling.

Typically, one sees sheets of closely packed cells with little intercellular substance. The cytoplasm of these cells is abundant and tends to be somewhat granular and basophilic; cytoplasmic outlines are distinct. The nucleus is characteristically round or oval and eccentric and may show the cartwheel arrangement of chromatin clumps. Occasionally two or even three nuclei are present within individual cells. In a series of myeloma cases, gradations are found to exist, apparently reflecting the maturity or degree of differentiation of the cells. At one end of the spectrum are tumors whose cells are so mature that they resemble those seen in inflammatory conditions, whereas at the other end of the spectrum, the cells are so very variable in size and shape that they simulate those of malignant lymphoma. Mitotic figures are unusual in the average myeloma. Amyloid is deposited in the tumor masses in approximately 10 percent of patients with myeloma. Occasionally this is in the form of small clumps of typically staining material, sometimes it thickens the walls of blood vessels, and large clumps of amyloid may be associated with a foreign body giant cell response. The presence of amyloid in a malignant tumor of bone is helpful evidence that the lesion is myeloma.

References

1. Macintosh DJ, Price CJG, Jeffree GM: Ewing's tumor. A study of behavior and treatment in forty-seven cases. J Bone Joint Surg 57B:331, 1975
2. Pritchard DJ, Dahlin DC, Dauphine RT, Taylor WF, Beabout JW: Ewing's sarcoma. A clinicopathological and statistical analysis of patients surviving five years or longer. J Bone Joint Surg 57A:10, 1975
3. Valls J, Muscolo D, Schajowicz F: Reticulum-cell sarcoma of bone. J Bone Joint Surg 34B:588, 1952
4. Dahlin DC: Is It Worthwhile to Differentiate Ewing's Sarcoma and Primary Lymphoma of Bone? Seventh National Cancer Conference Proceedings. Philadelphia, Lippincott, 1973
5. Boston HC Jr, Dahlin DC, Ivins JC, Cupps RE: Malignant lymphoma (so-called reticulum cell sarcoma) of bone. Cancer 34:1131, 1974
6. Kyle RA: Multiple myeloma. Review of 869 cases. Mayo Clin Proc 50:29, 1975

Round Cell Lesions—Radiology

Harold G. Jacobson

Round cell lesions include Ewing tumor, primary reticulum cell sarcoma of bone, and myelomatosis.

Ewing Tumor

Incidence and Clinical Features

Ewing tumor occurs primarily in the second and third decades of life, affecting males more than females (5:4). At least half the cases affect the spine, innominate bones, and major long bones (femora, tibiae, humeri). Clinical features include localized painful swelling. The child is usually quite ill on presentation.

Pathological Findings

A varied histologic pattern occurs. Fields of tumor cells of a uniform appearance in vacuolated cytoplasmic material are often observed. Cells are poorly delineated and contain round nuclei. When tissue undergoes degeneration, one may observe small polyhedral cells with clear-cut borders, containing small dark nuclei (pyknosis) with pale cytoplasm. Cells are frequently distributed around vessels (rosette formation). Necrosis, hemorrhage, and new vessel formation may be noted. New bone (nonmalignant) may form occasionally. It may resemble neuroblastoma metastases in infants and young children and reticulum cell sarcoma, myeloma, and anaplastic carcinoma metastases in older patients. Special staining for presence of glycogen granules (present in Ewing tumor) is helpful in differentiation.

Radiologic Features

Characteristically, there is a destructive lesion of bone with a permeating pattern and wide zone of transition. In the long bones, the diaphyses are principally affected. In flat bones, considerable reactive new bone formation may be present (Fig. 1). Characteristic periosteal reaction in the lesions of long bones is generally continuous, with Codman reactive triangles. Spiculated periosteal reaction is generally fine in contrast with that associated with osteosarcoma. A soft tissue mass is often a prominent feature.

General Comments

The prognosis is generally poor. Skeletal and pulmonary metastases occur frequently and often early. Diagnosis of Ewing tumor in patients under 5 years of age and over 30 years of age should be generally suspect. In patients under 5, the diagnosis is probably neuroblastoma, and in patients over 30, it is generally primary reticulum cell sarcoma of bone.

Primary Reticulum Cell Sarcoma of Bone

Incidence and Clinical Features

Primary reticulum cell sarcoma of bone occurs principally in the fourth, fifth, and sixth decades of life, affecting males twice as often as females. Approximately 35 percent of lesions occur around the knee—the proximal end of the tibia and the distal

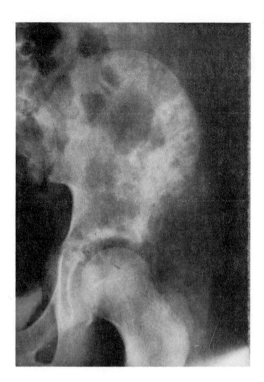 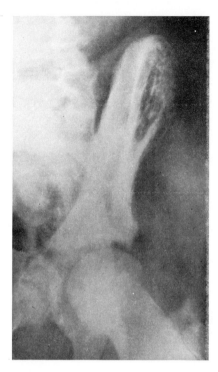

FIG. 1. Ewing tumor—iliac bone, showing evidence of reactive new bone formation.

end of the femur. Other long bones may be affected. Lesions in flat bones, particularly the innominate bones and spine, are not uncommon. Clinical features include localized pain and swelling. A pathologic fracture may be the cause of presentation.

Pathologic Findings

The characteristic histologic appearance shows sheets of round cells, each containing a fairly large, single, spheroid nucleus. Nucleoli may be present, and mitotic figures are common. Necrosis, secondary calcification, and reactive new bone formation are not uncommon. Demonstration of reticulum fibrils by special stains may be diagnostic.

Radiologic Features

Primary reticulum cell sarcoma may closely simulate Ewing tumor in both long and flat bones, although metaphyseal predilection in long bones is much more common than in Ewing tumor. Periosteal reaction and soft tissue masses are less extensive than in Ewing tumor. Lesions of flat bones may also show reactive new bone. Aggressive lytic lesions are not uncommon in long bones (Fig. 2). Reactive new bone formation may also occur in long bones even more frequently than in Ewing tumor.

General Comments

Patients with primary reticulum cell sarcoma of bone fare much better than those with Ewing tumor and also better than patients with the lymphoma type of reticulum cell sarcoma when bone is involved. Histologically, the differentiation between Ewing tumor and primary reticulum cell sarcoma of bone may be very difficult.

Myelomatosis (Including Plasmacytoma)

Incidence and Clinical Features

Myelomatosis affects males and females almost equally, usually in the sixth and seventh decades of life. Back pain is the most common presenting symptom. The patient may also experience weakness, loss of weight, and unexplained anemia. Pathologic fractures are common presenting manifestations. Collapse of a vertebral body with extradural block precipitating paraplegia may be the initial clinical finding. Laboratory findings are fairly specific, with a

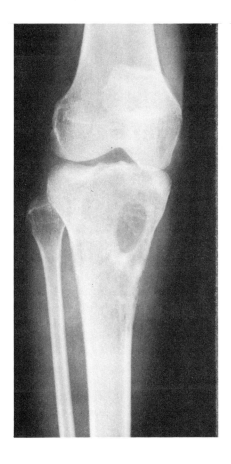
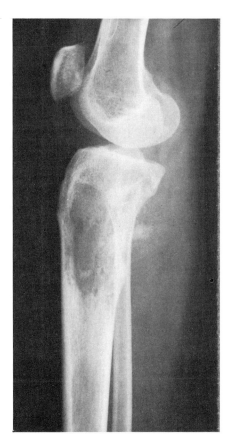

FIG. 2. Primary reticulum cell sarcoma—tibia, demonstrating lytic lesion.

characteristic M spot in the electrophoretic pattern of the serum. Hypercalcemia is common; Bence Jones proteinuria occurs in approximately 60 percent of patients. Usually increase in IgG is noted.

Pathologic Features

Myeloma tissue is highly cellular with a fairly uniform pattern of sheets of spheroid cells, usually containing eccentric nuclei. Tumor giant cells may be present. The condition may closely mimic neuroblastoma, Ewing tumor, anaplastic metastatic carcinoma, and anaplastic lymphoma. Reactive new bone formation is uncommon but may occur. Positive diagnosis can be established on finding plasma cells in excess of 10 percent in the marrow puncture of the sternum.

Radiologic Features

Lesions of plasmacytoma and multiple myeloma must be considered when the following characteristics are noted: *Plasmacytoma* is a lucent, grossly expanding lesion of the affected bone. It is most commonly noted in the spine and innominate bones, although major long bones are also affected (Fig. 3). Pathologic fractures are common. A soft tissue mass is frequently noted. Radiologic findings simulate metastatic deposits from renal or thyroid carcinoma, or even an aneurysmal bone cyst.

Radiologically, in multiple myeloma the skeleton may be completely normal. However, other findings may include:

1. Generalized and severe osteopenia without destructive lesions.
2. Discrete but disseminated lytic areas with widespread osteoporosis (Fig. 4).
3. Permeating mottled pattern of bone destruction, simulating other round cell malignancies.
4. Sclerotic form of myelomatosis. Although rare, this is being observed with increasing frequency. Sclerosis may be disseminated, simulating osteoblastic metastasis and myeloid metaplasia.

General Comments

The patient may present with primary amyloid disease followed by the development of myelomatosis. Approximately 15 percent of the patients

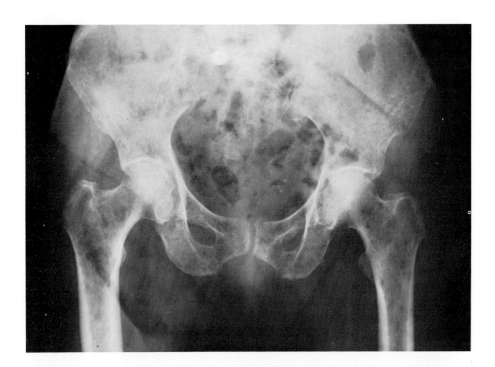

FIG. 3. Myelomatosis—pelvis (plasmacytoma).

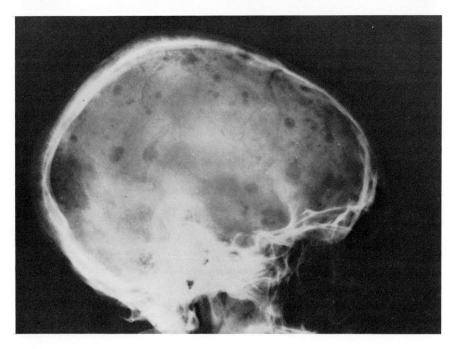

FIG. 4. Myelomatosis—skull (multiple myeloma). Note disseminated lytic areas with osteopenia.

with myelomatosis will develop secondary amyloidosis. There is also a relationship between Waldenström's macroglobulinemia and multiple myeloma, certain similar radiologic features may be noted.

Bibliography

Ackerman LV, Spjut HJ: Fascicle 4, Tumours of Bone and Cartilage, pp. 270-271. Washington, D.C., Armed Forces Institute of Pathology, 1962

Bhansall SK, Desai PB: Ewing's sarcoma: observation on 107 cases. J Bone Joint Surg 45A:541, 1963

Brown TS, Paterson CR: Osteosclerosis in myeloma. J Bone Joint Surg 55B:621, 1973

Bundens WD, Brighton CT: Malignant hemangioendothelioma of bone. J Bone Joint Surg 47A:762, 1965

Coley BL, Higinbotham NL, Groesbeck HP: Primary reticulum cell sarcoma of bone. Summary of 37 cases. Radiology 55:641, 1950

Dahlin DC, Coventry MB, Scanlon PW: Ewing's sarcoma: a critical analysis of 165 cases. J Bone Joint Surg 43A:185, 1961

Ewing J: Diffuse endothelioma of bone. Proc NY Pathol Soc 21:17, 1921

Griffiths DL: Orthopaedic aspects of myelomatosis. J Bone Joint Surg 48B:703, 1966

Heiser J, Schwartzman JJ: Variations in the roentgen appearance of the skeletal system in myeloma. Radiology 58:178, 1952

Ivins JC, Dahlin DC: Reticulum cell sarcoma of bone. J Bone Joint Surg 35A:835, 1953

Jacobson HG, et al: The vertebral pedicle sign. A roentgen finding to differentiate metastatic carcinoma from multiple myeloma. Am J Roentgenol 80:817, 1958

Sherman RS, Soong KY: Ewing's sarcoma: its roentgen classification and diagnosis. Radiology 66:529, 1956

Sherman RS, Snyder RE: The roentgen appearance of primary reticulum cell sarcoma of bone. Am J Roentgenol 58:291, 1947

Wilson TW, Pugh DG: Primary reticulum cell sarcoma of bone, with emphasis on roentgen aspects. Radiology 65:343, 1955

Osteosarcoma and Chondrosarcoma: Newer Variants and Pathologic Subclasses

D.C. Dahlin

Osteosarcoma of bone is defined as a malignant primary tumor of bone the malignant cells of which produce osteoid substance (Figs. 1,2). Although this material is relatively specific in appearance and most osteosarcomas are readily classified as such, sometimes the hyalinized matrix in a fibrosarcoma of bone is similar to the osteoid in fibroblastic osteosarcoma. Occasionally, it is difficult to differentiate the matrix substance produced in a chondrosarcoma from the osteoid present in a chondroblastic osteosarcoma. Although approximately 75 percent of osteosarcomas are type "ordinaire," it is important to recognize that there are several small groups of osteosarcoma that are different clinically and biologically from the ordinary type.[1] Recognition of these subtypes will be the main subject of this presentation. Table 1 lists the varieties of osteosarcoma described.

Table 1   Varieties of Osteosarcoma

Type of osteosarcoma	Number
Conventional	753
Others	268
In jawbones	66
In Paget's disease	30
In other benign conditions	8
Postradiation	35
Dedifferentiated chondrosarcoma	23
Multicentric	3
(?) Malignant fibrous histiocytoma	18
Telangiectatic	25
Low-grade intraosseous	10
Periosteal	14
Parosteal	36
Total	1021

Osteosarcomas of the jaws are different from standard osteosarcomas in three major ways. Patients with osteosarcomas of the jaws are older by from one to two decades than are patients with ordinary osteosarcomas. They are, on the average, lower in grade than are ordinary osteosarcomas. The usual osteosarcomas are grade 3 or 4 (high-grade Broders); osteosarcomas of the jaws are commonly grades 2 and 3. In fact, their low-gradeness sometimes causes the first biopsy specimen to be misinterpreted as benign. The third major difference from osteosarcomas in other sites is that those in the jaws are predominantly chondroblastic in at least 50 percent of cases. Many of these in the oral pathology literature are called *chondrosarcomas*, but the capability of those called *chondrosarcomas* to recur, metastasize, and kill is similar to that of those called *osteosarcomas of the jaws*. The consequence of these three characteristic differences is that osteosarcomas of the jaws are less apt to metastasize, and the chance of 5-year survival is approximately double that of osteosarcomas in conventional sites.

Osteosarcoma in Paget's disease of bone affected 30 patients in our series. The bones of predilection for osteosarcoma in Paget's disease are somewhat different from those that engender ordinary osteosarcoma. The patients are usually old, because Paget's disease does not begin in young people. The most compelling reason for recognizing this type of osteosarcoma as being different is that it is more lethal than is ordinary osteosarcoma. Patients with sarcoma in Paget's disease rarely are cured.

A few osteosarcomas arise in some benign condition such as benign osteoblastoma or fibrous

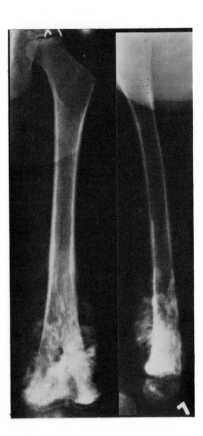

FIG. 1. Osteosarcoma in commonest location, the distal femoral shaft. This tumor in a 4-year-old has produced areas of lysis and sclerosis in the bone and has penetrated through the cortex to produce a "sunburst."

dysplasia. Although such tumors are exceedingly rare and comprised only eight instances in our experience, it is quite likely that they are different from ordinary osteosarcoma. They are readily recognized as different from the usual osteosarcoma.

Postradiation osteosarcoma provided 35 tumors in our group. In addition, the files of the Mayo Clinic contained 35 other malignant tumors, mainly fibrosarcomas, in radiated bone. The sarcomas that begin in radiated bone are often in sites that are difficult to treat adequately. The patients are usually middle aged or old, because they have developed the disease after radiation for something that began in adulthood. A most important consideration is that patients who develop their osteosarcomas after radiation for malignant tumors such as those of the uterus and breast are at risk because of two different malignant conditions. It seems only logical that patients whose osteosarcoma begins in radiated bone be sharply distinguished from the group of ordinary osteosarcomas.

The next group of cases, 23 in number, consists of osteosarcomas that developed in the course of chondrosarcomas. The dedifferentiated chondrosarcomas are different from ordinary osteosarcomas for several reasons. The bones of predilection are those of chondrosarcoma not osteosarcoma. Nearly all of the patients are middle aged or old, because they were old enough to have a chondrosarcoma before this malignant tumor developed by dedifferentiation. Patients who develop dedifferentiation of chondrosarcomas usually die of the disease, the mortality rate for the first 5 years being in the 80 to 90 percent range. Of 51 chondrosarcomas in our series to January 1, 1976, that showed dedifferentiation, 23 became osteosarcomas and the remainder became fibrosarcomas. The biologic problem for the patient whose chondrosarcoma has dedifferentiated is that of the highly malignant neoplasm that has supervened.

Multicentric osteosarcoma found at the onset of disease provided only three patients in our total experience. Patients who have sarcomas apparently arising in several bones simultaneously are distinctly different therapeutic problems than are those with the average osteosarcoma.

Malignant fibrous histiocytoma-like histology or features suggesting that that diagnosis may be valid are seen at least focally in a minority of osteosarcomas. Occasionally, this pattern dominates the histologic picture, but the tumor is recognizable as osteosarcoma, because there is osteoid produced by the malignant cells in regions of the tumor. It is not known at this time whether a prominent malignant fibrous histiocytoma-like appearance in an osteosarcoma imparts a special biologic capability. There is some suggestive evidence that such tumors may be more radiosensitive and more prone to metastasize to regional nodes than is the average osteosarcoma. It seems reasonable that a statistician should "red flag" osteosarcomas that show the pattern strongly suggestive of malignant fibrous histiocytoma so that we may possibly know in the future whether this histologic variant of osteosarcoma has clinically important differences from those of ordinary osteosarcoma.

Telangiectatic osteosarcoma had been called *hemorrhagic osteosarcoma* by Campanacci. We define this type of osteosarcoma as one that is lytic throughout; its radiographic appearance simulates that of Ewing's sarcoma at times. Characteristic tumors are bloody and necrotic, and viable tissue is often present only near the walls of the cavity they produce. The tumors may contain only small, even questionable, amounts of recognizable osteoid. The main problem in

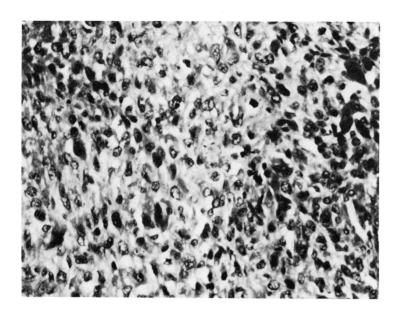

FIG. 2. Osteosarcoma with foci of spindling malignant cells and rounder cells associated with homogeneous osteoid. H&E, x250.

differential diagnosis by the histopathologist is aneurysmal bone cyst. The critical difference is that the mononuclear cells of telangiectatic osteosarcoma show hyperchromatism and nuclear irregularities that prove its malignant quality. Our experience with 25 "telangiectatic" osteosarcomas indicates that this tumor is probably worse than ordinary osteosarcoma. Twenty-three of the 25 patients have died as the result of their tumors. The twenty-fourth patient has had three operations for metastatic disease after the initial amputation. The twenty-fifth patient had probable metastasis to the lungs more than 5 years after amputation for the primary tumor.

Low-grade intraosseous osteosarcoma provided only ten of the total osteosarcomas at the Mayo Clinic. This type deserves separate categorization, because it is usually grade 1, is very difficult to diagnose, and has often been mistaken for fibrous dysplasia. Ordinarily, the roentgenogram indicates that this is a locally aggressive disease, and this finding aids the histopathologist. Even though it fulfills the criteria for the diagnosis of osteosarcoma in that malignant cells produce osteoid, its low gradeness correlates with a slow clinical evolution. We have seen many examples of 5-year survival for patients with this tumor even when they have been treated erroneously primarily.

Periosteal osteosarcoma comprised only 14 of our total cases of osteosarcomas. This is an extremely distinctive type of lesion. It characteristically occurs on the surface of the tibial shaft, although the tumor may occur on other bones. It typically consists of fusing lobules of heavily chondroid material. These tumors are usually small. The relatively radiolucent mass that they produce usually contains spicules of bone radiating from the underlying uninvolved cortex. Dr. Schajowicz[3] proposed that these tumors be called *periosteal chondrosarcomas* because of their prominent chondroblastic quality, but the matrix in parts of the centers of these lobules has the quality of osteoid, and we prefer to regard them as chondroblastic osteosarcomas. We have elected to regard arbitrarily any osteosarcoma that permeates deeply into and through the underlying cortex as an ordinary osteosarcoma. Periosteal osteosarcomas, as defined above, are characteristically at least one grade less active than the ordinary osteosarcomas. It has been aptly demonstrated that total removal of this tumor, often getting well around it without amputation, leads to a high possibility of survival. Only from 10 to 20 percent of the patients with this type of disease have died with metastases.

Parosteal osteosarcoma is the last type of osteosarcoma that we feel should be excluded from the group of osteosarcoma, type "ordinaire." They are much less apt to metastasize than are ordinary osteosarcomas, and the chance for 5-year survival is in the 80 to 90 percent range. This tumor is characteristically heavily ossified and has a marked predilection for developing on the posterior surface of the inferior part of the shaft of the femur. These tumors are characteristically low grade (Broders), either grade 1 or possibly grade 2. To qualify as a parosteal osteosarcoma, the tumor must be low grade and must be on the surface of the bone.

Chondrosarcoma should be separated from osteosarcoma because of its basic pathologic differences, which are reflected in vastly different clinical, therapeutic, and prognostic features (Fig. 3).

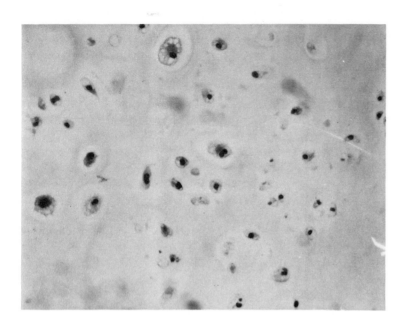

FIG. 3. Chondrosarcoma. The tumor is hypocellular but it contains a few large nuclei and scattered binucleate forms. H&E, x250.

The exact origin of chondrosarcomas is obscure, but the salient pathologic fact is that their proliferating tissue is cartilaginous throughout. Portions of these neoplasms may become myxomatous or calcified or even ossified. Osseous trabeculae, when present, result from maturation of the chondroid substance. When the malignant cells produce an osteoid lacework or osteoid trabeculae directly, even in small foci, the neoplasm has the clinical characteristics of osteogenic sarcoma and belongs in that category.

Chondrosarcoma usually has a slow clinical evolution. Metastases are relatively rare and often late in appearance. Accordingly, unlike osteosarcoma, in which prompt ablative surgical treatment is imperative because of early hematogenous dissemination, the basic therapeutic problem is prevention of recurrence by adequate control of the local lesion. Attainment of this goal demands adequate, frequently radical, early surgical treatment.

Secondary chondrosarcoma, when definitely recognizable as such, arises in a solitary osteochondroma or in one of the lesions in a patient with multiple osteochondromas or in a lesion in a patient with multiple chondromas. Such patients sometimes have the chondrodysplasia of Ollier. In the Mayo Clinic's series, there were 367 primary chondrosarcomas and 52 that arose in exostoses or in central chondromas. Several varieties of chondrosarcoma or chondrosarcomalike lesions are[2]:

1. Mesenchymal chondrosarcoma
2. Dedifferentiated cartilaginous tumors
3. Chondrosarcoma of skeleton of hands and feet
4. Cartilaginous tumors of soft tissue (usually in hands and feet)
5. Synovial chondromatosis
6. Cartilaginous tumors of larynx
7. Chondroid chordoma
8. Periosteal chondroma
9. Clear cell chondrosarcoma

Mesenchymal chondrosarcoma should not be lumped with the overall group of chondrosarcomas for the following reasons: mesenchymal chondrosarcomas are usually lethal; the likelihood of metastases is very high. Of the 15 patients with mesenchymal chondrosarcoma in the Mayo Clinic's group, 12 have died of their disease and another has known pulmonary metastases. The remaining two tumors have been resected recently, and their ultimate results are unknown. This tumor is very different from the ordinary chondrosarcoma histologically. It is recognized by its characteristic component of highly malignant small round cells similar to those of Ewing's sarcoma or hemangiopericytoma and its second component of chondroid islands, which usually contain cells with little anaplasia. The cartilaginous islands may be sparse or numerous and may undergo calcification or ossification. Approximately one-third of the mesenchymal chondrosarcomas reported in the literature have arisen in the somatic soft tissues.

Dedifferentiated cartilaginous tumors have been discussed above. Suffice it to reiterate that when a characteristically low-grade chondrosarcoma dedifferentiates and becomes a high-grade spindle cell sarcoma or osteosarcoma, its biologic capability is that of the lesion it has become.

Chondrosarcoma of the skeleton of the hands and feet is unusual. We recognized only 14 chondrosarcomas of the wrist and ankle bones or the bones distal to them in our material. These tumors did not differ histologically from chondrosarcomas in other sites. Considerable cellularity is permissible in chondromas if they appear radiographically benign in these sites. Chondrosarcomas of these bones show nuclear anaplasia characterized by increased nuclear size, irregularity of shape of nuclei, and multinucleation. The roentgenogram is extremely important in the recognition of malignant change. It nearly always shows the tumor's aggressiveness by cortical destruction over it and sometimes by invasion into the adjacent soft tissues. Approximately 90 percent of patients with chondrosarcomas of these distally located bones are cured by adequate removal, which is usually partial amputation of a hand or foot. Accordingly, this relatively good prognosis for chondrosarcomas in this location must be recognized. The difficulty in diagnosing chondrosarcomas in these small bones is accentuated by the fact that benign cartilaginous tumors are much more common in these locations than are the sarcomas. The cartilaginous tumors of the soft tissues of the hands and feet are relatively uncommon. Tumors with debatably chondroid features are not included. Such definitely cartilaginous tumors of the soft tissues are extremely rare in other parts of the body. They are nearly always of tenosynovial origin when they occur in the hands and feet. Like synovial chondromatosis in major articulations, the histologic features of these tumors may be worrisome to the pathologist, and the diagnosis of chondrosarcoma is often entertained. In spite of this ominous cytologic appearance, tumors of cartilage in these sites never metastasize distantly. They are benign in their evolution. We found metastases in none of 70 such tumors of the hands and feet; 59 of these were from our consultation files. Approximately 17 percent of the tumors gave rise to local recurrence, and recurrence developed more than once in a few of these. Nevertheless, they did not have a malignant clinical evolution. Roentgenographically, these tumors are distinctly extraskeletal, even though occasionally they erode into the nearby bone.

Synovial chondromatosis of a major articulation is another lesion that may be similar to chondrosarcoma histologically. Typically, patients with this monoarticular disease are young or older adults. The tumor masses in this condition characteristically occur in the synovium and produce polypoid extensions into the joint space. These may become loose and interfere with joint function and eventually produce secondary degenerative joint disease. The tumor mass often degenerates and calcifies or ossifies centrally and thus may show up in the roentgenogram. When the tumor masses are large, they may erode into the bone or adjacent soft tissues, thereby giving a false impression of invasiveness due to malignancy. The nuclei of the cartilage cells that characteristically occur in lobules are often large and multinucleated and histologically worrisome for malignancy. Malignant evolution of these cartilaginous proliferations in the synovium almost never occurs. If the pathologist, orthopedic surgeon, and radiologist can be assured that the process is of synovial derivation, they are almost certain that it is benign. Local recurrence is not uncommon, but the threat of metastasis practically never exists. In very rare instances, similar cartilaginous proliferations occur in sites away from major articulations such as in the ileopectineal bursa.

Cartilaginous tumors of the larynx deserve separate consideration. This tumor is unusual, but we have seen approximately 40 examples in the files of the Mayo Clinic. In our experience, such tumors have not given rise to metastases, even though many of them have cytologic features similar to those of low-grade chondrosarcoma in typical skeletal locations. The major threat for those that arise in the larynx is that they may produce death by airway obstruction. A few cartilaginous tumors of the larynx with distant metastases have been reported in the literature, but such an evolution is practically a medical curiosity.

Chondroid chordroma is not well understood generally. Approximately one-third of the chordomas that arise in the sphenooccipital region and that are roentgenographically characteristic of a chordoma in this location are similar histologically, at least in part, to cellular chondroma or low-grade chondrosarcoma. It is a paradoxical fact that chordomas at this location that simulate chondrosarcoma have a distinctly better prognosis than does an average chordoma. Recognition of this clinical feature is necessary in assessment of tumors in this location. Cartilaginous tumors of the skull at sites away from the sphenooccipital region are extremely unusual.

Periosteal chondroma occurs on the surface of a bone and ordinarily looks benign to the radiologist. This relatively radiolucent mass on the cortex is usually associated with a well-defined saucerlike defect. Cartilaginous tumors in this location are nearly always benign and can be cured by adequate local removal, even though histologically their cellularity, multinucleation, and increased size of nuclei make them difficult to differentiate from chondrosarcoma at times.

Clear cell chondrosarcoma[4] is a relative newcomer in the field of oncology. Only 9 examples of this disease were found in the Mayo Clinic's files. A recent report of 16 such tumors included 7 that had been sent in for consultation. The importance of this particular histologic type is that the tumors have been mistaken frequently in the past for osteoblastoma or even for chondroblastoma. These tumors nearly always begin in the very end of a major tubular bone. The available evidence indicates that it is impossible to cure them without total removal of the tumor, which often can be accomplished by some type of a local resection.

References

1. Dahlin DC, Unni KK: Osteosarcoma of bone and its important recognizable varieties. Am J Surg Pathol 1:61, 1977
2. Dahlin DC: Chondrosarcoma and its variants. Bones and Joints. I.A.P. Monograph No. 17: Baltimore, Williams & Wilkins, 1976, p 300
3. Schajowicz F: Juxtacortical chondrosarcoma. J Bone Joint Surg 59B:473, 1977
4. Unni KK, Dahlin DC, Beabout JW, Sim FH: Chondrosarcoma: clear-cell variant. A report of sixteen cases. J Bone Joint Surg 58A:676, 1976

Osteosarcoma and Chondrosarcoma: Radiology

Harold G. Jacobson

Osteosarcoma

Incidence and Clinical Features

Osteosarcoma usually presents in the first three decades of life, with a sex incidence of 3 males to 2 females. The appendicular skeleton is involved in about 90 percent of cases. At least half the cases occur about the knee. Any bone may be affected. In long bones the metaphyseal areas are mainly involved.

The patient usually complains of pain and swelling of a few weeks' duration. Low-grade pyrexia is common. Clinically, osteosarcoma resembles acute osteomyelitis. Pathologic fractures are common.

Pathologic Findings

Osteosarcoma presumably arises from primitive fibrous tissue or osteoblasts. The cellular pattern is extremely varied. The predominant tissue usually is either malignant osteoid, bone, cartilage, or fibrous tissue. The cellular spectrum varies from highly anaplastic, plump connective tissue cells with hyperchromatic nuclei to dense tumor bone with few stromal cells. Lesions usually show considerable neoplastic osteoid formation. New bone is frequently composed of non-neoplastic osteocytes and represents a response to the tumor. Lesions often are highly vascular, although angiography is not generally diagnostic.

Radiologic Features

Neoplastic osteoid tissue and cartilage appear radiolucent (Fig. 1), accounting for lytic defects with a wide zone of transition and cortical destruction. New bone formation causes irregular areas of increased density. Codman triangles of reactive periosteal bone are evident (Fig. 2). "Sunburst" spiculation is characteristic but relatively uncommon. Often, the epiphyseal plate acts as a temporary barrier to the spread of the tumor; however, soft tissue extension is common.

General Comments

Osteosarcoma is the most common primary malignant bone tumor, but it may be simulated by bizarre infections and unusual traumatic lesions. Metastases to the lungs may occur early and frequently are bone-forming. Metastases to the pleura also produce calcification and ossification. Multicentric osteosarcomatosis is a rare variant producing simultaneously-appearing blastic lesions often throughout the skull. Other forms include parosteal osteosarcoma, periosteal osteosarcoma, telangiectatic osteosarcoma and soft-tissue osteosarcoma.

Chondrosarcoma

Incidence and Clinical Features

Chondrosarcoma affects men twice as frequently as women, primarily after the fourth decade of life. Innominate bone, femur, tibia, or humerus is often involved, although the ribs, vertebrae, skull, and facial bones may be affected. Lesions of the long bones may be either diaphyseal or metaphyseal.

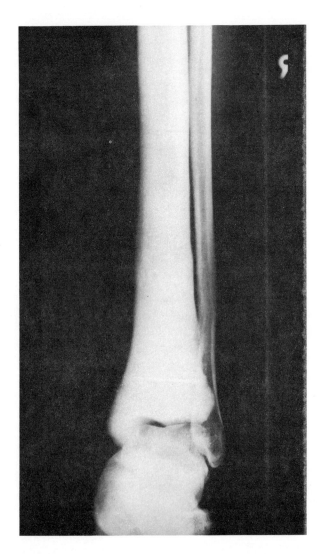 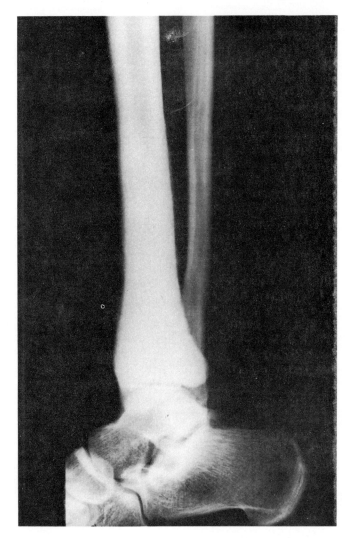

FIG. 1. Osteosarcoma—tibia. Note radiolucent area in lower portion.

There is often a long-standing history of the disease, with pain being the most frequent complaint. Constitutional signs and symptoms usually are absent early. A large localized soft tissue mass is a common feature.

Pathologic Findings

Chondrosarcoma arises from peripheral or central cartilaginous foci, with a tendency for tumors of peripheral origin to appear benign in contrast to central lesions. The tumor must be considered malignant if the tissue is hypercellular, if the cell nuclei are plump, if the cells have multiple nuclei, are hyperchromatic, and if giant cartilage cells are present. Even with firm criteria, differentiation between benign and malignant cartilage tumors may be difficult.

Radiologic Features

Chondrosarcoma may occur as a primary or secondary tumor arising from a cartilaginous precursor. It may be peripheral or central (Fig. 3) in location. Peripheral lesions may originate from osteochondroma or parosteal chondroma and show evidence of enlargement with no sharp margins of definition. A central lesion of a long bone, however, appears as a grossly destructive area with a large, well-defined soft-tissue mass. Endosteal new bone formation and endosteal scalloping are common, particularly in the tibia and femur (Fig. 4). Varying

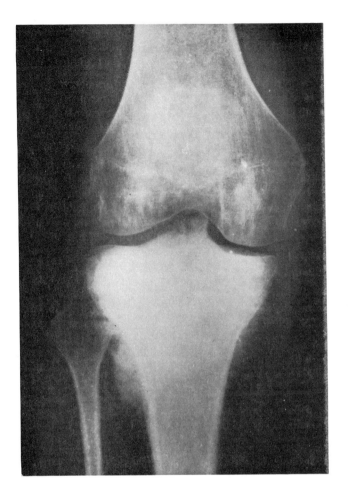

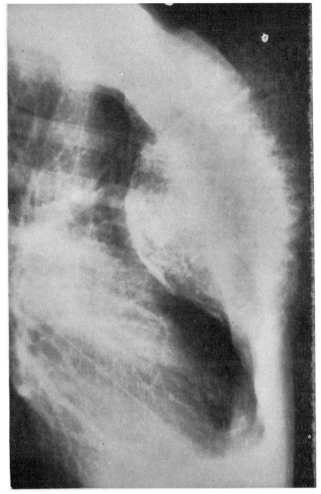

FIG. 3. Chondrosarcoma—sternum.

FIG. 2. Osteosarcoma—tibia. Note Codman triangles of reactive periosteal bone.

degrees of calcification occur. They are occasionally amorphous but frequently punctate or circular. Extensive cortical thickening is associated with organized periosteal reaction around the long bones. Chondrosarcoma of a flat bone tends to be grossly destructive but frequently is associated with varying amounts of reactive new bone, reminiscent of primary reticulum cell sarcoma.

General Comments

A large, soft-tissue mass is frequently observed, often larger than might ordinarily be anticipated. The soft-tissue mass may be the initial presenting radiologic feature, particularly in the pelvis. Skeletal metastases are very rare. Mesenchymal, juxtacortical, clear cell and dedifferentiated chondrosarcomata represent special variants.

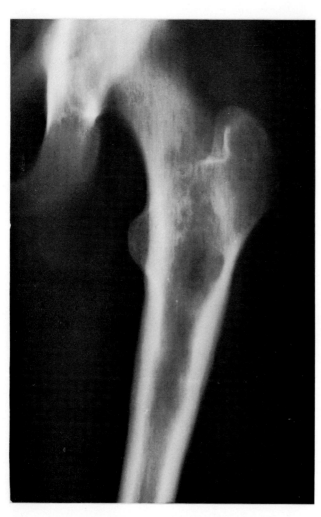

FIG. 4. Tomogram of femur demonstrating chondrosarcoma.

Bibliography

Murray RO, Jacobson HG: The Radiology of Skeletal Disorders: Exercises in Diagnosis, 2nd ed. Edinburgh, Churchill Livingstone, 1977

Chemotherapy of Malignant Bone Tumors

Robert H. DeBellis

The tumors that the orthopedic surgeon must concern himself with can be divided into three main groups—primary bone and cartilaginous tumors, soft tissue sarcomas often in close proximity to the skeleton, and metastatic tumors involving bone (Table 1). This chapter will concern itself with the first two groups only. Although there is a large variety of such tumors, numerically they represent less than 1 percent of all malignancies seen (excluding lymphomas and multiple myeloma). They are unique in that they affect a much younger population than do the more common carcinomas. As a group they result in approximately 2 deaths per million children under the age of 14 years and approximately 12 deaths per million young adults in the 15- to 19-year-old group. Although these are relatively small numbers of deaths, their impact on affected families is quite considerable.

Until recently, therapy beyond surgery, and in some instances radiotherapy, has been limited. In recent years, with the development of new agents and new combinations of existing agents, chemotherapy has played a progressively more important role in the management of these tumors.

Chemotherapeutic Agents Commonly Used in the Treatment of Bone Tumors—Mechanism of Action and Toxicity

Although over 30 agents are now clinically available, only a small number are used commonly in the treatment of primary bone tumors.

Table 1 Tumors of Bone

I. Soft tissue sarcomas—relatively uncommon, listed in approximate order of decreasing frequency (lymphomas and myeloma are excluded)

	Approximate survival	
	5-Year	10-Year
Liposarcoma	60%	56%
Fibrosarcoma	77%	71%
Rhabdomyosarcoma		
Pleomorphic	30–60%	10–40%
Alveolar, embryonal		50+%
Botryoidal		10–15%
Synovioma	30–35%	
Leiomyosarcoma	? poor	
Myxoma		
Mesenchymoma		
Malignant hemangiopericytoma		
Lymphangiosarcoma	(mean survival; 19 months)	
Malignant Schwannoma		
Extra Osseous Osteogenic Sarcoma		

II. Bone and cartilaginous sarcomas—less common than soft tissue sarcomas
  a. Osseous origin
    Osteogenic sarcoma — 5–20%
    Parosteal osteogenic sarcoma — >90%
    Chondrosarcoma — 25–50%
    Malignant giant cell tumor — 25%
  b. Nonosseous origin
    Ewing's sarcoma — 0–10%
    Fibrosarcoma
    Chordoma
    Angiosarcoma
    Liposarcoma
    Adamantinoma

III. Metastatic carcinomas—common

Alkylating Agents

Cyclophosphamide (Cytoxan, Endoxan).

Cyclophosphamide is an alkylating agent which, when enzymatically cleaved at the phosphorous-nitrogen linkage, forms a highly reactive agent which covalently binds to a number of biologically important groups on macromolecules such as amino, carboxyl, hydroxyl, sulfhydryl, and phosphate. Prior to the ring opening, cytoxan is relatively inert. Ring cleavage occurs primarily in the liver.

Cyclophosphamide can be administered orally in a continuous daily dose or intravenously at considerably higher doses intermittently.

Toxicity consists of gastrointestin, genitourinary, hematologic, and dermatologic side effects—the major effects being nausea and vomiting, leukopenia (with less common anemia and thrombocytopenia), alopecia, and hemorrhagic cystitis. The hemorrhagic cystitis can often be life threatening, but can be largely prevented by maintaining the patient's hydration coupled with frequent voiding.

Melphalan (Phenylalamine Mustard, Alkeran, L-PAM, Sarcolysin).

Melphalan is an alkylating agent having the same mechanism of action as cyclophosphamide. It is usually administered orally on a daily basis. Recently, an IV preparation has been made available which is used in an intermittent high dose schedule. To date, this route of administration has been limited to the treatment of multiple myeloma.

Toxicity is primarily hematologic (pancytopenia) and gastrointestinal (nausea and vomiting). There is evidence that patients with multiple myeloma treated with melphalan may be at greater risk for the development of monocytic leukemia.

Antibiotics

Actinomycin D (Dactinomycin, Cosmogen).

Actinomycin D is a cytotoxic antibiotic which forms a stable complex with DNA by intercolating in the minor groove of the DNA helix and specifically binding to guanine residues. It inhibits DNA-directed RNA synthesis, and at higher levels DNA-directed DNA synthesis.

Toxicity consists of gastrointestinal, hematologic, and dermatologic side effects. The gastrointestinal effects include nausea and vomiting, anorexia, diarrhea, proctitis, stomatitis, and chelitis. Alopecia is commonly seen, and the drug may enhance the toxic side effects of radiotherapy.

Adriamycin

Adriamycin is a cytotoxic antibiotic having a similar mechanism of action to that of actinomycin D resulting in an interference with nucleic acid synthesis. The toxic side effects of adriamycin are multiple. Myelosuppression, primarily leukopenia, occurs in over 60 percent of patients. Mucositis may be very severe with marked ulceration of the oral cavity and esophagus. Alopecia occurs in practically all patients with complete loss of scalp, axillary, and pubic hair which persists throughout the course of therapy. Nausea and vomiting occur and are usually progressively worse with subsequent courses of therapy. Chemical phlebitis may occur and extravasation of the drug during administration results in severe cellulitis, vesication, and tissue necrosis frequently requiring skin grafting. There is little evidence of spontaneous healing in affected tissues. Adriamycin (and daunomycin, the parent compound) has a unique potential cardiac toxicity. This correlates with the total dose of adriamycin given. At cumulative doses of <500 mg/M², only one nonfatal case of cardiomyopathy was found in 772 patients at risk. At doses >550 mg/M², the incidence increases to approximately 30 to 50 percent with 11 fatal cardiac complications in 86 patients reported (approximately 13 percent).

Vincristine (Oncovin)

Vincristine is a plant alkaloid having at least two distinct mechanisms of action. The first involves an interference with spindle formation during mitosis resulting in metaphase arrest. The second mechanism involves a block in the DNA-dependent RNA polymerase system resulting in an inhibition of RNA synthesis. Toxicity of vincristine is somewhat unusual in that it has little hematopoietic suppression. Thus, full doses of vincristine can be administered in the presence of moderately severe drug- or disease-induced pancytopenia. The main toxicity of vincristine is neurologic with loss of deep tendon reflexes and paresthesias of the extremities, constipation, and abdominal pain. Other neurologic changes such as hoarseness, ptosis, and diplopia can also be seen. Approximately 20 percent of the patients have significant reversible alopecia.

Methotrexate (Amethopterin)

Methotrexate competitively inhibits dihydrofolate reductase, thereby preventing the enzymatic conversion of folic acid to tetrahydrofolic acid and other structurally related compounds. Tetrahydrofolate, or activated folic acid, is an essential precursor in the one carbon donor system necessary for the biosynthesis of several purine and pyrimidine precursors of DNA + RNA. Thus, methotrexate indirectly interferes with nucleic acid synthesis.

The toxicities of methotrexate are multiple. Leukopenia and thrombocytopenia are common and megaloblastic anemia can be seen. Gastrointestinal side effects are also common with nausea and vomiting, diarrhea, and stomatitis. Diarrhea and stomatitis are indications to interrupt therapy. Pulmonary, hepatic, and dermatologic changes are occasionally seen. Scoliosis and osteoporosis with pathologic fractures have also been reported.[1]

Recently, "high dose methotrexate therapy with citrovorum factor rescue" (HDMTX$\bar{c}$CFR) has gained popularity in the treatment of many tumors, especially sarcomas. Doses of methotrexate as high as 7.5 g/M² have been used on a weekly schedule. Citrovorum factor (CF, leukovorin, folinic acid) is N^5 formyl tetrahydrofolic acid—the active derivative of folic acid. Since this intermediary is past the dihydrofolate reductase enzymatic steps, it is active in the presence of methotrexate and can be used to reverse the metabolic block induced by high levels of methotrexate. There are several requirements that should be met when HDMTX$\bar{c}$CFR is utilized:

1. Renal function with measurement of creatinine clearance should be evaluated prior to treatment.
2. Patients should be vigorously hydrated.
3. Alkalinization with oral diamox 250 mg b.i.d. should be carried out.
4. Methotrexate serum levels should be readily available and "rescue" should be continued until serum methotrexate levels are below 2×10^{-7} meters. Such treatment is very costly and should be used only by qualified oncologists in a research setting.

DTIC (Dimethyl Triageno Imidazole Carboxamide)

DTIC is a structural analogue of 5 amino imidazole-4-carboxamide, a precursor in purine biosynthesis. There are three postulated mechanisms of action: 1) inhibition of DNA synthesis by acting as a purine analogue, 2) action as an alkylating agent, and 3) interaction with SH groups. Toxicity includes nausea and vomiting and myelosuppression, primarily leukopenia and thrombocytopenia. A sometimes troublesome side effect involves severe pain and spasm along the course of the vein being injected. This can be particularly dangerous when DTIC is used in conjunction with adriamycin where venous stasis with pooling of the adriamycin may result in severe phlebitis and ulceration. Some patients experience a flu-like syndrome with fever, myalgias, and malaise. This occurs approximately 1 week after treatment with DTIC.

Chemotherapy of Osteogenic Sarcoma

With the demonstrations by Jaffe[2] using high dose methotrexate with citrovorum factor rescue, and by Cortes et al[3] using adriamycin, that 35 to 40 percent objective remissions of metastatic disease could be achieved, a great surge of activity developed in the treatment of osteogenic sarcoma. Out of these studies grew the numerous adjuvant programs of therapy that are now in vogue. Despite the apparent success of many of the adjuvant programs, the treatment of metastatic disease continues to be a major challenge to the medical oncologist, radiotherapist, and surgeon. Since the majority of patients with osteogenic sarcoma are given adjuvant chemotherapy as part of their primary treatment at the present time, subsequent chemotherapy at the time of recurrence will depend on what was used for adjuvant therapy initially.

The vast majority of reported studies have utilized adriamycin as a major component of the regimen, usually given to maximally cumulative doses that are considered safe. Therefore, at present adriamycin is rarely a drug used in the treatment of recurrent disease. High dose methotrexate with cirovorum factor rescue is the most commonly used regimen. Jaffe[4] has demonstrated that large doses of methotrexate given weekly can result in complete remissions even in patients previously given high dose methotrexate. Other drugs of value either singly or in combination include cyclophosphamide[5], vincristine[4-6], DTIC[7], actinomycin[7], melphalan[6], and cis-dichlorodiammine platinum (II) (DDP), an investigational chelated platinum compound with antitumor properties.[8] Many of the regimens listed in the discussion of adjuvant chemotherapy, which involve the drugs listed above, may be appropriate in the treatment of advanced disease. They should be used only by experienced chemotherapists familiar with doses, toxicities, etc. Several investigators have demonstrated the benefits of surgical resection of limited numbers of pulmonary metastases.[9-12]

Adjuvant Chemotherapy

Adjuvant chemotherapy involves the use of drugs in conjunction with surgery or radiotherapy in an attempt to improve the cure rate of the disease treated. The use of adjuvant chemotherapy is based on the assumption that micrometastases exist at the time of the primary definitive treatment. There are several requisites for successful adjuvant chemotherapy:

1. The disease being treated should have a poor prognosis following standard treatment.
2. Agents capable of producing objective remissions in advanced disease must obviously be available.
3. Treatment should be instituted as soon after diagnosis as is feasible.
4. Total excision of gross disease should be accomplished.
5. The patient must be free of distant metastases and appropriate work-up should be carried out to assure this.

The February 1978 issue of Cancer Treatment Reports (Vol 62, No 2) is devoted to the proceedings of the osteosarcoma study group meeting. The various adjuvant studies including adriamycin and amputation (Cortes et al, Fossati Bellani et al); high dose methotrexate with citrovorum factor rescue (Jaffe et al); CYVADIC (cyclophosphamide, vincristine, adriamycin, and DTIC) and CYVADACT (actinomycin substituted for DTIC) (Benjamin et al); CONPADRI (cytoxan, vincristine, adriamycin, and melphalan); CONPADRI-II (and high dose methotrexate with rescue); CONPADRI-III (modification of CONPADRI II with larger doses of adriamycin) (Sutow et al); en bloc resection rather than amputation ± chemotherapy (Marcove, Jaffe et al); and high dose cyclophosphamide (Shepp et al) are all summarized in this issue. It is difficult to draw firm conclusions from the multitude of studies. All are nonrandomized and use historical controls for comparison. The early successes of high dose methotrexate with leukovorin rescue are no longer as promising as they seemed. Rather than cure, treatment appears to result in a delay in recurrence and at 4 years recurrences are still being found with an overall relapse rate approaching that of historical controls.

The best results would appear to be with radical amputation plus full doses of adriamycin on a 3 day schedule q 28 days with therapy started within two weeks of surgery. Approximately a 60 percent 5 year disease-free survival is seen compared to a 20 percent 2 year disease-free survival in historical controls. Furthermore, few recurrences are seen beyond two years. The results of en bloc resection in attempts to preserve limb function are still too early to predict their success and further time will be required to evaluate these techniques.

In summary: the chemotherapy of primary and metastatic osteogenic sarcoma is an area of active research with ongoing evaluation of individual drugs plus combinations of drugs coupled with radical amputation or en bloc resection. Prophylactic radiotherapy appears to play no beneficial role in the treatment of primary disease.[13] Final standards of treatment are still to be determined. Until that time, all patients with osteogenic sarcoma should be actively involved in investigational studies by qualified specialists.

Chemotherapy of Ewing's Sarcoma

The chemotherapy of Ewing's sarcoma, like that of osteogenic sarcoma, is an area of active intensive research. Early studies with vincristine and cytoxan demonstrated that metastatic disease responded well to treatment. The addition of actinomycin D further increased response rates. With the development of adriamycin, a new powerful agent was added to the available list of agents active against Ewing's sarcoma. Due to the success of chemotherapy in the arrest of metastatic disease, patients with primary Ewing's sarcoma are being treated with complex programs of combined chemotherapy and radiotherapy in active, constantly evolving investigational studies. The program of the National Intergroup Study for Ewing's sarcoma typifies the programs being used in many centers. Following diagnosis, patients are treated with radiotherapy over a 6 week period to total doses of 4500 to 6500 rads. Concomitant with the radiotherapy, they are given weekly doses of vincristine and cytoxan. Toward the end of radiotherapy, they are randomized to receive bilateral pulmonary radiotherapy to total doses of 1500 to 1800 rads or adriamycin. Following a 6 week rest period, they are further treated with q3 month courses of actinomycin D (5 consecutive days followed by weekly vincristine and cytoxan on weeks 3 through 7). Six such courses are given. Such complex, time consuming regimens have resulted in promising long-term disease-free periods.[14-16] Best results appear to be achieved when either adriamycin or whole lung radiotherapy is used.[16] As was stated for patients with osteogenic sarcoma, it would seem imperative that all patients with Ewing's sarcoma be entered in active investigational programs if we are to make progress in the successful treatment of this disease.

Other Soft Tissue and Primary Bone Sarcomas

Because of unpredictability of the cure rates of most of these tumors and the poor prognosis seen with many of them (Table 1), it is felt that these patients should be entered in active investigational programs of adjuvant chemotherapy. If protocols are not available, a simple acceptable active program involves the use of adriamycin started as soon after diagnosis as is surgically safe. A schedule of 30 mg/kg/d × 3 q 4 wks × 6 doses is suitable and has been used with success to date in the treatment of a small group of patients with diagnoses of malignant Schwannomas, malignant fibrous histiocytomas, liposarcomas, malignant hemangiopericytomas, fibrosarcomas, and malignant giant cell tumors.

For metastatic disease a number of drugs or combinations of drugs can be used—basically resembling those used for osteogenic sarcoma.[17] However, remissions when they occur may be brief and the ultimate prognosis is poor.

References

1. Ragab A, Frech R, Vietti T: Osteoporotic fractures secondary to methotrexate therapy of acute leukemia in remission. Cancer 25:580, 1970
2. Jaffe N: Recent advances in the chemotherapy of metastatic osteogenic sarcoma. Cancer 30:1627-1631, 1972
3. Cortes EP, Holland JF, Wang JG, et al: Doxorubicin in disseminated osteosarcoma. JAMA 221:1132-1138, 1972
4. Jaffe N, Traggis D, Cassady JR, et al: Multidisciplinary treatment for macrometastic osteogenic sarcoma. Br Med J 2:1039-1041, 1976
5. Finklestein JZ, Hittle RE, Hammond GD: Evaluation of a high-dose cyclophosphamide regimen in childhood tumors. Cancer 23:1239-1242, 1969
6. Sutow WW, et al: Multi drug adjuvant chemotherapy for osteosarcoma: Interim report of the South West Oncology Group Studies. Cancer Treat Reports 62:265-269, 1978
7. Gottlieb JA, et al: Role of DTIC (NSC-45388) in the chemotherapy of sarcomas. Cancer Treat Rep 60:199-203, 1976
8. Ochs JJ, et al: Cis-dichlorodiamminoplatinum (II) in advanced osteogenic sarcoma. Cancer Treat Rep 62:239-245, 1978

9. Marcove RC, et al: Osteogenic sarcoma under the age of 21. A review of 145 operative cases. J Bone Joint Surg 52A:411-423, 1970
10. Martini N, et al: Multiple pulmonary resections in the treatment of osteogenic sarcoma. Annals Thoracic Surg 12:271-280, 1971
11. Rosen G, et al: Chemotherapy and thoracotomy for metastatic osteogenic sarcoma. A model for adjuvant chemotherapy and the rationale in the timing of thoracic surgery. Cancer 41:841-849, 1978
12. Mountain CF: Surgical management of pulmonary metastases. Postgrad Med 48:128-132, Nov 1970
13. Caceres E, et al: Adjuvant whole-lung radiation with or without adriamycin treatment in osteogenic sarcoma. Cancer Treat Rep 62:297-299, 1978
14. Gutierrez M, Marcove R, Rosen G: Four-drug chemotherapy in Ewing's sarcoma: Follow-up of prolonged disease free survival (Abstr). 12th Ann meeting of ASCO 17:268, 1976
15. Rosen G, et al: Curability of Ewing's sarcoma and considerations for future therapeutic trials. Cancer 41:888-899, 1978
16. Nesbitt M, et al: Abstract. Intergroup Ewing's sarcoma study (IESS: Results of three different treatment regimens). Proc Am Assoc Cancer Res 19:81, 1978

Bibliography

Block and Isacoff: Discussion of adjuvant chemotherapy principles. Semin Oncol 4:109-115, 1977

Cancer Chemotherapy Reports 6: No 1, July 1975. (Issue devoted to proceedings of the High-Dose Methotrexate Therapy Meeting, Dec. 1974)

Cancer Chemotherapy Reports 6: No 2, Oct. 1975 (Issue devoted to Drug Seminar on Adriamycin held Dec. 1974) Cortes et al. Review experience with osteogenic sarcoma in this issue pp 305-313

Cancer Treatment Reports: Proceedings of the Osteosarcoma Study Group Meeting. 62 No 2 pp 187-313, Feb. 1978. Entire issue devoted to osteogenic sarcoma with articles on adjuvant chemotherapy, use of Cis-platinum for advanced disease. Evaluation of transfer factor, adjuvant whole lung radiotherapy, en bloc resection for limb preservation, and consideration of prognostic factors

Holland & Frei (eds.) Cancer Medicine: Lea & Febiger 1973 Chapters on chemotherapy, soft tissue sarcomas and bone and cartilage tumors

Joseph WL, et al: Evaluated effect of doubling time on response to surgery. J Thorac Cardiovas Surg 61:23-32, 1971

Kaplan, et al: Radiotherapeutic alternative to radical surgery utilizing radiosensitizing agent and high dose MTX + CFR. Radiology 117:211-213, 1975

Shaeffer, et al: Experimental mouse model demonstrating efficacy of adjuvant chemotherapy vs radiotherapy. Radiology III: 467, 1974

Skipper HE: Discussion of adjuvant chemotherapy. Cancer, 41:936-940, 1978

Bone Tumors: Role of Nuclear Medicine

Philip M. Johnson

Scintillation imaging is a sensitive method for detecting focal or multifocal diseases of bone. Nearly all active skeletal lesions simultaneously stimulate regional blood flow and osteoblastic activity. These changes result in locally increased uptake of bone-seeking radiotracers that is detectable by scintillation imaging,[1] usually in advance of radiographic abnormalities. The positive bone scan thus reflects altered skeletal physiology regardless of stimulus, but its specificity in terms of differential diagnosis is low.

Development of the family of radiotechnetium-labeled phosphate compounds led to current high-information-density imaging with an absorbed radiation dose of only ~0.5 rad to the skeleton. Available evidence suggests that these agents react with skeletal hydroxyapatite crystals by "chemisorption".[1] Ectopic accumulations of hydroxyapatite that occur in myositis ossificans, soft-tissue metastases of osteogenic sarcoma, acute myocardial infarction, etc., are also demonstrable by skeletal imaging. The kidneys are the principal route of tracer excretion and together with the urinary bladder are normally visualized.

The superiority of the scan to radiography in detecting bony metastases is well established. In one representative study of 188 patients with proven malignancy, the scan was positive in 57 patients with negative radiographs, whereas only 3 patients had x-ray evidence of metastases not detected by scanning.[2] Thus, a negative bone scan virtually rules out the possibility of active skeletal disease. A false-negative scan may occur in certain situations—stable lesions such as calcified metastases; symmetrical or uniformly disseminated metastases; certain tumors, especially multiple myeloma,[3] that do not consistently evoke reactive changes; and anaplastic tumors in which bone is rapidly destroyed and reactive bone accretion is minimal. These lesions at times produce photon-deficient areas recognized as negative defects on the scan.

Primary Malignant Neoplasms of Bone

In the preoperative evaluation of primary skeletal malignancies, imaging is used to determine the proximal tumor margin, to identify "skip areas," and to detect polyostotic or multicentric involvement. In one series,[3] the scan was positive in 27 of 31 proven primary skeletal malignancies. The neoplasms included Ewing's sarcoma ($n = 12$), osteogenic sarcoma ($n = 8$), multiple myeloma ($n = 8$), fibrosarcoma ($n = 2$), and chondrosarcoma ($n = 1$). The four false-negative studies occurred in the group with multiple myeloma. There is close correlation between the results of imaging and radiography in staging osteogenic sarcoma.[4] The area of abnormal tracer uptake and the radiographic changes are usually concordant (Fig. 1). Osteoblastoma of the axial skeleton also concentrates bone-seeking radiotracers.[5]

Lymphoma commonly spreads to bone but may also arise as a primary skeletal malignancy. Benua et al[6] studied 60 patients with lymphoma, including 39 with Hodgkin's disease; the disease was advanced in the majority. In 21 patients, the scan was positive and x-ray studies were negative; in 8 others skeletal abnormalities were more extensive on scan than on x-ray. Gallium-67 citrate has affinity for both normal and abnormal lymphocytes.[7] Adler et al[8] demonstrated abnormal localization of gallium-67 in 10 of 12 patients with verified skeletal lymphoma. Given the appropriate clinical setting, the gallium-positive bony lesion likely is lymphoma.

Paget's disease is typically associated with intense uptake of radiotracers due to hyperemia and remodeling. In its systemic form, the disease often presents a characteristic pattern on imaging. Sarcomatous degeneration occurs in up to 3 percent of patients. However, the scan usually cannot distinguish this complication from pathologic fracture. Sarcoma

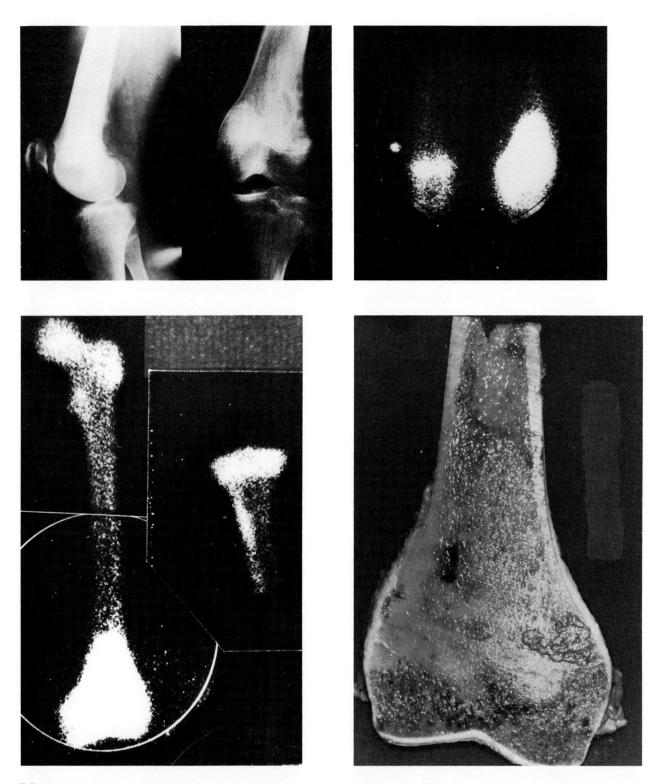

FIG. 1. Sclerosing osteogenic sarcoma in a 16-year-old male. Top, left: Anteroposterior and lateral radiographs show typical neoplastic changes in the left femoral metaphysis. Top, right: Preoperative scan of both knees demonstrates intense uptake in the left femoral metaphysis and part of the epiphysis. A radioactive landmark was placed lateral to the normal right knee. Technetium-99m diphosphonate was used. Bottom, left: Scan of the resected left femur and proximal tibia shows intense uptake confined to the lesion with no evidence of intraosseous spread. There is normally increased activity in the trochanters and femoral head. Bottom, right: Gross specimen, transsected. The findings on imaging and radiography correlate well with the tumor's actual extent.

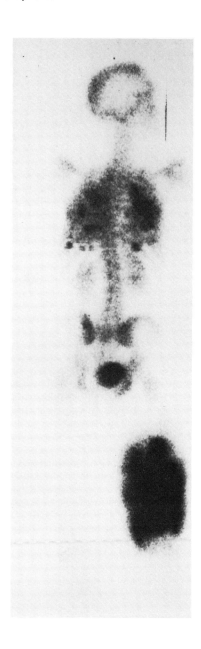

FIG. 2. Osteogenic sarcoma, metastatic. This child's tumor was locally advanced and had metastasized widely when first seen. Imaging with technetium-99m diphosphonate demonstrates avid uptake in the massive primary tumor of the left femur. Abnormal punctate and confluent localizations in the bones and soft tissues of the thorax are indicative of metastases. The left ischium also demonstrates increased uptake.

is suspected when localized intense activity extends beyond the margins of bone already involved with classic Pagetoid changes.[9] In one report, a photon-deficient area due to sarcomatous destruction of the ilium could not be differentiated from simple noninvolvement by the preexisting Paget's disease.[10]

In staging children with neuroblastoma, imaging is usually superior to radiography. However, bilateral metastases to the distal femora occur commonly. These lesions often display bilateral symmetry and also may be masked by normally intense activity in the adjacent epiphyses.[11] As a result, they may be impossible to detect by imaging. Therefore, workup should always include a complete radiographic survey as well as a complete radionuclide scan. Due to the frequency of calcification in neuroblastoma, the primary tumor is often visualized on imaging. Ewing's tumor is also associated with positive radiotracer uptake. The incidence of polyostotic involvement, detectable by imaging, may reach 30 percent.[3]

In addition to its role in preoperative staging and differentiating monostotic and polyostotic disease, imaging plays a useful role in assessing the result of treatment. It is important to note several interpretative pitfalls in the follow-up examination. Radiotracer uptake in viable bone diminishes after radiation therapy; the effect is dose-dependent and may be pronounced in intensity and duration. Iatrogenic injury of bone at biopsy or resection is associated with increased tracer localization for a period of months, as noted elsewhere in this volume (R.A. Fawwaz: Bone Trauma—Assessment by Scintigraphy). Active reossification during healing of a lesion is also associated with augmented uptake. These possibilities must be considered when assessing a persistent abnormality. Recurrence of tumor is suggested if the area of increased uptake at the operative site expands or intensifies, provided that infection and trauma (as from a prosthesis) are excluded as alternative causes.

Imaging allows detection of soft-tissue metastases that form malignant osteoid tissue (Fig. 2). Since the scan is nearly always more sensitive than radiography in detecting bony metastases, it is recommended for periodic routine follow-up and for evaluation of persistent skeletal pain. When the scan is positive and x-ray studies are negative, biopsy should be considered to establish etiology. If biopsy is contraindicated, then serial scans and radiographs may lead to definitive diagnosis.

Benign Neoplasms of Bone

Many of the benign skeletal neoplasms, such as cyst, osteochondroma, and enchondroma, are associated with normal or perhaps slightly increased uptake of tracer. This is to be expected, since the metabolic activity of these lesions seldom differs from that of adjacent normal bone. Although a scan-positive bone island has been reported,[12] this is clearly an exceptional case. Pathologic fracture of a benign tumor is of course associated with increased radiotracer uptake, as is active expansion of a benign lesion.

Osteoid osteoma may prove difficult to detect radiographically when located in the vertebrae, proximal femora, or other regions containing cancellous bone in large amounts.[13] This tumor is typically associated with intense radiotracer uptake; several verified cases of scan-positive, x-ray-negative lesions have been reported (Fig. 3). Imaging should be used to evaluate persistent atypical bone pain in children whose radiographs are negative or equivocal.[14] Eosinophilic granuloma is often, but not always, associated with increased localization of tracer.[13,15]

Fibrous dysplasia often causes intense tracer uptake confined to the radiographically demonstrable lesions. Increased uptake is also seen in leukemic infiltration and many of the reticuloses. In general, however, the bone scan is less useful than radiographic studies in assessing these conditions.

Summary

Correct clinical utilization of skeletal imaging in the primary benign and malignant skeletal neoplasms requires awareness of the method's capabilities and limitations. The general lack of diagnostic specificity of the positive bone scan must be remembered.

Focally increased uptake in bone distant from a known primary neoplasm may represent polyostotic disease or metastasis, but it can also result from an unrelated benign condition.[16] The message of the positive bone scan is that a local disturbance of skeletal physiology exists. This is an important message, because the bone scan is highly sensitive. In the majority of cases, however, the message is incomplete and the question of why the alteration arose must be answered by other methods.

References

1. Jones AG, Francis MD, Davis MA: Bone scanning: radionuclide reaction mechanisms. Semin Nucl Med 6:3, 1976
2. Pistenma DA, McDougall IR, Kriss JP: Screening for bone metastases. Are only scans necessary? JAMA 231:46, 1975
3. Shirazi PH, Rayudu GVS, Fordham EW: [18]F bone scanning: review of indications and results of 1,500 scans. Radiology 112:361, 1974
4. Goldman AB, Becker MH, Braunstein P, Francis KC, Genieser NB, Firoozhia H: Bone scanning—osteogenic sarcoma: correlation with surgical pathology. Am J Roentgenol 124:83, 1975
5. Martin NL, Preston DF, Robinson RG: Osteoblastomas of the axial skeleton shown by skeletal scanning—case report. J Nucl Med 17:187, 1976

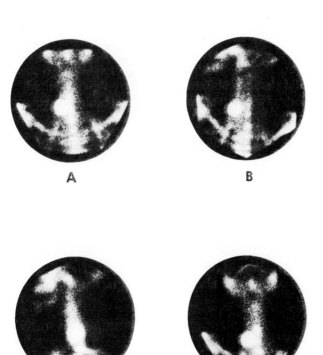

FIG. 3. Osteoid osteoma of cervical spine. The cause of this child's persistent neck pain remained unexplained for over a year despite extensive workup. Recent tomographs were suspicious but not unequivocal for a lucent lesion of the C-6 vertebra. Examination of the cervical spine after administration of technetium-99m diphosphonate showed a solitary collection of activity in the left side of C-6. A well-circumscribed osteoid osteoma was subsequently resected. A. Posterior image. B. Left posterior oblique image. C. Left lateral image. D. Posterior image with radioactive landmark in midline. (From Winter et al: Scintigraphic detection of osteoid osteoma. Radiology 122:177, 1977.) Reproduced with the permission of Radiology.

6. Benua RS, Laughlin JS, Lee BJ, Tilbury RS: Use of ^{18}F-sodium fluoride bone scans to determine the extent of disease in lymphoma. J Nucl Med 12:340, 1971
7. Merz T, Malamud L, McKusick K, Wagner HN Jr: Mechanism of ^{67}Ga association with lymphocytes. Cancer Res 34:2495, 1974
8. Adler S, Parthasarathy KL, Bakshi SP, Stutzman L: Gallium-67-citrate scanning for the localization and staging of lymphomas. J Nucl Med 16:255, 1975
9. Shirazi PH, Ryan WG, Fordham EW: Bone scanning in the evaluation of Paget's disease of bone. CRC Crit Rev Clin Radiol Nucl Med 5:523, 1974
10. McKillop JH, Fogelman I, Boyle IT, Greig WR: Bone scan appearance of a Paget's osteosarcoma: failure to concentrate HEDP. J Nucl Med 18:1039, 1977
11. Kaufman RA, Thrall JH, Keyes JW Jr, Brown ML, Zakem JF: False negative bone scans in neuroblastoma metastatic to the ends of long bones. Am J Roentgenol 130:131, 1978
12. Sickles EA, Genant HK, Hoffer PB: Increased localization of ^{99m}Tc-pyrophosphate in a bone island: case report. J Nucl Med 17:113, 1976
13. Gilday DL, Ash JM: Benign bone tumors. Semin Nucl Med 6:33, 1976
14. Winter PF, Johnson PM, Hilal SK, Feldman F: Scintigraphic detection of osteoid osteoma. Radiology 122:177, 1977
15. Lentle BC, Russell AS, Percy JS, Scott JR, Jackson FI: Bone scintiscanning updated. Ann Intern Med 84:297, 1976
16. Turner JW, Syed IB, Spencer RP: Two unusual causes of peripatellar nonmetastatic positive bone scans in patients with malignancies. Case reports. J Nucl Med 17:693, 1976

Angiography in the Management of Bone Tumors

William J. Casarella

Angiography has three major goals in the management of bone and soft-tissue tumors: (1) it aids in establishing the extent and blood supply of the lesion, (2) the differential diagnosis can be narrowed, and (3) preoperative embolization can aid in safe surgical removal. The third application has become more important as surgeons have become more willing to perform local resections and limb reconstructions in selected cases of bone or soft-tissue tumors.

The traditional first goal of angiography-localization of lesions is probably performed more easily by computed tomography or even routine radiographic tomography. This is especially true in hypovascular lesions, in which the size is generally underestimated by angiography. However, angiography can accurately demonstrate the invasive properties and intramedullary and soft-tissue extent of malignant lesions.

Some authors claim to have obtained a high degree of accuracy in distinguishing benign from malignant bone tumors by angiography alone. In our experience, this has been extremely difficult, and arteriography has not been able to replace biopsy in determining the benign nature of a lesion.

Various tumors and their hypervascular or hypovascular nature are listed in Table 1. In general, there is a tendency for the more malignant lesions to be hypervascular with considerable neovascularity. However, considerable overlap exists, and examples of very vascular benign tumors and avascular malignant lesions are present in most series. The marked neovascularity of aneurysmal bone cysts and giant-

Table 1 Various Tumors and Their Hypervascular or Hypovascular Nature

Hypervascular lesions
  Malignant
    Osteogenic sarcoma
    Fibrosarcoma
    Leiomyosarcoma
    Rhabdomyosarcoma
    Ewing's sarcoma
    Reticulum cell sarcoma
    Metastases—renal, thyroid, lung, some
      gastrointestinal lesions
    Fibrous histiocytoma
  Benign
    Giant-cell tumor
    Aneurysmal bone cyst
    Osteoid osteoma
    Chondroblastoma

Hypovascular lesions
  Malignant
    Chondrosarcoma
    Metastases—myeloma, melanoma, some
      gastrointestinal lesions
    Parosteal osteosarcoma
  Benign
    Osteoma
    Osteochondroma
    Fibrous dysplasia
    Bone cyst
    Chondroma
    Chondromyxoid fibroma
    Chronic osteomyelitis

cell tumors are examples of highly vascular benign tumors. Conversely, some chondrosarcomas may be very hypovascular and difficult to evaluate by angiography alone.

Clinical applications of selective embolization have multiplied dramatically during the past five years. The entire subject has been superbly reviewed by Grace et al. in 1976.

The first three clinical examples of bone tumor embolization were reported by Feldman et al. in 1975. In these cases and in most subsequent ones, particuate Gelfoam has been the material of choice for embolization. Our series has now been expanded to include 18 cases of bone and soft-tissue neoplasms with preoperative or palliative embolization.

Lesions of the pelvis, shoulder, or trunk in which tourniquet hemostasis is impossible are ideal candidates for embolization. Such a case is illustrated in Figure 1. The patient is a 45-year-old female with a leiomyosarcoma of the left buttock. Selective angiography demonstrates the highly vascular nature of the lesion. Embolization of the lesion with Gelfoam particles resulted in obliteration of the tumor's blood supply. At surgery, the lesion was easily resectable with the loss of less than one unit of blood.

Another example of preoperative embolization is demonstrated in Figure 2. This massively hypervascular lesion of the head of the fibula was caused by a giant-cell tumor in a 35-year-old male. The lesion was so vascular, with large dilated vascular lakes throughout, that attempts at local biopsy resulted in severe hemorrhage and loss of six units of blood. Following antegrade passage of a 6F catheter selectively into the feeding arteries derived from the popliteal artery, 1-mm cubes of Gelfoam suspended in contrast material were embolized into the lesion

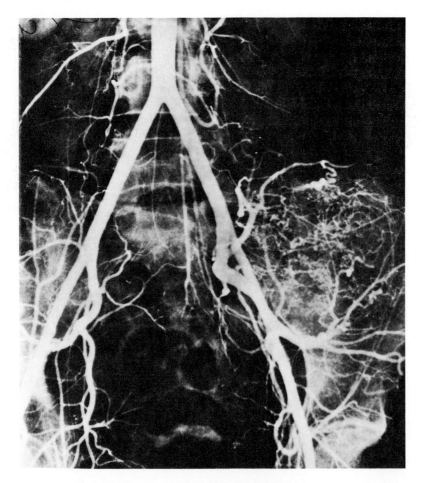
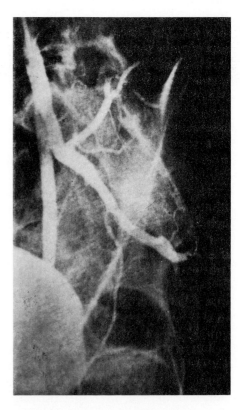

FIG. 1. Left: A 10-cm leiomyosarcoma of the left buttock with massive neovascularity from gluteal branches of the posterior division of the hypogastric artery. Right: Following embolization, the lateral view of a selective left common iliac study shows obliteration of neovascularity from branches of the gluteal arteries.

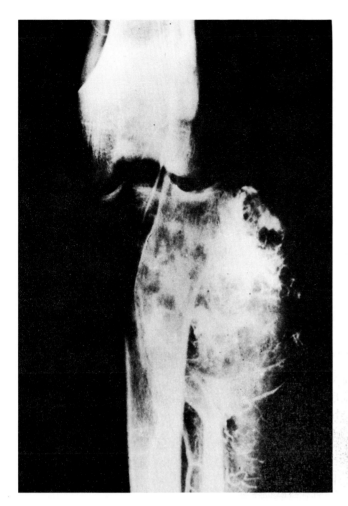 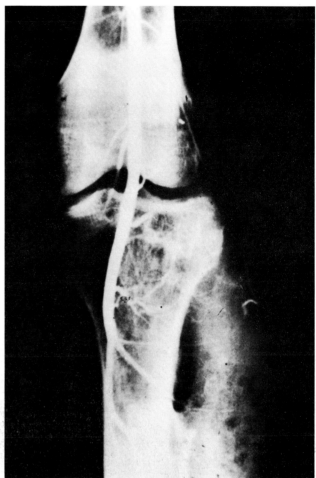

FIG. 2. A. An extremely vascular giant-cell tumor of the left fibula reveals huge vascular lakes on femoral arteriography. B. Following embolization of all feeding vessels, the lesion appears avascular and was ischemic at operation.

until angiographic hemostasis was achieved. Great care must be taken to direct the Gelfoam selectively into the lesion so that errant particles do not embolize the distal extremity. This can be achieved by using low-pressure hand injections with a 10-ml syringe and by discontinuing embolization when significant decrease in arterial flow is achieved. This prevents reflux of the particles downstream to the distal extremity. At surgery, the lesion was easily resected with minimal blood loss. This patient had a very similar lesion on the other side that was also embolized prior to its successful resection.

A slightly different application of tumor embolization is illustrated by the case in Figure 3. This 49-year-old man had metastatic adenocarcinoma to the right iliac crest. The lesion was extremely vascular, was pulsatile to examination, and was extremely painful. Radiotherapy had been ineffective. In this case, embolization was performed in an attempt at palliation. Following obliteration of the blood supply to the metastasis with Gelfoam particles, the pain was markedly improved and the patient remained free of symptoms for 3 months following embolization. Two other cases of painful bone metastases have been embolized in this fashion with some palliation achieved. This therapy should be reserved for use following radiotherapy, since rendering a lesion ischemic can itself decrease the effectiveness of radiation, which depends on tissue oxygenation for its tumoricidal properties.

Newer embolization techniques are being developed. Methacrylate glues, silicon polymers, and microfibrillar collagen suspensions are possible new embolization materials that may be more effective in the future.

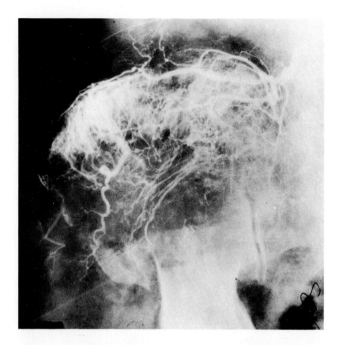 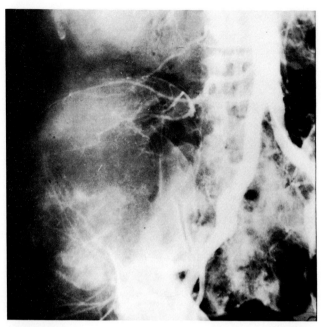

FIG. 3. Left: Selective lumbar arteriogram reveals a huge vascular mass of the right ilium. Right: Following embolization, the tumor vascularity is ablated. The lesion became much less painful, tender, and swollen.

Bibliography

Feldman F, Casarella WJ, Dick HM, Hollander BA: Selective intra-arterial embolization of bone tumors. Am J Roentgenol 123:130, 1975

Goldstein HM, Medezkin H, Ben-Menachen Y, Wallace S: Transcatheter embolization in the management of the cancer patient. Radiology 115:603, 1975

Grace DM, Pitt DF, Gold RE: Vascular embolization and occlusion by angiographic techniques as an aid or alternative to operation. Surg Gynecol Obstet 143:469, 1976

Hilal S, Mount L, Correll J, Wood EH: Therapeutic embolization of vascular malformations of external carotid circulation: clinical and experimental results. IX Symp Neuroradiologicum Göteborg, Sweden, August 1970

Steckel RJ: Usefulness of extremity arteriography in special situations. Radiology 86:293, 1966

Strickland B: Value of arteriography in diagnosis of bone tumors. Brit J Radiol 32:705, 1959

Viamonte M, Roen S, LePage J: Non-specificity of abnormal neovascularity in angiographic diagnosis of malignant neoplasms. Radiology 106:59, 1973

Yaghmai I, Zia Q, Shariat S: Value of arteriography in the diagnosis of benign and malignant bone lesions. Cancer 27:1134, 1971

Limb-Saving Resections of Malignant Bone Tumors with Autograft–Allograft Replacement

Harold M. Dick

The fatal biologic nature of primary malignant bone tumors has been well known through medical history. These include the osteosarcomas, chondrosarcomas, and malignant fibrous histiocytomas. Their rapid growth, late diagnosis, and blood-borne metastases have been well documented. Hemipelvectomy, disarticulation, and amputation have been the unhappy hallmark of treatment for these tumors until the past decade. During this same time period, more accurate pathologic study and aggressive primary surgical treatment have offered slow gains on the disappointing mortality rates from these tumors.

With the advent of more potent chemotherapeutic drugs such as adriamycin, high-dose methotrexate with citrovorum rescue, and other cytotoxic drugs, it appears that metastases may be controlled and possibly eliminated in some of these patients. With these adjuvant chemotherapeutic protocols as support, a renewed interest in radical en bloc resection has become an alternative approach to these lethal tumors. The rapidly expanding development of total joint replacements has also lent itself to the consideration of limb-salvage procedures. The technical feats of these resections has raised the need for replacement of the resected limb parts. The most successful replacement of metaphyseal–epiphyseal tumor resection has been total joint replacement.

For the diaphyseal long bone sections, massive allograft replacements have offered satisfactory results. The recent interest and development of successful techniques in the procuring, preparation, and storage of allografts has also enhanced our ability to offer a predictable limb-salvage procedure as a reasonable alternative to limb ablation.

The international terminology discussing various types of bone transplants (or any tissue transplant) is as follows:

Autograft refers to transplantation of a tissue or organ from one site to another in the same individual.
Isograft refers to a graft exchange between individuals of the same genetic background such as identical twins or inbred animal strains.
Allograft has replaced the old term *homograft* to describe transplantation from one individual to another of the same species but of different genetic background.
Xenograft has replaced the term *heterograft* to describe the transplantation from an individual of one species to another, i.e., cattle to human.

Orthopedic surgeons have been in the vanguard of tissue transplants, as recorded in 1668 by Job Van Meek'ren. Several other reports followed experimentally until in 1880 Mac Ewen described tibial bone grafting procedures. Albee in 1911 and several others continued to report their successful autografting. It soon was well established that autografts were safe, helpful, and convenient. They remain the mainstay of bone grafting in orthopedic surgery.

The massive resections required for adequate malignant tumor resection, however, does not lend itself to autograft replacement. For this reason, new interest in massive allografts was promoted by Volkov and Parish in 1966 and Ottolenghi in 1972. These successful reports were supported by recent reports of Wagner in 1975 and Mankin in 1976. Our experience from 1974 through 1978 was recently reviewed at the Columbia Presbyterian Medical Center Bone Tumor Service. Fourteen patients were treated during this time period with various size massive allograft and autograft replacements to substitute for limb-saving resection of long bones.

Technique of Allograft Replacement

The donor bone in all patients was obtained from the University of Miami Tissue Bank under the direction of Dr. T. I. Melanin. The donor bone is obtained within 6 hours of the donor's death under sterile operating room conditions. The donor must have been between 15 and 45 years of age, free of infection or neoplasm, and must not have received high doses of corticosteroids within the week prior to death. The long bone is cleared of all soft-tissue attachments. Cultures are taken of the long bone surfaces, marrow, cartilage, and donor blood. If cartilaginous surfaces are preserved, they are treated by immersion in dilute glycerol solution for 20 minutes. The graft is then wrapped in sterile gauze, x-rayed for defects, and then placed in refrigeration at 4 C for 24 hours to allow the glycerol to penetrate the cartilage cells. Mankin believes that glycerolization of cartilage during freezing prevents ice crystal formation and thus helps to maintain the viability of the chondrocytes. Freezing will help decrease the immunologic response and subsequent destruction of cartilage. After 24 hours of refrigeration at 4 C, the bone is stored at −100 C until needed. It is felt that

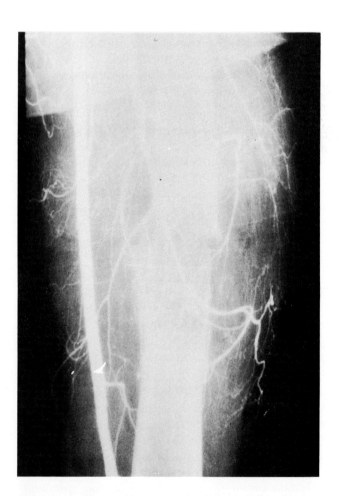

FIG. 1. Solitary bone metastases from hypernephroma diaphysis of femur.

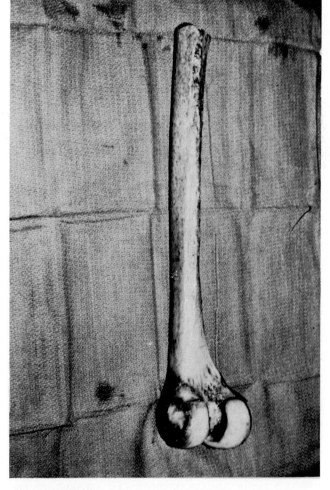

FIG. 2. Freeze-dried allograft, reconstituted with saline over 24 hours, ready for fitting to recipient defect.

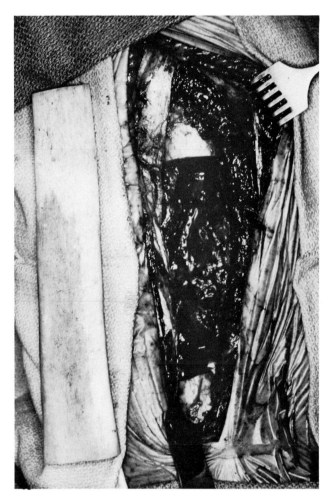

FIG. 3. Resection of diaphyseal tumor and allograft cut to size for replacement.

FIG. 4. Graft inserted to intercalary defect over an intramedullary rod.

immunogenicity of allograft bone is decreased when it is preserved at these low temperatures.

Prior to surgery, the patient receives a thorough clinical laboratory and radiographic study, including arteriography, bone scan, chest tomography, and computerized axial tomography scan to rule out metastatic disease and clearly delineate the extent of the bone tumor.

The allograft is selected of the proper size and shape to approximate the bone tumor. The majority of the allografts were diaphyseal and were not glycerinized but rather freeze-dried for easier transportation.

The freeze-dried grafts were reconstituted in their sterile containers 24 hours prior to surgery. The fresh-frozen glycerinized grafts were thawed in the operating room during the resection procedures. The tumor bone is removed en bloc with periosteum and soft tissue required to obtain an adequate tumor margin. The allograft is then cut to size, inserted in the defect, and internally fixed, usually with a combination of intramedullary fixation and cortical screws.

All patients had preoperative, intraoperative, and postoperative antibiotics. The lower extremity replacements were all placed in ischial weight-bearing braces in the postoperative period and maintained in those braces for a minimum of 18 months following graft insertion.

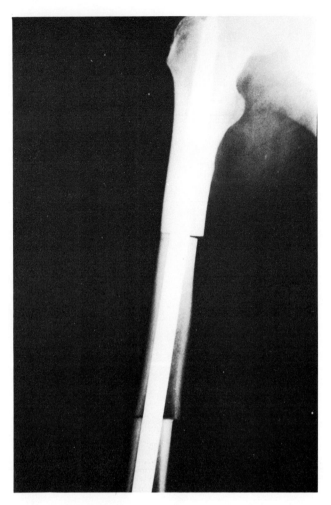

FIG. 5. Post operative x-ray demonstrating graft in place over intramedullary rod.

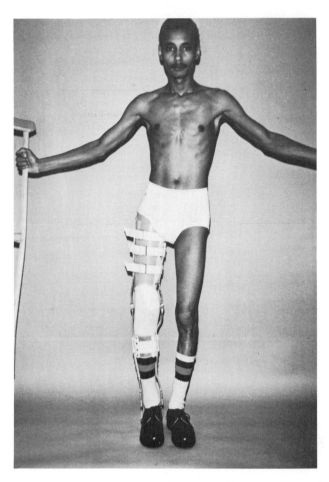

FIG. 6. Patient at 3 weeks' postoperative resection of osteosarcoma and allograft replacement using ischial weight-bearing brace for protection.

Results

The overall results of the procedure have been very gratifying (Tables 1–6). The two patients who died of their disease had advanced disease and proven metastases at the time of surgery. They both returned to a postoperative pain-free ambulatory state and died with no evidence of local disease at the original tumor sites. The young woman with a local recurrence at 2 years was treated by a shoulder disarticulation without recurrence or metastases 8 years later.

The single wound infection occurred in an adolescent who was severely immunosuppressed by preoperative chemotherapy using high-dose methotrexate and citrovorum rescue. She responded to surgical drainage and removal of the allograft. (Her resection was a hemisection of the posterior half of the distal femur for a juxtacortical malignant fibrous histiocytoma.)

We believe there is a real place for limb-saving procedures in treating malignant bone tumors. The improved techniques for replacement and adjuvant chemotherapy offer promise to those patients whose only treatment has been limb amputation in the past.

Table 1 Overall Results of Limb-Saving Procedures

Case number	Diagnosis	Age/sex	Graft site and size	Results	Complications
1	MFH	53/F	Femur, 20 cm	Satisfactory	Died of metastases of tumor—3 months postoperatively
2	MFH	50/M	Femur, 15 cm	Excellent	Results of graft at 2 years
3	Chondrosarcoma	19/F	Tibia, 12 cm	Excellent	Healed with good reconstruction
4	Juxtacortical osteosarcoma	24/F	Femur, 16 cm	Good	Good repair of graft
5	Metastatic cell	62/M	Femur, 20 cm	Excellent	Satisfactory graft incorporation
6	Chondrosarcoma	19/F	Femur, 10 cm	Excellent	Satisfactory graft incorporation
7	Osteosarcoma	22/M	Femur, 20 cm	—	Died of metastatic disease
8	Juxtacortical chondrosarcoma	14/M	Autograft and humerus prosthesis	Excellent	NED—6 years
9	Chondrosarcoma	22/M	Humerus prosthesis and autograft	Excellent	NED—2 years
10	Chondrosarcoma	19/F	Distal humerus autograft	Recent amputation	NED—8 years
11	Osteosarcoma	15/M	Femur, 25 cm	Excellent	NED—1 year
12	Juxtacortical osteosarcoma	70/F	Femur, 16 cm	Too early	NED—5 months
13	Malignant giant cell	27/F	Femur plus joint, 12 cm	Too early	NED—5 months
14	MFH	15/F	Femur, 25 cm	Post operative wound infection	NED—9 months

NED = no evidence of disease.

Table 2
Diagnoses

Osteosarcoma	6
Chondrosarcoma	4
Malignant histiocytoma	2
Giant cell tumor	1
Metastatic kidney carcinoma	1

Table 3
Internal Fixation Devices

Schneider rods	9
Humeral prostheses	2
Lottes nail	1
Rush rod	1
Supracondylar plate	1

Table 4
Location of Tumors

Femur	10
Humerus	3
Tibia	1

Table 5
Complications

Dead of disease (metastases)	2
Wound infection	1
Local recurrence (required amputation)	1

Table 6
Allograft Dimensions

Range	9–25 cm
Average	10 cm

Bibliography

Boyne PI: Review of the literature on cryo-preservation of bone. Cryosiology 4:341, 1968

Brown MD, Malinin TI, David PB: A roentgenographic evaluation of frozen allografts versus allografts in anterior spine fusions. Clin Orthop, 1977

Burwell RG: Editorials and annotations. Skeletal allografts for synovial reconstruction. J Bone Joint Surg 52B:10, 1970

Enneking WF: Histological investigation of bone transplants in immunologically prepared animals. J Bone Joint Surg 39A:597, 1957

Herndon CH, Chase SW: Experimental studies in the transplantation of en bloc joints. J Bone Joint Surg 34A:564, 1952

Langer F, Czitrom A, Pritzher WP, et al: The immunogenicity of fresh and frozen allogenic bone. J Bone Joint Surg 57A:216, 1975

Langer F, Gross AE: Immunogenicity of allograft articular cartilage. J Bone Joint Surg 56A:297, 1974

Malinin TI: University of Miami Tissue Bank, Transplant Proc., 8(Suppl):53, 1976

Mankin HJ, Fogelson FS, Thrasher AZ, et al: Massive resection and allograft transplantation in the treatment of malignant bone tumors. N Engl J Med 294:1247, 1976

Ohtolenghi CE: Massive osteo and osteoarticular bone graft. Technique and results of 52 cases. Clin Orthop 87:156, 1972

Pappas AM: Current methods of freezing and freeze drying. Cryosiology 4:358, 1968

Parish FF: Allograft replacement of all or part of the end of a long bone following excision of a tumor. J Bone Joint Surg 55A:1, 1973

Rish BL, McFadden JR, Penix JO: Anterior cervical fusion using homologous bone grafts: a comparative study. Surg Neurol 5:119, 1976

Sell KW, Billingham R, Russell P, Malinin T, Smith M (eds): Principles of Transplantation. Chicago, AMA, 1977

Smith RJ, Mankin HJ: Allograft replacement of distal radius for giant cell tumors. J Hand Surg 2:299, 1971

Musculoskeletal Applications of Computed Tomography

Harry K. Genant

Computed tomography (CT) is a new radiologic procedure now available to the radiologic diagnostician. Conventional radiography captures emergent x-rays by means of a screen–film system, with the resultant advantage of high spatial resolution but the shortcomings of overlapping anatomy and the inability to record the narrow differences in x-ray absorption between the various components of human soft tissues. Conventional tomography helps to eliminate the problem of overlapping structures but fails to distinguish subtle differences of tissue density. The much more sensitive system of CT affords detection and subsequent display of tissues of similar absorption characteristics and produces a transverse cross section of the part of the body being examined. The purpose of this discussion is to familiarize the radiologist with the potential orthopedic applications of this new diagnostic modality.

Study

To date, approximately 80 CT examinations for lesions of the musculoskeletal system have been performed at the University of California, San Francisco Medical Center using the EMI 5005 or the GE CT/T body scanners. Fifty-five of these cases have had surgical exploration or a sufficient follow-up interval to provide a basis for confirming or refuting the CT findings and diagnoses.

These cases have been reviewed and the data categorized as follows: (1) nature of the pathologic process, such as neoplastic or nonneoplastic; (2) location of lesion, such as extremity, limb girdle, or spine; and (3) type of tissue involved, such as bone, soft tissue, or both. In addition, these cases have been analyzed to determine the specific utility of CT with respect to establishing a diagnosis of the lesion, determining the extent of the process, and influencing the surgical or nonsurgical management of the patient. These judgments were based upon a composite of clinical, laboratory, and imaging data and were made in conjunction with referring clinicians.

The results are compiled in Table 1 and representative cases are shown in Fig. 1 through 3. Overall, CT was helpful in establishing a diagnosis in 46 percent of cases, in determining extent of the lesion in 78 percent, and in treatment planning in 78 percent. It was most useful in purely soft-tissue lesions of the extremity.

Table 1 Utility of CT in 55 Patients

Lesion category (total patients)	Percentage of patients		
	Diagnosis	Extent	Treatment
Disease process			
Neoplastic (38)	32	92	76
Nonneoplastic (17)	76	47	82
Location			
Extremity (25)	52	68	76
Limb girdle (21)	43	86	76
Spine (9)	33	89	89
Type of involvement			
Bone (13)	38	85	54
Soft tissue (22)	68	68	91
Both (20)	25	85	80

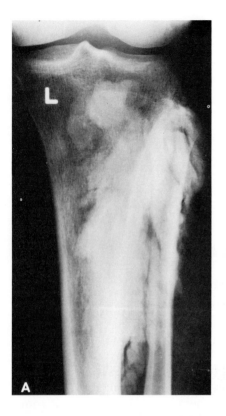
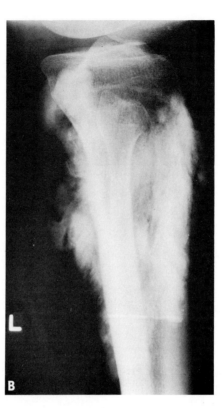
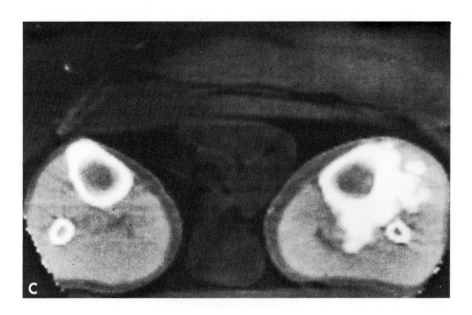

FIG. 1. Twenty-year-old male with painless mass in leg. (Top, left and right): Conventional radiographs demonstrate an ossified mass in the soft tissues surrounding the proximal tibia and fibula. Even with conventional tomography, the relationship of this mass to the underlying osseous structures could not be determined. (Bottom:) A CT scan defines precisely the intimate relationship between the ossified mass and the tibia. The image demonstrates cortical destruction but little medullary involvement, which is virtually diagnostic of paraosteal sarcoma and rules out myositis ossificans and osteosarcoma, the two other primary diagnostic considerations.

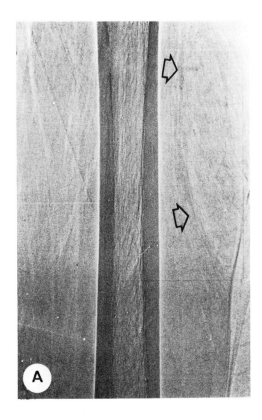
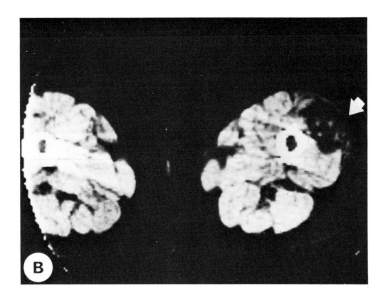

FIG. 2. Thirty-five-year-old female presented with a painful soft-tissue swelling in the thigh. (Opposite): A xeroradiograph shows a partially encapsulated mass (arrows) with lower density than surrounding tissues, suggesting lipoma. (Above): The CT scan demonstrates a somewhat lobulated mass of fatty density (arrow), which also contains elements of soft-tissue and calcium density. These features suggest a process other than simple lipoma. An angiolipoma was subsequently found surgically.

Discussion

Our early experience with CT scanning of the musculoskeletal system indicates that a more precise anatomic depiction of bone and soft-tissue lesions is afforded with this modality than has been possible by other means. It provides an accurate and detailed cross-sectional image of normal anatomy and shows the relationship of masses to normal structures. The encroachment of masses on vital structures, such as nerves or vessels, and actual invasion of structures such as bone and muscle are discernible. Preoperative evaluation is often clarified, which helps to establish the feasibility of surgical excision and the proper surgical approach. This is particularly evident for lesions in and around the pelvis and axial skeleton, where the transaxial view and the exceptional density resolution of CT scanning demonstrate the bone and soft tissue relationship to advantage. The capability of CT to obtain specific x-ray absorption coefficient values for a lesion renders it more accurate than conventional radiography in establishing the presence of water, fat, or calcium density. Furthermore, the serial assessment of malignant lesions following radiation therapy or partial surgical resection is enhanced, and the detection of recurrent disease is facilitated by CT.

The future contribution of CT to orthopedic diagnosis and management is unclear. The impact of this modality may be less in the musculoskeletal than in other anatomic areas, primarily because of the greater level of diagnostic accuracy of conventional radiography for the skeleton. Nonetheless, in many patients studied to date, substantial contributions have been demonstrated.

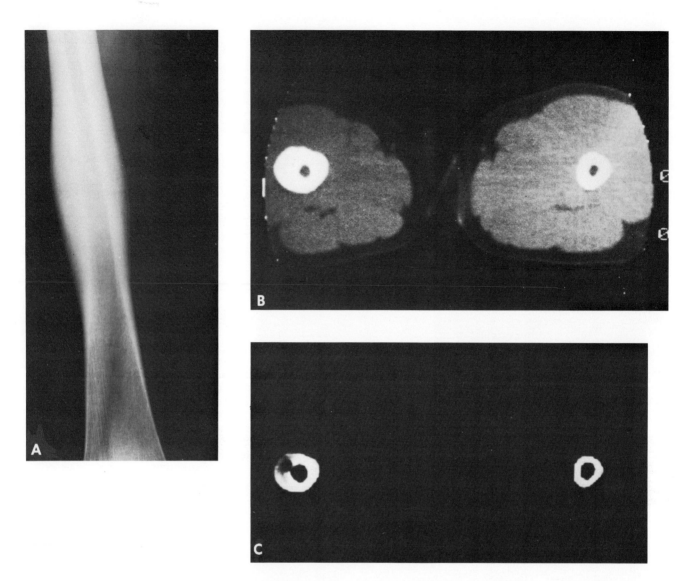

FIG. 3. Sixteen-year-old male presented with pain in the leg remotely associated with trauma. (Left): Conventional radiograph demonstrates primarily solid, but also slightly laminated, periosteal reaction in the midshaft of the right femur consistent with osteoid osteoma or sclerosing osteomyelitis. (Right): CT scans with window at two different settings demonstrate a central lucency at ten o'clock, highly suggestive of osteoid osteoma, and also demonstrates clearly the location for resection of the nidus.

Bibliography

Berger PE, Kuhn JP: Computed tomography of tumors of the musculoskeletal system in children. Radiology 127:171, 1978

Schumacher T, Genant HK, Korobkin MT, Bovill EG: Computerized tomography for diagnosis of musculoskeletal disorders. J Bone Joint Surg (in press)

Weinberger G, Levinsohn EM: Computed tomography in the evaluation of sarcomatous tumors of the thigh. Am J Roentgenol 130:115, 1978

Wilson JS, Korobkin M, Genant HK, Schumacher TM, Bovill EG: Clinical assessment of computed tomography for musculoskeletal disorders. Am J Roentgenol (in press)

Histiocytosis X: Pathology

D. C. Dahlin

Histiocytosis X includes a spectrum of conditions that ranges from the usually solitary and curable eosinophilic granuloma through the disseminated process that produces the Schüller–Christian syndrome to the fulminating and usually rapidly fatal variety known as Letter–Siwe's disease. Lesions of bone ordinarily dominate the pathologic picture, histiocytosis X being a disease of the reticuloendothelial system. Letter–Siwe's disease usually affects very young children, whereas the Schüller–Christian syndrome and the eosinophilic syndrome are seen most often in children and young adults. Practically any bone in the body may be affected, but there is a predilection for the skull.

Although a great variety of symptoms is produced, the commonest in patients with eosinophilic granuloma is a solitary painful focus and sometimes a palpable or visible mass. The triad of the Schüller–Christian syndrome classically includes exophthalmos (often unilateral), diabetes insipidus, and rarefied defects of the bones of the skull. A partial triad has the same significance if other evidences of dissemination such as anemia, splenomegaly, fatigability, weight loss, and lymphadenopathy are present.

A wide variety of symptoms is produced because of the wide variety of lesions that may be produced by this disease. These include discharge from the ear due to involvement of temporal bone, loosening or falling out of teeth secondary to lesions of the jaws, and the symptoms that might be produced by focal destruction of bone. Vertebral involvement may result in collapse of a vertebral body with resultant neurologic symptoms. Cutaneous manifestations, lymphadenopathy, and splenomegaly are most common in the progressive diffuse form of the disease. Pulmonary infiltration may become clinically important and, on rare occasions, is the most significant evidence of the disease. Diffuse pulmonary lesions of histiocytosis X, on the other hand, may occur in the absence of osseous lesions. Such patients are usually adults who are not seriously ill, may have episodes of spontaneous pneumothorax, and have an unpredictable clinical course.

Grossly, the lesional tissue is soft and it may be hemorrhagic, gray, pink, or yellow. The microscopic appearance is what links these three general conditions together (Fig. 1). The salient and pathognomonic feature consists of foci of proliferating histiocytic cells. These histiocytes frequently have ill-defined cytoplasmic boundaries and characteristically contain an oval or indented nucleus. Multinucleated histiocytes simulating benign giant cells of giant cell tumor of bone may be seen. Although chromatin clumping and nuclei are inconspicuous, mitotic figures are not uncommon. This has led to the occasional confusion of histiocytosis X with malignant tumors, especially reticulum cell sarcoma and even Ewing's sarcoma. Zones of necrosis are present in many of the lesions. Secondary chronic inflammatory infiltrate and fibrosis are often present in osseous or extraosseous lesions of histiocytosis X. The histiocytes may be swollen owing to cholesterol in their cytoplasm. Varying numbers of eosinophils, lymphocytes, and neutrophils are nearly always present. The characteristic clusters of histiocytes may be obscured by the zones of nonspecific subacute and chronic inflammatory cells and fibrosis.

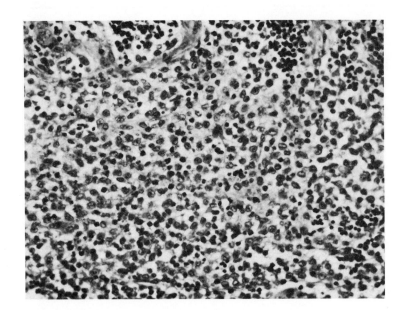

FIG. 1. Histiocytosis X of clavicle of a 9-year-old girl. Sheets of pale-staining histiocytes are interspersed with darker staining lymphocytes and a few eosinophiles. H&E, x250.

Bibliography

Enriquez P, Dahlin DC, Hayles AB, Henderson ED: Histiocytosis X: a clinical study. Mayo Clin Proc 42:88, 1967

Lichtenstein L: Histiocytosis X (eosinophilic granuloma of bone, Letter–Siwe disease, and Schüller–Christian disease). Further observations of pathological and clinical importance. J Bone Joint Surg 46A:76, 1964

Smith M, McCormack LJ, Van Ordstrand HS, Mercer RD: "Primary" pulmonary histiocytosis X. Chest 65:176, 1974

Histiocytosis X: Radiology

Frieda Feldman

The term *histiocytosis X* was introduced to encompass three major entities—Deosinophilic granuloma of bone, Hand–Schüller–Christian disease, and Letterer–Siwe disease—whose basic pathology consists of histiocytic infiltration of tissues in one or more sites of the body. The implication was that these disorders were merely variable expressions of the same basic abnormality and that they are clinically, etiologically, and pathologically related.

Classically, eosinophilic granuloma is thought of as the most benign and limited form of histiocytosis X, Hand–Schüller–Christian disease as the chronic disseminated form, and Letterer–Siwe disease as the fulminant and malignant form of the disorder. Apparent transformation from one variant to another has been noted, but the etiology of this group of diseases is still unknown.

Historical Review

Hand–Schüller–Christian disease was the first to be described by Hand in 1893 in a 3-year-old male with bone destruction, multiple visceral involvement (spleen, liver, kidneys), lymphadenopathy, exophthalmos, and polyuria.[28] In 1921, he noted similar patients[29] reported by Schüller, Christian, and Kay,[11,12,59] and Hand–Schüller–Christian disease then became a label applied to those cases with histiocytic infiltrates in multiple-organ systems, including bone, that have a slow, insidious onset and a chronic course throughout childhood.

Roland was subsequently credited with identifying the predominant cell type as a lipid-laden macrophage, later known as a histiocyte or foam cell.

Letterer–Siwe disease was initially described by Letterer in 1924[42] and then by Siwe in 1933.[62] Abt and Denenholz in their subsequent review of nine similar cases[1] initiated the label *Letterer–Siwe disease*.[63] This is the acute, fulminant, often rapidly fatal form of histiocytosis most common in children younger than 2 years of age. Although accompanied by histiocytic infiltration of many organs, it less commonly involves bone.

Eosinophilic granuloma was the last to be described, nearly simultaneously by Lichtenstein and Jaffe (1940),[48] Otani and Ehrlich (1940),[56] and Hatcher (1940). It was called *eosinophilic granuloma of bone* by Jaffe and Lichtenstein (1944)[31] and *solitary granuloma of bone* by Otani, who considered it a distinct entity.[55,56]

Originally defined as limited to bone and most commonly to a single bone, Farber (1941)[22] and Green and Farber (1942)[26] subsequently showed that multiple bones could be affected. Although their cases had no extraskeletal involvement, they suggested a relationship to Hand–Schüller–Christian disease and Letterer–Siwe disease and introduced the term *destructive granuloma*. Lichtenstein (1953)[45,46] then proposed *histiocytosis X* as a broad umbrella designation to emphasize the pathologic similarity of the three syndromes. Other designations such as *histiocytic granuloma*[58] continue to be coined.

Proponents of a unified concept maintain that the three clinical presentations are not sharply defined but are merely different manifestations of a single process. They continue to find the broad designation *histiocytosis X* useful and cite many cases with imprecise, overlapping features, as well as examples of transitional cases that have experienced a change in clinical course despite an underlying histologic similarity.

Microscopically, the three syndromes are fundamentally characterized by the proliferation of differentiated reticulohistiocytic elements of the granulomatous type accompanied by eosinophilic granulocytes and other inflammatory cells. Histiocytes may show secondary changes due to phagocytosis of blood pigment (hemosiderin) and cellular debris or of lipid, so that they have also been called *xanthoma cells* or *lipid-bearing foam cells*. Foam cells have been said to be more abundant in the Hand–Schüller–Christian syndrome, while in the Letterer–Siwe syndrome, histiocytic proliferation predominates along with variable numbers of eosinophils and other inflammatory cells.[58]

It has been noted that early lesions may be identified by larger numbers of eosinophils or actively phagocytic histiocytes, while older lesions have a preponderance of histiocytic foam cells, presumably due to necrosis. The latter is said to be more extensive in older lesions due to deficient circulation, which has, in turn, been related to hyperplasia of arterial endothelium with narrowing of the vascular lumen.[58] In still older lesions, a predominance of fibroblastic elements is associated with fibrosis and/or healing. Foam cells, however, have been found independent of necrosis. In addition, several lesions with varying histology and/or at various stages of development may exist in the same individual at different sites or at different times. A first biopsy may show an intense histiocytic proliferation, while the second, taken months or years later, may show a "typical" fibroxanthomatous lesion. Therefore, biopsies carried out consecutively on a particular case may exhibit transformation from a predominantly eosinophilic stage, to a histiocytic granulomatous stage, to a lipid granulomatous stage chiefly characterized by foam cells, to a stage composed of fibroblastic elements that have substituted the granulation tissue.[31,32,45–48,58] Intermediate types may occur especially in young children, e.g., about 3 years of age, in which the exact classification either as Hand–Schüller–Christian disease with conspicuous extraskeletal lesions or as milder, less acute forms of Letterer–Siwe disease may not be possible. Jaffe and Lichtenstein[31,47,48] in particular stressed this frequent overlap with transitional or intermediate types and the correlation of lesions with histologic maturation, thereby emphasizing the pathologic continuum or spectrum of the three diseases. Others believe[9,37,38] that the more benign variant may be differentiated from the more malignant on a histologic basis in that the former are distinguished by a mixture of eosinophils and histiocytes and the latter by diffuse histiocytic infiltration. However, most pathologists find the evolution of a particular case generally difficult, if not impossible, to predict without being apprised of the complete radiologic and clinical picture.

Not all authorities have ascribed to the unified concept[55] that has undergone recent reappraisal.[14,49] Some maintain that despite similar histologic patterns three separate entities are involved. Letterer–Siwe's disease in particular, with its associated extensive organ infiltration and high fatality rate, has been considered by some investigators as one of the malignant lymphomas and specifically as a malignant histiocytic lymphoma.[14] Eosinophilic granuloma, since it tends to remain stationary or indolent and may even spontaneously regress, is considered to be a benign disease. The proper classification and prognosis of Hand–Schüller–Christian disease remains in doubt in this scheme, with some investigators preferring to use the designation of *multifocal eosinophilic granuloma*.[49] Two designations or categories have been suggested for eosinophilic granuloma—unifocal and multifocal.

Unifocal eosinophilic granuloma in this scheme is defined as eosinophilic granuloma confined to one system, i.e., either bone or viscera, with one or multiple sites of involvement within that system, which is the only one involved, while *multifocal eosinophilic granuloma* is defined as eosinophilic granuloma involving multiple systems, i.e., the skeletal as well as nonosseous sites. In this classification, the term *multifocal eosinophilic granuloma* replaces the classic designation of *Hand–Schüller–Christian disease*. However, an explicit designation of *unifocal* or *multifocal eosinophilic granuloma* at any point in time is difficult, since not only may multiplicity not be recognized, but the number and extent of involved sites may increase over the years. Nevertheless, a distinction between Letterer–Siwe's disease on the one hand and eosinophilic granuloma (either unifocal or multifocal as in Hand–Schüller–Christian disease) on the other, has been advocated.[14,49,55,69]

Recently, however, ultrastructural studies have indicated that histiocytes in all three entities contain identical Langerhans cells that, in turn, contain cytoplasmic inclusion bodies called *Langerhans granules*. Langerhans cell granules have also been found in unrelated diseases and in normal as well as in abnormal conditions. They have been noted in the epidermis, dermis, lymph nodes, thymus, and other tissues and have been described in monocytic leukemia and reticulum cell sarcoma. However, studies have shown Langerhans cells to be unreactive with antilymphocytes, indicating that they are not of lymphatic or myelocytic origin.

Langerhans cells have been considered by some investigators to be the histiogenic precursors of the variants of histiocytosis X, while others consider them to be characteristic if not diagnostic of histiocytosis X.[17,23,24,43,54] Their presumed histiocytic origin and their purported phagocytic capabilities and mobility are held to account for their presence and frequent dissemination in all three syndromes in extracutaneous sites, including bone, as well as in lesions other than histiocytosis X. Their origin, however, is still controversial.

An immunoallergic hypersensitivity reaction to a still unknown infection, possibly viral, has been suggested as a cause of histiocytosis X and as an explanation for the therapeutic effect of antibiotics and steroids in some cases. Recently, Letterer–Siwe's disease was associated with an unusual reaction to a viral infection in an immunologic-deficient child[13] as well as with a familial opsonization defect.[60] However, other immunologic studies in histiocytosis X, using tests of lymphocyte and neutrophil function and immunoglobulin levels, revealed that though immunologic abnormalities were present during active illness, they were felt to be secondary to cell replacement by the underlying process.[41] An immunodeficiency disorder could not be found to explain the illness. HLA typing has not, to date, been useful in detecting those who might be at risk. There has been no evidence to support an underlying metabolic or lipid storage abnormality. The possible role of genetics has also been questioned. In view of certain clinical features, including its predilection for males and its rare occurrence in blacks, genetic mechanisms have been postulated as possible modifiers of environmental factors. A variety of histiocytic disorders have, in fact, been reported in several family members and in twins.[23,33,51] Juberg[33] postulated that "at least some instances result from a single autosomal recessive gene with slightly reduced penetrance." However, several salient questions remain unanswered. What determines the presence of a solitary lesion in some patients and a chronic unrelenting and sometimes fatal course in others? If multifocal (Hand–Schüller–Christian disease) and unifocal eosinophilic granuloma are etiologically related, what factors are operating in modifying their clinical expression? Could the wide spectrum of clinical presentation indicate that several etiologic factors are operative or that the disease is modified by host factors?

To date, none of these questions has been satisfactorily answered, so that etiology as well as the nomenclature continues to be disputed. Despite the continuing controversy, many clinicians continue to find the subdivision into three separate syndromes convenient, since certain overall differences in age of onset, clinical course, and prognosis permit the analysis of a considerable number of cases. Therefore, keeping the above-mentioned reservations in mind, as well as several of the newer names most frequently applied to the variants of histiocytosis X, the three syndromes will be discussed according to their classic designation.

Eosinophilic Granuloma of Bone

Single or multiple bones may be affected with no known extraskeletal involvement.

Incidence

Eosinophilic granuloma of bone comprises approximately 60 to 70 percent of the total number of cases of histiocytosis X. In Schajowicz and Sluttitel's series,[58] 76 of 106 cases had solitary bone lesions. However, most of their cases originated from orthopedic rather than children's hospitals, with resultant fewer examples of extraskeletal lesions.

Age and Sex

Ages ranged from 1 to 53 years in Schajowicz' series but only 27.7 percent of patients were over 20 years old (21 patients). It most commonly occurred between 1 and 15 years (62 percent), with a peak incidence between 5 and 10 years. Male to female ratios of 3:2 and 2:1 have been noted. If only solitary lesions are considered, male predominance may be higher—2.3:1.

Distribution

Solitary or multiple osseous loci may be radiographically apparent. Lesions most frequently involve the skull (70 percent) and femur in patients under 20. Lesions in patients over 20 were most common in ribs (67 percent), mandible (57 percent), clavicle (50 percent), scapula (50 percent), and skull (30 percent) in Schajowicz and Sluttitel's series.[58] In general, the skull (50 percent), bones of the trunk (25 percent), and bones of the proximal extremities (15 percent) are most frequently involved. In long bones, the diaphysis and metaphysis are predilected, while an epiphyseal locus is rare.

Clinical Features

Pain, warmth, and tenderness, usually localized to the area of the lesion, are the most common initial complaints and may lead to limp and/or soft-tissue atrophy when an extremity is affected. A palpable or visible mass overlying the osseous lesion is the most common presenting complaint in children, particularly when the skull or long bones are affected[21] (Fig. 2). Vertebral involvement may be associated with stiffness, pain, spasm, and neurologic findings, while pathologic fracture may lead to the initial discovery of long-bone and vertebral lesions.

Fever, chills, or weight loss is rare. The sedimentation rate may be moderately elevated. Eosinophilia may occasionally be noted in the peripheral white count, while normochromic anemia may be secondary to marrow replacement by histiocytic infiltration.

Hand–Schüller–Christian Disease

This syndrome is the disseminated form of histiocytosis X or multifocal eosinophilic granuloma and may involve the soft tissues as well as bone.

Incidence

Hand–Schüller–Christian disease comprises approximately 15 to 40 percent of the total number of cases of histiocytosis X.

Age and Sex

Ages have ranged from birth through the seventh decade, but the disease characteristically occurs between 5 and 10 years of age. In a review of 73 patients with multifocal disease over a 55-year period, only 15 patients were over 15 years of age.[21] A male preponderance was noted.

Clinical Features

Involvement of the reticuloendothelial system, i.e., the spleen, liver, lymph nodes, and bone marrow, is most commonly noted, although lungs, heart, brain, kidneys, gastrointestinal tract, and skin, as well as the musculoskeletal system, may be affected. In view of this multisystemic involvement in addition to bone, patients may present with a variety of complaints. Clinical manifestations that often have an insidious onset include chronic mastoiditis, otitis, cholesteatoma formation and hearing loss, diabetes insipidus, adrenal insufficiency, hypothyroidism, hepatosplenomegaly, exophthalmos, neurologic abnormalities, and skin and oral lesions, all or some of which may appear over a protracted period of time, which may be as long as several years. The triad of exophthalmos, diabetes insipidus, and skull defects classically associated with Hand–Schüller–Christian disease is found in less than 10 percent of cases, i.e., 6 of 129 cases of Cheyne[10] and 1 of 106 of Schajowicz and Sluttitel.[58]

Central nervous system involvement by Hand–Schüller–Christian disease has most frequently been related to infiltration and dysfunction of the hypothalamus and posterior pituitary. However, although diabetes insipidus is the most familiar endocrine disturbance, anterior pituitary hormone deficiencies have been documented.[5,40] Impairment in the hypothalamic–pituitary axis may also lead to impaired thyroid function. Recent reports suggest that extra-hypothalamic involvement is more prevalent than previously recognized, and lesions described in the cerebellum, brain stem, spinal cord, and cauda equina may be clinically reflected in the form of progressive cerebellar and pyramidal tract signs.[6,30,36] Extradural compression with signs and symptoms of radiculopathy have been evidenced by myelography.[20,68] Impairment of intellectual capacity has also been clinically manifested.

Of 117 cases of histiocytosis X studied at the Mayo Clinic, diabetes insipidus developed as a late complication in 21 patients with normal skull base morphology.[36] The oldest patient with diabetes insipidus was 49 years old at the time of onset.

Skin lesions are seen in approximately 30 to 50 percent of patients with Hand–Schüller–Christian disease and may be characterized by a greasy, scaly, papular eruption. It is most often confused with seborrheic dermatitis, since it is seen in areas typically involved with seborrhea, i.e., scalp; face; and retroauricular, axillary, and inguinal regions.[2,4,66] Infiltrations in the skin folds of the inguinal, retroauricular, and infragluteal regions may necrose and ulcerate with resultant deep longitudinal fissures. Cutaneous involvement occurred in 21 of 81 children and in 4 of 36 adults in the Mayo Clinic series.[20] Its occurrence correlated with extensive disease in very young children. Localized skin ulcers are more common in adults.

Oral lesions, in addition to gingival swelling and soreness, may occur. Oral mucosal ulcerations developed in 19 of 117 cases of Enriquez et al.,[21] usually in association with maxillary or mandibular lesions. There may be an associated loss of teeth. However, loss of teeth may not necessarily be accompanied by oral mucosal involvement.

Pulmonary involvement based on clinical, radiologic, or histologic evidence occurred in 19 of 177 patients in the same series.[21] Twelve were children and 7 were adults. At least 5 of the latter had residual impairment of respiratory function, while none of the surviving children did. Pulmonary disease in childhood does not seem to alter the prognosis unless it is associated with widespread disease, may not be accompanied by symptoms, and does not usually leave clinical evidence of sequelae.[8,21] In adults (patients over 20), it often produces symptoms despite a negative chest x-ray, may have disabling permanent sequelae, and has been known to be the sole cause of death.[8,21]

Letterer–Siwe Disease

This syndrome is the acute fulminant form of histiocytosis X, marked by widespread visceral infiltrates and, less commonly, by skeletal involvement.

Incidence

Letterer–Siwe disease constitutes approximately 10 percent of cases of histiocytosis X.

Age and Sex

The syndrome is most commonly manifested in children younger than 2 years. Congenital Letterer–Siwe disease has been noted, and cases in full-term stillborn infants have been reported.

Clinical Features

In addition to an age of onset of less that 2 or 3 years, criteria for diagnosing Letterer–Siwe disease, as established by Doede and Rappaport,[17] include clinical findings of diffuse nontender lymphadenopathy, hepatosplenomegaly, otitis media, and cutaneous manifestations. Purpura of the palms, a finding seldom seen in association with other skin diseases, is a poor prognostic sign and is often noted before demise. Fever, hemorrhage, anemia, and recurrent bacterial infections are frequent complications.

The course may be subacute but is most commonly acute and most commonly fatal. A rare case is protracted beyond 1 or 2 years of age, i.e., 5 of 96 cases reviewed by Doede and Rappaport.[11] Osteolytic lesions often do not develop or may not be evident.

Discussion

Multiple osseous lesions in all three syndromes predominated in patients under 15 years of age, with a predilection for the first 5 years. In those with multiple osseous lesions, with and without features of Hand–Schüller–Christian or Letterer–Siwe disease, the bones most frequently affected were the skull and femurs. Therefore, according to Schajowicz and Sluttitel,[58] in every case with a lesion in the skull, a radiograph of at least the femur is indicated and vice versa.

The prognosis is variable. An overall mortality approximating 30 percent has been noted with multiple areas of involvement. Patients with the classic triad or widespread or advanced visceral involvement, especially of the lungs, liver, or central nervous system, have a poorer prognosis.[3,21,52,61] Lahey[37,38] had directly related mortality rate to the number of organ systems involved, and in those with seven or more, the mortality rate was nearly 100 percent. The prognosis is also grave with anemia, which does not necessarily correlate with the extent of bone involvement detected roentgenographically.[21] Although most often related to marrow replacement,[10] its cause as well as that of the not infrequently associated leukopenia and thrombocytopenia is probably multiple.

The earlier the illness presents, the worse the outcome. In Lahey's series, 70 percent of children below 6 months of age at onset died, while 50 percent of those with onset before age 3 died.[37–39] It has also been noted that an earlier age of onset carried with it a greater likelihood of multiple sites of involvement, and the earlier the onset of disseminated lesions, the poorer the prognosis. Usually, dissemination of an initial solitary lesion tends to occur during the first few months and most frequently before 6 months. Several of Schajowicz and Sluttitel's patients had started with a solitary osseous lesion but developed multiple foci (with or without features of Hand–Schüller–Christian disease or Letterer–Siwe disease) after a few months and usually before 6 months.[58] However, they predicted a favorable prognosis in any case initially presenting with a solitary bone lesion if no new bone lesion or other focus appeared after 1 year. Cure was then said to be permanent, with conservative treatment consisting of curettage or radiotherapy.

Young age and evidence of soft-tissue involvement at presentation were also associated with a worse prognosis in Sims' series[61] in which 43 cases of histiocytosis X in children under age 12 were compiled over a 29-year period. Of these, 67 percent survived. The majority of deaths were associated with pulmonary involvement. Of the 67 percent that survived, 54 percent had detectable residual clinical disabilities. Survivors presented with more skeletal complaints. Every survivor had one or more bone

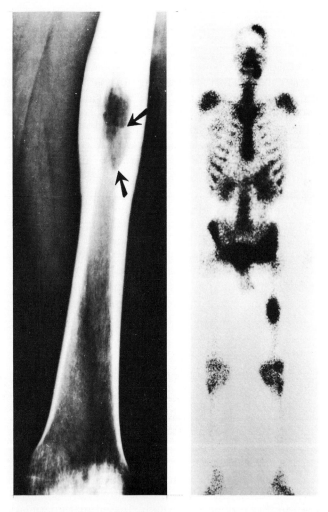

FIG. 1. Eosinophilic granuloma. This 21-year-old white male was diagnosed as having eosinophilic granuloma of bone based on a left supraacetabular biopsy 1 year prior to the current study. Left: An oval, centrally located area of osteolysis is seen within the diaphysis of the left femur. The lesion is well defined but has partially eroded the endosteal surface of the medial cortex. The defect is marginated by a beveled border (arrows), creating a three-dimensional or "hole within a hole" effect. An elliptically shaped localized area of cortical thickening is also noted medially. Right: A technetium-99m pyrophosphate bone scan. Multiple areas of abnormal tracer uptake predominate on the left side of the skeleton, including the skull, mandible, rib cage, ilium, and left midfemur. The latter area corresponded to the diaphyseal lesion in the figure on left.

lesions; however, they had fewer general features of ill health such as anorexia, failure to gain weight, skin manifestations, and lymphadenopathy.

In Lahey's experience,[37,38] excluding eosinophilic granuloma of bone, 63 percent of 59 patients suffered disability in the form of chronic active disease, small stature, diabetes insipidus, exophthalmos, vertebral compression, or lung fibrosis. Residual pulmonary fibrosis may affect carbon monoxide diffusing capacity despite a normal chest film. In older patients with lung involvement, dyspnea may coexist with a normal chest x-ray.[8,21] Lung involvement may lead to pulmonary fibrosis, emphysema, pneumothorax, and bronchiectasis, while major lung lesions may be associated with terminal dilatation of the right side of the heart.

Diabetes insipidus is seen in up to 20 percent of survivors and is related to previous hypothalamic and pituitary involvement. Pitressin therapy is usually effective. Small stature, in addition to resulting from vertebral collapse and scoliosis, may be related to growth hormone deficiency.[40,61] The effect of fibrous healing in areas of previously active disease may be manifest in clinical sequela such as pulmonary insufficiency, cirrhosis of the liver, and portal venous hypertension[27] when active disease is present. Liver involvement may be associated with histiocytic infiltration of the periportal areas. As the patient responds to treatment (chemotherapy), apparent hepatic healing with fibrosis may result in cirrhosis. Histologically, a dense macronodular cirrhotic pattern becomes evident that may lead to portal venous hypertension.[27] Gastrointestinal bleeding from esophageal varices may be an additional sequela. Portosystemic shunt procedures may prevent further hemorrhagic episodes and can provide long-term survival in children with portal hypertension.

Roentgenographic Findings

Long-bone lesions are usually rounded, ovoid, well-demarcated, lucent defects situated within the medullary cavity, which may be surrounded by slightly denser and occasionally scalloped margins. They are most common in the diaphysis or metaphysis, where they are often eccentrically placed. They are rarely situated in the epiphysis and infrequently appear as intracortical areas of rarefaction.

The lesions may enlarge within the medullary cavity, often causing localized expansion of the adjacent contour of the bone (Fig. 1). Expansion was observed in over 60 percent of affected long bones in one series[20] and was frequently seen in flat bones (Fig. 2). It may be accomplished by the erosion or thinning of the endostium or by localized cortical thickening due to periosteal as well as endosteal stimulation. Subtle erosion of the endosteal margin of the cortex may also be present in early lesions. This may progress to interruption or complete cortical dissolution with infiltration of the local soft tissues by friable

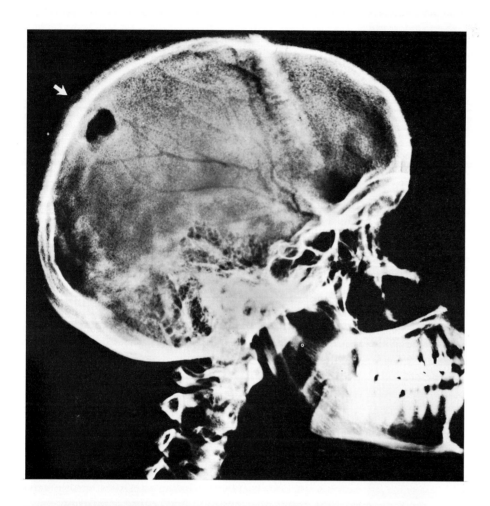
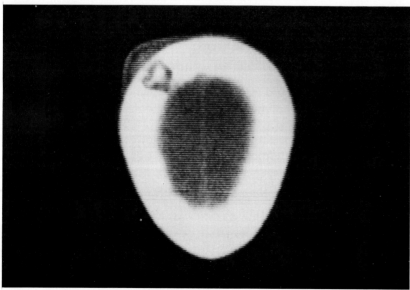

FIG. 2. Eosinophilic granuloma. This 20-year-old white female felt a "bump" on her head. There was no history of trauma. Top: A single, rounded, well-defined lytic area is seen within the calvarium. It has a characteristically "punched out" appearance. Unequal destruction of the inner and outer tables of the vault create a partially overlapping or beveled edge. There is suggestive evidence of slight localized expansion of the outer table (arrow) with an intact cortex. A tender mass was clinically palpable in this area. A large area of periodontal destruction was incidentally noted within the anterior half of the mandible, while a smaller area of osteolysis involves the opposing maxilla. The lower anterior teeth are so-called floating teeth, having no apparent bony mooring. Bottom: CT scan of the skull. A transverse section through the calvarium at the level of the lesion reveals the circular calvarial defect noted on the conventional lateral view. An additional bony density or "button" sequestrum is now seen projected within its radiolucent center. In retrospect, it is probably situated close to the inferior margin of the defect on the lateral view. The expansion of the outer table mass, as well as an associated soft-tissue mass, is now seen to better advantage. (Courtesy of Dr. Joel Budin, Hackensack, New Jersey.)

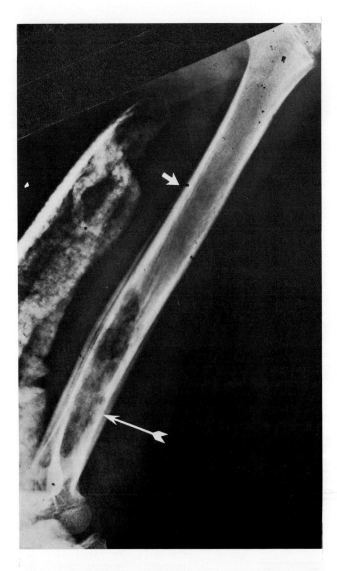

FIG. 3. Eosinophilic granuloma. This 7-year-old white male complained of pain and swelling of his left arm after minor trauma. An oblique view of the left humerus shows an ill-defined area of bone destruction centrally situated within the distal diametaphysis. The endosteum is focally eroded (large lateral arrow) as well as circumferentially thinned. A florid, multilayered periosteal reaction extends to the junction of the lower two-thirds and upper one-third of the humeral shaft (small medial arrow). A localized soft-tissue mass was not evident. (Courtesy of Dr. Jay Smith, Boston.)

material. A suppurative process may thereby be simulated on a histologic as well as a roentgenographic basis. Localized soft-tissue swelling and an associated mass may occasionally be noted on suitably exposed radiographs (Fig. 2).

Periosteal new bone formation, with or without fracture, is a common association of long-bone lesions. Active new bone formation may frequently be reflected in a positive bone scan (Fig. 1B) with or without concomitant roentgenographic evidence of neoostosis. Subsequent organization of a localized periostitis may appear as residual bony sclerosis and/or a fusiform area of solid cortical thickening often localized to the immediate site of the lesion (Fig. 1). Although marginal irregularity of localized lesions was thought to be associated with poor prognosis at one time[65] this has not been corroborated by additional experience.

Although an obvious periosteal reaction is a common association of long-bone lesions, it is relatively rare in flat bones. Among the flat bones, reactive marginal sclerosis about lytic lesions is most commonly observed in the pelvis and less commonly in the skull.

A periosteal reaction may be associated with an intramedullary lytic component with or without obvious cortical infringement. It may be florid and laminated in an "onionskin fashion," simulating a Ewing's sarcoma or an aggressive osteomyelitis (Fig. 3). An osteomyelitis may be further mimicked by a small fragment or "button of bone" often seen within the area of rarefaction, which has a sequestrumlike appearance (Fig. 2).

Lytic lesions may become manifest over a short period of time, i.e., 4 to 6 weeks after the onset of symptoms, and may be rapidly progressive.[20,65] The radiolucent defects may occur in crops or collections of holes. Occasionally, one larger area of rarefaction is seen to be associated with smaller peripheral satellite lesions. These two characteristics, multiplicity and rapid progression, may cause confusion with metastases, particularly if the lesions have ragged, irregular, or less well-defined margins than one is accustomed to seeing and are associated with visceral manifestations.

The skull is the most frequently affected flat bone, followed by the scapula, pelvis, mandible, and ribs. Skull lesions are radiolucent, of varying size, often multiple, and often described as being "punched out" in appearance (Fig. 2). They most commonly have well-defined but not necessarily sclerotic boundaries. Skull lesions tended to be anterior in some series and particularly prevalent in the frontal and facial bones.[65] The defects may appear to have beveled margins due to nonuniform growth and consequent unequal destruction of the inner and outer tables of the vault (Fig. 2). A three-dimensional or "hole within a hole" effect is thereby simulated due to the overlapping of the inner surfaces of the lesion. Calvareal lesions often contain fragments or "buttons" of intact bone within the confines of the circular area of destruction that may not be readily apparent

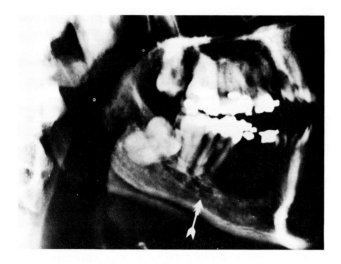

FIG. 4. Eosinophilic granuloma—lateral view of mandible in a 25-year-old male. Large areas of destruction are noted within the maxilla as well as in the mandible. These characteristically arise near the apices of the teeth. (arrow). Eventually, the bone surrounding one or several teeth, as well as the lamina dura, disappear and are replaced by radiolucent histiocytic granulation tissue. Less frequently, initial destruction may occur along the alveolar margin. Osteolytic areas may fill in completely during the healing phase. However, once teeth are shed, the mandible usually fails to regain its normal width.

on routine views unless seen en face or in proper perspective (Fig. 2). Calvareal lesions are commonly associated with a clinically palpable mass (Fig. 2), which may constitute the patient's initial complaint.

Maxillary and mandibular lesions may destroy the supporting alveolar bone, so that the radiopaque teeth appear to be "floating" in space (Figs. 2 and 4). Gingival lesions may further loosen the teeth. In the healing phase, the area of lysis may be completely reossified and appear regular. However, frequently, when teeth are shed, the mandible fails to regain its normal width.[20] Other lesions may arise in the mandible away from the alveolar margin. The disease may on occasion be routinely discovered by the dentist.

Otitis media and a history of a draining ear is one of the most frequent symptoms in any series and may or may not be associated with invasion or involvement of the temporal bone. Conversely, destruction of the skull base may be seen in the mastoid and petrous portions of the temporal bones without clinical sequelae. Lesions may also involve the skull base without involving the calvarium. Destructive lesions of the temporal bone are more common than those involving the sella turcica. The latter may or may not be associated with diabetes insipidus. The classic triad of exophthalmos and diabetes insipidus in association with skull lesions is an uncommon occurrence. Clouding of the ethmoidal air cells and sinuses may be related to an underlying lesion.

A predominance of flat-bone involvement, with the skull representing the commonest locus, has been documented in several series. The pelvis is next frequently involved, often just above the acetabular margin, with sclerosis rimming the superior border of the lesion. The ilium adjacent to the sacroiliac joint, the edge of the iliac wing, and the ischial pubic ramus are other sites of pelvic predilection.

A multilocular appearance is most often observed in association with lytic lesions in the pelvis as well as in the skull. This may again be attributed to the uneven progression of the advancing border of the destructive process, which erodes some portions of the inner cortical surface to a greater extent than adjacent portions. The ridged inner cortical surface then appears beveled when seen in tangent, while a multilocular or trabeculated appearance may be mimicked on en face views. This is a "pseudotrabeculated" appearance since bony septa have not been encountered histologically within the lesions.[20]

The ribs are another common site of flat-bone involvement. Rib lesions, too, tend to have a "punched out" appearance. They are infrequently associated with a florid periosteal reaction. Occasionally, a poorly confined, purely lytic, destructive process may destroy a major segment of an involved rib (Fig. 5).

The vertebral column is frequently involved, with its radiographic appearance depending on its phase of evolution (Fig. 6). Early lesions may consist of an osteolytic focus with or without compression. In one case,[53] the tip of the posterior spinous process of a vertebra was the early locus of a lytic lesion without compression.[53] The vertebral body per se is most often compromised by variable degrees of compression as an initial manifestation. The collapse may be symmetric, involving the entire body uniformly, or asymmetric. When asymmetric, anterior wedging with associated kyphosis may be noted. However, the transition from partial localized destruction to severe vertebral body compression may occur within a short time.

When severely compressed, the vertebral body appears as a uniformly wafer-thin disk, often referred to as a "vertebra plana." It has also been described as having a silver dollar or "coin on end" appearance (Fig. 6). The extreme vertebral body compression deformity was originally described by

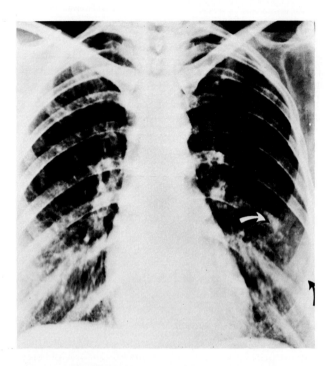

FIG. 5. Histiocytosis X—posteroanterior view of the chest. This 20-year-old white female was mildly dyspneic. A diffuse pulmonary parenchymal abnormality is present in the form of fine reticulonodular infiltrates which predominate at the bases. They have no predilection for any part of the lung. There are two clinical forms of pulmonary histiocytosis X—the primary form, which is a separate entity, and the pulmonary manifestations of the generalized disease. This patient had associated evidence of bone involvement. Note extensive destruction of a major segment of the left eighth rib (arrows) with no accompanying periosteal reaction or soft-tissue mass. Most patients with pulmonary involvement and generalized disease are below 20 years of age and have extrapulmonary lesions at the time abnormalities are detected on chest roentgenograms. They may be relatively asymptomatic. This patient had localized pain in the immediate area of the involved rib. (Courtesy Dr. Rubem Pochaczevsky, New York.)

Calvé (1925)[7] as an osteochondritis or osteochondrosis analogous to Perthe's disease and has been referred to as *Calvé's disease* and *osteochondritis vertebralis*, as well as *vertebra plana*. Osteoporosis and compression induced by steroids could mimic the vertebral body lesion. Compère et al.[15] were the first to associate the vertebra plana with eosinophilic granuloma. The affected body often appears dense due to compressed bony trabeculae and is commonly wider than normal in both its tansverse and antero-posterior diameters. It may project beyond the anterior margins of the adjacent vertebral body as well as posteriorly into the spinal canal for a short distance.

Kyphosis may occur, with minimal bulging of the paravertebral soft tissues. Although spinal involvement is usually confined to a single vertebra, two or more vertebrae may be affected (Fig. 6). When multiple bodies are involved, they are commonly adjacent. In Nesbit's series, 10 children had 35 vertebrae involved.[53] One case had 11, one case had 7, and one case had 5 vertebral lesions. Of the 35, 20 lesions were thoracic, 12 lumbar, and 3 in the cervical spine. Cervical involvement is the least frequent.[53,65]

The most common symptom relating to vertebral column lesions was pain. In two cases, the onset of back pain predated roentgenographic findings by 2 months and 6 months. However, only one patient had neurologic changes secondary to a vertebral lesion, while some patients were asymptomatic.

When two neighboring vertebral bodies are involved, the preservation of the intervertebral disk space (Fig. 6) as well as a lack of any adjacent soft-tissue mass are points that aid in differentiating eosinophilic granuloma from spondylitis due to infection. Paraspinal masses, when they do occur, are most often associated with focal hemorrhage and/or edema, and they tend to regress quickly and spontaneously. In addition to these highly characteristic roentgenographic findings, the usual absence of involvement of the pedicles or posterior elements is worthy of note and is a point of differentiation from lymphoma and/or metastatic disease.[35,53] The absence of neurologic findings despite involvement of many vertebral bodies is noteworthy.

The prognosis of vertebral involvement is good as supported by serial follow-up of 17 of 35 cases by Nesbit et al.[53] However, complete restitution in height seldom takes place. Partial reconstruction of vertebral height occurred in 13 of 17 cases during a 2- to 11-year interval. Despite the fact that the growth rate in repair is greater than that in normal adjacent vertebrae, the compensatory spurt is a limited process with resultant failure of full vertebral body height restitution (Fig. 6). Vertebra plana was defined by Nesbit et al. as a body whose thickness measured 2 mm or less.[53] New bone laid down within the vertebral body during recovery may give rise to a "bone within a bone" appearance (Fig. 6).

In addition to a total decrease in the height of a vertebral body, scoliosis of varying severity may constitute additional evidence of past vertebral involvement when the deformity or collapse remained asymmetric. Residual anterior protrusion of the vertebral body proper was seen after a 1-year follow-up in 2 of 6 lesions, and 3 of 6 lesions showed a

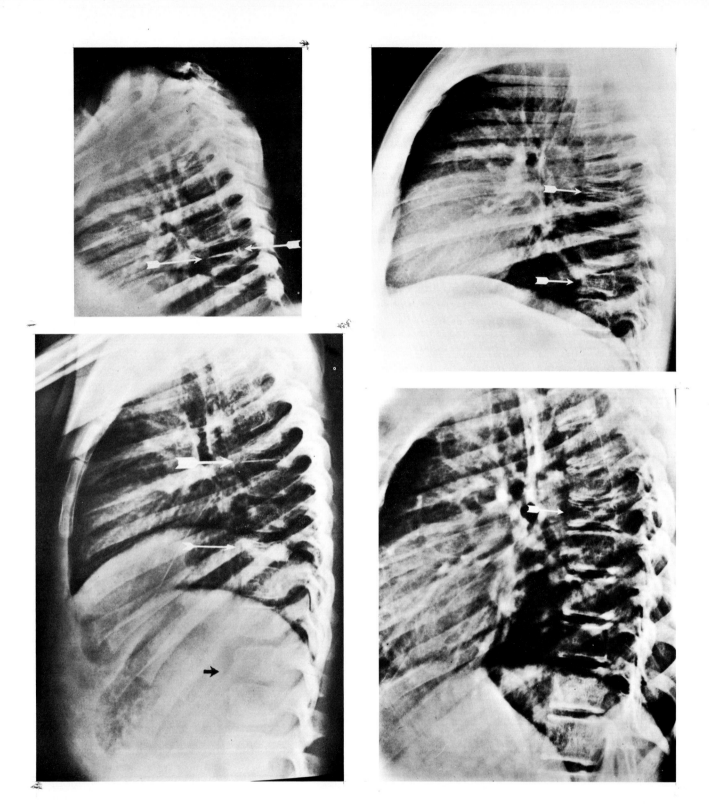

FIG. 6. Eosinophilic granuloma of the spine diagnosed in 2-year-old girl after the mother noted increasing irritability and sensitivity to touch in mid back. Top left: Lateral thoracic spine on admission. The T10 vertebral body is markedly compressed and has the classic vertebra plana or "coin on end" appearance of eosinophilic granuloma (arrows). Neighboring superior and inferior intervertebral disk spaces and articular endplates are intact and sharply defined. Bottom left: Lateral thoracic spine 3 years after. There has been partial reconstitution of the compressed vertebral body (small white arrow), which still has a residual increase in its anteroposterior diameter. Note involvement of L1 (black arrow) and T6 and T7. The last represented a new vertebra plana (large white arrow). Top right: Lateral thoracic spine 7 years after. Note increase in the height of T10. It is still shorter than vertebrae above and below. The severely compressed upper thoracic vertebral bodies noted in B have also undergone partial reconstitution. The templates of the previously collapsed vertebrae are still discernible (arrows), giving a "bone within a bone" appearance. Bottom right: Lateral thoracic spine 12 years after. There has been major reconstitution of the vertical heights of T6 and T10, with a lesser degree of reconstitution of T7 (arrow).

residual sclerotic area of increased density usually located near or at the end of the center of the vertebra.[53] Rarely, the involved vertebra may fuse to the vertebra above or below. Some degrees of deformity of the vertebral endplates may also occur before, during, and after recovery, with resulting irregularity of the superior and inferior vertebral body endplates, with or without subchondral sclerosis, which may in part be related to compressed trabeculae.

Characteristic roentgenographic findings in the vertebral column are variable degrees of vertebral collapse, and increased size of adjacent intervertebral disk spaces, absence of neighboring paraspinal soft-tissue mass, absence of posterior element involvement, varying degrees of restoration of vertebral height on comparison films, and the rarity of neurologic complications. Widening of the paravertebral shadows and soft-tissue swelling or prominence are more often apparent anterior to cervical vertebral body lesions. Considering the number of vertebral bodies at risk, the incidence of compression fracture is greatest in the lumbar spine.[20]

A generalized roentgenographic manifestation may take the form of widespread osteopenia, widened medullary spaces, and decreased cortical thickness of the long bones. The osteoporosis may simulate that associated with steroid therapy. However, it probably represents diffuse marrow replacement by the underlying process with little cortical involvement.[10]

Reactive marginal sclerosis is most commonly observed about long-bone lesions and is less commonly associated with lesions in flat bones. It is most often observed in association with pelvic lesions and is less commonly seen in the skull. It may also occur as a feature of spontaneous regression and/or as a response to treatment with lesions filling in by means of centripetal new bone formation. While one lesion is healing, however, others may be developing elsewhere in the skeleton. A mixed osteolytic and osteoblastic pattern may result. Rarely, lesions may be entirely sclerotic when first discovered, making diagnosis difficult unless other, more typical manifestations, in either bone or soft tissues, are evident (Fig. 7). This uncommon lytic and blastic pattern was recently noted in a 54-year-old woman with multifocal eosinophilic granuloma[34] and suggested metastatic disease. An extensive metastatic work-up was negative, and the diagnosis was made on the findings at open bone biopsy.

Extraskeletal lesions may involve almost every organ in the body and again may make metastatic disease a consideration. Hepatosplenomegaly may be a prominent roentgenographic feature.

The typical radiologic appearance of pulmonary involvement is that of bilateral, ill-defined reticulonodular infiltrations of varying sizes (Fig. 5). Multiple well-defined areas of radiolucency usually less than 1 cm in diameter having walls that can be traced through a nearly 360-degree circumference have also been described; they create a so-called honeycombed pattern.[8]

The lungs may be affected as part of a generalized involvement or as a separate entity. Generalized pulmonary involvement is more frequent in the pediatric age group and occurs in those who have widely disseminated disease. However, it rarely causes symptoms in this age group and almost invariably clears roentgenographically. By contrast, primary pulmonary histiocytosis is mainly seen in young adult males and quickly leads to severe disability.[8] Obstructive emphysema and spontaneous pneumothorax may be noted. However, 20 percent of patients have spontaneous pneumothoraces, which may occur with normal chest x-rays. Two of 12 patients with roentgenographic findings of pulmonary disease had hilar adenopathy. However, both had generalized histiocytosis.[8]

There is a predominance of flat-bone lesions, with the skull being the commonest affected site, in several series.[3,20,65] Although the disease may be manifested in any long bone, the femur is the most commonly affected. Involvement of bones distal to the knees and elbows is unusual, and it is particularly rare in the small bones in the hands and feet.

In general, the roentgenographic appearance of Hand–Schüller–Christian disease and/or multifocal eosinophilic granuloma of bone is similar to that of unifocal eosinophilic granuloma of bone, except that bone lesions are more numerous in the former and invariably involve the skull.

Treatment

Treatment has consisted of surgery, radiation therapy, steroids, chemotherapy, and even studied neglect, since spontaneous healing of eosinophilic granuloma involving solitary or multiple loci in bone has also been documented. In eosinophilic granuloma, surgically accessible bone lesions have been managed by curettage and bone grafting as needed to support structural integrity. A single curettage usually results in healing with a low recurrence rate. Even incomplete curettage has resulted in the disappearance of a bone lesion in a relatively short time. In two patients of Winkelman and Burgert, the lesions healed after surgery for biopsy alone.[67]

Radiotherapy has been used successfully in sur-

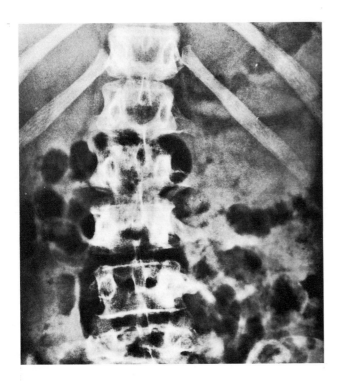 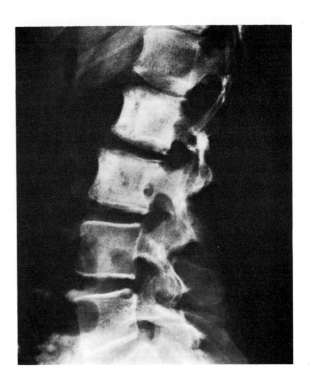

FIG. 7. Histiocytosis X. This 12-year-old white female complained of several months of low back pain. Left: Frontal view of lumbar spine. A scoliosis with a left-sided concavity is present. The left pedicles of L1, L2, and L3; the right pedicle of L4; and the L2 and L3 vertebral bodies appear dense. The last have a coarse trabecular pattern. Right: Lateral view of lumbar spine. The osteosclerosis predominantly involves the vertebral bodies. Radiolucent highlights within the centra of L1 and L2 create a mottled appearance. A biopsy of associated skin lesions revealed evidence of histiocytosis X. The patient is known to have had no other clinical complications after a 2-year follow-up but has not had additional roentgenograms. (Courtesy of Dr. Philip Sorabella, Ridgewood, New Jersey.)

gically inaccessible lesions, while steroids and chemotherapy have been advocated when multiple sites, either bony or visceral, have been involved and in those cases that do not respond to surgical or radiation therapy.

No single drug or group of drugs has emerged as a treatment of choice for induction or maintenance of chemotherapy. Vinblastine, 6-mercaptopurine, methotrexate, cyclophosphamide, chlorambucil, procarbazine, and danomycin have been used either singly or in various combinations. The results of drug trials by the Children's Cancer Study Groups A and B and the Southwest Oncology Group have shown that certain agents may be effective.[32,39,64] Chemotherapy has improved survival in multicentric disease. However, response may be slow and improvement gradual, so that in some instances long-term therapy is advised 6 to 12 months after there is no objective disease activity.[61] The response to therapy is influenced by the age of the patient, the type of organ dysfunction, and, according to some,[37,38] the histologic appearance of the tissue. A decreased response and increased mortality are observed in the youngest patients with a "malignant histologic pattern and multiple organ system involvement." Lahey has noticed response to chemotherapy in disseminated disease with a 71 percent survival in 59 of 83 children.[39] While 63 percent demonstrated some residual disability related to fibrosis of tissues or endocrine deficiency, 37 percent of the survivors were completely well. The following facts have been stressed:

1. Histiocytosis X may involve many tissues, and all patients presenting with a single sign or symptom should be examined for the presence of disease in other areas.

2. The location and extent of disease is of prognostic significance. The greater number of tissues or systems involved, generally the poorer the prognosis.[37-39,50] A knowledge of involvement or of dysfunction of particular organs or systems has particular prognostic significance, particularly when the liver, lung, or hematopoietic system are affected.

3. The situation is complicated by documented cases that have moved from one classic clinical category to another.[18] A solitary focus in bone, i.e., eosinophilic granuloma, may progress to both multicentric bone and visceral involvement indicative of classic Hand–Schüller–Christian disease. Likewise, cases with acute fulminant onsets typical of Letterer–Siwe disease have occasionally progressed to a chronic form and evolved into Hand–Schüller–Christian disease. Conversely, slowly progressive Hand–Schüller–Christian disease has been known to develop into a fulminant phase indistinguishable from Letterer–Siwe disease.

4. Another prognostic sign is age related. In the reticuloendothelioses, youthfulness is a liability. Before the age of 1 year, the most common syndrome is Letterer–Siwe disease, and after age 5, solitary eosinophilic granuloma becomes more common. Therefore, the youngest child may have the poorest prognosis.

Guidelines for management have been formulated that resemble those for patients with malignancy. A diagnosis should be based on histologic evidence and the patient evaluated as to the extent of disease. A scoring system has been suggested as a useful means of staging or assessing the severity of an individual case. Various staging methods have been proposed. Lucaya[50] divided patients into four groups: (1) disease in a single bone, (2) disease in two or more bones, (3) disease in bones and soft tissue, and (4) disease only in soft tissues. The highest mortality occurred in groups 3 and 4. The mortality of the first two groups was 4 percent and of the latter two groups 50 percent. Therefore, radiographs and radioisotopic scans to detect skeletal and visceral lesions become important.

Some of the suggested diagnostic studies, in addition to roentgenograms of the chest and bones and radioisotopic bone and visceral scans, include measurement of the integrity of the hypothalamic–pituitary axis by means of the evaluation of growth hormone reserve, gonadotropin secretion, adrenocortical and thyroid function, antidiuretic hormone, and plasma osmolality.

Summary

Cases of histiocytosis X or the reticuloendothelioses are not common occurrences. Cheyne (1971)[10] calculated on the basis of 34 cases of all ages that histiocytosis X occurred in about one in two million population per year in the Bristol area. By adjustment of yearly under-15 population figures in Sims's study,[61] one case of histiocytosis X may be expected per 350,000 children under age 12. Therefore, it is unlikely that an individual practice will afford more than a limited experience of the disease.

In addition, the clinical expressions of histiocytosis X may be so diverse that the patient may come to the initial attention of a host of different specialists. They may, furthermore, present over a protracted period of so many years, that a panoramic view of the underlying process may not be obtained by any one physician. Although an attempt was made to call attention to its involvement of multiple organ systems, chief emphasis has been placed on the clinical and roentgenographic diagnosis of skeletal lesions and related problems.

References

1. Abt AF, Denenholz EJ: Letterer–Siwe's disease. Am J Dis Child 51:499, 1936
2. Amardjazil Z, Serban-Aurel E, Konrad K: Histiocytosis X in an adult with skin and uncommon central nervous system involvement. Dermatologica 155:283, 1977
3. Avery ME, McAfee JG, Guild HG: The course and prognosis of reticuloendotheliosis (eosinophilic granuloma, Schüller–Christian disease and Letterer–Siwe disease): a study of 40 cases. Am J Med 22:363, 1957
4. Benisch B, Peison B, Carter H: Histiocytosis X of the skin in an elderly man. Am J Clin Pathol 67:36, 1977
5. Braunstein GD, Kohler PO: Pituitary function in Hand–Schüller–Christian disease. N Engl J Med 286:1225, 1972
6. Braunstein GD, Whitaker JN, Kohler PO: Cerebellar dysfunction in Hand–Schüller–Christian disease. Arch Intern Med 132:387, 1973
7. Calvé JA: Localized affection of the spine suggesting osteochondritis of vertebral body: clinical aspects of Pott's Disease. J Bone Joint Surg 7:41, 1925
8. Carlson RA, Hattery RR, O'Connell EJ, Fontana RS: Pulmonary involvement by histiocytosis X in the pediatric age group. Mayo Clin Proc 51:542, 1976
9. Case Records of Massachusetts General Hospital. Case 9. New Engl J Med 288:459, 1973
10. Cheyne C: Histiocytosis X. J Bone Joint Surg 53B:366, 1971
11. Christian HA: Defects in membraneous bones, exophthalmos and diabetes insipidus. An unusual syndrome of dyspituitarism; a clinical study. Contrib Med Biol Res 1:390, 1919
12. Christian HA: Defects in membranous bones, exophthalmos and diabetes insipidus; an unusual syndrome of dyspituitarism: a clinical study. Med Clin North Am 849, 1920
13. Clamon HN, Suvotte V, Githens JH, Hathaway WE:

Histiocytic reaction in dysgammaglobulinemia and congenital rubella. Pediatrics 46:89, 1970
14. Cline MJ, Golde DW: A review and re-evaluation of the histiocytic disorders. Am J Med 55:49, 1973
15. Compère EL, Johnson WE, Coventry MB: Vertebra plana (Calvé's disease) due to eosinophilic granuloma. J Bone Joint Surg 36A:969, 1954
16. Cutler LS, Krutchkoff D: An ultrastructural study of eosinophilic granuloma: the Langerhans cell—its role in histogenesis and diagnosis. Oral Surg 44:246, 1977
17. Doede KG, Rappaport H: Long term survival of patients with acute differential histiocytosis (Letterer–Siwe disease). Cancer 20:1782, 1967
18. Dutt AK, Lopez CG, Ganesan S, Dutt A: Acute disseminated histiocytosis X. A case with transition from eosinophilic granuloma of bone to Letterer–Siwe disease. Australas Ann Med 18:135, 1969
19. Eil C, Adornato BT: Radicular compression in multifocal eosinophilic granuloma. Successful treatment with radiotherapy. Arch Neurol 34:786, 1977
20. Ennis JT, Whitehouse G, Ross FGM, Middlemiss JH: The radiology of the bone changes in histiocytosis X. Clin Radiol 24:212, 1973
21. Enriquez P, Dahlin DC, Hayles AB, Henderson ED: Histiocytosis X. Mayo Clin Proc 42:88, 1967
22. Farber S: The nature of "solitary or eosinophilic granuloma" of bone. Am J Pathol 17:625, 1941
23. Feuerman EJ, Sandbank M: Histiocytosis X with skin lesions as the sole clinical expression. Acta Derm Venereol (Stockh) 56:269, 1976
24. Fridman B, Hanaoka H: Langerhans cell granules in eosinophilic granuloma of bone. J Bone Joint Surg 51A:367, 1969
25. Frisell E, Björksten B, Holmgren G, Angström T: Familial occurrence of histiocytoses. Clin Genet 11:163, 1977
26. Green WT, Farber S: "Eosinophilic or solitary granuloma" of bone. J Bone Joint Surg 24A:499, 1942
27. Grosfeld JL, Fitzgerald JF, Wagner VM, Newton WA, Baehner RL: Portal hypertension in infants and children with histiocytosis X. Am J Surg 131:108, 1976
28. Hand A Jr: General tuberculosis. Trans Pathol Soc Pa 16:282, 1893
29. Hand A: Defects of membranous bones, exophthalmos and polyuria in childhood: is it dyspituitarism? Am J Med Sci 162:509, 1921
30. Hewlett RH, Ganz JC: Histiocytosis X of the cauda equina. Neurology 26:472, 1976
31. Jaffe HL, Lichtenstein L: Eosinophilic granuloma of bone. Arch Pathol 37:99, 1944
32. Jones B, Kung F, Chevalier L, et al: Chemotherapy of reticuloendotheliosis, comparison of methotrexate plus prednisone vs. vincristine plus prednisone. Cancer 34:1011, 1974
33. Juberg RC, Kloepfer HW, Oberman HA: Genetic determination of acute disseminated histiocytosis X (Letterer–Siwe syndrome). J Pediatr 45:753, 1970
34. Kaufman A, Bukberg PR, Werlin S, Young IS: Multifocal eosinophilic granuloma (Hand–Schüller–Christian disease). Report illustrating Hand–Schüller–Christian chronicity and diagnostic challenge. Am J Med 60:541, 1976
35. Kaye JJ, Freiberger RH: Eosinophilic granuloma of the spine without vertebra plana. Report of 2 unusual cases. Radiology 92:1188, 1969
36. Kepes JJ, Kepes M: Predominantly cerebral forms of histiocytosis X. Acta Neuropathol 14:77, 1969
37. Lahey ME: Prognosis in reticuloendotheliosis in children. J Pediatr 60:664, 1962
38. Lahey ME: Histiocytosis X—an analysis of prognostic factors. J Pediatr 87:184, 1975
39. Lahey ME: Histiocytosis X—comparison of three treatment regimens. J Pediatr 87:179, 1975
40. Latorre H, Kenny FM, Lahey ME, et al: Short stature and growth hormone deficiency in histiocytosis X. J Pediatr 85:813, 1974
41. Leikin S, Puruganan G, Frankel A, Steerman R, Chandra R: Immunologic parameters of histiocytosis X. Cancer 32:796, 1973
42. Letterer E: Aleukämische Retikulose. Frankfurt Z Pathol 30:377, 1924
43. Lever WF, Schaumburg-Lever G: Histopathology of the Skin. Philadelphia, Lippincott, 1975
44. Levy JA: Auto-immunity and neoplasm: the possible role of C type viruses. Am J Clin Pathol 62:258, 1974
45. Lichtenstein L: Histiocytosis X. Arch Pathol 56:84, 1953
46. Lichtenstein L: Histiocytosis X: integration of eosinophilic granuloma of bone, "Letterer–Siwe disease," and "Schüller–Christian disease" as related manifestations of a single nosologic entity. Arch Pathol 56:84, 1953
47. Lichtenstein L: Histiocytosis X (eosinophilic granuloma of bone, Letterer–Siwe disease, and Schüller–Christian disease). J Bone Joint Surg 46A:76, 1964
48. Lichtenstein L, Jaffe HL: Eosinophilic granuloma of bone. Am J Pathol 16:595, 1940.
49. Lieberman PH, Jones CR, Dargeon HWK, Begg CF: A reappraisal of eosinophilic granuloma of bone, Hand–Schüller–Christian syndrome and Letterer–Siwe syndrome. Medicine 48:375, 1969
50. Lucaya J: Histiocytosis X. Am J Dis Child 121:289, 1971
51. McKusick VA: Mendelian Inheritance in Man, 4th ed. Baltimore and London, Johns Hopkins Univ Press, 1975
52. Mickelson MR, Bonfiglio M: Eosinophilic granuloma and its variations. Orthop Clin North Am 8:933, 1977
53. Nesbit ME, Kieffer S, D'Angio GJ: Reconstitution of vertebral height in histiocytosis X: A long term follow-up. J Bone Joint Surg 51A:1360, 1969
54. Nezelof C, et al: Histiocytosis X. Histogenic arguments for a Langerhans cell origin. Biomedicine 18:365, 1973
55. Otani S: A discussion of eosinophilic granuloma of bone, Letterer–Siwe disease and Schüller–Christian disease. Mt Sinai J Med NY 24:1079, 1957
56. Otani A, Ehrlich JC: Solitary granuloma of bone simulating primary neoplasm. Am J Pathol 16:479, 1940
57. Ransom JL, Murphy SB: Histiocytosis X: abnormal cerebrospinal fluid cytology in extrahypothalamic cen-

tral nervous system involvement. South Med J 70:1367, 1977
58. Schajowicz F, Sluttitel J: Eosinophilic granuloma of bone and its relationship to Hand–Schüller–Christian and Letterer–Siwe syndromes. J Bone Joint Surg 55B: 545, 1973
59. Schüller A: Über eingenartige Schadeldefekte im Jugendalter. Fortschr Roentgenstr 23:12, 1915–1916
60. Scott H, Moynahan EJ, Resdon RA, Harvey BAM, Sunthill JF: Familial opsonization defect associated with fatal dermatitis infections and histiocytosis. Arch Dis Child 50:311, 1975
61. Sims DG: Histiocytosis X. Follow-up of 43 cases. Arch Dis Child 52:433, 1977
62. Siwe SA: Die Reticuloendotheliose—ein neues Krankheitsbild unter den Hepatosplenomegalien. Z Kinderheilkd 55:212, 1933
63. Siwe S: The reticulo-endotheliosis in children. Adv Pediatr 4:117, 1949
64. Starling KA, Donaldson MH, Haggard ME, Vietti TJ, Sutaw WW: Therapy of histiocytosis X with vincristine, vinblastine and cyclophosphamide. The S.W. cancer chemotherapy study group. Am J Dis Child 123:105, 1972
65. Takahashi M, Martel W, Oberman HA: The variable roentgenographic appearance of idiopathic histiocytosis. Clin Radiol 17:48, 1966
66. Winkelmann RK: The skin in histiocytosis X. Mayo Clin Proc 44:535, 1969
67. Winkelmann RK, Burgert FO: Therapy of histiocytosis X. Br J Dermatol 82:169, 1970
68. Yamaguchi K, Yokoyama T, Morimatsu M: Involvement of the central nervous system in Hand–Schüller–Christian disease. Report of a case and discussion on the entity of the disease. Acta Pathol Jpn 22:363, 1972
69. Zinkham WH: Multifocal eosinophilic granuloma. Natural history, etiology and management. Am J Med 60:457, 1976

Giant-Cell Tumor and Its Variants

D. C. Dahlin

Giant-cell tumor of bone is a characteristic, benign neoplasm composed of poorly differentiated cells (Fig. 1).[1,2] The multinucleated giant cells apparently result from fusion of the proliferating mononuclear cells. Although benign giant cells are a constant and prominent part of these tumors, they are likely of less significance than are the mononuclear cells. These osteoclastlike giant cells, with or without minor modifications, occur in a host of pathologic conditions of bone.

Malignant giant-cell tumor is a separate and controversial subject. We employ the rule that this malignant tumor cannot be diagnosed with assurance unless evidence of ordinary benign giant-cell tumor exists within the lesion or has been demonstrated previously at the same site. Many osteosarcomas and even fibrosarcomas of bone contain benign giant cells, but there is much evidence that they are not related to benign giant-cell tumor. The important point is that an abundance of benign giant cells introduces the hazard of mistaking these malignant conditions for benign giant-cell tumor.

The 264 giant-cell tumors that we had seen prior to January 1, 1976, represented 4.2 percent of the total and 18.2 percent of the benign bone tumors. At least 60 percent of reported giant-cell tumors have occurred in women. More than 80 percent of the neoplasms are found in patients more than 19 years of age, and nearly all patients affected with giant-cell tumor have mature bone growth. This neoplasm occurs most often in the very end of a major tubular bone. More than half of our giant-cell tumors occurred above the knee. Histologically, typical examples are seen in various sites, including vertebrae above the sacrum, the sphenoid, the ribs, and small bones of the hands and feet.

The usual clinical presentation of giant-cell tumor is pain or swelling or both. Sometimes there is pathologic fracture. These tumors are characteristically soft, friable, and gray to red. Firmer portions may result, especially if there has been prior fracture or treatment. Necrotic portions or small cystic zones sometimes filled with blood may be present. Prominent brownish discoloration is common.

The proliferating and basic cell of these tumors has one round to oval or spindle-shaped nucleus. These nuclei are surrounded by an ill-defined cytoplasmic zone, and no discernible intercellular substance is produced in fields that are diagnostic of true giant-cell tumor. Mitotic figures are present in practically every lesion, and sometimes they are numerous. The nuclei, however, lack the hyperchromatism and variation in size and shape characteristic of sarcoma. The occasional osteosarcoma whose mononuclear cells are small may be difficult to differentiate from giant-cell tumor. In such cases the roentgenogram provides important evidence as to whether the tumor being assessed is benign or malignant. Osteoid and even bone spicules are common in giant-cell tumors, especially at their peripheries. Cartilaginous differentiation is unusual, and its presence is indicative of benign chondroblastoma, especially if it is disposed in discrete islands. Occasional tumors that simulate giant-cell tumors but have chondroid foci prove to be sarcomas with chondroid differentiation.

It must be remembered that histologically benign giant-cell tumor does, on rare occasions, produce metastasis, especially to the lung. Such metastases are histologically benign and are ordinarily quasimalignant; some of them have been proven to have limited

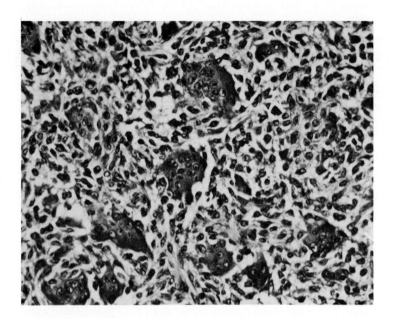

FIG. 1. Giant-cell tumor with multinucleated cells scattered among mononuclear forms, some of which are slightly spindle-shaped. They lack cytologic features of malignancy. H&E, x250.

growth potential even as metastases. This unusual phenomenon of metastases from benign giant-cell tumor occurs in only 1 or 2 percent of cases.

The main thrust of this chapter is to indicate that many other conditions contain benign giant cells and should not be confused with benign giant-cell tumor. These include the following:

Aneurysmal bone cyst
Giant-cell (reparative) granuloma
Benign chondroblastoma
Nonosteogenic fibroma (metaphyseal fibrous defect)
Benign osteoblastoma
Osteosarcoma
Parathyroid osteopathy
Miscellaneous
    Fibrous dysplasia
    Simple cyst of bone
    Others

Aneurysmal Bone Cyst

Aneurysmal bone cyst is probably the most important condition that must be differentiated from benign giant cell tumor. This nonneoplastic tumorlike lesion occurs in patients less than 20 years of age in about 75 percent of cases. It usually has its epicenter in the metaphysis rather than in the epiphysis of long bone. Unlike giant-cell tumor, this lesion has no known capability for spontaneous malignant transformation. Giant-cell tumor recurs in nearly 50 percent of cases after curettage, recurrences being a greater threat than with aneurysmal bone cyst. Some aneurysmal bone cysts are cured by less than total removal of the process. The roentgenogram may show a thin subperiosteal rim of bone over the region of expansion due to the tumor. Computerized tomography may emphasize that the lesion is well defined on the side within the bone.

Microscopically, the salient feature in most cases is the component of cavernomatous spaces, which may have been evident grossly. The walls of these spaces lack the features of normal blood vessels. Thin strands of bone are often present in the fibroblastic tissue of these walls. Benign osseous trabeculae are usually present, and sometimes the mineralizing elements have focal regions with a chondroid aura that is unusual in any other lesion of bone. Reconstruction of these cavernomatous spaces from curetted fragments may be difficult. The solid portions of an aneurysmal bone cyst may be fibrous, but they ordinarily contain a lacework of osteoid trabeculae somewhat like that in benign osteoblastomas. Benign giant cells are often conspicuous, at least in areas, thus accounting for the similarity of this lesion to true giant-cell tumor. The solid zones, which are often fibrogenic and somewhat edematous in appearance, resemble closely the lesion called giant-cell (reparative) granuloma of jaw bones. The most important histologic problem is to recognize that this lesion is benign. Although mitoses may be numerous in the spindle-cell areas, the nuclei lack anaplasia as manifested by hyperchromatism and irregularity in shape. The osteoid that shades from collagenous tissue of the predominantly spindle-cell regions is disposed in regular trabeculae. Regions with mononuclear cells producing no matrix as seen in a bona fide giant-cell tumor should not be present. Telangiectatic osteosarcoma often simulates aneurysmal bone cyst when viewed at low magnifications.

Giant-Cell Granuloma

Giant-cell (reparative) granuloma is a lesion that nearly always occurs in the jaws, although occasional characteristic examples are seen in other bones, especially the bones superior to the maxilla at the base of the skull. The lesional tissue usually contains numerous benign giant cells, but the stromal cells are fibrogenic. Their fibrogenic quality is prominent, and the matrix shades into the osteoid or bony trabeculae that are found in most of these lesions especially near their peripheries. Small blood spaces simulating those seen in aneurysmal bone cyst are common. Although the tumor may be quite destructive of the area that harbors it in the jaws, relatively simple but thorough removal nearly always results in cure. Spontaneous malignant transformation is unknown. A lesion of the jaws that looks exactly similar may occur in patients with hyperparathyroidism. Hyperparathyroidism should be expected in any adult patient with giant-cell reparative granuloma of the jaws, especially if it is recurrent. It is possible that true giant-cell tumor occurs in the jaws, but if it does occur, it is exceedingly rare, and I have been unable to identify with certainty any such lesion.

Benign Chondroblastoma

Benign chondroblastoma is most common in patients between the ages of 15 and 25 years. This is characteristically a small lesion, measuring only a few centimeters in diameter, and occurring in the epiphysis, most frequently of a major tubular bone. Although many benign chondroblastomas contain microscopic foci of calcification, only a minority have sufficient calcium within the lesion to have it reflected in the roentgenogram. Chondroid foci of variable number and sometimes difficult to find are the significant feature that differentiates this lesion from giant-cell tumor of bone. The mononuclear and multinucleated cells of benign chondroblastoma are very similar to those of giant-cell tumor. The lesion is more innocuous than is ordinary giant-cell tumor, and most of these tumors can be cured by relatively simple resection. Recurrence is somewhat unusual, and spontaneous malignant transformation is unknown. A few sarcomas have been produced in the region of chondroblastoma by the use of ionizing radiation.

Nonosteogenic Fibroma

Nonosteogenic fibroma (metaphyseal fibrous defect) is unusual in patients who are more than 20 years of age. It is characteristically seen in the metaphysis, where it produces an eccentric zone of rarefaction that is well delimited on its osseous side. Microscopically, the tumor is made up of fascicles of collagen-producing fibroblasts, which often contain hemosiderin and occasionally contain significant amounts of lipid. Scattered throughout the lesion are benign giant cells. These are usually less numerous than in giant-cell tumor and are often disposed in scattered aggregates. Metaphyseal fibrous defects are ordinarily easy to diagnose if one combines the roentgenographic features with the characteristic histology. The roentgenogram is so characteristic that unless the lesion is large and there is impending fracture of the bone that harbors it, treatment is not necessary. These tumors are most often found in the femur, tibia, and fibula, but characteristic metaphyseal fibrous defects may be found in any of the major tubular bones; they are rare in other bones.

Benign Osteoblastoma

Benign osteoblastoma sometimes contains many benign giant cells, and it may be confused with giant-cell tumor. The main characteristic of osteoblastoma is its capability of producing osteoid and even mineralizing bone throughout the lesion. The bony trabeculae that are commonly seen in giant cell tumor are less uniformly dispersed and often exhibit a parallelism that suggests they are a reparative phenomenon.

Osteosarcoma

Osteosarcoma contains benign giant cells in approximately 15 percent of cases. Occasionally, these cells are so numerous that an erroneous interpretation of giant-cell tumor may be made if the microscopic magnification is low. The roentgenogram aids in biasing one toward the diagnosis of malignancy, and the critical histologic feature is that the mononuclear cells of osteosarcoma exhibit the features of anaplasia with hyperchromatism and irregularity in size and shape. The quality of the osteoid produced in osteosarcoma may indicate strongly that the lesion is malignant.

Parathyroid Osteopathy

The osseous lesions of advanced hyperparathyroidism often contain numerous benign giant cells. They are typically present in a fibrogenic stroma in which are scattered trabeculae of immature bone. A significant difference from giant-cell tumor is that the lesion is apt to occur in unusual sites for giant-cell tumor, such as in the metaphyseal portion of major tubular bones or in the flat bones of the skull. The fibrogenic quality of the lesion and its unusual location should alert the histopathologist to the correct underlying process so that appropriate chemical determinations for hyperparathyroidism can be made.

Miscellaneous Diseases

A wide variety of benign and sometimes non-neoplastic processes in bone contain variable numbers of benign giant cells. Ordinarily, the background quality of the lesion and its roentgenographic interpretation indicate the correct diagnosis. An example of a benign process that may contain benign giant cells is fibrous dysplasia, especially if there has been prior treatment. The characteristic dense fibrous stroma of this lesion and its scattered irregularly shaped trabeculae of metaplastic bone are its hallmarks. Simple cysts of bone sometimes have a thick lining, and this lining may contain an abundance of benign giant cells. The centric quality of this metaphyseal lesion and the surgical findings usually indicate the appropriate diagnosis. Also in the miscellaneous category is the so-called giant-cell lesion that is usually found in the small bones of the hands and feet. This is a benign process that produces an area of rarefaction on the roentgenogram. It contains fibrogenic cells, often among trabeculae of bone. Their histology simulates that of giant-cell (reparative) granuloma of the jaws. It is likely that this benign process is related to some insult to the bone.

References

1. Jaffe HL: Giant-cell tumour (osteoclastoma) of bone: its pathologic delimitation and the inherent clinical implications. Ann R Coll Surg Engl 13:343, 1953
2. Dahlin DC, Cupps RE, Johnson EW Jr: Giant-cell tumor: a study of 195 cases. Cancer 25:1061, 1970
3. Goldenberg RR, Campbell CJ, Bonfiglio M: Giant-cell tumor of bone. An analysis of two hundred and eighteen cases. J Bone Joint Surg 52A:619, 1970

Skeletal Lesions Simulating Malignancy

Harold G. Jacobson

Skeletal lesions are often encountered in which the radiologic features suggest a malignant neoplasm. In many such instances, it is even true that the radiologic characteristics are definitive enough of a benign diagnostic entity so that a biopsy is not necessary, even though a malignancy is suspected. This concept is provocative, often disturbing, and frequently not accepted by orthopedic surgeons. However, a large number of such skeletal lesions exist. Biopsies in such instances may on occasion be painful, sometimes result in post-biopsy infection, and of course, are time consuming, expensive, and even emotionally traumatic. It is even more important to point out that cases have been observed in which an unnecessary biopsy in such a pseudomalignant skeletal lesion, which may properly be called in such instances a "leave-me-alone" lesion, e.g., pseudotumor of hemophilia (Fig. 1), has resulted in an erroneous diagnosis with serious consequences. Such an error obviously may lead to injudicious and disastrous treatment.

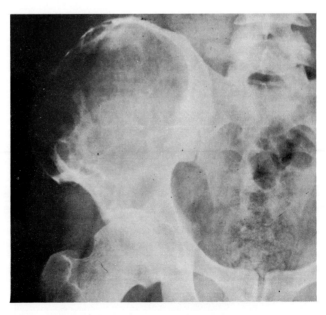

FIG. 1. Pseudotumor hemophilia—iliac bone.

When the radiologic features are definitive in a benign skeletal disorder mimicking a malignant process radiologically, biopsy *need not* and *should not* be performed.

Bibliography

Murray RO, Jacobson HG: The Radiology of Skeletal Disorders: Exercises in Diagnosis 2nd ed., Edinburgh, Churchill Livingstone, 1977

Cartilaginous Tumors and Tumorlike Conditions*

Frieda Feldman

Cartilaginous lesions are important because they are common, constituting the most prevalent form of bone tumor in several large series. They also present a difficult challenge for even the most experienced pathologist, since the line between their benign and malignant counterparts often cannot be finely drawn. The pathologist may be particularly insecure in the separation of the benign cartilaginous lesions from the so-called low-grade or well-differentiated chondrosarcomas, which constitute the large majority of chondrosarcomas. The distinction between benign and malignant is of additional importance within this family of lesions, since a greater potential exists for cure if malignant members are recognized early and treated definitively. At the other end of the spectrum lies a hazard with equally grim repercussions, i.e., that of "overdiagnosing" and "overtreating" the benign counterparts of this family of lesions. Furthermore, the high risk of local seeding of cartilaginous tumors, benign as well as malignant, frequently in the course of biopsy, leads to additional difficulty in terms of recurrence and future treatment.

The location and multiplicity of the lesions, as well as the age of the patient, play an inordinately important role in the classification of these lesions. Thus, difficulties in histologic interpretation and a poorly understood potential combine with a classically slow evolution to make cartilage lesions a difficult group to manage. More than most neoplasms, therefore, cartilaginous tumors of bone require the close cooperation of the pathologist, the surgeon, and the radiologist in achieving adequate treatment, and less of a tendency exists to rely solely on one discipline for definitive interpretation and/or evaluation.

In keeping with a primarily roentgenographic orientation, the simplest classification of a chondrogenic series of benign tumors and tumorlike lesions has been adopted (Table 1). Table 2 contains a number of terms that have been employed for these lesions.

Solitary Osteochondroma

A solitary osteochondroma is a cancellous bony projection with a cartilage cap, most frequently located in the metaphysis and occasionally in the diaphysis of a tubular bone. It may, however, be associated with any bone that is preformed in cartilage (Figs. 1 and 2).

Table 1 Benign Tumors and Tumorlike Lesions of Cartilaginous Origin (Location and Classification)

Peripheral	Metaphysis and/or diaphysis	Osteochondroma
	Epiphysis	Juxtacortical chondroma Dysplasia epiphysealis hemimelica
Central	Metaphysis	Enchondroma
	Epiphysis	Chondromyxoid fibroma* Chondroblastoma

*Rarely cortical in location.

---

*Abridged from Feldman F: Cartilaginous tumors and cartilage forming tumor-like conditions of the bone and soft tissues. In Diethelm L, Heuck F, Olsson O, et al (eds): Handbuchder medizinischen radiologie. Berlin, Springer, 1977.

Table 2 Nomenclature of Cartilaginous Tumors

Prefered Terminology	Synonyms
Solitary osteochondroma	Exostosis
	Enchondroma
	Osteocartilaginous exostosis
Multiple osteochondromatosis	Hereditary multiple exostoses
	Metaphyseal aclasis
	Diaphyseal aclasis
	Dyschondroplasia*
	Hereditary deforming dyschondroplasia*
Dysplasia epiphysealis hemimelica	Tarsoepiphyseal aclasis
	Benign epiphyseal osteochondroma
	Intra-articular osteochondroma of astragalus
	Carpal osteochondroma
Solitary enchondromas	Chondroma
Multiple enchondromatosis	Ollier's disease
	Dyschondroplasia*
	Hereditary deforming dyschondroplasia*
Juxtacortical chondroma	Paraosteal chondroma
	Periosteal chondroma
	Subperiosteal chondroma
	Eccentric chondroma
Chondroblastoma	Benign chondroblastoma
	Epiphyseal chondroblastoma
	Codman's tumor†
	Calcified giant cell tumor†
	Epiphyseal chondromatous giant cell tumor†
Chondromyxoid fibroma	Fibromyxoid chondroma

*Same name in use for both entities.
†Terminology no longer in use.

Osteochondromas are the commonest cartilaginous neoplasms as well as the commonest benign neoplasms of bone (Spjut et al., 1971). Solitary and multiple osteochondromas constituted 45 percent of 1025 benign bone tumors and 12 percent of all bone tumors studied by Dahlin (1967). Nearly 90 percent were solitary lesions.

Site

An osteochondroma may be related to any bone that develops by enchondral ossification, including the base of the skull. Long bones are predilected. The commonest locations are the lower femoral and upper tibial metaphyses with an incidence of 36 percent at the knee, 50 percent including the knee and proximal humerus (Dahlin, 1967), and 95 percent including all bones of the extremities (Spjut et al., 1971). The flat bones, including the scapula (Pratt et al., 1958; Wouters et al., 1974), ilium, clavicle (Pratt et al., 1958), sternum, spinous process of the vertebral body, the remainder of the vertebral column (Gokay and Bucey, 1955; Ilgenfritz, 1951; Jaffe, 1968; Meyerding, 1927; Peck, 1964), and the small tubular bones of the hands and feet are less frequent loci (Fig. 3). They are rarely associated with either the carpal (Heiple, 1961) or tarsal bones (Milch, 1954). Solitary osteochondromas of the distal phalanx of the first toe have been related to trauma and are not included in some series (Dahlin, 1956), despite the fact that they have similar radiologic and pathologic features.

Clinical Features

No significant sex predilection has been noted in some series (Spjut et al., 1971). In others (Dahlin, 1967), males predominate in a 3:2 ratio. The lesion has most commonly been discovered as a painless lump, usually noted between 2 and 21 years of age. Symptoms may be due to impingement on neighboring bones, nerves, or blood vessels, or they can be related to the irritation of a bursa, which may develop if the cartilaginous cap abuts a tendon or muscle.

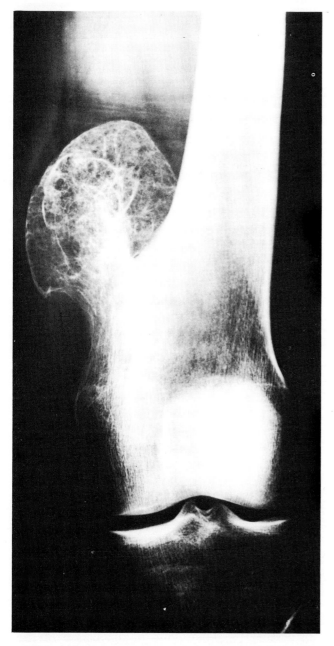

FIG. 1. Solitary benign osteochondroma in a 30-year-old male with a 15-year history of a painless mass in the left thigh. Exostosis extends beyond otherwise normally contoured femur. Note that the cortex of base of exostosis blends with that of the main shaft. The trabeculae of the host bone also extends into and blends with those of the exostosis, simulating pedunculated polyp on a narrow bony stalk. Exostoses most commonly point away from the epiphysis and joint. Although the bulbous periphery of the exostosis appeared smooth and sharply demarcated, its large size prompted excision.

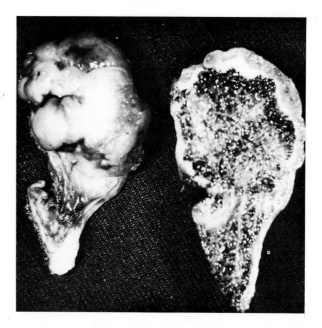

FIG. 2. Pedunculated osteochondroma. Left: Gross specimen, surface view. An irregular, smooth, glistening cartilage cap covers the bulbous end of the stalked exostosis. The narrow neck is seen inferiorly. The entire external surface of sessile tumor may be capped by cartilage. Right: Cut section. Note the fatty, cancellous bone with spongiosa extending into the neck of stalk. The cartilage cap is seen superiorly.

Accelerated growth of an exostosis with or without accompanying pain has also been related to trauma.

Although osteochondromas are infrequently noted in the vertebral column, serious complications have been encountered due to the proximity of important structures, such as the spinal cord, nerve roots, and cauda equina (Chiurco, 1970; D.M. Cohen et al., 1964; Fielding and Ratzan, 1973; Geschickter and Copeland, 1949; Gokay and Bucey, 1955; Ilgenfritz, 1951; Thomas and Andress, 1971). Compression of the carotid and subclavian arteries with obstruction of venous return has been reported. An odontoid osteochondroma was responsible for sudden death in one patient (Rose and Feketa, 1964). Malignant change has been estimated as occurring in less than 1 percent of cases and most commonly occurs in the form of a chondrosarcoma.

Roentgenographic Features

The osteochondroma appears as a bony protuberance extending beyond the cortical contour of

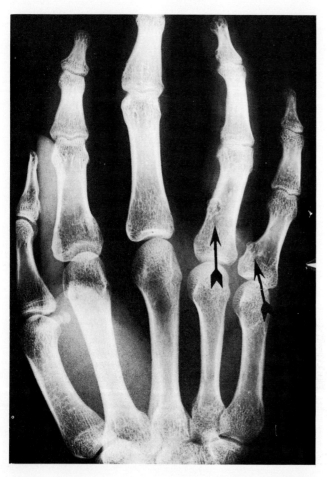

FIG. 3. Osteochondroma. Note the small nubbinlike exostoses with flaring of proximal shafts of proximal phalanges (arrows). Modeling deformity is usually noted at the level of exostosis.

the host or "mother" bone and is usually attached to it by a narrow neck simulating a pedunculated polyp on a stalk. Its trabeculae blend with those of the shaft via the base (Fig. 1). It most commonly points away from the epiphysis and neighboring joint (Fig. 1). Some osteochondromas are broad based with a plateaulike peripheral configuration, while others taper more sharply. Occasionally, they are sessile, small, or near nubbinlike excrescences. Even with the latter configuration, an osteochondroma is always suspect due to a localized failure of modeling, as evidenced by an abnormally wide shaft at the level of the exostosis.

The cartilage cap of the osteochondroma may not be apparent unless it is mineralized, but it is usually of a larger dimension than that appreciated on a roentgenogram, particularly in children. Since the cartilage cap becomes inactive with cessation of bone growth, it should be thin or atrophied in the adult. An unusually thick or growing cap may indicate malignancy in an adult, while mineral irregularly deposited along its periphery, or a sudden spurt in growth suggests malignancy in both young and old.

The visualized bony exostoses vary in overall size from less than 1 cm to over 10 cm. In adults, lesions over 8 cm are regarded with suspicion and should be carefully analyzed for the presence of chondrosarcoma (Spjut et al., 1971).

Histology—Gross Features

Whatever its size and shape, the exostosis is encased by a periosteum that covers and adheres closely to the irregular surface of the hyaline cartilage cap and is continuous with the periosteum of the adjacent cortical bone (Fig. 2). It is usually thick but may be delicate. If enchondral ossification is still in progress, a thin yellowish growth zone or plate can be identified on the undersurface of the cartilage cap (Fig. 2).

The younger the patient, the more prominent will be the cartilage cap, which varies from 1 to 3 mm and may be as large as 6 mm. Lichtenstein (1965) has noted that if the cartilage cap measures 1 cm or more in thickness, the possibility of chondrosarcoma should be seriously considered.

Since an exostosis grows by enchondral ossification of its proliferation cartilage, its enlargement should cease at puberty. It then tends to involute normally.

Multiple Osteochondromatosis

Multiple osteochondromas are inherited in an autosomal dominant mode and are transmitted to approximately half the offspring of affected parents. The disorder is transmitted by both parents. In some series it has a 2:1 male:female predilection. Growth disturbances are reflected in dramatic deformities. Shortness of the bones due to dissipation of the longitudinal growth force in a lateral direction, accompanied by curvature of the bones, adds to the "dwarfing" of the skeleton. The skull is rarely involved.

The lesions of multiple osteochondromatosis are generally more symmetrically distributed than those of multiple enchondromatosis which, though widespread, tend to predominate unilaterally, particularly in the extremities (Fig. 4).

Clinical Features

Lesions are usually first discovered at about 2 years of age. Most are detected prior to puberty; however, exostoses, particularly in cases with multiple osteo-

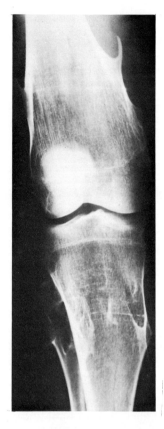
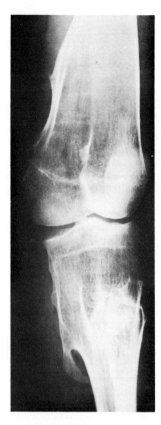
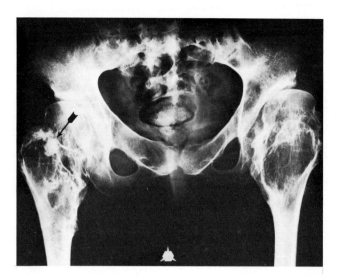
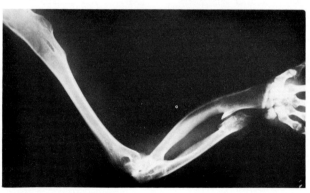

FIG. 4. Multiple osteochondromatosis. Top left: Anteroposterior view of the knees confirm widespread involvement and presence of multiple osteochondromatosis. Note the flared, flask-shaped bone ends studded with slender exostoses, most of which are directed away from the joint space. The cartilaginous caps are not mineralized and therefore not visualized. Top right: Anteroposterior view of the pelvis. Note the abnormally and asymmetrically modeled pubic bone with shallow acetabulae; flattened femoral heads; short, squat femoral necks partially obscured by the bulbous greater trochanters; and bilateral valgus deformities. The right-sided calcified exostosis seen on end simulates intramedullary enchondroma (arrow). Opposite: Right forearm. Bowed radius, foreshortened ulna, and multiple exostoses are typical of diaphyseal aclasis. Note the associated deformity of the proximal humeral shaft. The other arm was similarly affected.

chondromatosis, may appear for the first time in adult life. The dating of symptoms is often inexact. Pain is not a reliable criterion. It may be mild, intermittent, and insidious, not disturbing the patient sufficiently to seek attention. On the other hand, pain may occur acutely in relation to trauma with or without pathologic fracture.

Pain without fracture should arouse suspicion. Periodic follow-up studies including roentgenograms may be indicated. A slow but definite increase in size of one of the bony swellings after the end of the normal skeletal growth period may be the only clinical feature heralding the presence of a chondrosarcoma. Often this growth is so gradual that it may take years before the patient seeks advice unless pressure and/or impingement on neighboring soft-tissue structures such as vessels and nerves causes discomfort.

Roentgenographic Features

Frequently, the first site of involvement noted is a rib (Joseph and Fonkalsrud, 1972) or the vertebral margin of the scapula, since both areas are visualized on routine films of the chest. Occasionally an incidentally included proximal humeral shaft may furnish further corroborative evidence of multiple lesions. Since the bones about the knees are the most frequently involved, they should be examined when a

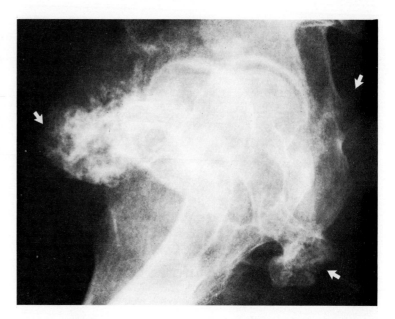
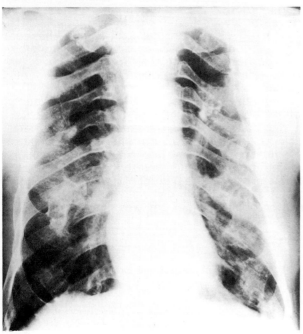

FIG. 5. Multiple osteochondromatosis in a 37-year-old male with chief complaint of changing bowel habits. Top left: Anteroposterior view of the abdomen. A huge abdominal mass (arrow) is seen originating in the pelvis, compressing and elevating the opacified urinary bladder. Subsequent barium enema was successful in solely filling the rectum. Top right: Lateral view of the pelvis. Numerous cauliflower-shaped exostoses arising from the iliac and ischial bones (arrows). The anterior pubic rim lesion had undergone sarcomatous transformation and had extended posteriorly, displacing the pelvic and abdominal contents. Partial pelvectomy afforded marked relief. The patient was alive and well 8 years later. Opposite: Posteroanterior view of the chest. The thorax is deformed by multiple exostoses arising from both proximal humeri and multiple ribs. Some of the osteochondromas simulate pulmonary parenchymal lesions as well as enchondromas.

diagnosis of multiple osteochondromatosis is considered on the basis of finding a single lesion elsewhere (Fig. 4). Pelvic exostoses are frequently prominent about the iliac surfaces in the region of the superior apophyseal attachments, and along the pubic rami and ischia (Figs. 4 and 5).

Growth disturbances in the form of skeletal shortening and bowing are dramatically reflected about the wrists and forearms (Fig. 4). When both the radius and the ulna are involved, the latter will grow disproportionately less, accounting for ulnar deviation of the hand in 30 percent of patients. Posterior dislocation of the radial head may also result from the disproportionate length of the radius.

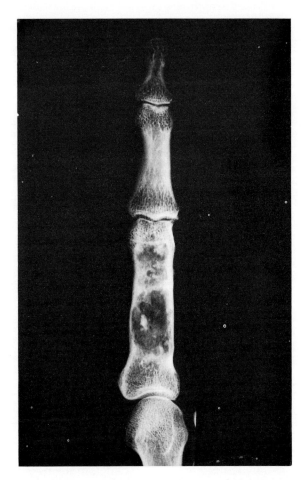

FIG. 6. Enchondroma—phalanx of the hand, sole site. Metaphyseal and diaphyseal lucent lesions represent cartilaginous foci, associated with coarsened central calcific deposits and focal endosteal scalloping resulting in an hour-glass configuration.

The incidence of chondrosarcomatous transformation in multiple osteochondromatosis, which earlier had been estimated at 5 percent (Ehrenfried, 1915) and then at 11 percent (Jaffe, 1943, 1968), is now considered to be closer to 20 percent (Dahlin, 1968; Jaffe, 1943) (Fig. 5).

Solitary Enchondromas

The benign cartilaginous neoplasms of bone are most frequently located within the medullary cavity and have no sex predilection. They are less common than exostoses and comprised just over 10 percent of 1025 benign bone tumors in Dahlin's 1969 series. This figure includes solitary and multiple enchondromas.

An enchondroma with conspicuous cortical expansion may be mistaken for an osteochondroma, just as an exostosis that is seen on end through the shaft of an affected bone may simulate an enchondroma. Analysis of the bone's interior and its external contour by means of multiple views, as well as the distribution of the visualized mineral within the lesion, will aid in distinguishing the two entities.

Site

More than 50 percent of all enchondromas are found in the small bones of the hands and feet (Fig. 6). Of these, 90 percent are in the hands (Dahlin, 1967; Spjut et al., 1971). Not infrequently, a large limb bone, such as the humerus, femur, and occasionally the tibia is involved (Fig. 7). Central chondromas of the ribs, sternum, and scapula are less common. The patella and vertebral column are rare sites.

Clinical Features

Enchondromas are most frequently incidentally discovered in asymptomatic patients in the second to fifth decades. Their observation in very young children is exceptional (Lichtenstein, 1965). Pain and swelling may be noted after pathologic fracture of an attenuated cortex, and occasionally after trauma per se.

Roentgenographic Features

Enchondromas most often appear as well-demarcated, round, or ovoid radiolucencies, centrally situated within the metaphysis of the middle and/or proximal phalanges of the hands or feet (Fig. 6). Diaphyseal lesions may be seen, most commonly, as a result of extension from a metaphyseal locus (Fig. 6). The epiphysis is not involved in the presence of an unfused growth plate and is usually not affected after fusion (Jaffe, 1968). The cortex is usually expanded in a symmetric, fusiform fashion by the enchondroma, but eccentric expansion may occur. Cortical thinning is related to gradual endosteal attrition by a slowly growing lesion whose lobular peripheral growth pattern contributes to an occasional hourglass or multilobulated configuration. The frequently cloudy or hazy appearance of the area of rarefaction serves as a clue to its chondroid composition, as well as to its mineral content. The latter is not always identifiable roentgenographically.

Although the evaluation of a primary chondrosarcoma per se and/or the reputed evolution of a chondrosarcoma from a benign enchondroma may be a

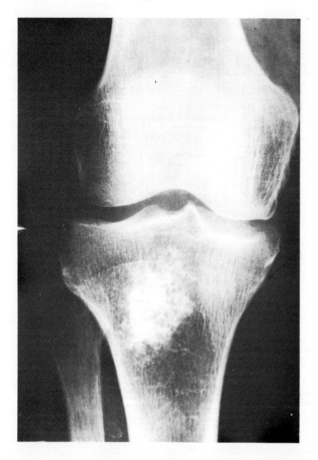 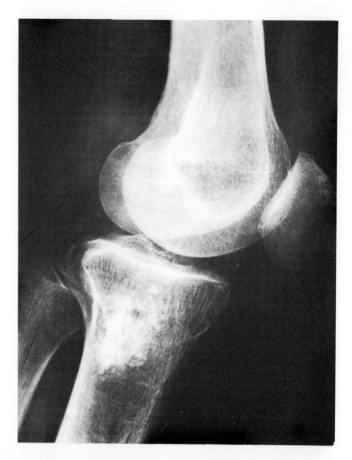

FIG. 7. Enchondroma—tibia.   Left: Anteroposterior view. A 58-year-old asymptomatic female had this rounded, heavily mineralized, centrally located metaphyseal lesion.   Right: Lateral view further defines coarse flocculant calcifications scattered uniformly throughout the lesion. Normal trabeculae abut its somewhat irregular contours. Curettage yielded normal cartilage tissue.

slow process, serial roentgenograms, when available, may be of inestimable value in confirming a change in the character or aggressive nature of a lesion. Any centrally located cartilaginous lesion, and particularly one situated within a long bone close to the trunk, that exhibits a sudden or gradual change in its size, configuration, or margination, perforates the cortex, or appears to have its mineral content dissipated or "eaten away," should be suspected of being or having become malignant. This is particularly true if the lesion is harbored by an older individual complaining of pain.

Size has occasionally been used as a criterion for malignancy, with cartilaginous lesions larger than 3 to 4 cm being suspect. This suspicion is enforced as the locus of the tumor approaches the body axis. The closer the lesion is to the trunk, the greater is the degree of inherent biologic malignancy. Therefore, chondromatous lesions situated within the proximal shafts of long bones, the pelvis, and the thoracic cage (including the ribs and sternum) larger than 3 to 4 cm should be regarded with particular suspicion. Conversely, site, as identified radiographically, may also be of importance in confirming an impression of benignity, since, as a rule, both malignant degeneration of a solitary enchondroma and primary chondrosarcoma in the phalanges, metacarpals, or metatarsals per se rarely occur (Shellito and Dockerty, 1948).

In general, the lesions of younger patients show a greater degree of cellularity despite their benignity. This is particularly true of biopsies from phalangeal enchondromas and those from cases of multiple enchondromatosis at any age and from any site. Therefore, the age of the patient and the site of the lesion from which the biopsy was obtained are important in the histologic evaluation of cartilage tumors. Generally speaking, primary chondrosarcomas of the skeleton are uncommon in the young, while primary chondrosarcomas of the small bones of the extremities are rare at any age. Secondary

Cartilaginous Tumors and Tumor-Like Conditions

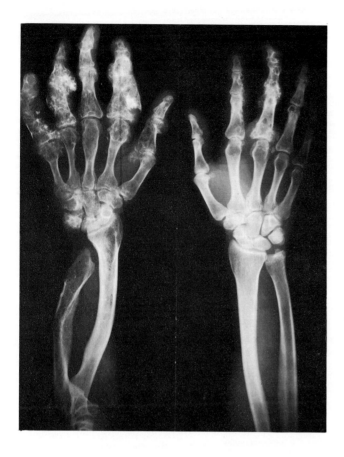

FIG. 8. Multiple enchondromatosis. Posteroanterior view of both hands and forearms. The marked deformities are due to multiple expanding heavily mineralized lesions in both hands. Note the diaphyseal aclasis. Deformities are similar to those associated with osteochondromatosis. The left side is more severely involved. Note the complete disruption of the lateral cortical contour of both proximal and mid phalanges of left fifth finger. The cartilaginous mass had proceeded to grow beyond confines of cortex, as noted by new calcific deposits as well as cortical disruption. A chondrosarcoma was resected from the left fifth finger.

chondrosarcomas deriving from solitary enchondromas of the small bones are even rarer. Both malignant degeneration of solitary enchondromas and primary chondrosarcomas are more frequent in the long bones and more frequent in adults.

Multiple Enchondromatosis

Multiple enchondromatosis is a rare abnormality. Unlike multiple osteochondromatosis, no hereditary or familial influence has been demonstrated. In contrast to a solitary enchondroma, it represents a widespread anomaly of skeletal development and was referred to as a dyschondrophasia by Ollier (1900) (Fig. 8).

Clinical Features

The appearance of multiple enchondromata in early childhood is followed by progressive skeletal deformity resulting from the failure of normal enchondral ossification. In some cases, skeletal involvement becomes stabilized after puberty. In others it is progressive, with the development of monstrous deformities.

Early recognition generally parallels widespread involvement. The condition usually becomes evident at about 2 years of age. Phalangeal swelling, bowing of the long bones, growth discrepancies between the radius and ulna associated with restricted forearm motion, and ulnar deviation of the hand may be noted early (Fig. 8). The forearm and wrist deformities are similar to those present in many cases of multiple osteochondromatosis, and both entities have been referred to by the common name, or hereditary deforming chondrodysplasia.

Retarded growth of the lower extremities may be striking, particularly when one lower limb is more heavily affected (Fig. 9). At the age of 2 or 3 years, the discrepancy in lower-limb length may approach 2 to 4 cm, and in young adults it may be as much as 25 cm (Fig. 9). A compensatory scoliosis and sharp pelvic tilt are frequent. The hazard of malignant transformation in enchondromatosis approaches 50 percent in some series. This again emphasizes the importance of a clear distinction between instances of skeletal enchondromatosis and multiple osteochondromatosis in which the purported risk is considerably less.

Roentgenographic Features

The enchondromas are identified as multiple rarefactions involving the shafts of the tubular bones, including those of the hands and feet. The lesions are most often rounded or ovoid, and since normal modeling does not take place, bone ends become bulbous, asymmetrically expanded, and generally deformed. Affected limb bones are always short. Linear or rounded radiolucencies representing columns or islands of cartilage are most commonly seen in the flat bones of the pelvis.

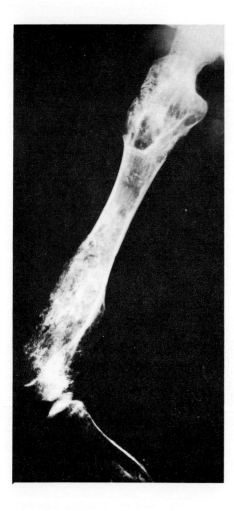
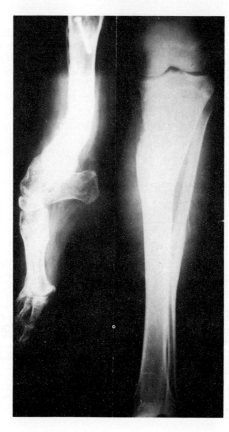

FIG. 9. Multiple enchondromatosis in a teenaged patient with chief complaint of recently increasing leg length disparity. Marked shortening of the right lower extremity was noted at an early age. Left: Right femur. Club-shaped expansion of the distal and proximal femur is due to multiple medullary enchondromas. Heavy mineralization is seen, with cartilaginous lesions contained within the medullary cavity. Mineral is therefore contained within the confines of the medullary cavity rather than external to it. This contrasts with peripheral mineralization seen in the cartilaginous caps of osteochondromas. Right: Note the marked disparity in leg length with deformity of the contralateral left proximal tibia.

Chondrosarcomatous transformation has most frequently occurred in middle age and may or may not be heralded by pain. Roentgenographically, one looks for a sudden change in size, configuration, or cortical contour, as well as extension of the lesion beyond cortical confines (Fig. 8).

Dysplasia Epiphysealis Hemimelica

This developmental disorder has most often been described as an asymmetric cartilaginous overgrowth of one side of one epiphysis or of a carpal or tarsal bone. Fairbank (1956) used the term *hemimelica* to graphically describe the limited locus of the lesions. They are, however, not always hemimelic or solitary. The study of Kettelkamp et al. (1966) of 167 lesions in 57 patients revealed multiple lesions in more than two-thirds of the cases.

Dysplasia epiphysealis hemimelica is distinct from the epiphyseal hypoplasias, since it is a disorder of excessive growth, and it therefore shares much in common with hyperplasias of other segments, i.e.,
enchondromatosis and osteochondromatosis. Its limited distribution, however, precludes its being classified as a true bone dysplasia (Rubin, 1964).

Site

Distribution is characteristically on one side of the limb and most commonly in the lower limb. The talus (Fig. 10) and distal femoral and tibial epiphyses are the most common sites (Keats, 1957; Trevor, 1950). Upper limb involvement has also been reported (Fairbank, 1956; Kettelkamp et al., 1966). The medial side is affected twice as often as the lateral. Multiple lesions are usually confined to one extremity.

Clinical Features

Pain is not a prominent feature. It is most often related to restricted motion due to the presence of the local cartilaginous growth or to occasional fracture, as in multiple osteochondromatosis or multiple enchon-

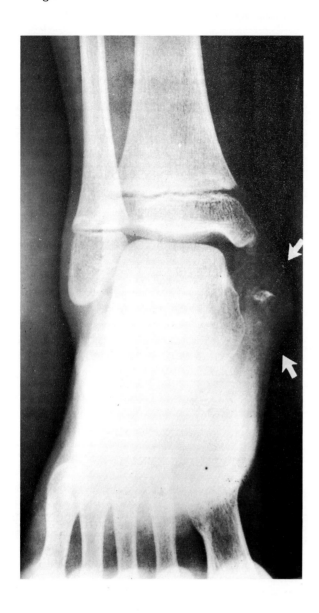

FIG. 10. Dysplasia epiphysealis hemimelica in a 10-year-old male. The lesion is associated with the medial aspect of the talus, which is affected twice as often as the lateral aspect. Note the multiple irregular radiopacities that appear to be separate from the talus itself. As the lesion matures, however, the cartilaginous mass continues to ossify and will eventually fuse and blend with neighboring bone. The final appearance may be that of an exostosis.

dromatosis. Locking or catching of the knee joint rarely occurs. Growth appears to be unaffected, and shortening of the limb is unusual.

Roentgenographic Features

The only apparent abnormality in infancy may be metaphyseal widening or deformity. Ossification centers of affected bones usually appear prematurely. Later, numerous and irregular multicentric radiopacities may occur adjacent to one side of, but not necessarily connected to, the affected bone. Therefore, an irregularly shaped, mineralized mass may be noted, which appears isolated from the main epiphysis of a long bone or from an ossification center of a carpus or tarsus (Fig. 10). As the lesion matures, the mass fuses and blends with the texture of the affected neighboring bone. The final appearance may be similar to that of an exostosis.

Juxtacortical (Periosteal) Chondroma

Juxtacortical chondromas are usually solitary, benign cartilage lesions that develop in the periosteum and/or immediate paraosteal connective tissue of a bone. They characteristically erode their underlying cortical bed forming a depression or declivity in the underlying cortex. The latter is not penetrated and serves to separate the lesion from the medullary cavity. Lichtenstein and Hall (1952) classified this lesion as a distinctive, benign, cartilaginous tumor.

Clinical Features

These lesions are most frequently found in young or middle-aged adults, have no sex predilection, and occur most frequently in the phalanges (Fig. 11) of the hands and feet (Lichtenstein and Hall, 1952; Meyer, 1958).

Roentgenographic Features

Juxtacortical chondromas present as an eccentric, semilunar, cortically oriented radiolucency that appears to have eroded the cortex from without. The erosion, which has a well-marginated sclerotic base separating it from the medullary cavity, has a crater or cuplike configuration. A buttress of periosteal new bone may be present, particularly at the proximal end of the defect (Jaffe, 1968). They usually are 1 or 2 cm and rarely exceed 3 or 5 cm in their greatest dimension.

Its peripheral cartilaginous component may be ill defined or not roentgenographically visible. However, it may be sprinkled with fluffy or coarse granular calcifications. Occasionally, and particularly if the cartilaginous mass is large, it may be outlined by a thin radiopaque shell representing a calcified or ossified capsule.

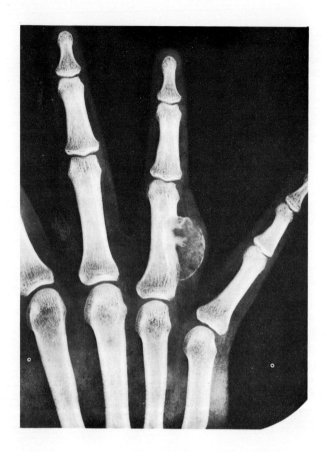

FIG. 11. Juxtacortical (periosteal) chondroma—hand. The lesion again involves a common site, the phalanx of the hand. The eccentric, cortically oriented, cartilaginous mass is well circumscribed both medially and laterally. A well-marginated sclerotic base separates the cortically oriented lesion from the medullary cavity. The cartilaginous mass is peripherally defined by a thin radiopaque calcific shell.

More than 500 cases have been reported in the world literature, and many of them are single case reports or small series of cases. The femur, humerus, tibia, and tarsal bones, in descending order, harbor nearly 70 percent of the tumors.

Long-bone lesions affect the epiphysis in the overwhelming majority of cases, although lesions may extend into the adjacent metaphyses (Table 3). The apophyses or secondary centers of ossification, such as those for the humeral tubercle, the femoral lesser and greater trochanter (Fig. 13), and the acromion constitute occasional sites. The lesion is rarely solely metaphyseal.

Age and Sex Distribution

There is an approximate 2:1 male to female ratio in several large series—81:42 (Dahlin and Ivins, 1972), 44:25 (Schajowicz and Gallardo, 1970), 116:48 (Salzer et al., 1968), 33:23, 15:10 (Huvos and Marcove, 1973)—with ages varying from 3 to 73 years. Chondroblastoma rarely occurs over the age of 30 (Dahlin and Ivins, 1972). More than 60 percent of patients were in their second decade at the time of diagnosis. In Schajowicz and Gallardo's (1970) series, 75.3 percent were in the second decade of life and 88.5 percent were 5 to 25 years old. In the Netherlands Committee on Bone Tumors (1966) series, 95 percent were 5 to 25 years old, while Salzer et al. (1968) noted 71 percent of 171 patients in the second decade.

Symptoms

The symptoms are nonspecific. Pain is the most characteristic complaint, generally of a few months' duration, but in some instances noted for 1 to 8 years (Dahlin and Ivins, 1972). Local swelling, ordinarily of a few months' duration, was noted by approximately 10 percent of patients, usually in areas not obscured by a great deal of overlying soft tissue. Stiffness or limitation of motion of the adjacent joint were occasionally experienced. However, associated effusion described in 21 of the 69 cases of Schajowicz and

Course

Juxtacortical chondromas, like other benign cartilaginous tumors, may recur if incompletely excised (Nosanchuk and Kaufer, 1969). Nests of cartilage cells within the periosteum and adjacent sclerotic bone several millimeters from the main mass have been observed. Since they, too, may serve as nidi for recurrence, en bloc resection is usually attempted.

Chondroblastoma

The lesion was given its present name and was first regarded as an independent clinicopathologic entity by Jaffe and Lichtenstein (1942). The presence of multilobulated giant cells within chondroblastomas had led to considerable confusion in the semantics of this lesion (Table 2).

Prevalence and Site

Chondroblastomas comprise approximately 1 percent of benign primary bone tumors in several reported series (Fig. 12) (Kunkel et al., 1956; Schajowicz and Gallardo, 1970; Spjut et al., 1971).

Cartilaginous Tumors and Tumor-Like Conditions

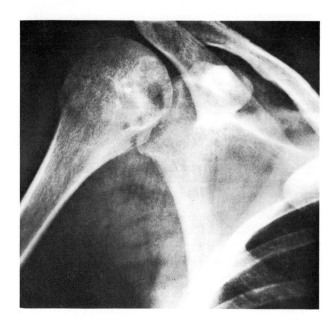

FIG. 12. Chondroblastoma humerus. This 15-year-old female was admitted with painful right shoulder. Note the rounded, well-defined, eccentric, well-marginated, radiolucent lesion with coarser mineral deposits within it. The tumor is separated from adjacent spongiosa by a rim of bony condensation.

Gallardo (1970) was rarely recorded in the Mayo Clinic Series (Dahlin and Ivins, 1972).

Roentgenographic Findings

Since gross specimens showing an entire tumor in its setting are seldom available, our knowledge of the tumor's in situ appearance, development, and effect on surrounding bone is largely gleaned from roentgenograms. The characteristic roentgenographic appearance is that of a small (less than 4 cm in greatest diameter), round or ovoid, well-defined, eccentric, epiphyseal rarefaction most commonly involving the epiphysis of a long bone (Dahlin and Ivins, 1972; Huvos and Marcove, 1973; McLeod and Beabout, 1973; Plum and Pugh, 1958; Sherman and Uzel, 1956; Spjut et al., 1971).

The epiphyseal tumor is usually separated from the adjacent spongiosa by a sclerotic rim or well-defined margin of bony condensation. In the majority of lesions, the size of the central defect has ranged from 1.5 to 4 cm, and the epiphyseal line was most commonly open. These tumors may originate in a secondary ossification center such as the apophysis of the greater trochanter or greater tuberosity of the humerus with metaphyseal extension. Stippled calcification has been noted in 50 percent of some series and in 25 percent of others (McLeod and Beabout, 1973). Flocculent "fluffy" or "popcorn ball" calcifications were noted on roentgenograms of more than half of Schajowicz and Gallardo's (1970) cases.

Although chondroblastomas abut the articular cartilage, it is unusual for them to disrupt it. Larger lesions, in addition to extending into the adjacent metaphyses, may expand laterally as well and may occasionally erode the overlying cortex, producing a rounded extraosseous soft-tissue mass (Dahlin and Ivins, 1972). Xerography and/or angiography may serve to better define or identify any associated soft-tissue component.

Microscopic Features

As seen by light microscopy, the cells considered to be chondroblasts, which give chondroblastoma its distinctiveness, are usually polygonal or rounded with

Table 3 Chondroblastoma—Sites of 692 Lesions*

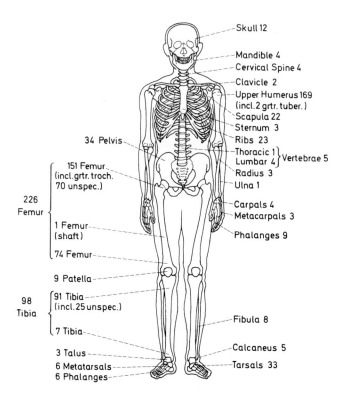

*692 Lesions in 691 Cases (1 case had 2 sites). Data from Buraczewski et al., 1957; Cares, 1971; Coleman, 1966; Dahlin and Ivins, 1972; Fechner and Wilde, 1974; Garavanis and Giansanti, 1971; Huvos and Marcove, 1973; Levine and Bensch, 1972; McBryde and Goldner, 1970; Mangini, 1964; Netherlands Committee on Bone Tumors, 1966; Neviaser and Wilson, 1972; Oppenheim and Boal, 1965; Piepgras et al., 1972; Balzer et al., 1968; Schajowicz and Gallardo, 1970; Sherman and Uzel, 1956; Spjut et al., 1971; Sundaram, 1966; Tanghe and Martens, 1969; Varma and Gupta, 1972; Wellman, 1969; Wiesniewski et al., 1973; Witwicki et al., 1969.

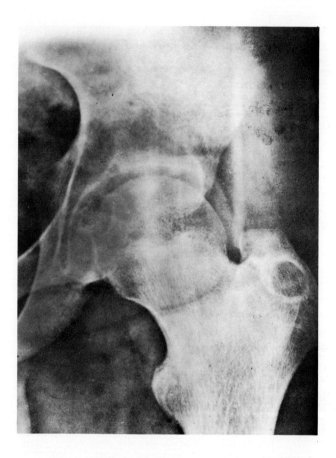

FIG. 13. Chondroblastoma in a secondary center of ossification. The apophysis of the greater trochanter is the site of this rounded lucent lesion bounded by well-defined sclerotic border.

a relatively large nucleus (Fig. 14). Nucleoli are common. The cells are commonly polyhedral in shape, characteristically uniform, closely packed, and separated by a scanty interstitial matrix that is occasionally frankly chondroid in its makeup. A delicate latticelike intercellular calcification resembling a picket fence or chicken wire may be seen.

Treatment and Results

Since in the vast majority of cases chondroblastoma exhibits a benign clinical behavior, conservative therapy has usually been advocated. Curettage, with or without bone grafting, or local excision, preferably en bloc together with an uninvolved zone of normal tissue as appropriate to the size and site of the lesion, and curettage combined with radiation therapy have all been employed (Aegerter and Kirkpatrick, 1968; Dahlin and Ivins, 1972; Huvos and Marcove, 1973; Huvos et al., 1972). Some authorities stress that radiation should not be employed in view of its inherent risk and the generally good results obtained with conservative therapy. Radiation has been advocated when a surgical approach is impossible, after local aggressive recurrence, and when implantation has occurred within the joint space (Dahlin and Ivins, 1972; Schauwecker et al., 1969).

Several case reports (Case Records of Massachusetts General Hospital, 1964; Dahlin and Ivins, 1972; Huvos and Marcove, 1973; Huvos et al., 1972) document aggressive recurrence with neighboring soft-tissue involvement. This is felt to be related to spillage and seeding of tumor cells during the course of the initial surgical intervention. This eventuality may even necessitate amputation, despite the fact that true malignant transformation has not occurred (Kahn et al., 1969; Sundaram, 1966).

Huvos et al. (1972) reported recurrence rates among 25 patients treated with different modalities (excluding amputation, block excision, and irradiation) on a 3-year basis, since all but one of the recurrences occurred within 3 years of treatment. A 25 percent recurrence rate for curettage plus packing with bone chips, as opposed to a rate of 60 percent with curettage alone, suggested that bone packing was beneficial. Patients treated by cryosurgery also had a lower recurrence rate than those who had curettage alone, but the difference was not statistically significant due to the small number of cases in each group.

Curettage effected cure in about 90 percent of 125 cases of benign chondroblastoma reviewed by the Mayo Clinic (Dahlin and Ivins, 1972). Good results were obtained by conservative means in all cases involving the extremities. Treatment consisted chiefly of curettage with or without bone grafting in the Mayo Clinic patients, i.e., 38 of the 125 cases reviewed.

Although chondroblastoma generally has an essentially benign clinical behavior, a few cases with a more aggressive or malignant course have been documented. Several of these, however, share a common denominator of previous radiotherapy.

Chondromyxoid Fibroma

This benign tumor of bone was established as an entity in 1948 when it was first described by Jaffe and Lichtenstein (1948). It occurs less frequently than chondroblastoma. It constitutes less than 1 percent of all benign plus malignant bone tumors (Dahlin, 1956, 1967; Feldman et al., 1970; Goldanich, 1957; Iwata

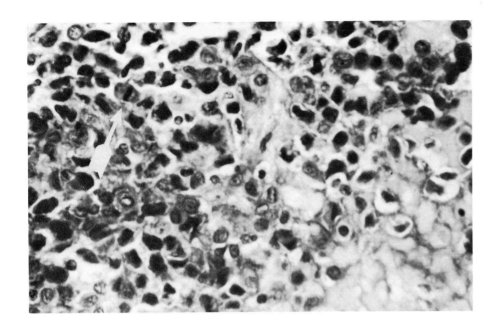

FIG. 14. Chondroblastoma photomicrograph (X50). Note the chondroid matrix at the lower right. Some cells have disappeared. Viable cells are pholyhedral and remain individually distinct. A mitotic figure is seen above left (arrow). There is sufficient atypism to raise histologic suspicion of malignancy. Lesion was biologically benign. (Courtesy Austin D. Johnston)

and Coley, 1958; Rahimi et al., 1972; Schajowicz and Gallardo, 1971). The incidence in Schajowicz and Gallardo's (1971) series was somewhat lower than 1 percent for all tumoral and tumorlike lesions and somewhat less than half of all chondroblastomas observed in the same time period.

Most authors have emphasized the distinct histologic characteristics as well as the benign clinical behavior of this lesion, despite recurrences (Dahlin, 1956; Frank and Rockwood, 1969; Jaffe and Lichtenstein, 1948; Schajowicz and Gallardo, 1976; Spjut et al., 1971; Turcotte et al., 1962), a more aggressive course in the young (Ralph, 1962; Scaglietti and Stringa, 1961), and a few reports of malignant transformation (Aegerter and Kirkpatrick, 1968; Iwata and Coley, 1958), some of which have not been universally accepted. Others have separated off histologic subgroups from chondromyxoid fibroma, such as myxoma of bone (Bauer and Harrell, 1954; Scaglietti and Stringa, 1961) and fibromyxoma of bone (Marcove et al., 1964), which have also not been universally accepted.

An interrelationship between chondromyxoid fibroma and chondroblastoma (Dahlin, 1956; Ralph, 1962; Willis, 1967) has also been postulated with both tumors presumably arising from epiphyseal plate cells, one lesion growing toward the metaphysis and the other toward the epiphysis. Individual cases with microscopy reminiscent of both lesions have been cited. Jaffe (1956, 1968), however, did not accept this as evidence for a morphologic interrelationship, stating that "although one may encounter sporadic microscopic fields in some chondromyxoid fibromas in which groups of cells resemble those of benign chondroblastoma, one does not see spotty calcification within these tissue fields, such as are characteristic of benign chondroblastoma."

A small number of cases of chondromyxoid fibroma have reputedly undergone malignant transformation. However, a review of the original biopsy material by other authorities has not always supported this contention. Lichtenstein (1965) has described a small number of atypical chondromyxoid fibromas as well as chondroblastomas that have behaved in a more aggressive fashion locally. However, Dahlin has noted that the evidence overwhelmingly indicates that the risk of malignant transformation of "bona fide chondromyxoid fibroma," unless radiation is empolyed, is so slight that radical treatment is unnecessary (Rahimi et al., 1972).

Recognition of this uncommon lesion is therefore important, not only from the standpoint of its histologic and clinical features, which on occasion may be misleading and ominous, but from a roentgenographic standpoint, since it is one of those bone lesions in which the roentgenogram may be decisive in tempering a histologic impression of malignancy.

Age and Sex

Ages of patients have reputedly ranged from 4 to 79 years (Salzer et al., 1968). Fifty-seven percent of patients were less than 30 years old. It is believed (Salzer et al., 1968) that a slow clinical evolution and a subtle unappreciated onset of symptoms account for the lesions discovered in older people. As opposed to chondroblastoma, which has a male predominance, chondromyxoid fibroma has shown no significant sex predilection.

Clinical Features

Compaints were nonspecific. In the majority of cases, pain, most often mild, local, and intermittent had been present from months to years prior to consultation. Most were aware of swelling and/or a mass, in view of the lesion's commonly peripheral location, i.e., about the knee. Occasional tenderness to palpation, limp, and slight limitation of motion were also noted. However, in one series of five children 5 to 10 years of age, pain, swelling, and restricted motion developed within 3 to 6 weeks, leading to speculation that the lesion may be more aggressive in the young (Scaglietti and Stringa, 1961).

Pathologic fracture per se was a rarity. It was noted in one case in our experience (Feldman et al., 1970) and in nine other cases (Herfarth, 1932; Lettin, 1963; Seth and Rao, 1964; Sideman et al., 1960-61; Turcotte et al., 1962). Four pathologic fractures are mentioned in the recent Mayo Clinic review that may have been included in our previous compilation (Rahimi et al., 1972). Therefore, either 10 or 14 pathologic fractures have been noted among 295 cases, an incidence of 3 to 4 percent.

Sites of Localization

Sites of localization as noted in several major series are shown in Table 4. A chondromyxoid fibroma is most commonly located in the metaphyseal region of a large tubular bone of a lower limb at a variable distance from the epiphyseal line and often in contact with it. The preferred site appears to be the upper tibia. It is of interest that Schajowicz and Gallardo (1971) had no case involving the distal femur, a frequent site in most series. This tumor has only recently been recognized in such sites as the mandible (Schutt and Frost, 1971), sternum (Teitelbaum and Bessone, 1969), rib (Goorwitch, 1951; Ramini, 1974), and vertebral column.

Roentgenographic Features

Roentgenograms most commonly suggest a benign lesion whose appearance varies depending on whether a long, small tubular or flat bone is involved (Figs. 15 and 16) (Feldman et al., 1970; Murphy and Price, 1971; Ramini, 1914). In a long tubular bone, the lesion typically appears as a round or oval metaphyseal radiolucency whose long axis is directed along

Table 4 Fibroma—Sites of 363 Cases*

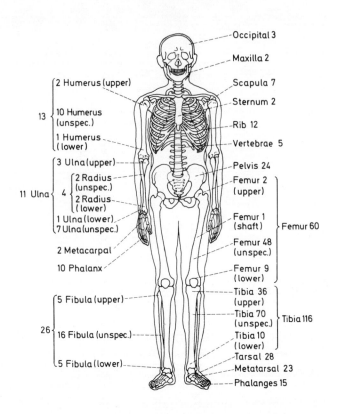

*Data from Dutt et al., 1969; Feldman et al.,1970; Frank and Rockwood,1969; Mikulowski and Østberg,1971; Murphy and Price,1971; Netherlands Committee on Bone Tumors, 1966; Pisar, 1969, Rahimi et al., 1972; Ramini, 1974; Ryall, 1970; Schajowicz and Gallardo, 1971; Schutt and Frost, 1971, Spjut et al., 1971.

the axis of the bone. They have ranged in size from 1 to 10 cm, but in long bones, they are most commonly 3 × 2 × 2 cm. When the lesion involved a small tubular bone, it generally occupied its entire width. In flat bones, the tumor has the same generally benign features but tends to be larger (Dutt et al., 1969; Feldman et al., 1970; Pisar, 1969; Rahimi et al., 1972). The largest lesion in our series (Fig. 16) was in an iliac bone and measured 11.5 × 13 cm (Feldman et al., 1970). Another tumor in an iliac bone measured 12 × 8 × 6 cm (Rahimi et al., 1972).

Only rarely and in advanced lesions does the tumor cross the unfused growth plate. In the series of Rahimi et al., (1972) 12 of 76 lesions abutted the growth plate. The lesion occasionally abuts the adjacent diaphysis, but (Turcotte et al., 1962) a solely diaphyseal location is rarely observed, e.g., in 3 of 207 cases in our series. Three tumors were noted as having a "definite diaphyseal location" in the review of Rahimi et al. (1972).

The eccentrically situated defect is typically well demarcated along its internal border. Frequently, a

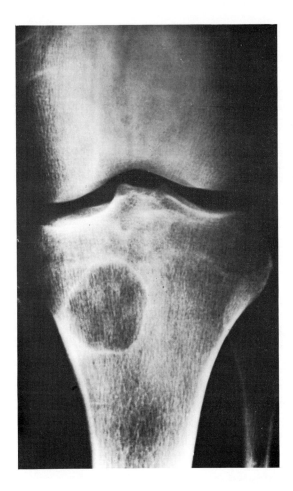 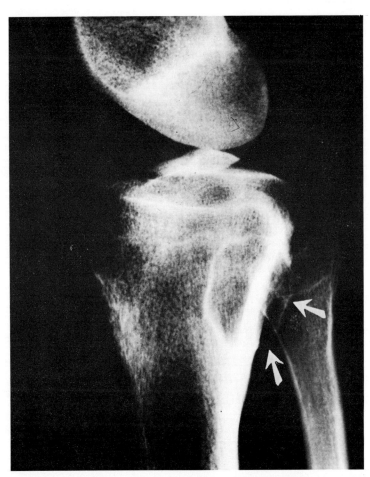

FIG. 15. Chondromyxoid fibroma—tibia. Left: Anteroposterior view. 21-year-old female with chief complaint of 4 months of pain and 2 weeks of swelling of left knee. Right: Lateral view. A clearly demarcated metaphyseal lesion expands and exquisitely thins the cortex (arrows). Note the lack of intralesional mineralization.

rim of sclerotic bone, which may be pronounced, separates the lesion from the remainder of the marrow cavity. Occasionally, a scalloped internal border simulates the curvilinear medial margin most often associated with nonossifying fibroma. The outer cortical surface may be expanded in a fusiform manner, but it is usually well defined. However, the expansion may be so marked that the lesion is externally delimited by an attenuated exquisitely thinned cortex that is barely discernible. Although this apparent lack of external margination or "bite out of bone" appearance is localized, it may on occasion mimic an aneurysmal bone cyst or a malignant tumor (Ralph, 1962).

Cortical layering or other evidence of active periosteal new bone formation as well as a Codman's triangle is unusual (Feldman et al., 1970; Rahimi et al., 1972; Salzer et al., 1968). Spjut et al. (1971) note that periosteal reaction is not seen unless a fracture is present. Rarely, apparently uninvolved bone a short distance beyond the visible confines of the lesion may show cortical thickening. The usual benign roentgenographic features of the tumor, therefore, vary in bones of different size and types.

Although the lesion has frequently been described as having a "soap bubble" appearance or as being trabeculated, in most instances, the roentgenographic impression of trabeculation is a false or misleading one. The bony septa that appear to be compartmentalizing or dividing the lesion are in fact pseudotrabeculae, which do not traverse the tumor but are a reflection of its lobulated periphery grooving the cortex with which it comes in contact. Therefore, intralesional trabeculae are simulated by the ridged or corrugated endosteal bone seen on end through the usually radiolucent, unmineralized lesion. If one were to tomograph a particular lesion, no bony septa would be delineated within the tumor substance.

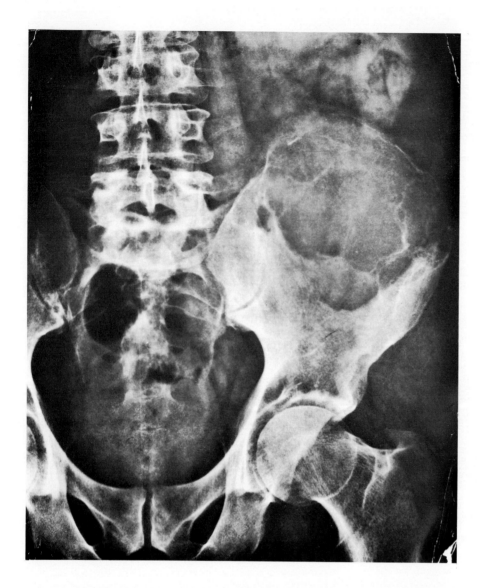

FIG. 16. Chondromyxoid fibroma—pelvis. Anteroposterior view. A large lytic lesion literally balloons the superior iliac cortex, which appears interrupted superior-medially. The periosteum, however, was preserved at surgery. Note the sharp inferior demarcation of the tumor. No mineral was noted within the lesion either roentgenographically or pathologically.

In our previous study (Feldman et al., 1970), it was noted that tumoral calcification has rarely been observed either roentgenographically (5 of 207 cases —2 percent), or microscopically (14 of 207 cases with two said to show concomitant ossification). Schajowicz and Gallardo (1971) reported only one case in which areas of spotty calcification were radiographically visible. In a recent review of 76 cases, Rahimi et al., (1972) affirmed the rarity of roentgenographically visible calcification by adding a single case. However, they noted microscopic evidence of calcification in 27 percent, an unusually high figure. Jaffe and Lichtenstein (1948), in their original report on this lesion and Lichtenstein (1965) more recently continued to emphasize that microscopically evident calcification or ossification is only an occasional and inconspicuous feature.

Microscopic Features

The basic microscopic pattern is that of areas of spindle-shaped, ovoid, or stellate cells, without cytoplasmic borders, loosely dispersed within a myxoid and occasionally chondroid intercellular matrix. The lesion tends to be demarcated into pseudolobules by narrow vascularized curving bands of more compact tumor cells. It is the increased concentration of cells at the periphery of the lobules that is of extreme importance in identifying this neoplasm. At the more cellular periphery, the cells are often spindle-shaped (Fig. 17) and fibroblastic and occasional mitotic figures may serve to simulate fibrosarcoma.

In addition to the problematic overlap between the two benign lesions of chondroblastoma and chondro-

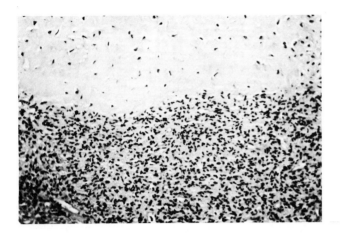 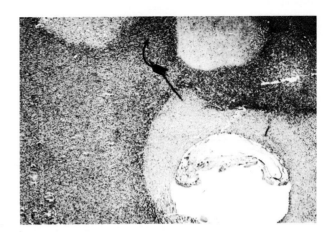

FIG. 17. Left: Chondromyxoid fibroma, photomicrograph (X150) showing cellular peripheral area (top) and chondroid area (bottom). Right: Photomicrograph (X150) showing two adjacent areas of chondroid tissue inferiorly divided into characteristic lobules (arrow). Lesions tend to be demarcated into pseudolobules by narrow vascularized curving bands of compact tumor cells. Note the blood vessels in the upper field.

myxoid fibroma and the small number of atypical chondromyxoid fibromas reported, chondromyxoid fibroma per se may also be confused histologically with chondrosarcoma, myxosarcoma, myxoma, giant-cell tumor, and chondroblastoma. In view of the broad spectrum of histologic features with which it may be associated, roentgenograms take on added significance in diagnosing this tumor, particularly where problem cases are concerned. The ultrastructure of chondromyxoid fibroma has just begun to be documented (Tornberg et al., 1973) (Fig. 18).

Treatment, Recurrence, Malignant Transformation

The treatment of choice is complete excision, when practical, with curettage and packing with bone chips as an alternative. Jaffe and Lichtenstein (1948) originally knew of no recurrence and noted that "even with incomplete removal, spontaneous regression of the remnants follow." In our review of 207 cases, 17 of 26 recurrences were attributed to incomplete tumor removal (Feldman et al., 1970). Among the 24 cases diagnosed and treated at the Mayo Clinic and separated from the 52 seen in consultation, curettage was the initial therapy in 15 tumors, 6 of which recurred (Rahimi et al., 1972). Four recurred within 2 years, one within 6.5 years, and one at 9 years post curettage. Various combinations of cautery, grafting, and irradiation employed as adjuncts did not obviate recurrence. Of the 14 recurrences that developed among the entire group of 76 patients, 11 occurred within 2 years, 2 after 3 years, and 1 after 6 years. Soft-tissue involvement by lobules of tumor contiguous with or adjacent to recurrent lesions in bone were seen in a few cases.

The tendency for local recurrence after curettage seems to be higher in young children (Ralph, 1962; Scaglietti and Stringa, 1961). Although the tumor is usually easily separated from the surrounding bone, its lobulated periphery and friability may contribute to its incomplete removal. A small lobule extending from the main mass into a groove or pocket of bone may be overlooked and may serve as a nidus for future recurrence. This circumstance, rather than the natural tendency for the tumor to recur, could also explain the recurrences with apparent multicentric foci (Dahlin, 1956; Feldman et al., 1970).

The role of radiation therapy was difficult to evaluate or assess, since it was most often used as an auxiliary postoperative measure. Although 12 cases had recieved radiotherapy in the Mayo Clinic series, details as to tumor dose, timing, equipment used, or quality of radiation were unknown in 5 cases (Rahimi et al., 1970). The value of radiotherapy to a predominantly benign lesion composed of tissue unlikely to be radiosensitive has been questioned and is generally felt to be contraindicated for surgically accessible lesions.

The malignant potential for chondromyxoid fibroma has been contested in the literature. Rahimi et al. (1972) noted that, "malignant transformation is extremely improbable and we have found no case in which malignant change has been convincingly documented." Schajowicz and Gallardo (1971) had no case in their series and also note that, "it appears to be very exceptionable."

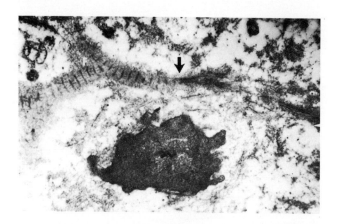 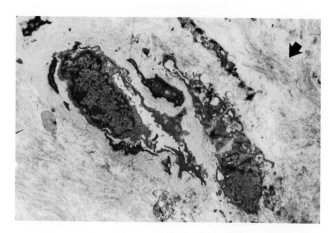

FIG. 18. Left: Chondromyxoid fibroma, electron micrograph (X30,000). A collagenous fiber bundle (F) having distinctive major banding periodicity of 973Å is seen in the intercellular matrix. The intraperiod bands are also visible in many of these structures, and loose aggregations of fibril 100 to 200Å in diameter (arrow) are also present. Right: Electron micrograph (X5500). Cells from the fibrous component. Note the collagen fibrils in the upper right field (arrow). (Courtesy of Austin D. Johnston)

Nevertheless, isolated cases have been reported. Three of these (Filippone, 1966; Seth and Rao, 1964; Witwicki and Dziak, 1967) were felt by Rahimi and Dahlin to contain insufficient documentation or none at all.

In the series of Rahimi et al. (1972), the only case in which malignant evolution was observed was in a patient who developed an anaplastic fibrosarcoma at the site 6 years after radiation therapy for chondromyxoid fibroma. They felt, that, "All evidence indicates that the risk of malignant transformation of bona fide chondromyxoid fibroma, unless radiation is employed, is so slight that it need not lead to unnecessary radical treatment." They also point out that "paradoxically because of the fear of unnecessary radical treatment for benign tumors, pathologists have often considered, or made the diagnosis of chondromyxoid fibroma when the lesion in question was actually a chondrosarcoma, with disastrously inadequate treatment resulting" (Rahimi et al., 1972).

The majority of recorded cases have responded well to simple operative measures. However, most authorities agree that complete removal by a resection that includes the tumor bed is necessary to prevent the relatively high rate of recurrence seen after curettage alone. The efficacy of complete resection has been documented by Schajowicz and Gallardo (1971) and is also attested to by the success of such treatment in recurrent tumors.

The most rewarding approach to all these cartilaginous lesions, i.e., the approach that yields the happiest results, heavily depends on a synthesis of biologic, clinical, pathologic, and radiologic data, any one of which may be decisive at a particular time and for a particular cartilaginous lesion.

References

Aergerter E: Radiol Clin North Am 8:215, 1970

Bauer WH, Harrell A: J Bone Joint Surg 36A:263, 1954

Chiurco A: Neurology 20:275, 1970

Cohen DM, Dahlin DC, MacCarty CS: Proc Mayo Clin 39:509, 1964

Dahlin DC: Cancer 9:159, 1956

Dahlin DC: Bone Tumors, 2nd ed. Springfield, Illinois, Thomas, 1967

Dahlin DC, Ivins JC: Cancer 30:401, 1972

Dahlin D, Swedlow M: 34th Semin Neoplasms Bone, Chicago, Am Soc Clin Pathol, 1969

Dutt AK, Dhillon DS, Din O: Med J Malaya 24:71, 1969

Fairbank HAT: J Bone Joint Surg 38B:237, 1956

Feldman F, Hecht HL, Johnston AD: Radiology 94:249, 1970

Fielding JW, Ratzan S: J Bone Joint Surg 55A:640, 1973

Filippone JJ: J Am Pediatr Assoc 56:237, 1966

Frank WE, Rockwood CA Jr: South Med J 62:1238, 1969

Geschickter CF, Copeland MM: Tumors of Bone, 3rd ed. Philadelphia, Lippincott, 1949

Goldanich IF: I Tumori Primitivi dell'Osso. Bologna, Societa per Azione Poligrafici il Resto del Carlino, 1957, p 197

Gokay H, Bucey PC: J Neurosurg 12:72, 1955

Goorwitch J: Dis Chest 20:186, 1951

Heiple KG: J Bone Joint Surg 43A:861, 1961

Herfarth H: Arch Klin Chir 170:283, 1932

Huvos AG, Marcove RC: Clin Orthop 95:300, 1973

Huvos AG, Marcove RC, Erlandson RA, Mike V: Cancer 29:760, 1972

Ilgenfritz HC: Am Surg 17:917, 1951

Iwata S, Coley BL: Surg Gynecol Obstet 107:571, 1958

Jaffe HL: Bull Hosp Joint Dis 17:20, 1956

Jaffe HL: Tumors and Tumorous Conditions of the Bones and Joints. Philadelphia, Lea and Febiger, 1968

Jaffe HL, Lichtenstein L: Am J Pathol 18:969, 1942

Jaffe HL, Lichtenstein L: Arch Pathol 45:541, 1948

Joseph WL, Fonkalsrud EW: Am Surg 38:338, 1972

Keats TE: Radiology 68:558, 1957

Kettelkamp DB, Campbell CJ, Bonfiglio M: J Bone Joint Surg 48A:746, 1966

Kettelkamp DB, Dolan J: J Bone Joint Surg 48A:329, 1966

Kunkel MG, Dahlin DC, Young HH: J Bone Joint Surg 38A:817, 1956

Lettin AWF: Proc R Soc Med 56:10, 1963

Lichtenstein L: Bone Tumors, 3rd ed. St. Louis, Mosby, 1965

Lichtenstein L, Hall JE: J Bone Joint Surg 34A:691, 1952

Marcove RC, Kambolis C, Bullough PG, Jaffe HL: Cancer 17:1209, 1964

McLeod RA, Beabout JW: Am J Roentgenol 118:464, 1973

Meyer R: Br J Radiol 31:106, 1958

Meyerding HW: Radiology 8:282, 1927

Mikulowski P, Östberg G: Acta Orthop Scand 42:385, 1971

Milch RA: Am J Surg 87:134, 1954

Murphy NB, Price CHG: Clin Radiol 22:261, 1971

Netherlands Committee on Bone Tumours: Radiological Atlas of Bone Tumours, Vol. 1. The Hague, Mouton; Baltimore, Williams and Wilkins, 1966

Nosanchuk JS, Kaufer H: J Bone Joint Surg 51A:375, 1969

Oller M: Lyon Med 93:23, 1900

Peck JH Jr: J Bone Joint Surg 46A:1379, 1964

Pisar DE: Tex Med 65:52, 1969

Plum GE, Pugh DG: Am J Roentgenol 79:584, 1958

Pratt GD, Dahlin DC, Ghormley RK: Surg Gynecol Obstet 106:536, 1958

Rahimi A, Beabout JW, Ivins JC, Dahlin DC: Cancer 30:726, 1972

Ralph LL: J Bone Joint Surg 44B:7, 1962

Ramini PS: J Neurosurg 40:107, 1974

Rose EF, Feketa A: Am J Clin Pathol 42:606, 1964

Rubin P: Dynamic Classification of Bone Dysplasia. Chicago, Year Book Medical, 1964

Ryall RDH: Br J Radiol 43:71, 1970

Salzer M, Salzer-Kuntschik M, Kretschmer G: Arch Orthop Unfallchir 64:229, 1968

Scaglietti O, Stringa G: J Bone Joint Surg 43A:67, 1961

Schajowicz F, Gallardo H: J Bone Joint Surg 52B:205, 1970

Schajowicz F, Gallardo H: J Bone Joint Surg 53B:198, 1971

Schauwecker F, Weller S, Klumper A, Anlauf B: Bruns Beitr Klin Chir 217:155, 1969

Schutt PG, Frost HM: Clin Orthop 78:323, 1971

Seth HN, Rao BDP: Indian J Pathol Bacteriol 7:112, 1964

Shellito JG, Dockerty MB: Surg Gynecol Obstet 86:465, 1948

Sherman RS, Uzel AR: Am J Roentgenol 76:1132, 1956

Sideman S, Sarrafian S, Topouzian LK: Northwestern Univ Med School Quart Bull 34-35:346, 1960-61

Spjut HJ, Dorfman HD, Fechner RE, Ackerman LV: Tumors of Bone and Cartilage. Atlas of Tumor Pathology, 2nd Ser, Vol 5. Washington, Armed Forces Inst Pathol, 1971

Teitelbaum SL, Bessone L: J Thorac Cardiovasc Surg 57:333, 1969

Thomas ML, Andress MR: Br J Radiol 44:549, 1971

Tornberg DN, Rice RW, Johnston AD: Clin Orthop 95:295, 1973

Trevor D: J Bone Joint Surg 32B:204, 1950

Turcotte B, Pugh DG, Dahlin DC: Am J Roentgenol 87:1085, 1962

Willis RS: Pathology of Tumours, 4th ed. London, Butterworths, 1967

Witwicki T, Dziak A: Wiad Lek 20:2211, 1967

Wouters HW, Szepesi K, Kullman L: Arch Chir Neerl 26:63, 1974

The Upper Cervical Spine*

J. William Fielding, William R. Francis, and Richard J. Hawkins

Correct roentgenographic diagnosis of patients with injured necks may be crucial to safe management, but the absence of roentgenographic findings does not preclude damage, even severe damage, to the spinal cord or vertebral arteries. Normal variations, epiphyseal lines, unique vertebral architecture, and congenital and developmental anomalies all may serve to confuse the x-ray interpretation of the injured neck.

The incidence of cervical spine injury in accidental deaths may not be fully appreciated, since postmortem examinations of individuals dead on arrival at hospitals do not always include the cervical spine. In a unique postmortem radiologic investigation of traffic accident victims dead on or shortly after arrival at a large metropolitan (Buffalo) hospital, Alker[1] found that 50 to 200 (25 percent) had traumatic lesions of the cervical spine, largely in the upper segments. Many of these could by themselves have been responsible for the death of the patient, but more obvious injuries to the chest, abdomen, and head might normally be accepted as the cause of death. Bohlman[6] reported the inadvertent production of neurologic deficits or death in 11 of 300 patients with multiple injuries who had unrecognized neck fractures and whose necks were moved during the course of emergency management. This further emphasizes that serious neck injury can be hidden, expecially in patients with multiple injuries.

Short of a bullet wound or similar trauma, it is difficult to injure the cervical spine by direct violence. Blockey and Purser[5] considered mandibular fractures to be the most common injury seen in association with fractures of the odontoid. Nachemson[18] noted that 40 percent of patients with odontoid fractures had associated head injuries such as large scalp wounds, cerebral contusions, and skull fractures. Evidence of head or facial trauma with or without multiple injuries, particularly in the comatose patient, should alert the physician to rule out "hidden" cervical spine injury.

Radiologic Technique

It is often impossible to obtain a thorough x-ray assessment of the patient with multiple injuries, particularly if the patient is unconscious or uncooperative. Most obvious serious problems, however, can usually be diagnosed or ruled out. The type of study will depend on the clinical appearance. The suggested first films should include lateral, anteroposterior, open-mouth, and oblique views plus an "overpenetrated" lateral view.

Routine Views

The lateral view generally provides valuable immediate information because it will show most traumatic lesions. If the patient is unable to stand, a cross-table lateral view will suffice. Care must be taken to attempt to visualize down to the upper border of T1; generally, traction on the wrists will depress the

---

*Adapted from Fielding JW, Hawkins RJ: Roentgen diagnosis of the injured neck. A.A.O.S. Instructional Course Lectures, Vol 25. St. Louis, Mosby, 1976, p 149.

shoulders enough to visualize T1. If this maneuver fails, a swimmer's or Twining view can be attempted.[15] Even with these maneuvers, it may not be possible to visualize T1 or the lower cervical spine.

Anteroposterior and open-mouth views can be obtained with the patient supine by placing the film beneath the neck on the stretcher. An open-mouth view can be obtained if the patient is cooperative. The x-ray beam may have to be redirected caudally or cranially to project the overlying maxillary teeth or other obstructions away from the odontoid. It is occasionally helpful to use a prolonged exposure with the patient repeatedly opening and closing his mouth (Otanello's technique).

An overpentrated view allows better visualization of bony injuries at the expense of soft-tissue definition. These routine views are of better quality with the patient erect. Satisfactory films, however, can usually be obtained without moving the patient from a stretcher. Common causes for errors in diagnosis include poor radiographs and failure to visualize the upper border of T1.

Normal Variations

Prevertebral Soft Tissue (Retropharyngeal Soft Tissue)

The space between the cervical spine and the pharynx in the region of the third cervical vertebra has been estimated to be a maximum of 5 mm in adults and about two-thirds of the thickness of the second cervical vertebra in children. An increase in width after trauma is presumptive evidence of hemorrhage or edema from fracture. The retropharyngeal soft-tissue swelling should be observed on a lateral roentgenogram taken at rest during quiet respiration. Since the pharynx is attached to the hyoid bone, any action displacing it forward will "artificially" increase the width of the retropharyngeal shadow. It has been demonstrated by Ardran and Kemp[3] that in inspiration the pharyngeal wall is close to the vertebrae, while in forced expiration there is a marked "physiologic" increase in the size of the shadow. Anything that displaces the hyoid bone and larynx forward, such as emitting a high-pitched sound as in crying, will "physiologically" increase the width of the shadow, a fact that should be considered when interpreting the film (Fig. 1).

Atlantal–Dens Interval

The atlantal–dens interval (ADI), a term introduced by MacEwen and Hensinger, should be 3 mm or less in the adult, even with the neck flexed.[11,13,19,20] In children less than 8 years of age, the distance has been reported to be as much as 4 mm in flexion; it was

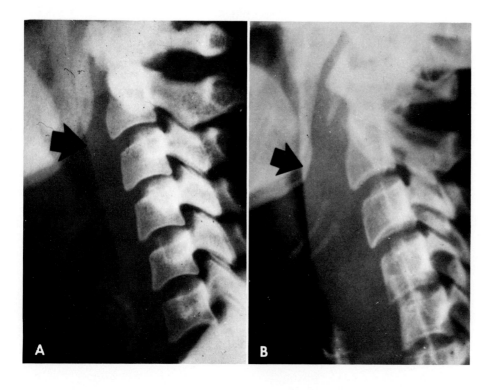

FIG. 1. Roentgenogram of a 45-year-old woman during quiet respiration. A. Enlarged retropharyngeal soft-tissue space. B. Markedly enlarged retropharyngeal soft-tissue space. (From Fielding JW: Selected observations on the cervical spine in the child. In Ahstrom JP Jr (ed): Curr Pract Orthop Surg 5. St. Louis, Mosby, 1973)

postulated that the increase was caused by the greater ligamentous laxity thought to be present in children. Until anatomic studies of children are available to document this, it cannot be stated with accuracy whether this increased space is caused by the increased amount of cartilage in children or by the reported ligamentous laxity.

Space Available for Cord

The space available for the cord (SAC), or diameter of the spinal canal,[19] is the narrowest distance between the posterior edge of a vertebral body and the anterior edge of the posterior vertebral arch. In children and adults, it should be greater than 14 mm from C1 to C7.[20] The width of an odontoid is roughly equal to the width of the spinal cord at the C1 level and is therefore a guide to the SAC. If the SAC is less than the width of the odontoid, the cord may be compromised. A small posterior ring of C1 may accentuate this problem.

Overriding of Anterior Arch of Atlas

A mistaken impression of odontoid hypoplasia may be given by the lateral extension x-ray film of a very young child, because the anterior arch of the atlas slides upward and protrudes beyond the ossified part of the dens to lie against the unossified tip. Cattell and Filtzer[8] reported this finding in 14 of 70 (20 percent) children aged 1 to 7 years. This is essentially the same age group showing a physiologic increase in the ADI, which was thought to be caused by ligamentous laxity but in reality may be caused by a cartilage space (Fig. 2).

Cervical Spine in the Child*

The injured child's neck often becomes a diagnostic enigma, baffling the most astute physician. "Epiphyseal variations, unique vertebral architecture, incomplete ossification, and hypermobility of the cervical spine may all cause uncertainty when interpreting the cervical roentgenograms of a child with a history of neck injury, pain and stiffness. There is an understandable tendency to conclude that certain of these normal variations represent actual injury."[8]

---

*This discussion is reprinted, with minor modifications, from Fielding JW: Selected observations on the cervical spine in the child. In Ahstrom JP Jr (ed): Curr Pract Orthop Surg 5: St. Louis, Mosby, 1973.

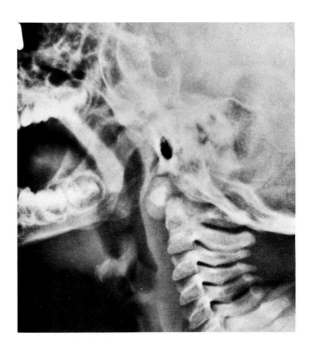

FIG. 2. In an extension x-ray film of a very young patient, the anterior arch of the atlas may slide upward to protrude beyond the ossified part of the dens, giving a mistaken impression of odontoid hypoplasia. (From Fielding JW, Hawkins RJ: Roentgen diagnosis of the injured neck. A.A.O.S. Instructional Course Lectures, Vol 25, St. Louis, Mosby, 1976)

In neck injuries, unlike extremity injuries, contralateral comparison views cannot be made. The problem is further compounded by the limited ability of the frightened child with a painful neck to cooperate in obtaining satisfactory roentgenograms.

Epiphyseal Development and Variations

A thorough knowledge of normal epiphyseal development is mandatory in interpreting a young child's roentgenograms. Epiphyseal plates are smooth, regular, and in predicted locations, and they have subchondral sclerotic lines. Fractures are irregular, without sclerosis, and generally in unpredictable locations. The exact dates of epiphyseal appearances and disappearances cannot always be committed to memory, but knowledge of their location is helpful. The first two cervical vertebrae are unique in their development; the remaining five are essentially uniform.[4,7]

Atlas

At birth, the atlas is composed of three ossification centers—one for the body and one for each of the

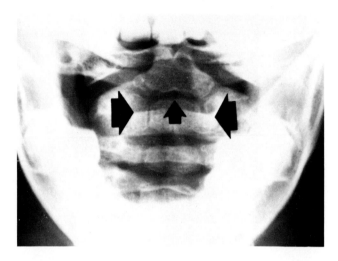

FIG. 3. The epiphysis at the base of the odontoid process is shown by the small arrow. This is well below the level of the superior articular facet of the axis. Synchondroses between the body of the axis and neural arches are shown by the large arrows. Just above this are synchondroses between the odontoid process and the neural arches. The odontoid process therefore surmounts the body of the axis and is sandwiched between the neural arches. The epiphyseal line and the synchondroses combine to form an H. (From Fielding JW: Selected observations on the cervical spine in the child. In Ahstrom JP Jr (ed): Curr Pract Orthop Surg 5. St. Louis, Mosby, 1973)

two neural arches. The center for the body, occasionally bifid, is not usually present at birth. It usually appears during the first year of life, and in rare cases it remains absent throughout life. In such cases, the anterior ring of the atlas may close by fusion of the neural arches anteriorly.

The neural arch usually closes by the third year, although it can remain open. Rarely, there may be an independent ossification center appearing in this region. The posterior ring of the first cervical vertebra may develop partially or remain completely absent throughout life. The neurocentral synchondroses binding the neural arches to the body are best seen in the open-mouth view. They close by the seventh year of life and should not be mistaken for fractures.

The ligament crossing the foramen for the vertebral artery rarely, of ever, calcifies in children. However, it calcifies in about 12 percent of adults to form the so-called ponticulus posticus enclosing foramen arcuale (posterior atlantoid foramen) through which passes the vertebral artery.

Axis

The developing axis, perhaps the most unusual of all the cervical vertebrae, and perhaps of all the body's bones, has four ossification centers at birth—one for each neural arch, one (occasionally two) for the body, and a fourth for the odontoid process. In the anteroposterior open-mouth view, the odontoid process seems to fit like a cork in a bottle, sandwiched between the neural arches. It surmounts the body of the axis and is separated from it by a synchondrosis, or the epiphyseal plate of the odontoid process. Below this are the synchondroses between the body and the neural arches. The epiphyses and synchondroses combine to form an H.

The epiphysis of the odontoid does not run across the apparent base of the structure, which appears to be at the level of articular processes of the axis; it lies well below this level. In the adult, a persistent epiphyseal line is not seen at the base of the odontoid process, where a fracture would be anticipated, but within the axis body and well below the level of the articular facets of the axis (Fig. 3).

The odontoid joins the neural arches and the body of the axis between 3 and 6 years of age, at essentially the same time that the body joins the neural arches. Therefore, no epiphysis or synchondrosis should be present in the axis in the open-mouth view of a child older than 6 years.

The ossification center of the inferior vertebral ring of the second cervical vertebra should cause little difficulty. It ossifies during the late years of childhood and fuses with the body at approximately 25 years of age.

Cattell and Filtzer[8] indicate that the basilar epiphysis of the odontoid may persist up to age 11 as a decidely narrow, occasionally sclerotic line resembling an undisplaced recent fracture.

The odontoid process is probably the most unusual ossicle in the spine. In utero, it develops two independent ossification centers on either side of the midline, usually during the fifth fetal month. These fuse by the seventh fetal month. Persistence of this line should not be interpreted as a fracture. The tip of the odontoid process is not ossified at birth, and it has a V-shaped appearance. A small ossification center known as the summit ossification center appears at its tip about at age 3 to 6 years and usually fuses with the main portion of the odontoid by age 12 years. Its persistence is referred to as an ossiculum terminale (Fig. 4).

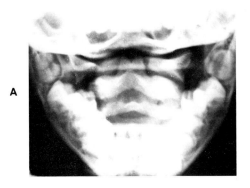
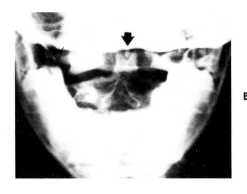
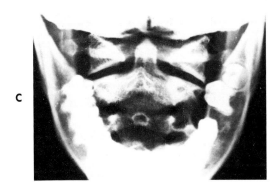

FIG. 4. **A.** Bifid tip odontoid process in a 2-year-old child. **B.** Terminal ossification center in a 5-year-old child. **C.** Ossification center is fused with odontoid process at age 12. (From Fielding JW: Selected observations on the cervical spine in the child. In Ahstrom JP Jr (ed): Curr Pract Orthop Surg 5. St. Louis, Mosby, 1973)

Anomalies of the Cervical Spine

Anomalies of the cervical spine may be confused with bone injury. Hypoplasia and os odontoideum may cause instability of the atlantoaxial complex with potential neurologic sequelae or even death.[16,21] Odontoid aplasia is extremely rare and must include that part of the odontoid contributing to the upper part of the main body of the axis. The most common form of hypoplasia has a short "peg" of odontoid projecting just above the level of the lateral facet articulations (Fig. 5).

Os odontoideum consists of a rounded ossicle, usually posterior to the anterior arch of the first cervical vertebra or occasionally lying within the foramen magnum; it is separated by a wide gap from the hypoplastic remnant of the odontoid.[21] It usually moves with the anterior arch of C1. In both hypoplasia and os odontoideum there may be a hypertrophied anterior arch of the atlas and an underdeveloped posterior ring (Fig. 6).

Occipitocervical synostosis, partial or complete, is a congenital union between the atlas and the base of the occiput and may be associated with basilar impression with proximal positioning of the odontoid. Other anomalies occasionally present include condylar hypoplasia, aplasias, irregular segmentation, spina bifida, Klippel–Feil syndrome, and isolated congenital fusions, particularly at the C2–C3 level.

Specific Injuries

Since most cervical injuries are the result of indirect violence, the direction of the force may provide a clue to the area damaged.

Atlantooccipital Dislocation

Atlantooccipital dislocations are usually bilateral and are rarely reported, probably because they are often incompatible with life. There are scattered reports of survivors.[14] The lateral x-ray film shows displacement of the occipital condyles from the superior facets of the atlas, and usually there is marked retropharyngeal swelling.

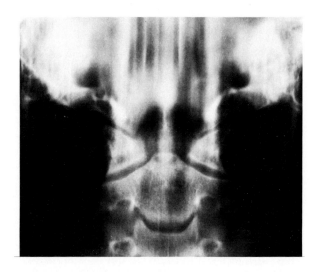

FIG. 5. Odontoid hypoplasia. The hypoplastic dens usually projects just above the level of the superior facets of the axis. (From Fielding JW, Hawkins RJ: Roentgen diagnosis of the injured neck. A.A.O.S. Instructional Course Lectures, Vol 25, St. Louis, Mosby, 1976)

C1–C2 Ligament Disruption

A biomechanical study of adult cadaver specimens at St. Luke's Hospital in New York indicates that up to 3 mm of anterior displacement of the atlas on the axis implies that the transverse ligament is intact. If the displacement is from 3 to 5 mm, the transverse ligament is ruptured. If the displacement exceeds 5 mm, the transverse ligament has ruptured and the accessory ligaments (alars, capsular, etc.) are stretched and partially deficient.[13]

Isolated traumatic transverse ligament ruptures are rare and usually follow severe trauma (Fig. 7).

Rheumatoid arthritis, anomalies of the odontoid, and odontoid fractures are more common causes of atlantoaxial instability.

Atlantoaxial Rotary Dislocation

Complete rotary dislocation of the atlantoaxial joint is rarely reported, since it appears to be incompatible with life. In this condition, the skull and atlas are rotated approximately 90 degrees in relation to the axis. Thus, in the routine anteroposterior radiographic projection, there is a lateral view of the skull and atlas and an anterior view of the axis.

Atlantoaxial Rotary Displacement

Atlantoaxial rotary displacement occurs predominantly in childhood and adolescence and is probably one of the most common causes of short-lived torticollis in this age group. Most patients recover either spontaneously or with minimal treatment. It is thought that this condition occurs most commonly by a "locking" of the facet joints at some point in the normal excursion of rotation.[9] The mechanism of the locking is obscure. The onset may be spontaneous, follow an upper respiratory tract infection, or follow minor or (rarely) major trauma. The patient has a torticollis deformity; the head is rotated to one side and tilted to the opposite side with a slight flexion, associated with pain, muscle spasm, and marked resistance to movement. Very rarely, the pain subsides but the deformity and decreased neck motion persist. Under these conditions it is best described as atlantoaxial rotary fixation.[22]

The radiologic features of rotary displacements are sometimes difficult to demonstrate, partly because of pain, difficulty in positioning, and difficulty in x-ray interpretation. Cineradiography has been valuable in studying this problem. This technique is of little help in the acute stages of rotary displacement, because pain precludes the motion necessary for a cineradiographic study. Therefore, the diagnosis at this stage is based more on history and physical examination. In longstanding cases, the pain has usually subsided and the deformity is fixed; cine-

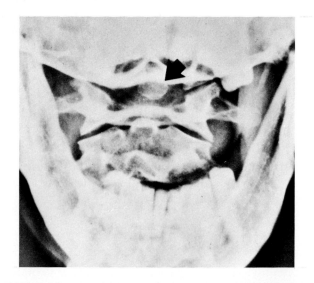

FIG. 6. Os odontoideum. The small rounded ossicle usually is just posterior to the anterior arch of the atlas and is separated by a wide gap from the hypoplastic remnant of the odontoid. (From Fielding JW, Hawkins RJ: Roentgen diagnosis of the injured neck. A.A.O.S. Instructional Course Lectures, Vol 25, St. Louis, Mosby, 1976)

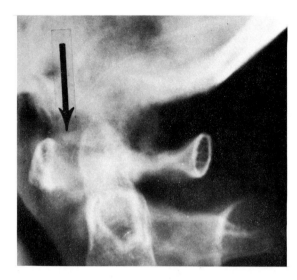

FIG. 7. Traumatic disruption of the transverse ligament with increased atlantal–dens interval (From Fielding JW, Hawkins RJ: Roentgen diagnosis of the injured neck. A.A.O.S. Instructional Course Lectures, Vol 25, St. Louis, Mosby, 1976)

radiography will then usually demonstrate the atlas and the axis rotating as a fixed unit, confirming the diagnosis of atlantoaxial rotary fixation.

The following are some radiologic features of atlantoaxial rotary displacements or torticollis from any cause. It should be noted that these features can be reproduced by a normal individual placing his head in the torticollis position. Hence, the clinical findings of pain and spasm assume important roles in diagnosis.

1. The lateral mass of C1 that has rotated forward appears wider and closer to the midline (medial offset), while the opposite lateral mass is narrower and away from the midline (lateral offset) (Fig. 8A).
2. One of the facet joints, usually the facet of C1, that has moved backward may be obscured by apparent overlapping (Fig. 8A)
3. On the lateral projection, the wedge-shaped lateral mass of the atlas that has moved forward lies anteriorly (where the oval arch of the atlas normally lies) (Fig. 8B).
4. The posterior arches of the atlas fail to superimpose because of the head tilt and may suggest assimilation of the atlas to the skull, because tilting may cause the skull to obscure C1.
5. In the anteroposterior projection, tomography may show the lateral masses of the atlas in different coronal planes and may erroneously suggest the absence of one atlantal mass, which has rotated to a different plane (Fig. 8C).
6. Since about 50 percent of neck rotation occurs at C1–C2, the spine of C2 is not deviated from the midline (indicating rotation) until greater than 50 percent of head rotation has occurred.[12] It deviates to the left with right rotation and vice versa. With lateral tilt, however, even minimal rotation of the cervical spine occurs below C1 to a much greater degree than with head rotation. The best indication of this rotation is the bifid spine of C2. It deviates to the right with left tilt and vice versa. Therefore, with tilt in one direction and rotation in the opposite direction as might occur in torticollis from any cause, the spine of the axis and the chin would be on the same side of the midline, because there is more rotation of C2 in head tilting than in head rotation (Fig. 8D).

Flexion–extension stress films are suggested to rule out possible anterior displacement of the atlas due to transverse ligament deficiency, which occasionally is seen in association with rotary displacements (Fig. 9).

Atlas Fractures

In 1920, Jefferson [17] classified atlas fractures into anterior arch, posterior arch, and ring fractures.

ANTERIOR ARCH FRACTURE Anterior arch fractures are rare; they are usually comminuted and minimally displaced. Tomograms are often necessary for documentation.

POSTERIOR ARCH FRACTURE Posterior arch fractures are the most common of atlas fractures and are probably the result of hyperextension with compression of the posterior arches of the atlas between the occiput and the axis pedicles. Displacement is usually mild; hence, the fracture may be overlooked. Oblique views usually best demonstrate the fracture. Retropharyngeal soft-tissue swelling is not usually present.

RING FRACTURE Ring fractures are the classic Jefferson fractures and usually result from a direct blow on the vertex of the head. The forces are dissipated laterally, fracturing the atlantal arches, usually where they join the lateral masses, and spreading the lateral masses laterally. With greater force, the transverse ligament may rupture, allowing even greater displacement. The typical radiologic finding is bilateral, usually symmetrical, overhang of the lateral masses of the atlas and associated widening of paraodontoid spaces. Many variations of overhang of the lateral atlantal masses and widening of the paraodontoid spaces occur with normal tilt or rota-

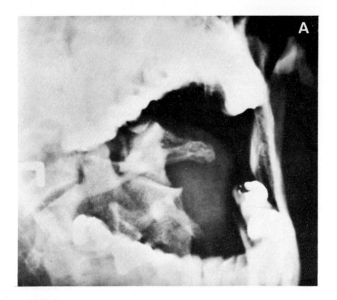
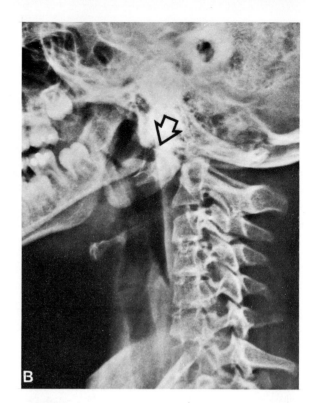
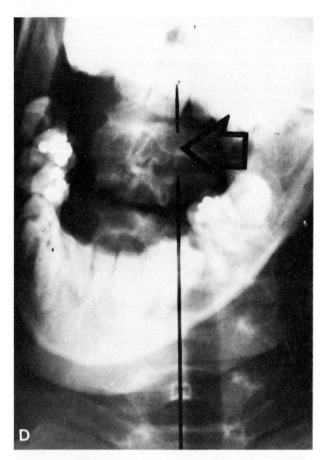
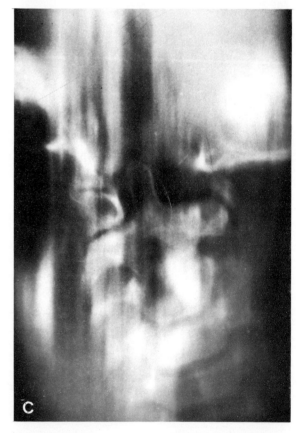

FIG. 8. Rotation of the atlas on the axis, which may occur in torticollis regardless of cause. A. Lateral mass of C1 that has rotated forward appears wider and closer to the odontoid (medial offset), whereas the opposite lateral mass appears narrower and away from the midline (lateral offset). Note that one facet joint appears obscured because of overlapping. B. The lateral mass has rotated to appear as a wedge-shaped mass lying anteriorly. C. Tomographic cut showing the well-outlined lateral mass of C1 on one side and the apparent absence of lateral mass on the opposite side. The "absent" lateral mass was seen on deeper cuts. D. In torticollis from any cause or even with normal tilt in one direction and rotation in the opposite, the spine of the axis and the chin are on the same side of the midline. (A and B from Fielding JW, Hawkins RJ: Roentgen diagnosis of the injured neck. A.A.O.S. Instructional Course Lectures, Vol 25. St. Louis, Mosby, 1976. C and D from Fielding JW: Selected observations on the Cervical Spine in the Child. In Ahstrom JP Jr (ed): Curr Pract Orthop Surg 5. St. Louis, Mosby, 1973)

The Upper Cervical Spine

tion or are of a developmental nature, but these are generally asymmetrical and unilateral. Retropharyngeal soft-tissue swelling is usually present (Fig. 10).

Odontoid Trauma

Children

Odontoid "fractures" in children younger than 6 years are uncommon; in reality they are epiphyseal slips.[10,11] The open-mouth anteroposterior view may demonstrate a normal-appearing epiphyseal line; the degree of anterior or posterior displacement can only be seen in the lateral projection.

Adults

Odontoid fractures in adults usually result from a force directed to the head. The force may dictate the direction of displacement; it is usually anterior or posterior. The open-mouth and lateral views usually show the fracture. Tomograms may be necessary, and flexion–extension films may show displacement, revealing a fracture not thought to be present. Repeat exposures with redirection of the x-ray beam in Otanello's technique may also be helpful.

Anderson and D'Alonzo[2] have classified odontoid fractures into three types. Type I is an avulsion of the upper part of the odontoid (which is rare). Type II is a fracture through the base of the odontoid at or below the level of the superior articular facets of the axis (the most common type). Occasionally in type II injury the odontoid is not displaced but is slightly angulated; this angulation may result in a toggle effect on flexion–extension films. Type III is a fracture of the body of the axis; these fractures may not be displaced and require tomograms for clarification. A small bone chip separated from the anteroinferior rim of the axis at the point of rupture of the anterior longitudinal ligament may be a clue to the type III injury. The importance of the classification lies in the fact that the type II fractures may be unstable and frequently lead to nonunion (36 percent of type II fractures treated conservatively in Anderson's series), and type III occur through the cancellous bone of the axis body, are usually stable, and heal (Fig. 11). The instability associated with odontoid fractures may be insidious, and follow-up x-ray films in cases of nonunion should include flexion–extension films.

Nonunion of the odontoid may be confused with os odontoideum.[20] In nonunion, the fracture line is narrow and at or below the level of the superior

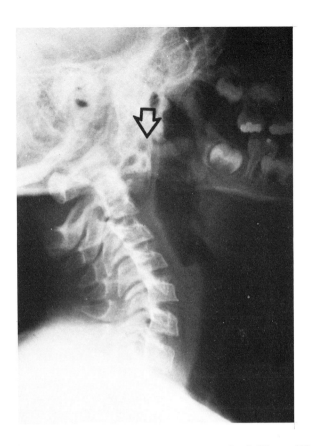

FIG. 9. Marked anterior displacement of C1 on C2 found in a patient with atlantoaxial rotary fixation. (From Fielding JW, Hawkins RJ: Roentgen diagnosis of the injured neck. A.A.O.S. Instructional Course Lectures, Vol 25, St. Louis, Mosby, 1976)

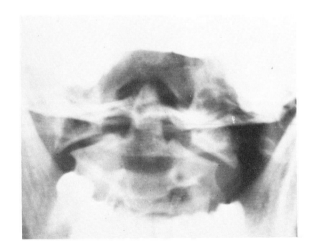

FIG. 10. Classic Jefferson fracture showing bilateral lateral overhang and associated widening of the paraodontoid spaces. These features are symmetric. (From Fielding JW, Hawkins RJ: Roentgen diagnosis of the injured neck. A.A.O.S. Instructional Course Lectures, Vol 25, St. Louis, Mosby, 1976)

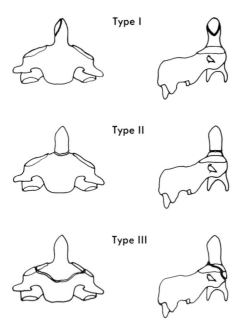

FIG. 11. Anderson's classification of odontoid fractures. Type I fracture is in the tip of the odontoid. Type II fracture is through the base of the odontoid; fracture line is at or below the level of the superior articulating facets of the axis. Type III fracture involves the main body of the axis. (From Anderson LD, D'Alonzo RT: J Bone Joint Surg 56A:1663, 1974.)

articular facets. The odontoid is of normal configuration and size. In os odontoideum, the ossicle is smaller than the normal odontoid (usually one-half its size), round, and separated from the hypoplastic odontoid by a wide gap. The remnant hypoplastic odontoid projects like a hill, continuing the upward slope of the superior articular facets of C2 (Fig. 11).

Conclusion

Appreciation of normal variation, epiphyseal architecture, and congenital and developmental anomalies will aid in diagnosing injuries to the cervical spine. With few exceptions the neck has generally attained an adult form in children over 8 years of age, and normal epiphyseal lines are no longer a diagnostic problem. Increased prevertebral or retrotracheal soft tissue, variations in alignment, teardrop fractures, and increases in interpediculate distances should alert the examiner to rule out more ominous cervical spine damage. The examiner should look for the radiologic signs of instability while keeping these radiologic patterns in mind.

References

1. Alker GJ: Postmorten radiology of head and neck injuries in fatal traffic accidents. Personal communication, 1975
2. Anderson LD, D'Alonzo RT: Fractures of the odontoid process of the axis. J Bone Joint Surg 56A:1663, 1924
3. Ardran GM, Kemp FH: The mechanism of changes in form of the cervical airway in infancy. Med Radiogr Photogr 44(2):26, 1968
4. Bailey DK: The normal cervical spine in infants and children. Radiology 59:712, 1962
5. Blockey NJ, Purser DW: Fractures of the odontoid process of the axis. J Bone Joint Surg 38B:794, 1956
6. Bohlman H: Personal communication, 1974
7. Caffey J: Pediatric X-ray Diagnosis. Chicago, Year Book Medical, 1967
8. Cattell HS, Filtzer DL: Pseudo-subluxation and other normal variations in the cervical spine in children. J Bone Joint Surg 47A:1295, 1965
9. Coutts MB: Atlanto-epistropheal subluxation. Arch Surg 29:297, 1934
10. Ewald FC: Fracture of the odontoid process in a 17-month-old infant treated with a halo. J Bone Joint Surg 53A:1636, 1971
11. Fielding JW: The cervical spine in the child. In Ashrom JP Jr: Curr Pract Orthop Surg 5. St. Louis, Mosby, 1973
12. Fielding JW: Cineroentgenography of the normal cervical spine. J Bone Joint Surg 39A:1281, 1957
13. Fielding JW, Cochran GVB, Lawsing JF III, Hohl M: Tears of the transverse ligament of the atlas: a clinical and biomechanical study. J Bone Joint Surg 56A:1683, 1957
14. Gabrielson TO, Maxwell JA: Traumatic atlantoaxial dislocation. With case report of a patient who survived. Am J Roentgenol 97:624, 1966
15. Garber JN: Fracture and fracture–dislocation of the cervical spine. American Academy of Orthopedic Surgery Symposium on the Spine. St. Louis, Mosby, 1969
16. Gwynn JL, Smith JL: Acquired and congenital absence of the odontoid process. Am J Roentgenol 88:424, 1962
17. Jefferson G: Fractures of the atlas vertebrae: report of four cases and review of those previously recorded. Br J Surg 7:407, 1920
18. Nachemson A: Fracture of the odontoid process of the axis. Acta Orthop Scand 29:185, 1958
19. Rothman RH, Simeone FA: The Spine, Vol 2. Philadelphia, Saunders, 1975
20. von Torklus D, Gehle W: The Upper Cervical Spine. New York, Grune, 1972
21. Wollin DG: The os odontoideum. J Bone Joint Surg 45A:1459, 1965
22. Wortzman G, Dewar FP: Rotary fixation of the atlantoaxial joint: rotational atlanto-axial subluxation. Radiology 90:479, 1968

Fractures and Dislocation about the Shoulder

Murray K. Dalinka

Fractures of the Proximal Humerus

Fractures of the proximal humerus are more common in females and occur more frequently in the older age group. They are often secondary to a fall on the outstretched, pronated upper extremity in osteoporotic individuals.[1]

Neer[2,3] has described the four-part classification based on displacement of one or more of the major segments of the proximal humerus. This classification is practical, reproducible, and may serve as a guide to both treatment and prognosis. The four major segments defined by Neer are:

1. Articular segment of the humerus or anatomic neck
2. Greater tuberosity
3. Lesser tuberosity
4. Surgical neck or shaft

This fracture classification is not dependent upon the number of fracture fragments, but upon segmental displacement. One-part fractures are those in which the fracture fragments are displaced less than 1 cm or angulated less than 45 degrees. These fractures are held together by an intact rotator cuff and periosteum. They make up approximately 85 percent of fractures of the proximal humerus and can usually be treated with protection and functional exercises.

The two-part fracture has one of the major fragments displaced by at least 1 cm or angulated more than 45 degrees. Greater tuberosity displacements of more than 1 cm are indicative of a torn rotator cuff. Lesser tuberosity fractures are frequently associated with posterior dislocations of the shoulder. Fractures through the head or anatomic neck are rare but have a high incidence of avascular necrosis.

Fractures through the surgical neck are frequently displaced anteriorly and medially because of the pull of the pectoralis major muscle (Fig. 1). They may be associated with brachial plexus injuries. Most two-part fractures, except occasional unstable surgical neck fractures and fractures of the greater tuberosity, can usually be treated closed.[2]

Three-part fractures consist of a displaced fracture through the surgical neck of the humerus and one displaced tuberosity. These frequently require operative intervention, as anatomic reductions are difficult to perform by closed methods.

A four-part fracture is one in which the four major segments are separated. The articular segment loses its soft-tissue attachments and blood supply; therefore, there is a very high incidence of avascular necrosis. These fractures are frequently treated with immediate prosthesis.

Epiphyseal injuries of the proximal humerus are rare, heal well, and are usually treated closed. They are often Salter II injuries, and since 80 percent of humeral growth is proximal, slight residual shortening is not uncommon.

Fractures of the Clavicle

Clavicular fractures make up approximately 5 percent of fractures about the shoulder, and the majority occur in children.[3] The fractures are frequently subdivided according to which third of the clavicle is involved. Approximately 80 percent of these fractures are midclavicular and heal without

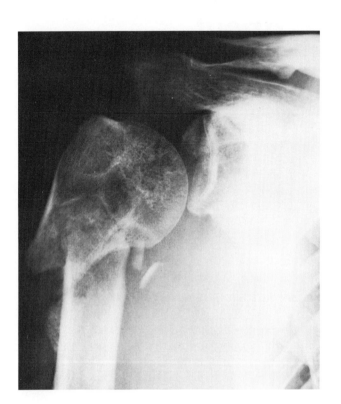 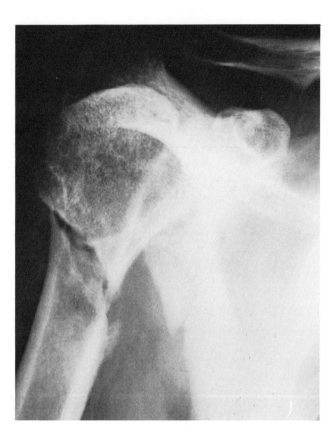

FIG. 1. Two-part proximal humeral fracture. (Left) Two-part fracture of proximal humerus with medial displacement of humeral shaft secondary to pull of the pectoralis major muscle. Note multiple fracture fragments in this two-part fracture. (Right) Same patient. Note associated nondisplaced fracture of scapula.

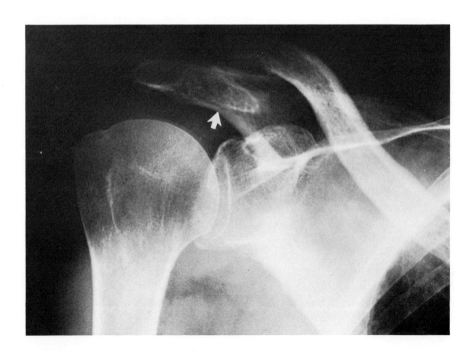

FIG. 2. Clavicular fracture with inferior displacement of distal clavicle. The coracoclavicular ligaments are attached to the distal fragment (arrow).

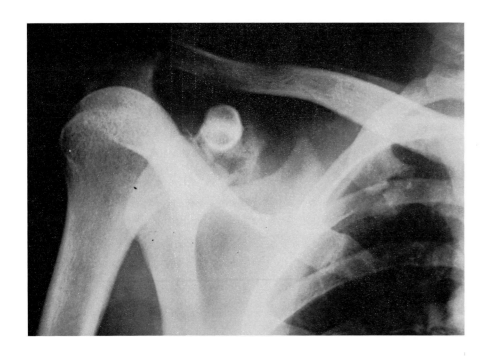

FIG. 3. Fracture at base of coracoid process associated with dislocation of acromioclavicular joint. Note additional fracture of third posterior rib and chip from distal clavicle.

significant problems. Fractures of the inner third of the clavicle are uncommon and account for approximately 5 percent of clavicular fractures. These may represent epiphyseal injuries, as the medial clavicular epiphysis does not close until approximately 25 years of age.

Fractures of the outer third of the clavicle can be subdivided into two types.[4] In the type one fracture, the ligaments are intact and there is no significant displacement. This makes up approximately three-quarters of outer clavicular fractures. In the type two fractures, the ligaments are attached distally but detached medially. The fractured clavicle is displaced inferior medially with frequent interposition of the trapezius muscle and may require open reduction (Fig. 2).[3,4]

mioclavicular separations (Fig. 3). These fractures may be difficult to see on routine views but are frequently visualized on a supine view with 30 to 35 degrees of cephalad angulation.[5] Coracoid fractures must be differentiated from an ununited epiphysis, which usually fuses between 15 and 18 years of age.

Acromion fractures result from a downward blow on the shoulder and may lead to avulsion of the brachial plexus. Fractures of the neck of the scapula are uncommon and are usually impacted without significant displacement.

Fractures of the glenoid are associated with approximately 20 percent of shoulder dislocation.[3] They are best seen on axillary views of the shoulder. Vigorous contractions of the triceps muscle, which can occur with throwing, may cause avulsion injuries to the inferior aspect of the glenoid.[6]

Fractures of the Scapula

Fractures of the body and spine of the scapula usually occur secondary to severe direct trauma and are frequently associated with other more serious injuries. The fractures are usually comminuted, but because of strong muscular attachments, there is frequently no significant displacement. The complications of scapular fractures are injury to the axillary nerve or brachial plexus, but these are rare.

Fractures of the coracoid process are unusual as isolated findings and may be associated with acro-

Dislocations

Dislocation of the shoulder is the most common dislocation in the body. Most dislocations about the shoulder are anterior dislocations of the glenohumeral joint. These accounted for 84 percent of dislocations about the shoulder in one large series.[7] Acromioclavicular dislocations made up 12 percent of Rowe's cases, sternoclavicular dislocations accounted for 2.5 percent, and posterior dislocations for only 1.5 percent.

Anterior Dislocations of the Glenohumeral Joint

The anatomy of the glenohumeral joint is such that the glenoid has one-third the articular surface of the humeral head, leading to an unstable joint with a large range of motion. The deeper the glenoid socket is, the less gliding motion there is, and the greater the tendency for subluxation. The joint capsule has twice the surface area of the humeral head and is loose, redundant, and synovial lined. It is reinforced anteriorly by the glenohumeral ligaments. The subscapularis bursa is an outpouching of synovium from the joint between the mid and superior glenohumeral ligaments. The glenoid labrum is a rim of fibrous tissue about the glenoid fossa formed by the interconnecting and joining of the periosteum of the glenoid, the hyaline articular cartilage of the glenoid fossa, the anterior capsule, and the synovium. The axillary nerve is inferior to the glenohumeral joint and the neurovascular bundle anterior medial to it.

Anterior dislocations are usually described by the anatomic location of the humeral head. They may be subdivided into subcoracoid (Fig. 4), subglenoid, subclavicular, intrathoracic, or luxatio erecta types. The subclavicular, intrathoracic, and luxatio erecta dislocations are rare.[8]

When the humerus dislocates anteriorly, its posterior lateral margin is frequently driven against the glenoid rim, which produces a compression fracture called the Hill-Sachs deformity (Fig. 5). This deformity may be demonstrated on internal rotation or axillary views, although occasionally special projections may be necessary for its demonstration.[8] The dislocated humerus frequently detaches the cartilaginous labrum and capsule from the glenoid rim, occasionally with an avulsed fragment of bone. This cartilaginous defect, the Bankhart lesion, and the detached capsule may be demonstrated on double-contrast arthrography, as can the occasional associated tear of the rotator cuff. The younger the patient is the greater are the chances of recurrent dislocation. If the glenoid rim is fractured, the recurrence rate is approximately 95 percent. Other predispositions to recurrent dislocation are the size of the capsular deformity, the Hill-Sach's lesion, and the normal range of lateral motion. The greater the normal range, the greater the incidence of recurrent dislocation.[8]

Posterior Dislocations of the Shoulder

Posterior dislocations of the shoulder are rare and frequently overlooked. In one series,[9] they made up 24 percent of unrecognized dislocations. Convulsive seizures are the most common cause of posterior dislocations, and when posterior dislocations are bilateral, they are almost always caused by seizures.[10]

The x-ray findings on the "routine" internal and external rotation views may be deceptively normal (Fig. 6). The shoulder is in extreme internal rotation,

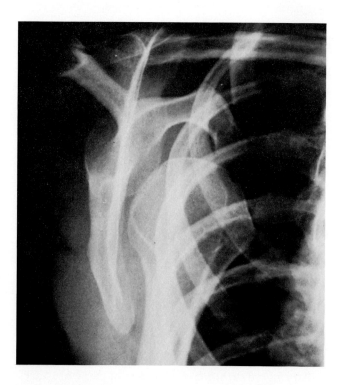

FIG. 4. Subcoracoid dislocation of shoulder. (Left) Subcoracoid dislocation of shoulder on external rotation view. (Right) Same patient. Scapula Y view revealing the humeral head anterior and beneath the coracoid process.

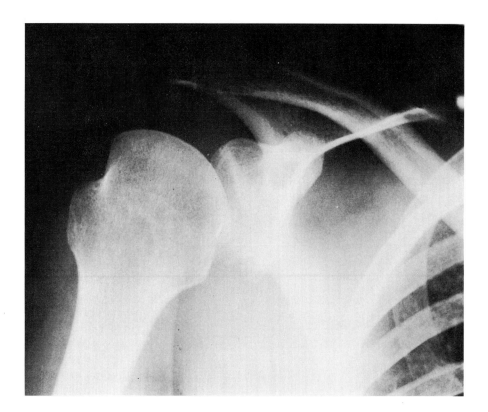

FIG. 5. Hill–Sach's deformity. Note the posterior lateral defect of the humeral head on the internal rotation view. Also note fragment of bone beneath the glenoid rim, the osseous portion of a Bankhart lesion.

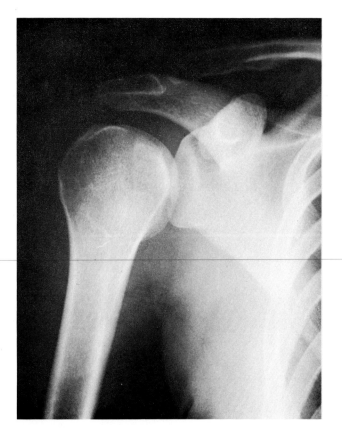 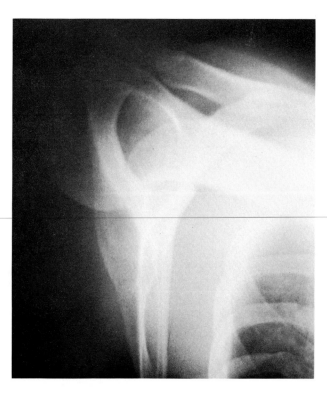

FIG. 6. Posterior dislocation of the shoulder. Left: The shoulder is in internal rotation but appears normal. Right: Lateral scapula (scapula Y) view. Note head of humerus over acromion process indicating posterior position. Compare to Fig. 4.

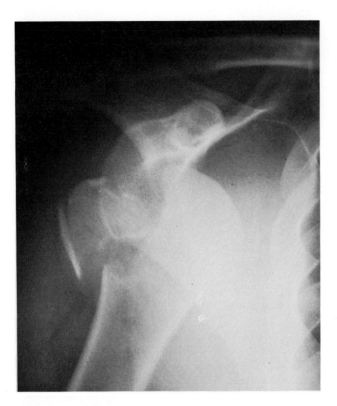

FIG. 7. Two-part anterior fracture dislocation of humerus. The greater tuberosity is displaced and comminuted, and the humerus is anteriorly displaced beneath the coracoid. This displacement indicates a longitudinal tear of the rotator cuff.

with the lesser tuberosity in sharp profile. The normal lateral prominence of the greater tuberosity is not seen. A vertical line may be seen paralleling the humeral head. This represents a compression fracture and not the bicipital groove. This defect is the counterpart of the Hill–Sachs deformity and is caused by the posterior rim of the glenoid compressing the cancellous bone between the anatomic neck and lesser tuberosity. The normal eliptical shadow representing the overlap of the humeral head and the glenoid is, of course, not present.[8] An avulsion fracture of the lesser tuberosity should lead one to search for a posterior dislocation.[8]

Rubin et al[11] described the use of the scapula Y projection in suspected dislocations. This is a 60-degree anterior oblique view of the shoulder or lateral view of the scapula. The coracoid process and acromion join with the scapula spine, resembling the letter Y. In posterior dislocations the humerus is projected over the acromion (Fig. 6) and in anterior dislocations beneath the coracoid process or glenoid (Fig. 4). We use this projection routinely in all cases of shoulder trauma.

The axillary view of the shoulder shows posterior dislocations clearly, but it is frequently difficult to obtain in patients with acute posterior dislocations who are often unable to abduct the arm. The author feels that the transthoracic lateral view is often technically unsatisfactory and does not advocate its use.

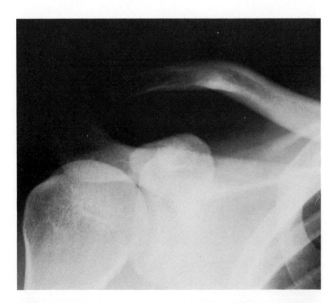
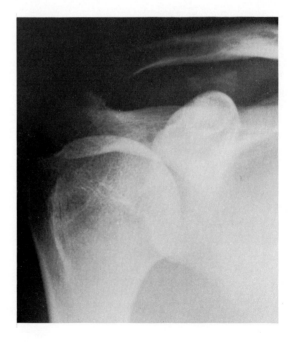

FIG. 8. Acromioclavicular dislocation. Above: Note clavicle is higher than acromion. Distance between acromion and clavicle is increased as is coracoclavicular distance. Right: Same patient following immobilization. Note calcification in area of coracoclavicular ligaments.

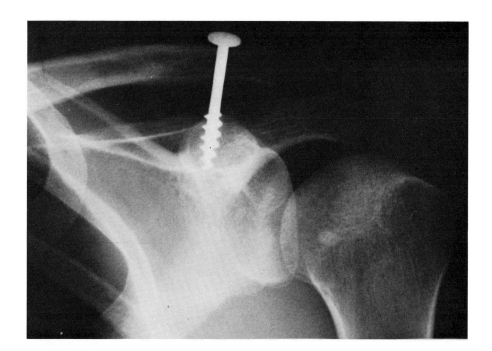

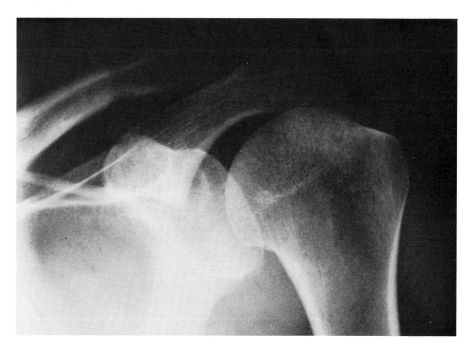

FIG. 9. Acromioclavicular dislocation. Top: Screw fixation of clavicle to coracoid process in treatment of dislocation. Bottom: Calcification in coracoclavicular ligaments following removal of the screw.

Fracture Dislocations

In fracture dislocations, the humeral fracture is almost always displaced with the articular surface outside the joint. True dislocations must be differentiated from pseudosubluxations in which the humerus is displaced inferiorly by hemarthrosis and poor muscle tone.[12]

Anterior fracture dislocations are frequently associated with displacement of the greater tuberosity (Fig. 7). These dislocations do not usually displace the capsule, and hence, recurrent dislocation is infrequent. In posterior fracture dislocations, the lesser tuberosity is frequently displaced.[8] Neer's classification[2,3] can, of course, be extended to include fracture dislocations.

Acromioclavicular Dislocations

Acromioclavicular dislocations may lead to widening of the acromioclavicular joint or elevation of the distal clavicle. Films with the patient erect and holding a 10-pound weight taken with 15 degrees of cephalad angulation may be necessary together with films of the opposite shoulder for purposes of comparison. These films may demonstrate joint widening or clavicular elevation. The clavicular elevation represents widening of the coracoclavicular distance. An increase of the coracoclavicular distance by 5 mm or a distance greater than 50 percent larger than the opposite side indicates a true acromioclavicular dislocation[8] (Fig. 8). The coracoid process may be fractured with this injury[3] and should be carefully evaluated, since one of the more popular surgical repairs for acromioclavicular dislocations consists of fixation of the coracoid process to the clavicle (Fig. 9).[5] Calcification in the coracoclavicular ligaments may be seen on followup examinations (Figs. 8 and 9). Injury to the joint may lead to osteolysis of the distal clavicle, even when the original injury was relatively benign.[13]

Sternoclavicular Dislocations

The sternoclavicular joint is the least stable major joint in the body. The obliquity of the articular surfaces is such that only one-half the clavicle articulates with the angle of the manubrium. An intraarticular disk or miniscus absorbs the impact between the bony surfaces.

Anterior and posterior sternal–clavicular ligaments reinforce the joint, with the posterior being much stronger than the anterior. Lateral displacment is prevented by the interclavicular ligament. The rhomboid or costoclavicular ligament connects the medial aspect of the clavicle to the anterior first rib.

Anterior dislocations are more common than posterior and usually result from indirect trauma. Posterior dislocations are frequently secondary to direct trauma, and approximately 25 percent have complications such as injury to the great vessels, trachea, or esophagus. Posterior dislocations may on occasion be spontaneous and secondary to degenerative arthritis.[8]

Routine radiography of the sternoclavicular joint is difficult. Rockwood[8] has advocated an anteroposterior view with 40 degrees of cephalad angulation taken at a distance of 60 inches. In this projection, the anterior dislocations are superior and posterior dislocations inferior. Lee and Gwinn[14] have advocated a view originally described by Henig. This projection is performed with the patient recumbent and the tube parallel to the tabletop 30 inches from the sternoclavicular joint. The central ray is directed along the clavicular axis with a grid cassette perpendicular to the opposite shoulder. This view is helpful in the evaluation of the sternoclavicular joint, particularly when compared to the opposite side.

References

1. Heppenstall RB: Fractures of the proximal humerus. Orthop Clin North Am 6:467, 1975
2. Neer CS: Displaced proximal humeral fractures. J Bone Joint Surg 52A:1077, 1970
3. Neer CS: Fractures about the shoulder. In Rockwood and Green (eds): Fractures, Vol 1. Philadelphia, Lippincott, 1975, p 585
4. Heppenstall RB: Fractures and dislocations of the distal clavicle. Orthop Clin North Am 6:477, 1975
5. Protass JJ, Stampfli FV, Osmer JC: Coracoid process fracture diagnosis in acromioclavicular separation. Radiology 116:61, 1975
6. Bowerman JW, McDonnell EJ: Radiology of athletic injuries: baseball. Radiology 116:611, 1975
7. Rowe CR: Prognosis in dislocations of the shoulder. J Bone Joint Surg 38A:957 1956
8. Rockwood CA Jr: Dislocations about the shoulder. In Rockwood CA, Green DF (eds): Fractures, Vol 1. Philadelphia, Lippincott, 1975, p 624
9. Schulz TJ, Jacobs B, Patterson RL Jr: Unrecognized dislocations of the shoulder. J Trauma 9:1009, 1969
10. Shaw JL: Bilateral posterior fracture-dislocation of the shoulder and other trauma caused by convulsive seizures. J Bone Joint Surg 53A:1437, 1971
11. Rubin SA, Gray RL, Green WR: The scapular Y: A diagnostic aid in shoulder trauma. Radiology 110:725, 1974
12. Markham DE, Rowland J: The shoulder joint—Is it dislocated? Apparent dislocation of the shoulder joint. Clin Radiol 10:61, 1969
13. Levine AH, Pais MJ, and Schwartz EE: Posttraumatic osteolysis of the distal clavicle with emphasis on early radiologic changes. Am J Roentgenol 127:781, 1976
14. Lee FA, Gwinn JL: Retrosternal dislocation of the clavicle. Radiology 110:631, 1974

Angiography of Trauma—Diagnostic and Therapeutic

William J. Casarella

The role of angiography in acute trauma is twofold—to identify the site and extent of possible vascular or organic injury and to stop significant bleeding by angiographic hemostasis. Our purpose is to describe the major indications for angiography and the characteristic radiologic findings. In addition, we will review the indications for interventional angiography in achieving hemostasis secondary to significant injury.

Vascular Trauma

The most significant trauma induced vascular injury is laceration of the thoracic aorta. This lesion is life threatening and requires emergency aortography when suspected. The usual inciting injury is a rapid deceleration to the thorax caused either by an automobile accident or a severe fall. The usual site of aortic injury is at the insertion of the ligamentum arteriosum in the transverse arch. Laceration of the aorta may also occur at the aortic root or the origin of the right innominate artery.

Approximately 7500 deaths occur annually in the United States secondary to automobile accidents associated with rupture of the thoracic aorta. Ten to twenty percent of these patients live long enough to be brought to hospitals for potential repair. If gone unrecognized, 90 percent of these patients will die of delayed aortic rupture within 4 months. When correctly diagnosed in time, 80 percent can be successfully repaired by surgery. In view of these statistics, it is mandatory to carefully examine all patients with severe chest injuries for possible aortic rupture. Widening of the mediastinum does not occur in all cases of aortic injury. As many as 25 percent of significant aortic tears may not be associated with detectable mediastinal widening. Furthermore, only one-half of all trauma patients with widened mediastinums will prove to have compromised aortas.

Because of the severe consequences of missing an aortic laceration, we feel that emergency aortography should be performed whenever there is any clinical suspicion of it. Besides widening of the mediastinum, a falling hematocrit, unexplained hypotension, changing peripheral pulses, or medial displacement of a nasogastric tube should alert one to the diagnosis.

In performing thoracic aortography in this situation, we prefer to use the femoral route, inserting a soft, pigtail-shaped catheter and monitoring arterial pressure via the catheter tip at all times. When the region just distal to the ligamentum arteriosum is reached, a floppy J-tipped guide wire is passed around the aortic arch and then followed by the catheter. The catheter is positioned in the aortic root 3 cm above the aortic valve for the injection of contrast material. It is essential to obtain two views of the aorta in either the anteroposterior and lateral or RAO and LAO positions before clearing the aorta. Lesions may be obvious or quite subtle, and are usually best seen in the lateral or LAO position. An example of an angiographically subtle intimal tear is seen in Fig. 1. The anteroposterior projection (Top) appears normal. However, on the lateral view (Bottom), an intimal tear is seen at the level of the ligamentum arteriosum. In our experience, traumatic aortic lacerations have always occurred at this site unless the patient had an anomalous right subclavian artery that resulted in a tear at the origin of the anomalous vessel.

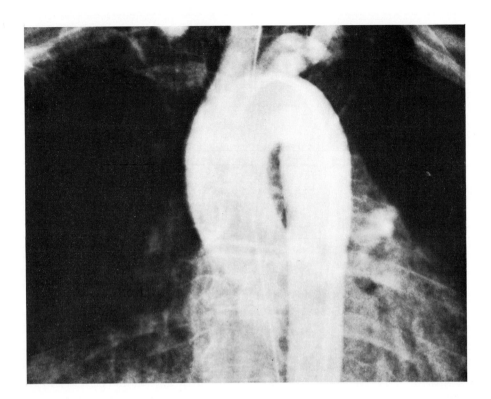
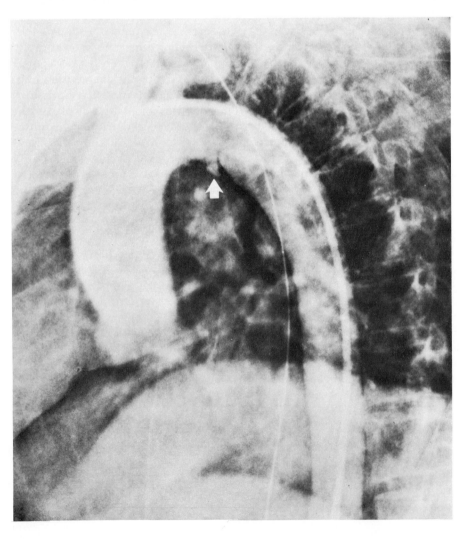

FIG. 1. A thoracic aortogram in the anteroposterior and lateral projections reveals an intimal tear at the level of the ligamentum arteriosum only in the lateral projection (arrow).

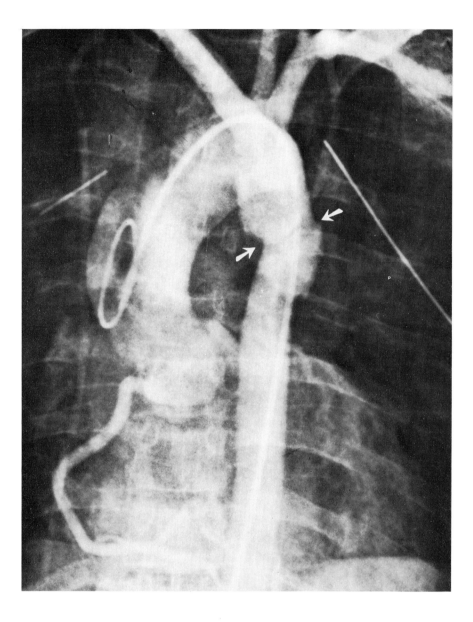

FIG. 2. An unusual example of traumatic aortic rupture in a 25-year-old man. The transection is in the descending aorta immediately distal to the origin of an aberrant right subclavian artery (arrows).

Frequently, patients with this lesion will have other, more obvious, but less hazardous injuries as well. The major role of the radiologist is to alert the orthopedic or general surgeons to the possibility of life-threatening central injuries before the clinician's attention is distracted by more obvious extremity injuries.

In penetrating injuries of the extremities or the neck, there is considerable controversy regarding the indications for angiography (Fig. 2). Although it is our own experience as well as that of Pochaczevsky and others that significant arterial injury is always accompanied by physical findings of significant hematoma, decreased pulses, severe bony fracture, or audible bruits, some authors claim that as many as 30 percent of significant arterial injuries may have negative physical examinations. In addition, arteriography is not 100 percent sensitive in detecting arterial injuries. Many of the reports of significant vascular injuries without abnormal physical findings are associated with battlefield injuries suffered from high-velocity rifle bullets, which cause severe shock waves in soft tissue and can damage vessels at points far removed from their path through the body.

In penetrating injuries of the extremities, arteriography should be performed when there is a history of preexisting arterial disease in order to guide surgical exploration and potential bypass as well as local repair if there is a large hematoma, a decrease in pulse or bruit present, or if there is unreliable history or physical examination.

Clearly, injuries to major vessels such as the superficial femoral or brachial artery need to be carefully explored and repaired. Bleeding from a small branch of the profunda femoris can be safely embolized at the time of angiography to achieve hemostasis.

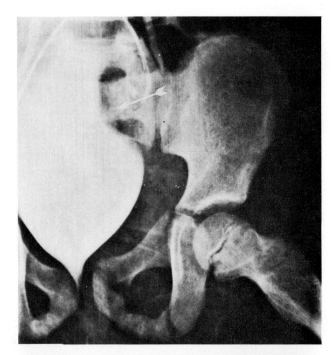
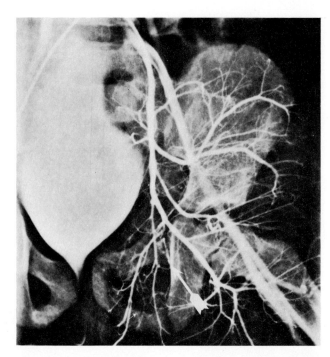

FIG. 3. This 5-year-old boy suffered multiple pelvic fractures following a fall from a second floor window. The bladder is deviated to the right by a large hematoma (top, left). A selective left internal iliac arteriogram (top, right) reveals a linear streak of extravasation of contrast from the obturator artery at the level of the superior pubic ramus (arrow). Following embolization (left), the obturator artery is occluded and hemostasis achieved.

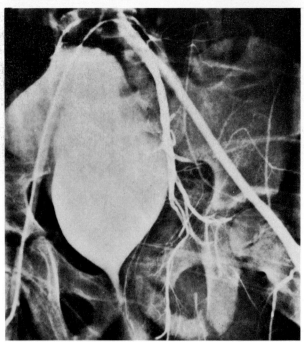

Pelvic Hematomas

One of the most useful applications of angiography in trauma is the management of pelvic hematomas by selective embolization of the hypogastric arteries (Fig. 3). This technique was introduced by Ring and his colleagues and has been rapidly accepted in many centers around the world.

In pelvic fractures that involve the superior and inferior pubic rami, the obturator and internal pudendal arteries are at risk, as they course through the obturator foramen. These arteries are frequently torn, and the injury results in massive pelvic hematomas. When surgery is required to control hemorrhage, precise ligation of the bleeding point is very difficult because of the surgeon's inability to find the artery in the massive hematoma. Although it would be ideal to be as selective as possible in embolizing the specific bleeding vessel, the hypogastric artery can be embolized with almost complete impunity as long as it is not a source of major collateral to the colon or lower extremity. In fact the major pitfall of the technique is rebleeding due to immediate recruitment of collaterals from the contralateral internal iliac. Therefore, it is our routine to perform selective injections of both internal iliacs after embolization to the affected side in order to verify adequate hemostasis. When properly performed with selective embolization with gelfoam particles, a 90 percent success rate can be expected in controlling pelvic hemorrhage.

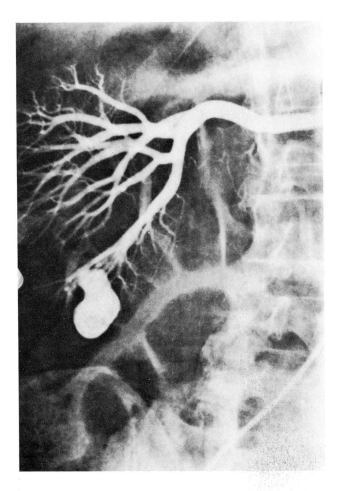 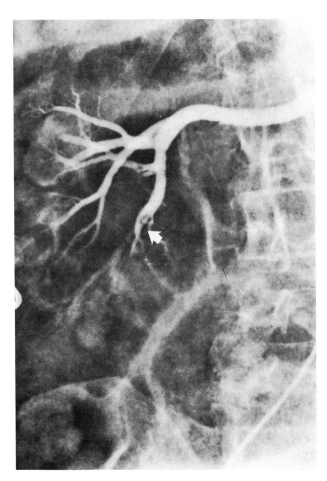

FIG. 4. Left: Massive extravasation from the lower pole of the right kidney is seen following lithotomy. Right: Life-threatening hemorrhage was prevented by subselective catheterization and embolization of the lower pole renal artery (arrow).

Visceral Trauma

Angiography remains the most sensitive radiologic technique for the detection of major traumatic injuries to the abdominal viscera. The other major added advantage of angiographic evaluation of the abdominal viscera is the potential for hemostasis by means of selective embolization.

Renal trauma is clearly defined by angiography, but the technique is unnecessary unless surgery is contemplated to control hematuria or retroperitoneal hemorrhage. Selective embolization with autologous blood clots to control renal hemorrhage was first described as a means of treating hematuria secondary to percutaneous renal biopsy. Subsequently, the technique has gained wide use for management of blunt and penetrating renal trauma and postoperative complications of renal surgery (Fig. 4). Autologous clot has become a popular embolic substance in the kidney because it is relatively short-lived and will usually lyse in 24 to 72 hours. By this time, hemostasis is usually achieved. We have performed renal embolizations exclusively with gelfoam particles in arteriovenous fistulas secondary to biopsy, gunshot wounds of the cortex, avulsion of segmental renal artery branches (Fig. 5), and in postoperative bleeding. In none of our cases has there been significant loss of renal cortex or resultant hypertension due to ischemia. The major factor in performing successful renal embolization is superselective catheter position with precise delivery of the emboli. This allows for maximum sparing of normal parenchyma and prevents the formation of collaterals that can restart the bleeding or result in ischemia to an embolized area of normal parenchyma.

In the spleen, angiography has proven to be an accurate predictor of significant trauma. The angiographic signs of splenic rupture are extravasation of contrast material, arteriovenous shunting, avascular intrasplenic mass, extrasplenic subcapsular mass,

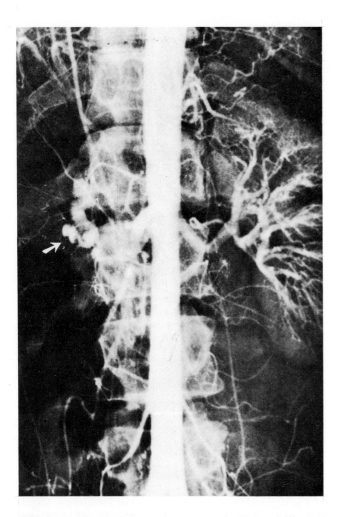 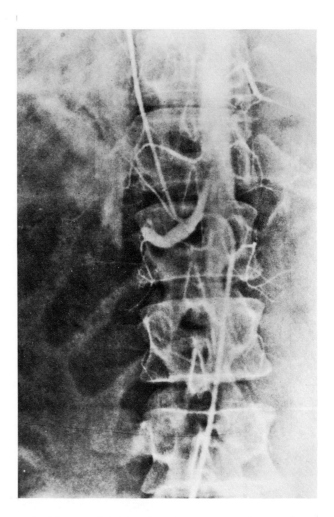

FIG. 5. Left: Avulsion of the renal artery with retroperitoneal hemorrhage (arrow) is demonstrated in a midstream aortogram in a 26-year-old girl who attempted suicide by jumping from a third-floor window. Right: Embolization of the renal stump (arrow) resulted in hemostasis from the renal artery.

and irregular splenic contour. The first three signs are the most reliable. Selective high-volume splenic artery injections performed in two views are necessary to rule out splenic trauma.

Since it is desirable to leave in the spleen because of its immunological activity, embolization of the spleen in trauma appears to be an attractive alternate to splenectomy. However, selective splenic embolization has been associated with an unacceptable number of splenic abscesses and is not recommended under usual circumstances. However, if surgery is risky and the patient can be carefully observed for development of left upper quadrant abscess, splenic embolization can be useful and also can spare functional splenic tissue.

Angiography is a useful diagnostic tool in the evaluation of blunt and penetrating hepatic trauma. The consequences of significant hepatic trauma include subcapsular hematoma, parenchymal hematoma or contusion, laceration of hepatic vessels, arteriovenous fistuals, pseudoaneurysm, hematobilia, traumatic bile cyst, and obstruction of the inferior vena cava or bile duct. Many of these lesions are diagnosable on angiography.

Isolated case reports of successful embolization of cases of traumatic hematobilia, arteriovenous fistulas, and pseudoaneurysms have appeared in the literature. Embolization should definitely be attempted in cases of significant hepatic trauma even if the patient is to be explored. Selective hemostasis may be easier for the angiographer than for the surgeon.

Other types of soft-tissue trauma associated with significant hemorrhage are subject to angiographic diagnosis. But perhaps the most useful contribution of angiography in the management of the trauma patient is its ability to deliver hemostatic emboli to the precise bleeding site and thereby save difficult surgery in an already traumatized patient.

Bibliography

Berk RN, Wholey MH: The application of splenic arteriography in the diagnosis of rupture of the spleen. Am J Roentgenol 104:662, 1968

Chuang VP, Reuter SR: Selective arterial embolization for the control of traumatic splenic bleeding. Invest Radiol 10:18, 1975

Enge I, Aakhus T, Evensen A: Angiography in vascular injuries of the extremities. Acta Radiol Diagn 16:193, 1975

Grace CM, Pitt DF, Gold RE: Vascular embolization and occlusion by angiographic techniques as an aid or alternative to operation. Surg Gynecol Obstet 143:469, 1976

Kalish M, Greenbaum L, Silber S, Goldstein H: Traumatic renal hemorrhage treated by arterial embolization. J Urol 112:138, 1974

Levin DC, Watson RC, Sos TA, Baltaxe HA: Angiography in blunt hepatic trauma. Am J Roentgenol 119:95, 1973

Paton BC, Elliott DP, Taubman JO: Acute treatment of traumatic aortic rupture. J Trauma 11:1, 1970

Pochaczevsky R, Mufti MA, LaGuerre NJ, Bryk D, Kassner EG, Richter RM, Levowitz B: Angiography of penetrating wounds of the extremities: Help or hindrance? J Can Assoc Radiol 24:354, 1973

Ring EJ, Athanasoulis C, Waltman AC: Arteriographic management of hemorrhage following pelvic fracture. Radiology 109:65, 1973

Ring EJ, Waltman AC, Athanasoulis C: Angiography in pelvic trauma. Surg Gynecol Obstet 139:375, 1975

Silber S: Renal trauma: treatment by angiographic injection of autologous clot. Arch Surg 110:206, 1975

Wilson RF, Arbubu A, Bassett RS, Walt AJ: Acute mediastinal widening following blunt chest trauma. Arch Surg 104:551, 1972

Ankle Fractures

Murray K. Dalinka

The evaluation of traumatic injuries to the ankle is usually more complex than a simple description of the fracture lines or the position of the fracture fragments. An understanding of the mechanism and pathologic anatomy of ankle injuries will enable the radiologist to thoroughly evaluate the radiographs and render a meaningful report to his orthopedic colleagues.

The author believes that all ankle injuries should be x-rayed in four projections—standard anteroposterior lateral, mortise views (with the foot in 15 to 20 degrees of internal rotation), and an external oblique with the heel elevated and the toes dropped. The external oblique or "poor lateral" view separates the posterior malleolus from the fibula, allowing one to visualize the third malleolus.[1] In the mortise view, the medial articular surface of the talus should be tangent to the X-ray beam. The concave insertion of the deep deltoid fibers should be lateral to the medial articular surface.[2]

There is considerable disagreement concerning the value of stress radiography particularly in the unanesthetized patient. Staples[3] found inversion stress unreliable; in a series of 27 patients only 1 patient had a positive inversion stress radiograph, and in 10 patients, the study could not be performed at all because of pain.

Arthrography of the ankle is a simple, accurate procedure for the precise diagnosis of ligamentous injury. It must be performed within a week of the injury, otherwise blood clot and fibrin will lead to false negative examinations. Further discussion of ankle arthrography is included elsewhere in this volume.

In children, comparison views may be helpful, particularly for doctors who do not deal primarily with pediatric patients. These views can be performed as a routine or when in doubt.

Ankle Fractures in Children

In children, the growth plate, or physis, is the area of weakness, and while epiphyseal injuries are common, ligamentous injuries are almost unknown. The ligaments are attached to the epiphysis, and hence epiphyseal separations are more common than fractures.

The classification of Salter and Harris (Fig. 1) is simple, widely accepted, and easily applicable to ankle injuries in childhood.[4] In the Salter–Harris classification, the type 1 injury occurs through the zone of provisional calcification. The entire epiphysis may be separated or the growth plate widened, but the prognosis is usually excellent because the germinal layer is intact. The periosteum may be intact or torn on one surface. The radiographs may appear deceptively normal. Point tenderness is usually present, and soft-tissue swelling is frequent. These injuries are usually the result of an abduction or inversion force. This injury is uncommon in the tibia but is the most common fibular injury in the child (Fig. 2). Rang[5] recommends the use of a cast for 3 weeks and feels that followup films are unnecessary, while Salter[4] states that the diagnosis can be confirmed by periosteal new bone formation 2 weeks following the injury.

Type 2 injuries include a small metaphyseal fragment, the Thurston–Holland sign, as well as the epiphyseal separation. Since the separation is through the zone of provisional calcification, and since the germinal layer remains with the epiphysis,

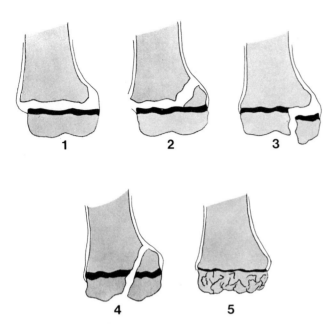

FIG. 1. Salter–Harris classification. Note that type 5 fracture frequently does not have any abnormality in the ossified epiphysis.

injury, as it frequently leads to premature cessation of growth in the area of fracture.

Type 5 injuries are compression injuries of the growth plate, usually on the medial side. This injury may be difficult to recognize. It leads to premature growth arrest and, fortunately, it is rare.

Adults

The early classifications and descriptions of ankle injuries were frequently inadequate because of the failure to emphasize the importance of ligamentous injuries, which can be thought of as fracture equivalents.

Lauge–Hansen[7] formulated a classification of ankle injuries derived from a detailed study of recently amputated extremities. His classification was based upon the position of the foot and the direction of the deforming force. Each injury was divided into various stages that were dependent upon the magnitude and the blood supply to the epiphysis has not been disturbed and the prognosis is excellent. The fracture is produced by lateral displacement, leaving the periosteum intact on the side of the metaphyseal fragment. If the fracture is grossly displaced, ischemia of the foot may be seen. Plantar flexion and eversion are thought to be the mechanism, and a green stick fibular fracture is frequently associated with the tibial injury.[5]

The type 3 epiphyseal injury represents an intraarticular epiphyseal fracture. This type of epiphyseal injury occurs most commonly in the ankle, where it is usually seen in the lateral aspect of the tibial epiphysis (Fig. 3). This is thought to be a lateral rotation or shearing injury and is called a Tillaux fracture.[5] The medial portion of the tibial epiphysis fuses at 13 to 14 years of age. The lateral portion is attached to the fibular metaphysis by the anterior tibial fibular ligament. The pull of this ligament avulses the lateral or anterior lateral portion of the tibial epiphysis, which remains open. This condition has a good prognosis despite the damage to the growth plate, since growth has almost ceased. The fracture is usually nondisplaced, but if displaced it should probably be treated with open reduction.[6]

A type 4 epiphyseal injury is a fracture across the epiphysis, including the growth plate, with a small metaphyseal fragment. Fortunately, it is a rare ankle

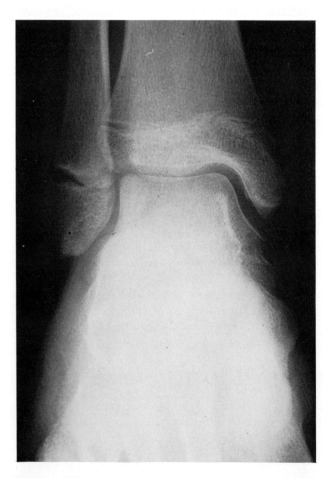

FIG. 2. Anteroposterior view of ankle demonstrating type 1 epiphyseal injury with widening of epiphyseal plate of fibula.

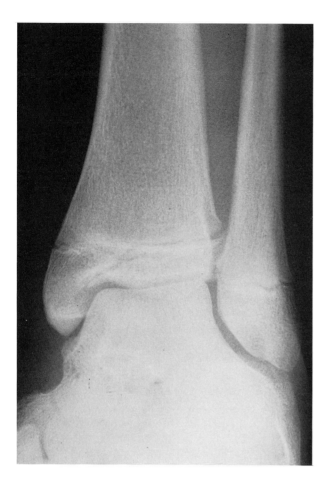

FIG. 3. Anteroposterior view of ankle demonstrating type 3 fracture through distal tibial epiphysis. Note difference in appearance between medial and lateral portions of epiphyseal plate.

1. Primary force: external rotation
   a. Isolated
   b. External rotation plus abduction
   c. External rotation plus adduction
   d. External rotation plus vertical compression

2. Primary force: abduction
   a. Isolated
   b. Abduction plus external rotation
   c. Abduction plus vertical compression

3. Primary force: adduction
   a. Isolated
   b. Adduction plus external rotation
   c. Adduction plus vertical compression
      i. Talus dorsiflexed
      ii. Talus plantar flexed

4. Primary force: vertical compression
   a. Isolated
   b. Vertical compression plus external rotation
   c. Vertical compression plus abduction
   d. Vertical compression plus adduction

The ankle is a complex hinge joint allowing only dorsal and plantar flexion. It is closely related to the subtalar joint, where supination and pronation of the foot occur. The talus sits between the tibia and fibula, with the collateral and syndesmotic ligaments reinforcing the osseous structures.

Fractures secondary to ligamentous avulsion are horizontal, while those secondary to impact of the talus are oblique and frequently comminuted.[8] The direction of the forces determines the obliquity of the fracture line. Ligamentous injury is frequently associated with ankle fractures and, in fact, represents a fracture equivalent. Cedell[9] stated that total rupture of one or more syndesmotic ligaments occurs in greater than 90 percent of malleolar fractures. Displaced malleolar fractures are indicative of ligamentous injuries.

duration of the force. He felt that the sequence of injury was constant and pathognomonic for the force applied, and when the force was terminated, incomplete stages resulted. This classification was not easily clinically applicable, and hence, many simpler classifications based on the mechanism of injury are in use.

The understanding of the mechanism is crucial to the orthopedist who reverses the injury forces in his reduction. It is of importance to the radiologist so that his report is comprehensive and meaningful. Although most of our orthopedic colleagues do not classify ankle injuries in any particular fashion they all feel that an understanding of the forces involved is important, particularly in the performance of closed reductions.

The author has familiarized himself with Wilson's[8] classification (Table 1) which will be used here. This is based upon the direction of the primary force and the addition of secondary forces.

Patterns of Ankle Injuries

External Rotation Injuries

The roentgenographic findings are dependent upon the duration and degree of the force and vary from soft-tissue swelling to a trimalleolar fracture. An oblique fibular fracture directed from anterior inferior to posterior superior is frequent and often comminuted along the posterior cortex (Fig. 4). The foot is usually in pronation, and an avulsion injury causes a deltoid tear (fracture equivalent) or a

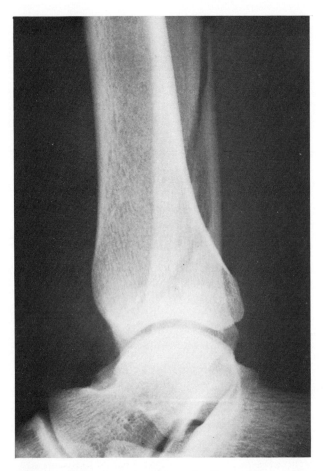

FIG. 4. Lateral view demonstrating oblique fibula fracture running from anterior inferior to posterior superior typical of an external rotation injury.

transverse fracture of the medial malleolus at or below the articular surface of the tibia. Fracture fragments are usually displaced inferiorly by the pull of the deltoid ligament. The force then stresses the anterior tibiofibular ligament, either tearing the ligament or fracturing the fibula obliquely. The fibular fracture may occur at any level, depending upon the degree of syndesmotic disruption. Hence the entire fibula must be evaluated in external rotation injuries (Figs. 5 and 6). The rotating fibula may then cause a small posterior malleolar fracture. External rotation and abduction is the commonest mechanism of ankle injury. If the foot is in supination rather than pronation, the interosseous ligaments are usually spared.

Abduction Injuries

Abduction injuries also cause horizontal medial malleolar fractures or deltoid ligament ruptures. The fibular fractures below they syndesmosis, or the syndesmosis itself is ruptured (Figs. 7 and 8). With syndesmotic rupture, the fibula may fracture above the syndesmosis. The fibular fracture is usually a short oblique fracture, frequently with comminution of the lateral cortex.[8] If external rotation occurs with abduction, the fibular fracture is usually higher and/ or more oblique. The Dupruytren fracture disloca-

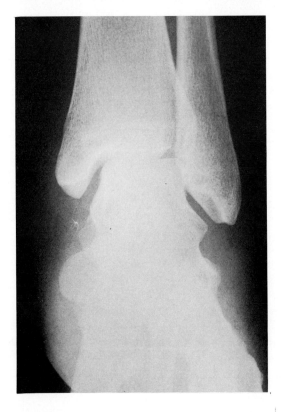
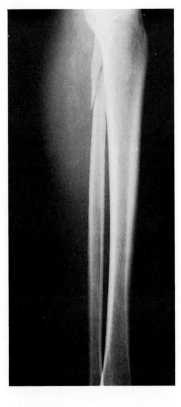

FIG. 5. Left: Anteroposterior view of ankle showing diffuse soft-tissue swelling without fracture. Right: Lateral view of tibia and fibula in the same patient demonstrating fracture of proximal fibula. This represents the Maisonneuve fracture. A rupture of the syndesmosis must occur with this injury, and there may be associated ligamentous injuries or malleolar fractures. The mechanism of this injury is external rotation.

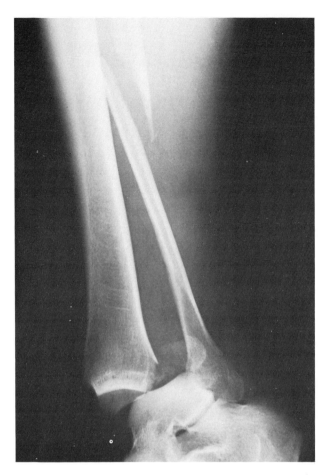

FIG. 6. Oblique view showing high fibular fracture with syndesmotic rupture, fracture of posterior aspect of medial malleolus and lateral talar displacement. This represents an injury secondary to external rotation and abduction.

first, anatomic reduction of the medial malleolus occurred without incident.

Adduction Injuries

These injuries are frequently associated with horizontal fractures of the distal fibula at or below the articular surface (Fig. 10). There is frequently a vertical fracture of the medial malleolus extending above the articular surface, and this may be associated with a fracture of the lateral talar dome (Fig. 11). There may be comminution of the articular surface predisposing to posttraumatic arthritis. Adduction injuries do not cause diastasis, and posterior marginal fractures are uncommon[8] (Fig. 12).

Vertical Compression Injuries

POSTERIOR MARGINAL FRACTURES These fractures are of two types—those without significant vertical compression and those involving a large portion of the articular margin. Small posterior marginal frac-

tion (Fig. 9) consists of a horizontal fracture of the medial malleolus or its equivalent (torn deltoid), a complete tear of the syndesmosis, and a high fibular fracture. It is an unstable injury produced by abduction and lateral rotation and is probably best treated with open reduction.[10] The lateral fibular cortex may by comminuted with abduction fractures. Diastasis is more common with abduction injuries, as the entire syndesmosis is torn. This may be associated with small posterior avulsions of the tibia or fibula and occasionally with depression of the lateral tibial articular surface. Yablon et al[11] felt that adequate reduction of the lateral malleolus was the key to stable reduction of abduction external rotation bimalleolar fractures. Their cadaver experiments demonstrated that fixation of the medial malleolus followed by forcible reduction of the lateral malleolus resulted in stretching of the lateral ligaments. After discontinuance of the force, the talus went back to its displaced position. When the lateral malleolus was reduced

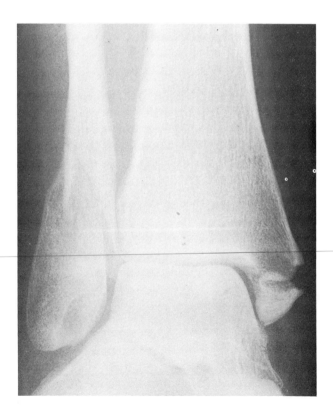

FIG. 7. Anteroposterior view demonstrating horizontal fracture of medial malleolus and oblique fracture of lateral malleolus below the syndesmosis. This is an abduction injury.

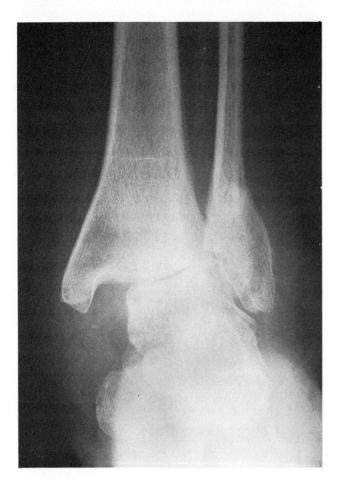

FIG. 8. Oblique view demonstrating widening of medial aspect of mortise secondary to rupture of deltoid ligament (fracture equivalent) and fibular fracture below the syndesmosis.

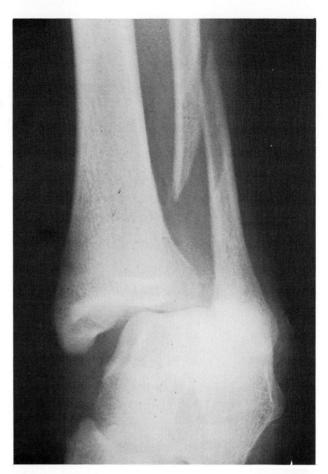

FIG. 9. Oblique view demonstrating widening of medial aspect of ankle mortise, syndesmotic rupture, fibular fracture above the syndesmosis, and lateral talar displacement. This is an abduction external rotation injury called Dupruytren's fracture dislocation. This is very similar to the fracture shown in Fig. 6, both representing injuries secondary to abduction and external rotation.

tures may be caused by external rotation injuries with or without an abduction component (Fig. 13), but large fragments require vertical compression in addition to external rotation. Posterior marginal fractures rarely occur as isolated injuries. One must examine the entire fibula to exclude a proximal fibular fracture (the Maisonneuve type of external rotation injury).[8] "Isolated" posterior marginal fractures frequently have associated tears of the anterior tibial fibular ligament. If the posterior articular fragment is large (one-quarter to one-third of the articular surface), there is frequent instability of the ankle and a high incidence of posttraumatic arthritis. This is thought to be an indication for open reduction, which yields better results.[12]

ANTERIOR MARGINAL FRACTURES These fractures usually occur in the dorsiflexed foot and are more common as isolated injuries than the posterior marginal fractures. They may be comminuted.

SUPRAMALLEOLAR FRACTURES The supramalleolar, or Malgaigne's fracture of the ankle is a fracture of the distal 4 cm of the tibia above the ankle joint and is almost always associated with a fibular fracture (Fig. 14). It frequently results from high impact injuries in the direction of axial compression. It is commonly open and severely comminuted with frequent complications, particularly when it extends into the tibial articular surface.[13]

The anatomic reduction of ankle fractures is extremely important, as small amounts of lateral displacement result in traumatic arthritis and stable joint reconstruction is necessary to prevent ankle instability.[2,9] Ramsey et al[14] recently demonstrated that as little as 1 mm of lateral displacement reduces the area of tibial talar contact by 42 percent. The primary indication for open reduction is the inability of maintaining an anatomic reduction of the tibia and talus. Other common indications for operative treatment include ligamentous avulsion fractures, displaced malleolar fragments and displaced large posterior tibial fragments.[9] Comminution is the major contraindication.

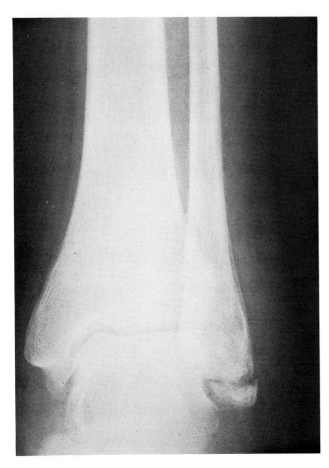

FIG. 10. Healing adduction injury showing horizontal fracture of lateral malleolus below the joint line. Considerable soft tissue swelling is present as is periosteal reaction.

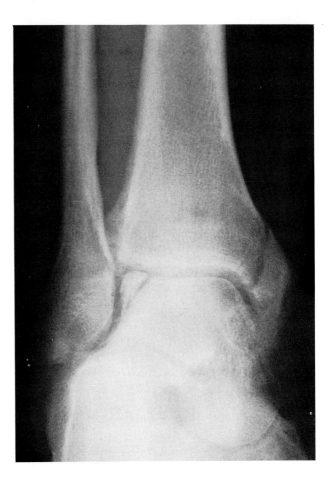

FIG. 11. Vertical (adduction) fracture of medial malleolus associated with fracture of lateral talar dome.

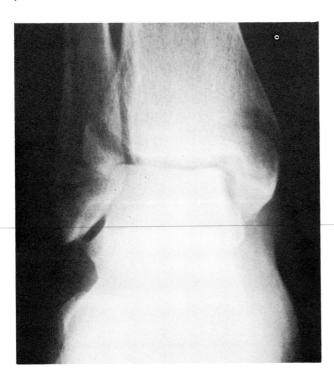

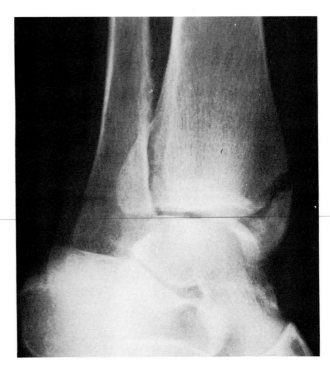

FIG. 12. Left: Fracture of medial and posterior malleolus on anteroposterior view. Right: Same patient. Oblique view demonstrating malleolar fracture running obliquely from the junction of the malleolus and lateral tibial roof (plafond). This is an adduction injury associated with vertical compression.

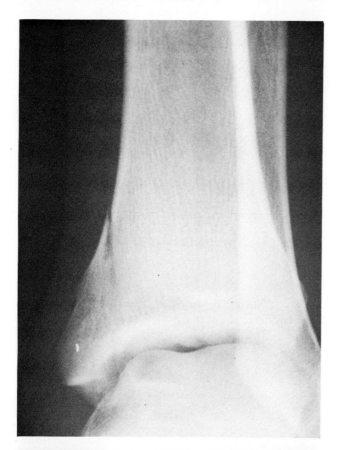

FIG. 13. Oblique view of ankle showing oblique fracture of posterior malleolus. These fractures when isolated are almost always associated with tears of the anterior tibial fibular ligament.

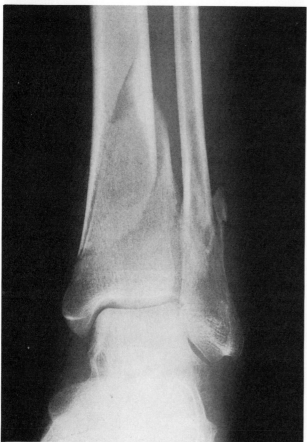 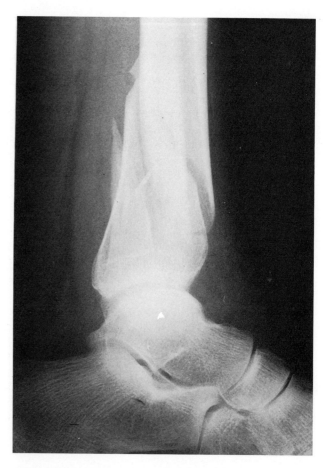

FIG. 14. Malgaigne's fracture. Left: Anteroposterior view of oblique fracture of distal tibia and comminuted fibular fracture not entering the articular surface. Right: Lateral view of same patient. This represents an injury secondary to a high-velocity axial compression force.

References

1. Mandell J: Isolated fractures of the posterior tibial lip at the ankle as demonstrated by an additional projection, the "poor" lateral view. Radiology 101:319, 1971
2. Georgen TG, Danzig LA, Resnick D, Owen CA: Roentgenographic evaluation of the tibiotalar joint. J Bone Joint Surg 59A:874, 1977
3. Staples OS: Ruptures of the fibular collateral ligaments of the ankle. J. Bone Joint Surg 57A:101, 1975
4. Salter RB: Injuries of the ankle in children. Orthop Clin North Am 5:147, 1974
5. Rang M: The Growth Plate and Its Disorders. Baltimore, Wilkens and Williams, 1969
6. Molster A, Soreide O, Solhaug JH, Raugstad TS: Fractures of the lateral part of the distal tibial epiphysis (Tillaux or Kleiger fracture). Injury 8:260, 1977
7. Lauge-Hansen N: Fractures of the ankle. II. Combined experimental surgical and experimental roentgenologic investigations. Arch Surg 60:957, 1950
8. Wilson FC: Fractures and dislocation of the ankle. In Rockwood Jr CA, Green DP (eds): Fractures, Vol 2, Philadelphia, Lippincott, 1975, p 1361
9. Cedell CA: Ankle lesions. Acta Orthop Scand 46:425, 1975
10. Colton CL: The treatment of Dupuytren's fracture—dislocation of the ankle. J Bone Joint Surg 53B:63,1971
11. Yablon IG, Heller FG, Shouse L: The key role of the lateral malleolus in displaced fractures of the ankle. J Bone Joint Surg 59A: 169, 1977
12. McDaniel WJ, Wilson FC: Trimalleolar fractures of the ankle. Clin Orthop 122:37, 1977
13. Lee CK, Hansen HT, Weiss AB: Supramalleolar fracture of the ankle (Malgaigne's fracture). Am Surg 43:589, 1977
14. Ramsey PL, Hamilton W. Changes in tibiotalar area of contact caused by lateral talar shift. S Bone Joint Surg 58A:356, 1976

Bone Trauma: Assessment by Scintigraphy

Rashid A. Fawwaz

With the advent of technetium-99m-labeled phosphates, bone scintigraphy has become a useful adjunct for evaluation of bone trauma. The physical properties of technetium-99m allow for superior resolution with currently available imaging devices, and radiation exposure following systemic administration of this radionuclide is low.

Although an abnormal pattern on bone scintigraphy can be diagnostic per se, more often the findings are nonspecific. The value of bone scintigraphy lies in its sensitivity, and it should be utilized either as a screening procedure or to assess progress of disease after diagnosis is established.

Methods

Technetium-99m tin hydroxyethylidene diphosphonate (99mTc-Sn-EHDP) or Technetium-99m tin methylene diphosphonate (99mTc-Sn-MDP) are currently the radiopharmaceuticals of choice for skeletal imaging.[1,2] Following intravenous administration of 200 microcuries per kilogram of either agent, imaging is performed at 2 to 3 hours. Spot imaging of the area of interest is usually sufficient, although occasionally, whole-body scintigraphy is recommended (e.g., battered child, involvement of paired structures).

Quantitative bone imaging requires interfacing of computer and camera. Count rates are obtained over the region of interest and are compared to count rates obtained over normal bone. The derived ratio then forms a baseline for objective evaluation of progression of disease through sequential imaging.

Because technetium-99m-labeled diphosphonates are excreted via the kidneys, adequate hydration and frequent voiding are imperative to a minimized radiation dose to the patient.

Fractures

Scintigraphic abnormalities are demonstrable within 24 to 48 hours after trauma, when increased radiotracer uptake is noted both at the fracture site and in surrounding normal bone.[3] The latter change is attributed to a generalized nonspecific hyperemia, which occurs as the reparative process in bone is initiated. Subsequently, radiotracer localization becomes more focal and limited in extent to the exact location of the fracture.[4] In fractured bone that has healed with normal alignment, radiotracer distribution usually returns to normal in 9 to 12 months after trauma.[5] However, when anatomic continuity is not fully restored, increased radiotracer uptake at the site of fracture may persist indefinitely due to continuous bone remodeling.[6]

Scintigraphic manifestations of trauma are generally nonspecific. Occasionally, however, the abnormal scintigraphic pattern can be diagnostic. Focally increased radiotracer uptake involving several adjacent ribs in a linear distribution is characteristic of multiple fracture. This is demonstrated in Figure 1.

There are four main areas where scintigraphy can play an important role in the evaluation of bone fracture. The first involves the detection of small or incomplete fractures, such as those secondary to stress, which can be difficult to diagnose radiographically, as demonstrated in Figure 2. The second pertains to evaluation of fracture age. As depicted in Figure 3, scintigraphy of the pelvis in this patient, who had recently sustained left pelvic trauma, revealed intense radiotracer concentration in the left iliac bone. Radiography also demonstrated fracture of the right hip of indeterminate age and significance. Scintigraphy of the right hip was normal, indicating that this lesion was old and had stabilized. The third area involves the assessment of the integrity of bone

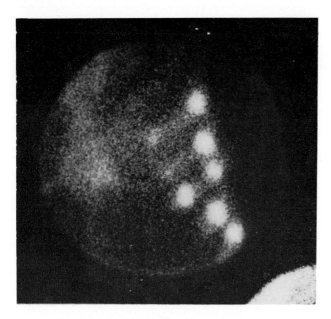

FIG. 1. Limited bone scintigraph of the anterior left hemithorax. Note focally increased radiotracer uptake involving several adjacent ribs in a linear distribution. Radiography confirmed the presence of multiple rib fractures.

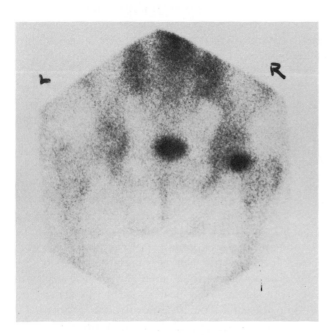
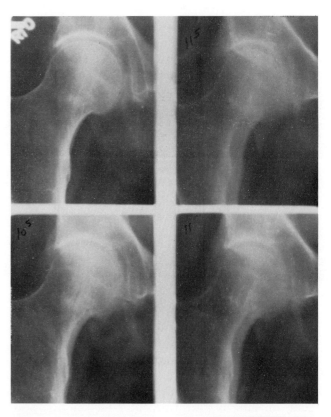

FIG. 2. Left: Limited bone scintigraph of the posterior pelvis in a patient with medial right thigh pain. Note intense radiotracer uptake in right femoral neck. Plain radiography was equivocal. Right: x-ray tomography of the right femoral neck of the same patient. Note impacted stress fracture of the right femoral neck.

structures adjacent to fracture. In femoral neck fractures, for instance, compromised blood supply to the femoral head is exemplified by reduced or absent radiotracer concentration in this region. However, when collateral circulation is established, increased accumulation of radiotracer in the femoral head occurs due to enchanced blood flow and osteoblastic activity associated with bone repair. This is demonstrated in Figure 4. Thus, reduced or absent radiotracer uptake in the femoral head early after femoral neck fracture, while indicative of compromised curculation, does not necessarily carry a bad prognosis. However, failure to detect radiotracer uptake in the femoral head later on is an ominous sign. The fourth area relates to detection of nonunion and its differentiation from delayed healing. Preliminary studies indicate that decreased radiotracer uptake at the fracture site at a time reparative bone formation is

Bone Trauma

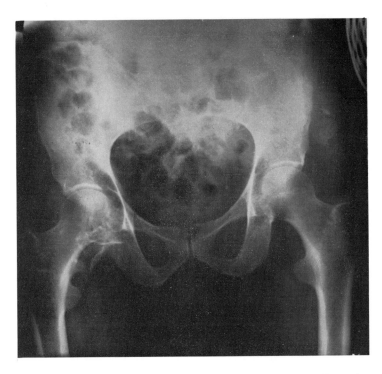
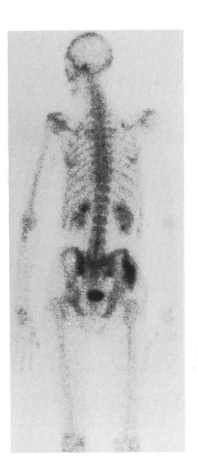

FIG. 3. Anterior whole-body bone scintigraph (right) and a plain radiograph of the pelvis (left) in a patient with recent left hip trauma. Radiography demonstrates left iliac fracture and evidence of trauma to the right acetabular region, the latter of indeterminate age. Scintigraphy shows intense localization of radiotracer in the left iliac bone but normal radiotracer concentration in the right hemipelvis, indicating that the right acetabular fracture is old and stabilized.

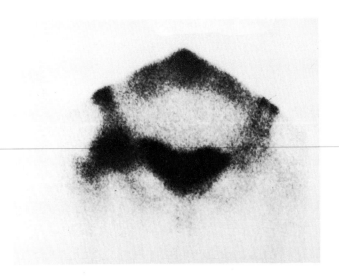

FIG. 4. Limited bone scintigraph of the anterior pelvis in a patient with right femoral neck fracture and aseptic necrosis of the right femoral head diagnosed radiographically. Note intense radiotracer accumulation in the right femoral head.

expected to be intense is suggestive evidence for nonunion. Discontinous accumulation of radiotracer at the fracture site is another sign indicative of nonunion. Scintigraphic changes in nonunion precede radiographic evidence by as much as 3 to 6 weeks.[7]

Iatrogenic Trauma

Operative procedures such as low-friction arthroplasty and insertion of internal fixation devices result in trauma to bone leading to increased radiotracer uptake lasting approximately 6 months.[8] Persistent increased uptake lasting more than 6 months raises the possibility of osteomyelitis or instability of the prosthetic or internal fixation device. Figure 5 demonstrates the scintigraphic changes observed in loosening of a prosthetic device. Differentiation of osteomyelitis from an unstable prosthesis or internal fixation device is difficult. Occasionally, clarification

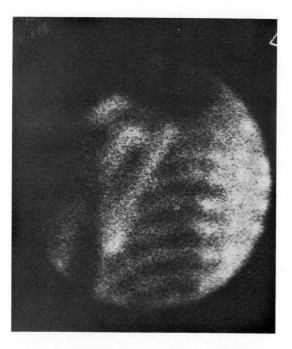 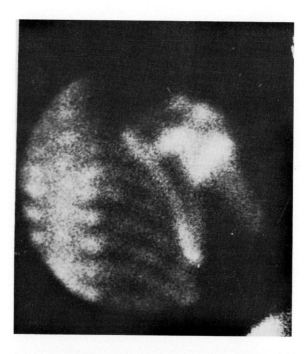

FIG. 5. Limited bone scintigraphs of the posterior humeri in a patient who underwent arthroplasty of the right humeral head 18 months earlier. Note increased radiotracer uptake adjacent to prosthesis (right) relative to the contralateral normal side (left). Instability of prosthesis was confirmed surgically.

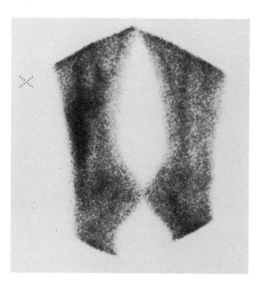 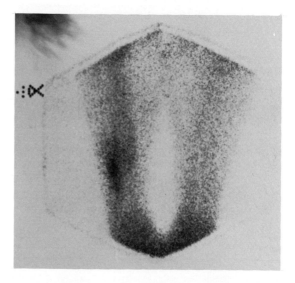

FIG. 6. Limited bone scintigraph of the distal tibial shafts (left) and a corresponding radiogallium image obtained 2 days later (right) in a patient with an internal fixation device inserted in the right tibia 4 months earlier. A mismatch between radiogallium and radiolabeled diphosphonate distribution is evident. Intense radiogallium concentration in right lateral tibial shaft was considered indicative of infection. This was subsequently confirmed surgically.

can be made by reexamination with gallium-67 citrate. Radiogallium normally localizes faintly in bone and also has an affinity for leukocytes. Thus, when infection is present, localization of gallium should be more intense than localization observed with the technetium-99m-labeled phosphate compounds.[9] This is demonstrated in Figure 6.

In the first 6 months after surgical manipulation of bone, scintigraphic detection of pathology is difficult and requires quantitative sequential bone imaging. When bone abnormality is present, the ratio of counts over involved bone relative to counts over normal bone increases with time. Normally, a gradual decrease in this ratio is expected to occur with time.

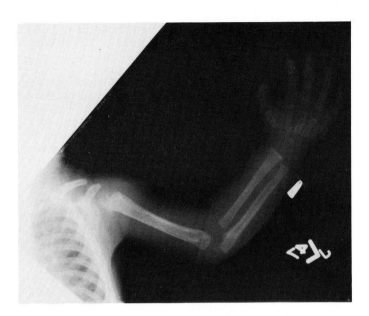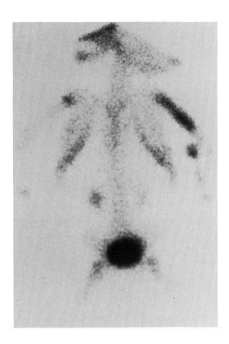

FIG. 7. Plain film of the upper extremity of a battered child (left) and a whole-body bone scintigraph of the same child (right). Note that while radiography is normal, scintigraphy reveals multiple areas of bone injury.

The Battered Child

Bone trauma need not be associated with fracture to produce an abnormal scintigraph. Trauma resulting in periosteal injury is also associated with increased radiotracer uptake in the traumatized area. These abnormalities may prove difficult to detect by x-ray as shown in Figure 7. In battered children, bone scintigraphy may be critical in that it may provide the only legal method that permits hospitalization.

Osteoarthritis

Osteoarthritis is characterized by increased radiotracer uptake on both sides of the joint. A problem of critical importance in osteoarthritis is the detection of superimposed disease such as osteomyelitis. In this regard, the combined use of bone and gallium scintigraphy may prove fruitful as previously discussed.

Indirect Trauma

In patients who sustain injury leading to limited use or total disuse of an affected skeletal structure, as with nerve damage or immobilization, several abnormal patterns are observed on bone scintigraphy. Alteration in weight-bearing results in emergence of new stress points, which are perceived as areas of increased radiotracer concentration.[6] Skeletal structures immobilized for prolonged periods often exhibit increased radiotracer uptake.[9] This is attributed to relative increase in bone blood flow due to diversion of blood from the muscular compartment.[10] In paralyzed patients, metastatic ossification is not uncommon and is reflected on bone scintigraphy as generalized increased radiotracer uptake in the soft tissues of the involved region. Quantitative sequential imaging is important in the management of patients with myositis ossificans, since therapeutic intervention other than physiotherapy is not usually contemplated until the turnover of calcium in affected soft tissues is markedly reduced. Figure 8 demonstrates a sequential study whereby radiotracer uptake in metastatic ossification had not subsided over a period of 1 year.

Summary

The use of bone scintigraphy for evaluation of osseous trauma is a well-established procedure. Static and quantitative bone scintigraphy combined with radiogallium imaging permit detection of a wide variety of traumatic bone disorders difficult to evaluate by other noninvasive techniques. Examples of the value of radionuclide imaging techniques include evaluation of bone fracture, iatrogenic bone trauma, the battered child, osteoarthritis, and indirect bone trauma.

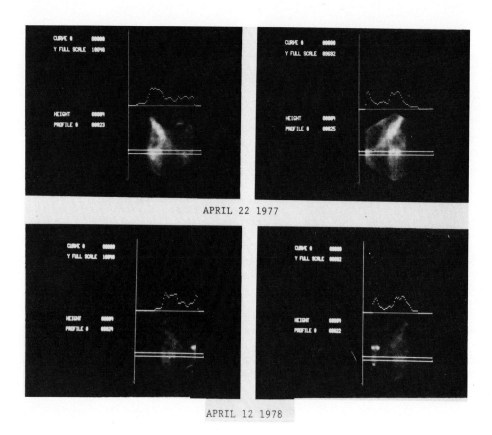

FIG. 8. Two quantitative bone scintigraphic studies performed in a patient with myositis ossificans of the proximal thighs obtained a year apart. The images on the left side represent the right hemipelvis and images on the right the left hemipelvis. Areas of interest are enclosed between the solid white lines. A profile count of this slice is shown in the upper portion of each image. Note that the ratio of radiotracer uptake in the soft tissues of the thigh relative to normal pubic bone is unchanged with time.

References

1. Castronovo FP, Callahan RJ: New bone scanning agent: 99m-Tc-labeled 1-hydroxyethylidene-1, 1-disodium phosphonate. J Nucl Med 13:823, 1972
2. Subramanian G, McAfee JG, Blair RJ, et al: Technetium-99m-methylene disphosphonate—a superior agent for skeletal imaging. J Nucl Med 16:744, 1975
3. Ray RD, Aouad R, Kawabata M: Experimental study of peripheral circulation and bone growth. Clin Orthop 52:221, 1967
4. Fordham EW, Ramachandrian P: Radionuclide imaging of osseous trauma. Semin Nucl Med 4:411, 1974
5. Marty R, Denney JD, McKamey M, et al: Bone trauma and related bone disease: assessment by bone scanning. Semin Nucl Med 6:107, 1976
6. Bell EG, Mahon DF: Bone. In Handmaker H, Lowenstein J (eds): Nuclear Medicine in Clinical Pediatrics. Publishing Science Group, 1975, p 144
7. Stevenson JS, Bright RW, Dunson GL: Technetium 99m phosphate bone imaging; a method for assessing bone graft healing. Radiology 110:391, 1974
8. Campeau R, Hall M, Miale A: Detection of total hip arthroplasty complication with Tc-99m polyphosphate. J Nucl Med 17:526, 1976
9. Lisbona R, Rosenthal L: Observation on the sequential use of 99m-Tc phosphate complex and 67-Ga. Radiology 123:123, 1977
10. Brookes M: The Blood Supply of Bone. London, Butherworks, 1971

Orthopedic Considerations of Trauma to the Growing Skeleton

John R. Denton

In the area of fractures, the adage still holds true that "children are not small adults." The diagnosis, treatment plan, and prognosis are very different in the growing child, with his open growth plates and excellent healing capacity, and in the adult, who has completed growth and has a lesser healing ability and an increased propensity for joint stiffness. In this chapter, the problems of fractures in the growing skeleton will be discussed from the standpoints of diagnosis, treatment, healing, and growth disturbances.

Diagnosis

Fractures, particularly nondisplaced ones, through the growth plate may be difficult to define because of the radiolucency of the growth cartilage. Fractures that involve the growth plate should be classified by the Salter–Harris method.[1,2] An advantage of this classification (Fig. 1) is that some idea of prognosis is implicit within it.

Congenital and developmental anomalies may also make diagnosis of the injury difficult. For this reason, a radiograph of the *other* (I emphasize "the other" because it is not always "the normal" extremity) areas of the body will be of benefit. Such anomalies in the cervical spine, elbow, and knee regions are among the most common.

Unusual alignment and radiographic measurements different from those in the adult are common in children. The cervical spine is again an excellent example of this problem.[3] Accessory bones, particularly in the feet, may masquerade as avulsion fractures.

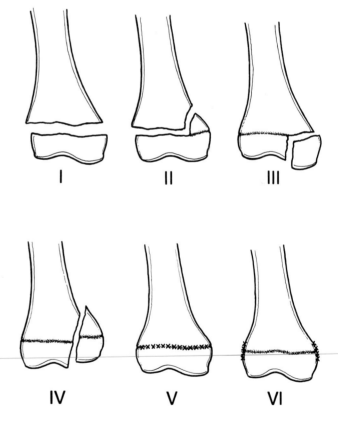

FIG. 1. Salter I: fracture through the growth plate only; Salter II: fracture through the growth plate and a portion of the metaphysis; Salter III: fracture through the growth plate and the epiphysis; Salter IV: fracture through the metaphysis, growth plate, and the epiphysis, (note that types III and IV are intraarticular injuries); Salter V: a crushing injury to the growth plate; Salter VI: a fracture of the perichondral ring surrounding the epiphysis.

The secondary centers of ossification are notorious for being obstacles to diagnosis, and the most frequent site is the elbow. With its six secondary centers, some of which are bipartite or multicentric, and their varying ages of appearance, the child's elbow is truly an area of challenge. Unfortunately, it is also one of the most frequently injured areas, and an improper diagnosis and treatment can lead to a poor result. Because of their common occurrence and the admonition of Dupuytren to "Pity the young surgeon with a difficult elbow fracture," some emphasis will be placed on this subject. I believe the following outline will be of benefit to radiologist and surgeon alike:

PRINCIPLES

1. Appearance of secondary centers of ossification (years)[4]

K	capitellum	1
I	(internal) medial epicondyle	4
T	trochlea (multicentric)	8
E	(external) lateral epicondyle	12
R	radial head	4
O	olecranon[2]	10

    Varies with sex and the individual, but the order of appearance is constant
2. "Joint above and joint below"
3. Proper x-rays
    a. To include the *other* elbow
    b. To include the ipsilateral elbow in displaced shaft fractures
4. Put your finger on each fragment and identify it

TYPES OF FRACTURES

1. Supracondylar fracture
    Most common
    The two complications:
    a. Volkmann's ischemic contracture—catastrophe
    b. Cubitus varus—cosmetic
    Plan for a displaced fracture:
    a. Reduction is the best form of decompression
    b. Overhead skeletal traction if any question
    c. General anesthesia for initial treatment
2. Salter IV fracture of the lateral condyle (elbow)
    Recognize
    Very common
    Surgery if *any* displacement
    Malunion probably not of serious functional consequence
3. Transcondylar fracture (rare)
    Neither joint nor growth plate involved
4. Salter II fracture (rare)
    Medial displacement
    Good prognosis
5. Medial epicondyle fracture
    Very common
    Ulnar nerve problems
    Often associated with a dislocation
    Consider surgery if:
    a. Fragment within the joint
    b. Ulnar nerve symptoms
    c. Greater than 1-cm displacement
6. Radial *neck* fracture
    Retrograde blood supply to head
    Up to 30 to 40 degrees angulation acceptable
    Reduce and fix, not excise, the radial head
7. Dislocations
    Direction of dislocation
    Look for fractures
    Check post reduction x-rays without plaster
    Early motion
8. Monteggia fracture—dislocation
    Make the diagnosis
    Usually anterior
    May be radial
    Rarely posterior in children
    Closed treatment
    Posterior interosseous nerve injury

Treatment

A few basic principles will be emphasized, as treatment in detail is not within the scope of this discussion, but I feel the radiologist should have some idea of what the surgeon is trying to accomplish. As a general rule, children's fractures do not require surgery to reduce and fix them, because more deformity is acceptable in a child due to his remodeling potential; healing is usually not a problem; bed rest and joint immobilization do not present the same hazards as in the adult; and the consequences of a postoperative infection, especially if the growth plate is involved, can be extremely serious and long lasting. There are exceptions to these rules and they are (1) anatomic restoration of the growth plate and joint surfaces is essential; (2) a neurologic or vascular problem that needs bony reduction and stabilization; (3) open (compound) fractures are debrided and irrigated; (4) displaced femoral neck fractures, including slipped capital femoral epiphyses, of any degree; (5) marked displacement of the fracture, particularly when not in the plane of the adjacent joint motion; (6) a multisystem problem in which bone stabilization would be of benefit to the patient's total situation; (7) certain pathologic fractures. Despite the length of this list, probably less than 10% of closed fractures in children require surgery.

Healing

The marvelous healing capacity of a long bone in a child cannot be overemphasized, and the younger the child the greater this ability. Healing time is much more dependable and faster than in the adult. It has been said that the long bone of a child will heal "if the bone ends are in the same room." The two parameters that are of concern in long-bone fractures are shortening at the fracture site and malunion, that is, deformity due to angulation (usually varus/valgus) and/or rotation. These two deformities are not well corrected by future remodeling.

Growth

The above characteristic—indeed it is the definition of the child in orthopedic parlance, "A child is a person whose growth plates are open"—is a major point to consider in children's fractures. It is truly a double-edged sword in that growth can correct many fracture deformities to a remarkable degree. The amount that can be corrected is dependent upon (1) the age of the child, (2) the proximity of the fracture to the growth plate, and (3) whether the deformity is in the plane of motion of the adjacent joint. An example of an ideal situation would be a young child with a deformity in the distal radial metaphysis in the flexion–extension plane.

On the other hand, if the growth plate is injured and a growth disturbance occurs, then serious and *progressive* deformity may ensue. The two general types of disturbance are longitudinal and angulatory. A longitudinal disturbance may produce an overgrowth or an undergrowth of the limb. Obviously, this is of much greater import in the lower extremity. The Salter V and VI types of injuries are the most likely to cause an undergrowth. The overgrowth problem is not clear in its etiology but is thought to be due to an increased blood supply to the growth plate in response to interruption of the interosseous blood supply at the fracture site, which is usually in the diaphysis. Children in the age group 2 to 10 years are most likely to have an overgrowth problem. With this in mind, displaced femoral shaft fractures in that age group are usually left with 0.5 to 1.5 cm of shortening in traction to allow for the later probable overgrowth.[5,6] Children outside of that age group should have no override in traction. It is not clear why this phenomenon does not occur in older children, even during their growth spurt. This problem usually does not progress beyond 2 years after the time of injury.

The undergrowth problem may be due to complete closure of the growth plate or a partial closure. The problem is then progressive, as the child is growing normally from the noninjured side.

The angulatory disturbance may be either varus or valgus. The angulation is usually caused by a premature closure or bony bridge across one side of the growth plate creating a "tethering" situation. This is most commonly seen following a Salter IV fracture, often with an incomplete reduction, which allows part of the plate to be bridged by bone. Even in cases in which the fracture is anatomically reduced, a deformity still may occur due to the initial trauma that injured the growth plate cells at that time. The problem with angulatory and undergrowth disturbances is that they tend to be progressive as long as the child is growing, so that the younger a child is, the more serious the problem because of the longer period of improper growth remaining before reaching skeletal maturity, at which time growth will naturally stop altogether.

Unfortunately, the Salter V and Salter VI injuries are not radiographically evident on initial x-rays. The damage is done at the cellular level, so no fracture or displacement is appreciated. Indeed, the midshaft of the bone may be fractured and the distal femoral growth plate injury may not even be suspected, but it may turn out to be the major injury in the form of a growth disturbance. For this reason, the parents of a child with a long-bone fracture, particularly in the lower extremity, must be cautioned about the possibility of a growth disturbance from the outset. The child should be followed for at least 2 years post injury to adequately assess growth of the extremity. Such injuries are usually the result of high-energy violence, such as a fall from a height or a motor vehicle accident.

Figures 2 and 3 show a 10-year-old male who is 7.5 cm short in the right femur due to an injury at age 5 years. Note that the femoral shaft fracture has healed, but the distal femoral growth plate has closed, probably due to a Salter V injury. The problem is not only the present shortening, since 7.2 cm more additional shortening can be predicted[7] as the noninjured normal left femur continues to grow.

Figures 4 and 5 show a 10-year-old male with a progressive right-ankle varus deformity due to a Salter III or IV injury at age 8 years. Note the bony bridge medially, which tethers the normal growth of the lateral side into a varus deformity. The normal left ankle has had a correctional osteotomy for the right ankle diagrammed on it.

Both of these problems are caused by a growth disturbance and are a major factor in children's fractures but, of course, are nonexistent in adults.

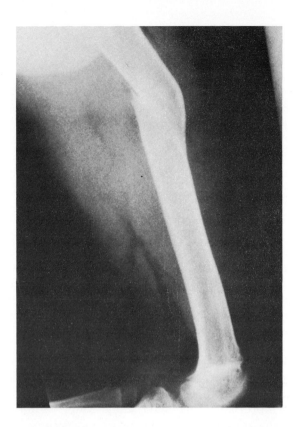 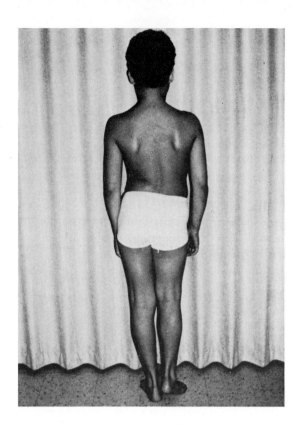

FIG. 2,3. A 10-year-old male who is 7.5 cm short in the right femur due to an injury at age 5 years.

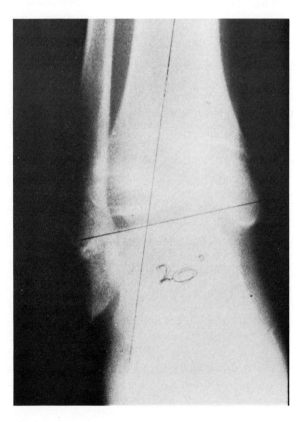 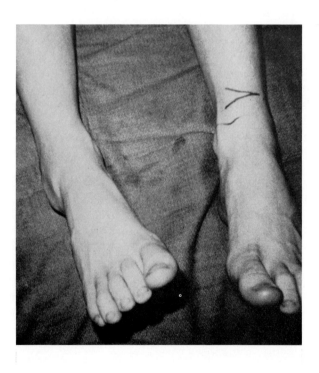

FIG. 4,5. A 10-year-old male with a progressive right-ankle varus deformity due to a Salter III or IV injury at age 8 years.

Summary

This chapter has discussed the salient features of fractures in children, including diagnostic difficulties, classification, treatment principles, healing properties, and the effects of the phenomenon of growth.

References

1. Salter RB, Harris WR: Injuries involving the epiphyseal plate. J Bone and Joint Surg 45A:587, 1963
2. Personal communication with R.B. Salter
3. Cattel HS, Filtzer D: Pseudosubluxation and other normal variations in the cervical spine in children. J Bone Joint Surg 47A:1295, 1965
4. Caffey J: Pediatric X-Ray Diagnosis, Vol 2. Chicago, Year Book Medical Publishers, 1972 p 882
5. Salter RB: Textbook of Disorders and Injuries of the Musculoskeletal System. Baltimore, Williams and Williams, 1970, p 414
6. Tachdjian MO: Pediatric Orthopedics, Vol 2. Philadelphia, Saunders, 1972, p 1702
7. Anderson M, Green WT, Messner MB: Growth and prediction of growth in the lower extremities. J Bone Joint Surg 45A:1, 1963

Arthrography of the Ankle, Elbow, and Wrist

Harvey L. Hecht

Ankle Arthrography

Ankle arthrography is not as frequently performed as is knee or shoulder arthrography. Many orthopedists still feel that clinical evaluation and plain radiographs with stress suffice to make the diagnosis of an acute ankle "sprain."[1] Furthermore, even if a torn ligament is diagnosed, the orthopedist will frequently treat it conservatively; thus, more detailed evaluation is not considered necessary. Recently, however, reports of early surgical repair of ligamentous injuries have revived interest in the arthrogram.[1-4] Ankle arthrography has always been helpful in the diagnosis of chronic recurrent ankle sprain or the unexplained "painful ankle."[5-8]

Method

Prior to an arthrogram, anteroposterior lateral, and oblique films of the ankle are obtained for both technique and evaluation of loose bodies and osteochondritis dissecans. The ankle to be examined is first prepared by washing the skin with soap and water. Hexachlorophene 3 percent or a similiar antibacterial soap is usually used. Povidone iodine or benzalkonium chloride 1:750 is applied. The ankle is then draped and placed under a fluoroscope. The fluoroscopic controls are also draped. The best site for injection is anterior medial just lateral to the curve of the medial malleolus around the talus.[9] The anterior tibial artery should be palpated at this level and avoided. The injection site is between the tendon of the tibialis anterior and the extensor hallucis longus.[9] Using a 25-gauge needle, 1 ml of lidocaine 1 percent is injected into the site. This needle is then replaced by a 20-gauge 1½-inch needle. A 10-ml glass syringe is recommended rather than a plastic syringe because of the ease of injection when using a glass syringe. Upon entering the joint, a small amount of clear yellow joint fluid can be seen bubbling back through the needle. Injection of 3 to 5ml of saline is then performed. This material is used as a test to see if one is in the joint. If one is not in the joint, a great deal of resistance is felt. Another method of testing whether one is within the joint space proper is to use air. Again, 3 to 5ml of air is injected into the joint. If the needle is properly placed, the ankle joint will swell as the air is injected and collapse as air is withdrawn or comes back into the syringe, pushing the plunger back with it. Once there is free communication between the syringe and the joint as tested by saline and/or air, 5 to 8ml of methylglucamine diatrizoate (Renografin 60 percent) is injected. Once the contrast is seen to be within the joint fluoroscopically, the needle is withdrawn. Anteroposterior, lateral, and oblique projections are then obtained.

Anatomy

The lateral ligaments of the ankle (Fig. 1) are most frequently injured. The most commonly injured ligament is the anterior talofibular ligament.[10] It originates from the anterior surface of the distal fibula and inserts on the neck of the talus.

The calcaneofibular ligament is the strongest lateral ligament. It originates from the posterior and distal end of the fibula, traverses posteriorly, inferiorly, and slightly medially to insert on the

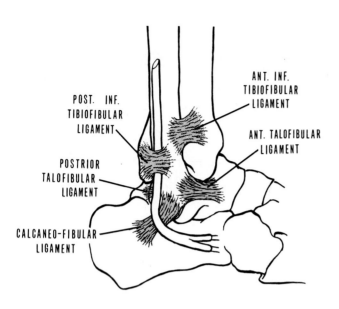

FIG. 1. Ligaments of the ankle. After Fordyce and Horn.[1]

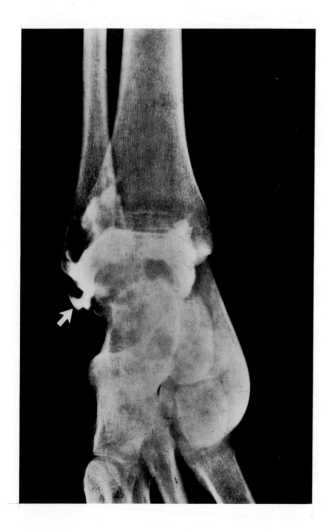

FIG. 2. Fifteen-year-old female with lateral instability 6 months after ankle injury. (Referred by Lawrence Lefkowitz, M.D., Norwalk, Connecticut.) Anteroposterior view showing leakage of contrast under and lateral to the fibula (arrow).

superior portion of the os calcis.[10,11] It is connected to the tendon sheath of the peroneus longus and brevis muscles.

The smallest of the lateral ligaments is the posterior talofibular. It starts from the medial and distal aspect of the fibula, traverses up and medially to the talus, and attaches to the fibula.

Two other ligaments are slightly higher in the ankle and are called the anterior tibiofibular ligament and the posterior tibiofibular ligament. The anterior tibiofibular ligament starts from the distal tibia posteriorly, inferiorly, and laterally and inserts on the anterior and lateral aspect of the distal fibula.[10,11] The posterior tibiofibular ligament extends from the posterior lateral angle of the tibia to the posterior aspect of the distal fibula. Anteriorly, it touches the interosseous membrane.[1,11]

The medial side of the ankle, which is less frequently injured, is covered by the deltoid ligament. It starts from the medial malleolus of the tibia and stretches downward to the talus and the os calcis.[10,11]

Radiographic abnormalities

When the anterior talofibular ligament tears, contrast will leak laterally underneath the distal fibula. This is best seen in the anterior posterior projection (Fig. 2). In the lateral projection, the contrast will leak anteriorly and beyond the anterior recess of the ankle joint (Fig. 3).

Tears of the calcaneofibular ligament allow the leakage of contrast into the peroneal tendon sheaths (Fig. 4). Contrast also flows downward below the talus to overlie the os calcis when viewed in the lateral and oblique projections (Fig. 5). Combined tears of these two ligaments are common, and contrast may leak out along both the anterior and lateral aspects of the joint.

Tears of the anterior tibiofibular ligament show contrast escaping upward between the tibia and fibula. Normally, a small amount of contrast does escape into the recess between the tibia and fibula (Fig. 6), but in a tear of the anterior tibiofibular ligament, the contrast is seen to extend much higher.

The posterior tibiofibular and posterior talofibular ligaments are rarely involved with isolated tears but may be found with calcaneofibular ligament tears

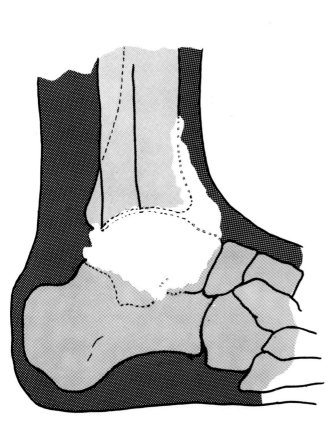 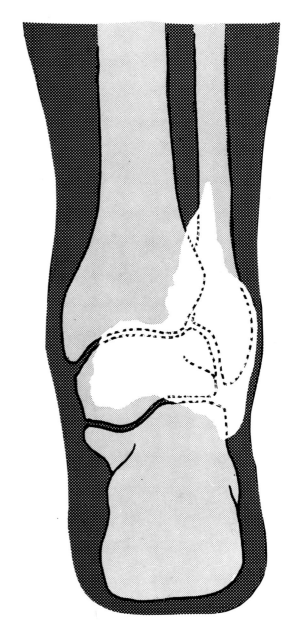

FIG. 3. Diagramatic representation of a large tear of the anterior talofibular ligament. Contrast leaks (above) anterior to the normal recess and (right) lateral to the fibula.

(Fig. 7). The deltoid ligament is sometimes torn. Contrast leaks medially outside of the medial recess.

Although some authors have felt that the peroneal tendon sheaths may occasionally fill normally,[9] filling of these sheaths usually indicates a calcaneofibular ligament tear. However, filling of the tendon sheaths of the tibialis posterior, flexor digitorum longus, and flexor hallucis longus, which pass posterior to the medial malleolus, is normal (Fig. 8).

Thus, in the chronically painful ankle and in the acute ankle injury, arthrography may help diagnose many ligamentous tears and allow for better surgical treatment of these injuries. In the search for a loose body, arthrography has not been applied to the ankle. The technique would be different and involve injection of only 1 to 2ml of contrast with 6 to 7ml of air. This has not been done with frequency, however.

Elbow Arthrography

The elbow arthrogram was introduced in Sweden and has been investigated by Arvidsson and Johansson[12] mostly in patients with fractures or dislocations of the elbow. Similarly to ankle and shoulder arthrography, leakage of contrast out of the joint at specific sites indicates a tear of the ligament or joint capsule. More recently, using tomography and double-contrast technique, an attempt has been made to visualize loose bodies and osteochondritis dissecans.

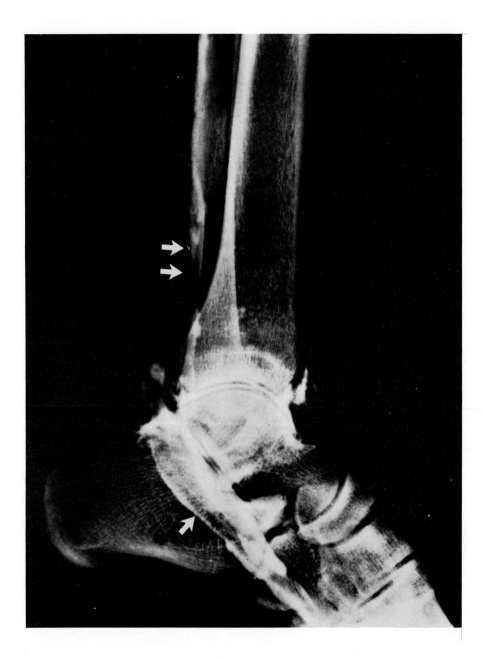

FIG. 4. Twenty-eight-year-old male with chronic ankle swelling and pain. (Referred by Alan Weisel, M.D., Stamford, Connecticut.) Contrast seen extravasating into the peroneal tendon sheath (arrow) from a calcaneofibular ligament tear. Normal filling of a tibialis posterior tendon sheath (double arrows).

This has proved successful.[13,14] Arthrography has also been used for evaluating patients with rheumatoid arthritis.[15,16]

Method

After obtaining preliminary anteroposterior, lateral, and oblique radiographs of the elbow with comparison films in children, the patient is positioned supine with the elbow to be examined closest to the arthrographer. First, after washing the skin with soap and water, povidone iodine or another appropriate antiseptic is applied. The elbow is properly draped. Using a 25-gauge ½-inch needle, 1 ml of lidocaine 1 percent is injected on the lateral side of the elbow just proximal to the radial head at the radiohumeral articulation (Fig. 9). This can be readily palpated. Using fluoroscopy with the elbow in the extended anteroposterior position and leaving the needle in place, one can see if the direction of the needle is satisfactory. The 25-gauge needle may then be replaced with a 22-gauge 1½-inch needle. Frequently, the ½-inch needle will be adequate in length to puncture the joint. When entering the joint, fluid will leak out. To test that the needle is in the joint, 3 to 4 ml of saline or 3 to 4 ml of air may be injected. Using glass syringes allows for the easy manipulation of the plunger. It is important that the needle be held at the skin surface so that it will remain within the confines of the joint, since it has advanced only a short distance. Following the test injection and when one is certain that there is free communication between the joint and the syringe, 0.5 ml of methylglucamine diatrizoate (Renografin 60 percent) is injected with 6 ml of air. The needle is then withdrawn and the elbow is exercised. The patient is turned into the prone position. Lateral films of the elbow flexed and anteroposterior films with obliques are obtained. Tomography in the lateral and anteroposterior projection is performed. The technique employed in the tomography utilizes a three-phase unit and linear tomographic motion (GE Telegem) with 34 to 40 kv 20 MAS. The sweep is 2 seconds. Cuts were obtained from 2.5 cm to 6.0 cm in both anteroposterior and lateral planes.

Anatomy

The articular surfaces of the elbow joint are confined by a capsule that is thickened medially and laterally into the ulnar collateral and radio collateral ligaments, respectively.[11] The ulnar collateral, or medial, ligament is composed of three bands—anterior, posterior, and transverse. The anterior band starts from the anterior portion of the medial epicondyle and attaches to the medial margin of the coronoid process. The posterior band attaches to the lower and posterior part of the medial epicondyle and runs to the medial portion of the olecranon. The transverse band crosses from the olecranon to the coronoid process.

The radial collateral, or lateral collateral, ligament runs from the lateral epicondyle to the annular ligament around the radius and the margin of the radial notch.

Three joint recesses are seen on the arthrogram. The coronoid recess is anterior, the olecranon recess posterior and the periradial recess annular around the neck of the radius (Fig. 10). The borders of the recesses are all smooth, except for the anterior border of the coronoid recess, which is wrinkled in the flexed lateral projection, and the border next to the medial collateral ligament, which is usually irregular.[12,13].

A portion of the articular cartilage of the ulna is

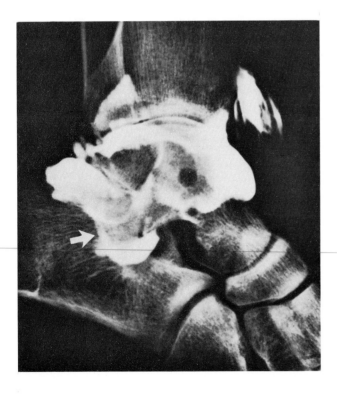
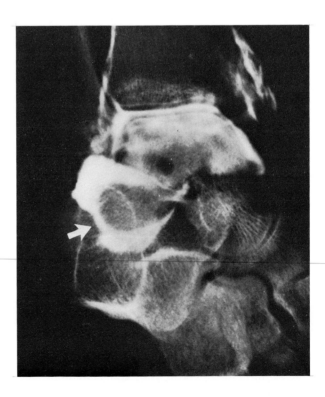

FIG. 5. Twenty-three-year-old female presenting with chronic pain and swelling of the lateral aspect of the ankle. (Referred by Alan Weisel, M.D., Stamford, Connecticut.) Tear of the calcaneofibular ligament seen in the lateral (left) and oblique (right) views (arrow).

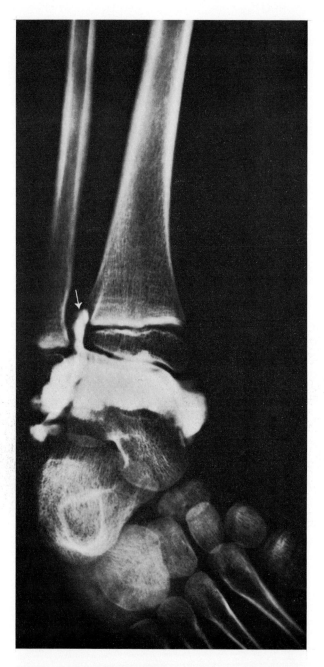

FIG. 6. Seven-year-old male who injured right ankle one year ago and still has pain. (Referred by Alan Weisel, M.D. Stamford, Connecticut.) Contrast extending superiorly into the normal recess between the tibia and fibula (arrow).

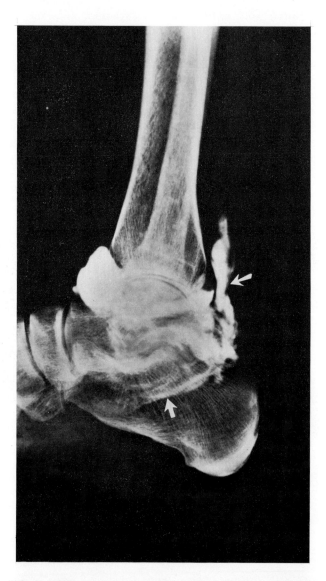

FIG. 7. Fifty-seven-year-old male with 4 month history of pain and swelling on lateral side of ankle. (Referred by Peter W. Hughes, M.D., Stamford, Connecticut.) Calcaneofibular and posterior talofibular ligament tear with contrast extending up posteriorly and down into the peroneal tendon sheath (arrow).

absent in the trochlear notch[11,13] (Fig. 11). It should not be mistaken for a defect. The remainder of the articular cartilages are all smooth (Fig. 12). The articular cartilage of the capitellum is present on the anterior and inferior surfaces only. The trochlear has cartilage on its anterior, inferior, and posterior surfaces.

Radiographic abnormalities

Leakage of contrast on the medial side of the elbow joint usually occurs after fracture of the head or neck of the radius. It suggests a tear of the joint capsule and the medial collateral ligament.[12] The annular

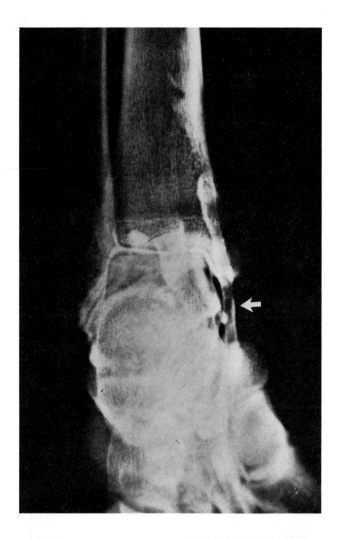

FIG. 8. Normal filling of the tendon sheath of the tibialis posterior passing posterior to the medial malleolus in anteroposterior view (arrow). The patient had a torn calcaneofibular ligament (Fig. 4). (Referred by Alan Weisel, M.D., Stamford, Connecticut.)

ligament of the radius may also be torn, allowing contrast to leak into the annular recess.

Two major types of dislocation of the elbow result in two types of ligamentous capsular tears. In posterior dislocation, the anterior capsular recess is stretched, and leakage of contrast occurs through an anterior tear.[12] In lateral dislocation, a rupture of the medial ligament allows for leakage of contrast at that site. A fracture of the coronoid process can give rise to capsular tears. The injury is usually the result of abduction or hyperextension and therefore causes medial ligament tears.

In patients suffering from a locked elbow or clicking elbow, loose bodies or osteochondritis dissecans must be considered. Plain films will sometimes reveal calcified loose bodies. The arthrogram will show cartilagenous loose bodies and chondral defects (Fig. 10). Defects have been demonstrated such as osteochondritis dissecans of the capitellum, hypertrophic synovium over the capitellum, interarticular coronoid fossa loose bodies, and olecranon fossa loose bodies.[13]

Wrist Arthrograms

Arthrography of the wrist has been most useful in rheumatoid arthritis[17-22] by discovering early cartilagenous changes that are not revealed on routine films. In addition, in the traumatized wrist or the "sprained" wrist, arthrography may help make a more specific diagnosis of ligamentous tear rather than a general diagnosis of a "sprain." For example, a tear of the articular disk or of the ligaments separating the radiocarpal joint from the intercarpal joints may easily be diagnosed by arthrography.[18]

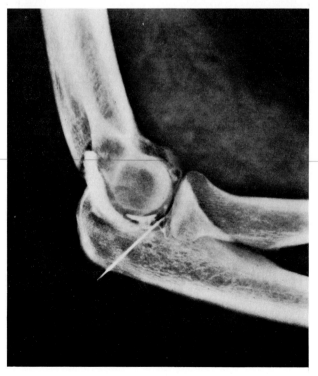

FIG. 9. Elbow arthrogram showing the needle placement between the radius and capitellum of the humerus. Note the olecranon recess posteriorly. (Courtesy of Helene Pavlov, M.D., and Robert Freiberger, M.D., Hospital for Special Surgery, New York.)

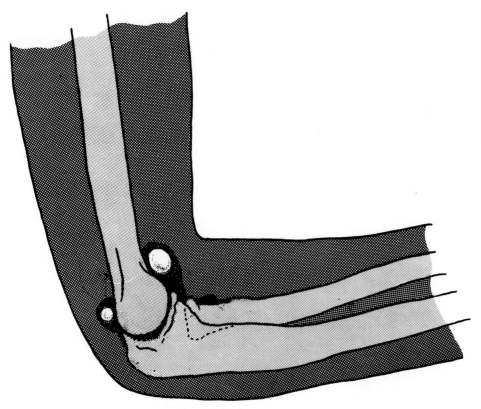

FIG. 10. Diagram showing cartilagenous loose bodies within the coronoid recess anteriorly and the olecranon recess posteriorly. The peri-radial recess can also be seen. This patient had a 3-year history of pain and limited range of motion. After radiographs in Eto et al.[13]

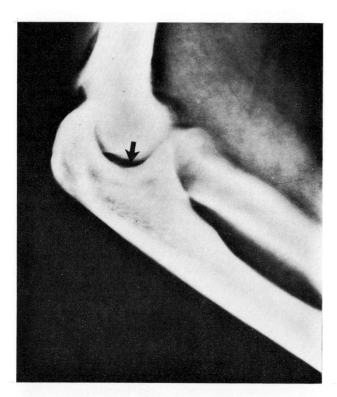

FIG. 11. Slight depression on the ulna, where there is an absence of cartilage (arrow). (Radiographs from Helene Pavlov, M.D., and Robert Freiberger, M.D., Hospital for Special Surgery, New York.)

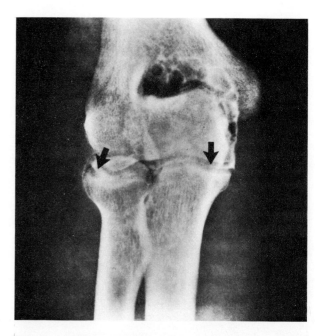

FIG. 12. Anteroposterior elbow arthrogram showing smooth cartilage of the trochlea and capitellum (arrow). (Radiographs from Helene Pavlov, M.D., and Robert Freiberger, M.D.)

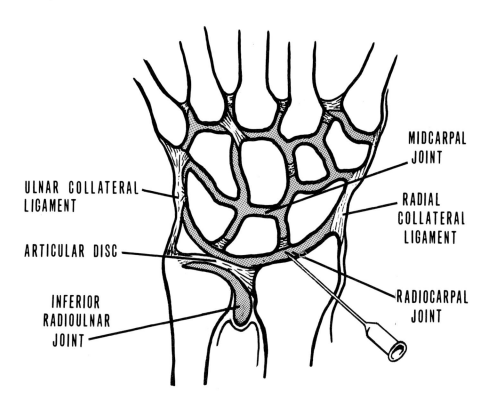

FIG. 13. Needle insertion in preparation for wrist arthrography. Diagram after Ranawat et al.[19]

Method

Preliminary anteroposterior, lateral, and oblique films are obtained. After washing the overlying skin of the hand and wrist with soap and water, the patient is placed in the prone position and the wrist is placed on a small triangular sponge with the palm down. The heel of the triangular sponge is at the wrist and the apex is at the fingertips. After preparation of the skin with povidone iodine, the wrist is draped. Using a 25-gauge ½-inch needle, the wrist is entered between the distal articular border of the radius and the proximal row of carpal bones close to the navicular (Fig. 13). An injection of 1 ml of lidocaine 1 percent is used. A 22-gauge 1½ inch needle replaces the 25-gauge needle, and using fluoroscopy as a guide, the needle is introduced dorsally and angled under the radial lip.[17,22] After puncture of the joint, 2 to 3 ml of saline is injected. If the injection is intraarticular, fluid will rapidly return through the needle. Using a glass syringe, 2 to 3 ml of air is injected. When one is in the joint, the syringe plunger will move smoothly back and forth as air is injected from and released back into the syringe. Once one is unequivocally in the joint, 3 ml of methylglucamine diatrizoate (Renografin 60 percent) is injected. Inject contrast only when you are certain you are in the joint.

Anatomy

The radiocarpal joint fills with contrast and is seen as a thin, continuous, even line across the joint.[18] No contrast enters the inferior radioulnar cavity, which lies proximal to the radiocarpal cavity separated from it by the articular disk (Fig. 13). The triangular fibrocartilage or disk attaches to the base of the ulnar styloid and to the radius on its ulnar side. The interosseous intercarpal ligaments prevent the extension of contrast, and the proximal carpal row is not outlined. A prestyloid recess in front of the ulna styloid and a pisiform triquetral cavity[17,22] exists separated from the radiocarpal cavity by fibers of the ulna collateral ligament.[22]

Radiographic Abnormalities

Leakage of contrast into the inferior radioulnar joint implies damage or tear of the triangular fibrocartilage or disk. Markedly displaced radial and ulnar fractures may tear the disk. Fractures of the ulnar styloid itself may tear the attachment of the disk (Fig. 14). Usually, however, both the radius and ulna are torn together, and the disk and its attachment are intact.[18]

In cadaver studies, communication with the in-

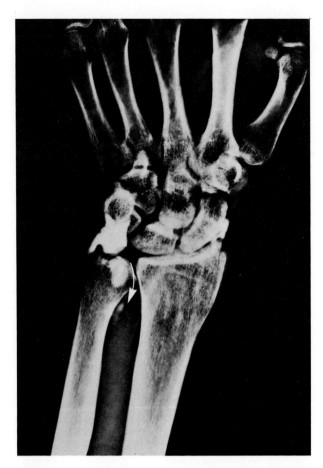

FIG. 14. Thirty-five-year-old male with pain and clicking left wrist. (Referred by Robert Orth, M.D., Stamford Connecticut.) Wrist arthrogram showing leak of contrast into the inferior radioulnar joint (arrow).

ferior radioulnar joint is demonstrated in from 7 to 60 percent of cases.[17,18,23,24] Arthrography in adult subjects usually fails to show such communication, however, unless there is a tear present. These wide discrepancies are thought to be related to the older age of cadavers, since tears and degenerative changes increase with age.[22]

Radiocarpal to midcarpal connections can also be seen in dissection of cadavers. Once again, a tear has to be presumed if a communication is found on an arthrogram. Dissection of elderly cadavers shows a 36 percent incidence of communication across the lunate triquetrum ligament and a 40 percent incidence of communication across the lunate navicular ligament.[22,24]

Radiocarpal to pisiform triquetral cavity communication normally exists. Lewis noted 34 percent incidence in cadavers.[24] It can be traumatic or degenerative.[22] Midcarpal to carpal–metacarpal communication is very rare.

In rheumatoid arthritis, a corrugated arthrographic pattern is found.[17,19,20,22] The proliferation of synovium into villae creates a ridged nodular pattern on the arthrogram in 90 percent of such patients.[12] Communication between the radiocarpal and radioulnar joint is found in approximately 70 percent of patients. A radiocarpal to midcarpal connection was also found in 70 percent of patients.[17] Communication between the wrist joint and flexor tendon sheaths is found in approximately 21 percent of patients.[17] A combination of one or more of the above was present in 77 percent.[17] Lymphatic filling is also frequently found (42 percent).[22]

Communication between joints is caused by destruction of the triangular cartilage and ligaments. The carpal–metacarpal joint of the thumb was never visualized, however. These abnormal findings may be seen before bone alterations occur in patients with rheumatoid arthritis.

Erosions of the ulnar styloid are well known in rheumatoid arthritis. Marked synovial proliferation and the intimate relationship of the ulnar styloid to the prestyloid recess accounts for the styloid erosion.[21,22]

Similar changes—that is, the corrugated patterns, interjoint communication, and lymphatic filling—are seen in ankylosing spondylitis and would be expected in Driller's disease and psoriasis.[22]

Patients with posttraumatic arthritis and neurotrophic joints show synovial irregularity without corrugation. Frequent intercompartment communication and tendon visualization without lymphatic filling is seen.[22]

Wrist arthrography seems to be useful in planning synovectomy for patients with rheumatoid arthritis. It may aid in the diagnosis of the patient with a posttraumatic wrist by defining specific ligamentous tears.[25] It may also be useful in differentiating types of arthritis prior to bony changes.[22]

Acknowledgments

The author wishes to thank Renald VonMuchow for the artist's drawings, Michael Carlin for the radiographic reproductions, and Bonnie Scarvey for help in preparing the manuscript.

References

1. Fordyce AJ, Horn CV: Arthrography in recent injuries of the ligaments of the ankle. J Bone Joint Surg 54B:116, 1972
2. Staples OS: Result study of ruptures of lateral liga-

ments of the ankle. Clin Orthop 85:50, 1972
3. Staples OS: Ruptures of the fibular collateral ligaments of the ankle. Result study of immediate surgical treatment. J Bone Joint Surg 57A:101, 1975
4. Speigel PK, Staples OS: Arthrography of the ankle joint: Problems in diagnosis of acute lateral ligament injuries. Radiology 114:587, 1975
5. Anderson KJ, LeCocq JF, Clayton ML: Athletic Injury to the fibular collateral ligament of the ankle. Clin Orthop 23:146, 1962
6. Meherez M, El Geneidy S: Arthrography of the ankle. J Bone Joint Surg 52B:308, 1970
7. Olson RW: Arthrography of the ankle: its use in the evaluation of ankle sprains. Radiology 92:1439, 1969
8. Percy EC, Hill RO, Callaghan JE: The "sprained" ankle. J Trauma 9:972, 1969
9. Callaghan JE, Percy EC, Hill Ross O: The ankle arthrogram, J Can Assoc Radiol 21:74, 1970
10. Kaye JJ, Bohne WHO: A radiographic study of the ligamentous anatomy of the ankle. Radiology 125:659, 1977
11. Goss CM (ed): Gray's Anatomy, 28th ed. Philadelphia Lea and Febinger, 1966
12. Arvidsson H, Johansson O: Arthrography of the elbow joint. Acta Radiol 43:445, 1955
13. Eto RT, Anderson PW, Harley JD: Elbow arthrography with the application of tomography. Radiology 115:283, 1975
14. Del Buono MS, Solarino GB: Arthrography of the elbow with double contrast media. Ital Clin Ortop 14:223, 1962
15. DeSeze S, Debeyre N. Djian A, et al: The elbow joint. Int Symp Radiol Aspects Rheum Arthritis. Amsterdam, Excerpta Med Int Congr Ser 61, 1964 p 115
16. Weston WJ: Lymphatic filling during positive contrast arthrography in rheumatoid arthritis. Australas Radiol 13:368, 1969
17. Harrison MO, Freiberger RH, Ranawat CS: Arthrography of the rheumatoid wrist joint. Am J Roentgenol 112:480, 1971
18. Kessler I, Silberman Z: An experimental study of the radiocarpal joint by arthrography. Surg Gynecol Obstet 112:33, 1961
19. Ranawat CS, Freiberger RH, Jordan LR, et al: Arthrography in the rheumatoid wrist joint. A preliminary report. J Bone Joint Surg 51A:1269, 1969
20. Ranawat CS, Harrison MO, Jordan LR: Arthrography in the wrist joint. Clin Orthop 83:6, 1972
21. Resnick D: Rheumatoid arthritis of the wrist: why the ulnar styloid. Radiology 112:29, 1974
22. Resnick D: Arthrography in the evaluation of arthritic disorders of the wrist. Radiology 113:331, 1974
23. Liebolt FL: Surgical fusion of the wrist joint. Surg Gynecol Obstet 66:1008, 1938
24. Lewis OJ, Hamshere RJ, Bucknill TM: The anatomy of the wrist joint. J Anat 106:539, 1970
25. Coleman HM: Tears of fibrocartilage disc at the wrist. J Bone Surg 38B(proc):782, 1956

Shoulder Arthrography

Murray K. Dalinka

In the shoulder joint, stability has been sacrificed for increased motion. The joint capsule arises from the glenoid rim and inserts on the anatomic neck of the humerus. The humeral head has three times the articular surface of the glenoid cavity. These anatomic factors give rise to an unstable joint, and the surrounding soft tissues are necessary for enhancing stability. This renders the shoulder particularly prone to soft-tissue injury, namely, dislocations (these are discussed elsewhere in this volume) and rotator cuff tears.

Patients with rotator cuff tears may present with a chronically painful shoulder.[1] Plain films may be nonspecific or show suggestive but nondiagnostic findings.[2] In acute injuries, they are frequently normal. Clinically, there may be no significant muscular atrophy or weakness on abduction. A relatively normal range of motion may be present with little or no tenderness about the greater tuberosity. The "drop arm" test may be negative.

Arthrography of the shoulder is an accurate, safe, and simple technique of considerable value in the diagnosis of rotator cuff tears. Other soft-tissue abnormalities, such as adhesive capsulitis and capsular detachments secondary to previous dislocations, can also be demonstrated.[3,4]

Technique

Shoulder arthrography can be performed by either positive[4,5] or double-contrast techniques.[6,7] Erect filming is a necessary adjunct to the double-contrast method.

The shoulder is shaved and then prepped with antiseptic solution, and a keyhole drape is placed over the joint. The author then places a sterile lead grid over the shoulder to localize the injection site via coordinates[8] (Figs. 1 and 2). The entire procedure is performed under fluoroscopic control, but filming is done with the overhead technique. The shoulder is infiltrated with local anesthetic, with the patient in the anteroposterior position. A 20-gauge spinal needle is then advanced to the joint while injecting small amounts of lidocaine; entrance to the joint can usually be determined by the differences of resistance during the injection.[4] When a small amount of contrast material is injected, it flows away from the needle edge when the needle is in the joint.[2]

With the single-contrast technique, 12ml of a mixture of Renografin 60 percent and 1 percent lidocaine is injected. Utilizing double-contrast technique, 4 to 5ml of contrast are used along with 10ml of room air.[7]

Routine postarthrographic radiographs consist of anteroposterior views in internal and external rotation and an axillary view. When utilizing double-contrast technique, erect external and internal rotation views are added to the routine views. If these films are normal, the shoulder is exercised for approximately 5 minutes and repeat filming is performed. Small tears may not be demonstrated until after the postexercise films.

We feel that if single-contrast studies are performed, the Renografin should be diluted with local anesthetic. Renografin is a hypertonic solution, and its dilution decreases postarthrographic pain without decreasing the diagnostic accuracy of the study.

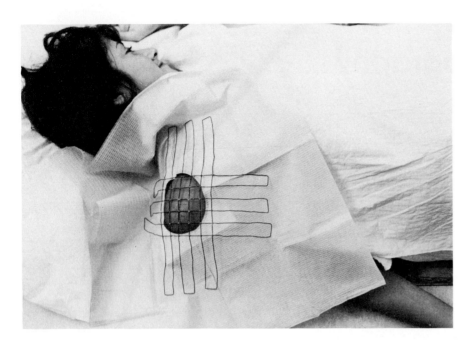

FIG. 1. Sterilizable lead grid placed over shoulder for localization of injection site.

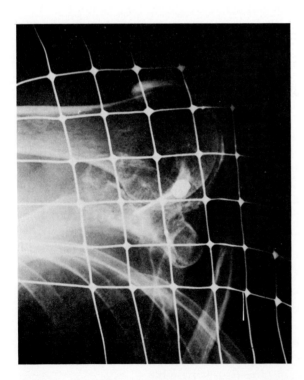

FIG. 2. Film taken to demonstrate aid of grid in localization. Note contrast has flowed away from needle site and is external to humeral head and below acromion representing a torn rotator cuff. (From Dalinka MK: A sample aid to the performance of shoulder arthrography. Am J Roent 129:942, 1977. With the permission of American Journal of Roentgenology.)

Normal Arthrogram

Single Contrast

The smooth articular cartilage over the humeral head is well seen as is the biceps muscle within the superior aspect of the shoulder joint (Fig. 3). The synovial sheath of the biceps is usually visible within the bicipital groove, and contrast may be seen leaking into the upper arm, particularly following exercise (Fig. 4). We and others feel that this is a normal finding, probably related to increased intraarticular pressure and joint distention.[4]

The axillary and subscapularis recesses are normally visualized. The subscapularis bursa appears as a tonguelike recess beneath the coracoid process, which is deep to the muscle. Overdistention may lead to dissection of contrast adjacent to the bursa. The axillary fold is located at the inferior joint margin with an indentation between it and the subscapularis recess. In external rotation, the subscapularis bursa is small and, contrast extends to the anatomic neck of the humerus (Fig. 3). In internal rotation, the subscapularis bursa is better seen, but the axillary recess is almost obliterated (Fig. 3). The axillary view demonstrates the glenohumeral space, glenoid labrum, bicipital tendon, and subscapularis recess. On the air contrast studies, the cartilaginous labrum is seen to better advantage, as is the cartilage about the glenoid[4] (Fig. 5). When the rotator cuff is abnormal, both edges of the torn muscle or tendon may be demonstrated.[7]

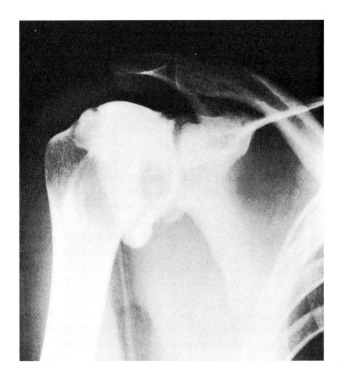 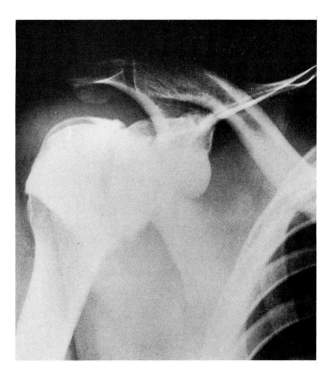

FIG. 3. Normal shoulder arthrogram. Left: External rotation. Note the bicipital tendon sheath in the bicipital groove. A small subscapularis recess and axillary fold are present. The joint capsule extends to the anatomic neck of the humerus. Right: Internal rotation. The axillary recess is obliterated. The subscapularis bursa is well seen beneath the coracoid. The lateral aspect of the humeral articular cartilage is well seen.

Abnormal Arthrograms

Complete Tear of the Rotator Cuff

When the rotator cuff musculature is completely torn, there is filling of the subdeltoid–subacromial bursa (Fig. 6). Contrast material is seen beneath the acromion process and above and lateral to the greater tuberosity (Fig. 7). Normally, the rotator cuff separates the subacromial–subdeltoid bursa from the shoulder joint. With a torn cuff, there is a communication between the glenohumeral joint and the bursa, and they are both filled with contrast agent. This can be seen on the axillary view, since the bursa projects over the humeral neck, which is normally devoid of contrast media (Fig. 8).

The bursal size does not correspond to the size of the tear. Small tears may not be seen until after exercise. With long-standing tears or atrophy of the musculature, there may be a communication with the acromioclavicular joint (Fig. 9). With double contrast and erect filming, the rotator cuff is coated both superiorly and inferiorly when torn. The thickness of the musculature and the size of the tear can both be demonstrated[7] (Fig. 10).

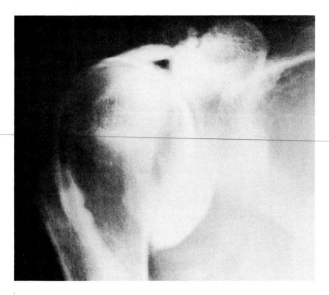

FIG. 4. Contrast extending along bicipital tendon in after-exercise film. Note dilution of contrast in delayed film.

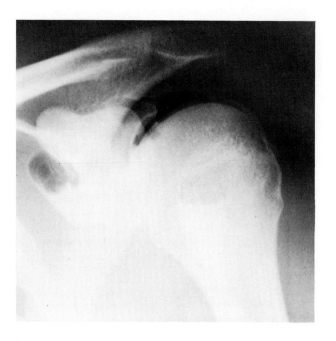

FIG. 5. Internal rotation erect film with double contrast demonstrating the superior portion of the cartilagenous labrum and the cartilage over the humeral head. The gas lateral to the glenoid labrum coats the undersurface of the rotator cuff.

shoulder, the axillary and subscapularis recesses may become continous (Fig. 11). The size of the pouch is best demonstrated on internal rotation and indicates the degree of capsular disruption. The Bankhart lesion or defect in the cartilagenous labrum may be demonstrable particularly with double-contrast techniques. Reeves[5] described an abnormality in the normal triangular appearance of the glenoid labrum on axillary views. In our experience, this has not been of value.

Partial Cuff Tears

In patients with partial tears of the rotator cuff, the arthrogram will be normal unless the abnormality occurs on the undersurface of the tendon.[4] In these cases, the contrast material will enter the tendonous defect, but no communication with the bursa exists.

In patients with recurrent dislocations of the

Adhesive capsulitis

In patients with adhesive capsulitis, the joint capsule is thickened and its capacity decreased. The bicipital tendon is poorly seen, and the recesses are small or partially obliterated.[4,9] The pressure on the injecting syringe is increased and only small amounts of contrast may be injected.

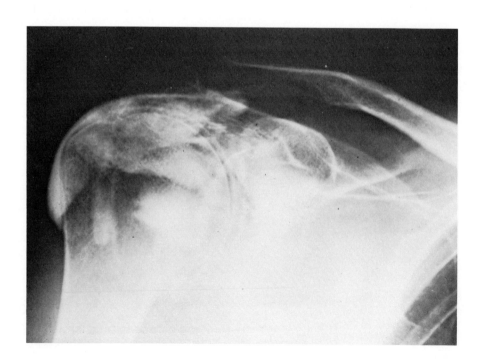

FIG. 6. Torn rotator cuff. The subacromial subdeltoid bursa is filled just beneath the acromion and lateral to the humeral head.

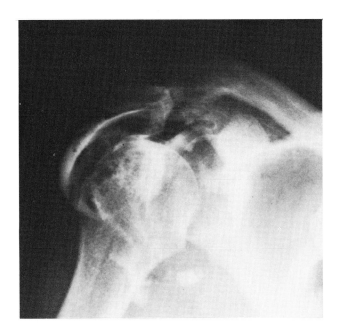

FIG. 7. Torn rotator cuff. The small subacromial subdeltoid bursa is filled. The lucency beneath the bursal shadow represents the rotator cuff.

Other Findings

Arthrography can demonstrate the status of the articular cartilage in patients with ischemic necrosis. Synovial chondromatosis and other synovial tumors can also be shown, as can giant synovial cysts. We have also seen filling of the paraarticular lymphatics, which we feel is a nonspecific finding probably secondary to inflammation (Fig. 12).

Summary

Shoulder arthrography is a valuable procedure in the evaluation of patients with suspected rotator cuff tears or chronic shoulder pain. In addition, adhesive capsulitis and soft-tissue damage from recurrent anterior dislocation can also be demonstrated.

FIG. 8. Torn rotator cuff, axillary view. Large subacromial–subdeltoid bursa projecting over neck of bursa that is normally devoid of contrast.

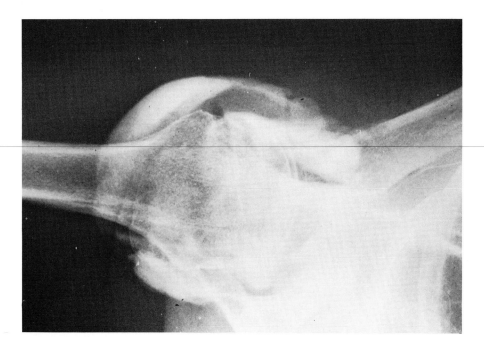

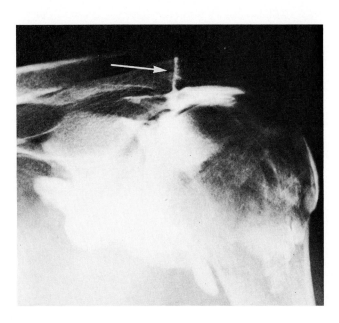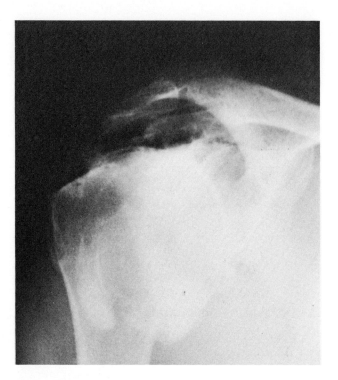

FIG. 9. Torn rotator cuff with contrast extending into acromioclavicular joint (arrow).

FIG. 10. Air contrast study showing torn rotator cuff with space between edges representing size of tear. The thickness of the remaining musculature is easily appreciated.

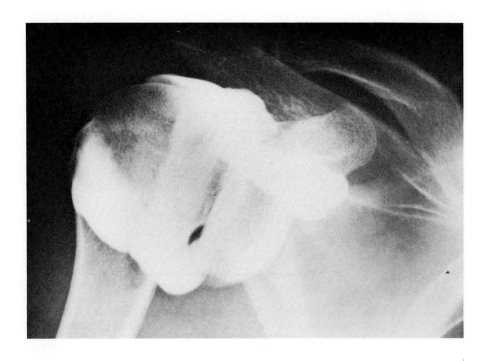

FIG. 11. The very large convexity between the axillary and subscapularis recesses represents the abnormally large capsule in this patient with previous anterior dislocation of the shoulder.

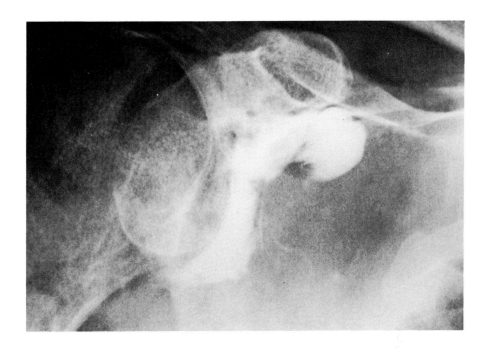

FIG. 12. Lymphatic filling about the shoulder seen on single-contrast arthrography.

References

1. Neviaser JS: Ruptures of the rotator cuff of the shoulder. Arch Surg 102:483, 1971
2. DeSmet AA, Ting YM: Diagnosis of rotator cuff tear on routine radiographs. J Can Assoc Radiol 28:54, 1977
3. Kummel BM: Arthrography in anterior capsular derangements of the shoulder. Clin Orthop 83:170, 1972
4. Killoran PJ, Marcove RC, Freiberger RH: Shoulder arthrography. Am J Roentgenol 103:658, 1968
5. Reeves B: Arthrography of the shoulder. J Bone Joint Surg 48B:424, 1966
6. Preston BJ, Jackson JP: Investigation of shoulder disability by arthrography. Clin Radiol 28:259, 1977
7. Ghelman B, Goldman AB: The double contrast shoulder arthrogram: evaluation of rotary cuff tears. Radiology 124:251, 1977
8. Dalinka MK: A simple aid to the performance of shoulder arthrography. Am J Roentgenol 129:942, 1977
9. Den Herder BA: Clinical significance of arthrography of the humeroscapular joint. Radiol Clin (Basel) 46:185, 1977

Arthrography of the Knee

R. H. Freiberger

Arthrography is a highly accurate method for assessing the state of the meniscal cartilages, articular cartilages, and cruciate ligaments. The double contrast method using a small quantity of radiopaque water soluable contrast agent plus air with spot filming is preferred.

An image amplified fluoroscope permitting spot filming on a fractional mm focal spot is essential. The focal spot ideally should be 0.3 mm, but 0.6 mm is acceptable. A focal spot larger than this will not produce adequate pictures. A reciprocating grid in the fluoroscopic tower is optimal; a stationary fine line grid is acceptable. A restraining device to permit distraction of the knee is necessary. It may consist of a sling fastened to the side of a table or other holding device that can be placed just above the knee. Any of the commercially available arthrogram injection trays are acceptable.

Technique

The knee is scrubbed with povidone iodine solution and draped. A 22 gauge needle is inserted under the midpoint of the patella, either medially or laterally. If fluid is present, it must be aspirated as completely as possible. Then 20 ml of air, 3 to 5 ml of a meglumine 60 percent contrast agent, and another 20 ml of air are injected and the needle withdrawn. The patient walks a few steps to distribute contrast agents and is then placed prone on the x-ray table, and filming commences. The overlapping of the femoral condyles serves as a guide to make certain that a complete examination is carried out from the most anterior to the most posterior portion of each meniscus. The cruciate ligaments are examined in a lateral position and traction applied to the proximal tibia to stretch the anterior cruciate ligament. A second examination consists of having the patient sit with a pillow forced into the popliteal space to push the tibia forward. An over-penetrated, sharply coned lateral view is taken.

Meniscal Anatomy

The medial meniscus has a complete peripheral attachment; therefore, contrast agent in the meniscus represents a tear. The medial meniscus is normally wider posteriorly than anteriorly.

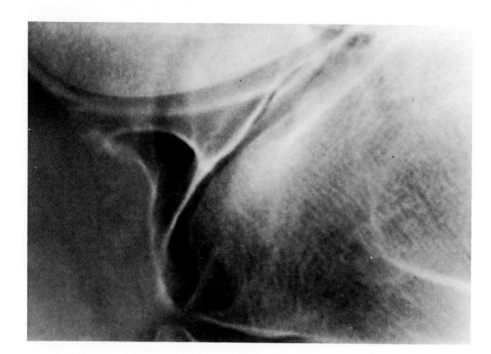

FIG. 1. Lateral meniscus posteriorly. The complex attachments of the posterior portions of the lateral meniscus are shown. The popliteal bursa is filled by air, bordered superiorly by a thick superior attachment of the meniscus and inferiorly by a thin inferior attachment. The lateral bulge into the lumen of the popliteal bursa represents the popliteal tendon, which is attached to the capsule.

The lateral meniscus is more circular than the medial meniscus and its most anterior and posterior portions cannot be examined arthrographically. Posterolaterally, the attachment of the lateral meniscus is incomplete because of the popliteal tendon and its bursa (Fig. 1). Knowledge of the attachments of the lateral meniscus in this region, and in the normal variants, is necessary for arthrographic interpretation. Since the popliteal bursa communicates with the joint, contrast agents seen peripherally, posterolaterally to the lateral meniscus, do not indicate a meniscus tear or detachment.

Meniscus Tears

Meniscus tears can be roughly divided into vertical, horizontal, and complex.

Discoid Lateral Meniscus

The discoid meniscus, as the name implies, is disc shaped in appearance and lacks the central "hole" of the normal meniscus. This abnormality is almost exclusively lateral, and often leads to tears and disability.

Cystic Meniscus

Cysts of the menisci are common laterally, but rare medially. They are ganglion-like structures located at the periphery of the meniscus and are often associated with a horizontal tear within the meniscus which allows partial opacification of the cyst.

Total Knee Replacement

Howard A. Kiernan

History

Gluck, in 1890 suggested joint replacement with an ivory component, cementing the component in place with a bone cement of plaster of Paris. He actually carried out replacement of the temporomandibular joint.

Metallic interposition arthroplasties were not introduced until the 1940s, when Campbell used a titanium plate. The Massachusetts General group used a femoral replacement made of vitalium in 1953, and McKeever introduced the tibial prostheses in 1960. But the first total knee replacement as we know it was the Waldius hinge, which was first used in 1957. The introduction of methyl methacrylate, high-density polyethylene, stainless steel, and chromium and cobalt alloys ushered in a new era in joint replacement surgery.

For any joint replacement surgery to be successful, it must relieve pain and ensure motion and stability. These three criteria—pain, motion, and stability—are the criteria by which replacements are judged, because an arthrodesis will provide pain relief and stability, and therefore, a joint replacement must be superior to an arthrodesis. The current joint replacement falls into three groups: (1) simple resurfacing procedures, (2) semiconstrained devices, and (3) totally-constrained devices.

Gunston, in 1962, introduced the polycentric knee. This is an example of a resurfacing procedure. It simulates the normal anatomy of the knee, attempts to reproduce normal knee motion, and retains the cruciate ligament. In 1970, Freeman and Swanson introduced their prosthesis (Fig. 1). In 1973, the Mayo Clinic group introduced the geometric prosthesis (Fig. 2). These two prostheses are examples of the semiconstrained devices, in which the cruciate ligaments are sacrificed and the knee motion is not exactly anatomical. The Shiers prosthesis introduced in 1970 (Fig. 3) and the Waldius prosthesis first introduced in 1957 are examples of the fully constrained, or hinged, prosthesis.

Biomechanics of the Total Knee

The design of any total joint replacement must include the normal tibiofemoral motion that occurs during daily activity, the normal articular surface motion, and the loads applied to the knee joint surface. We must also consider the gait alterations that occur in diseases. Morrison has determined the tibial–femoral motion necessary for normal activities. Walking on level ground requires 67 degrees of flexion, 7 degrees of total adduction–abduction motion, and 13 degrees of rotation. Stair climbing requires 83 degrees of flexion, and sitting requires 93 degrees of flexion. Indeed tall people require almost 110 degrees of flexion to sit. The adduction–abduction motion that takes place in normal knees is to alter the load on the tibial plateaus and thereby protect them from abnormal wear. Rotation in the flexed knee is possible, and approximately 13 degrees of rotation takes place in normal walking. Tibiofemoral rotation in part compensates for any position in relation to foot placement and is therefore essential in normal gait, especially activities such as stair climbing. The surface velocity of the femur to the tibia should be tangent to the tibiofemoral contact point. In abnormal knees, this is not true, and the surface velocity is almost perpendicular to the tibiofemoral contact point. This will significantly contribute to wear and loosening. Morrison also calculated the loads applied to the knee during

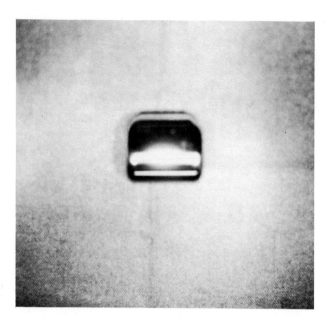

FIG. 1. Freeman–Swanson femoral component

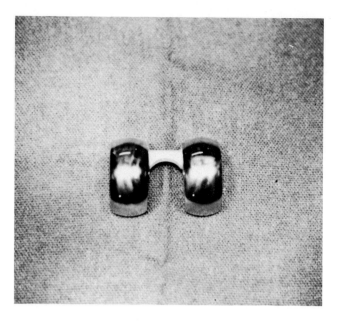

FIG. 2. Geometric femoral component

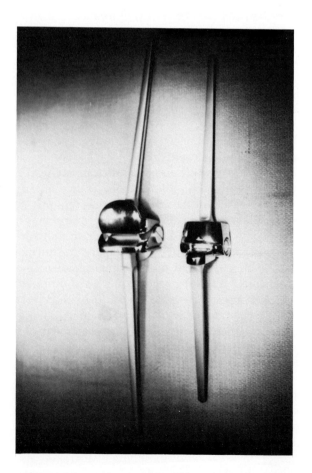

FIG. 3. Left: Shiers' hinge. Right: Waldius hinge

normal activities, and they are well in excess of what we previously thought. For instance, in level walking, the loads applied to the knee are about 3.97 times body weight. Walking down stairs, the load is about 3.83 times body weight, while walking up stairs, it is about 4.25 times body weight. These loads are maximum at heel strike and push offs. However, the tangents of knee motion are altered in rheumatoid and osteoarthritis. In normal gait, the knee is fully extended at heel strike and then flexes to 20 degrees as foot flat is reached. The knee then goes into extension at push off. Sixty degrees of flexion is needed during the swing phase. This is important, because the tibiofemoral contact area normally shifts forward and backward, enlarging the area to which the load is applied. In osteoarthritis, there is a marked loss of stance space flexion, which substantially increases the load on the knee. In rheumatoid arthritis, grossly abnormal abduction–adduction and rotation take place in addition to loss of stance space knee flexion, and this increases the load. Ideally, stance phase flexion should be restored by resurfacing procedures. Frequently it is not, and persistent flexion deformities increase the load on the tibiofemoral contact area. A 20-degree flexion contraction may increase the load by a factor of 2.6 times body weight. Varus and valgus deformities also increase the load in the corresponding compartment. Therefore, loss of stance space flexion, flexion deformity, and varus–valgus malalignment do increase the force per unit area on the implant. The amount of flexion at stance space depends upon extensor lag, quadriceps strength, and hip deformities. This may have a greater effect than the fixed deformities. Morrison has actually calculated the loads:

Level walking	3.12 body weight
Walking up ramp	3.97 body weight
Walking down ramp	3.95 body weight
Walking down stairs	3.83 body weight
Walking up stairs	4.25 body weight

Design Considerations in Total Knee Prostheses

Size

The knee is a subcutaneous joint and cannot tolerate large components. For instance, the Waldius and Shiers prostheses are large units. The Shiers is smaller and therefore better tolerated.

Stability

The constrained prostheses are unicentric and cannot duplicate normal knee motion. Stress is therefore transferred to the prosthetic bone interface, threatening the fixation. Nonconstrained or condylar units have not yet reached the ideal, because they sacrifice the anterior cruciate ligament, which tends to increase instability and impart large additional loads on the knee articular surface. This may contribute to wear and loosening of the prosthesis if there is no good means of fixation.

Fixation

The cancellous bone of the proximal tibia does not provide a good surface for implant fixation. So far, the one-piece tibial device with an intramedullary skin provides the best fixation designed. Additionally, instability in varus–valgus overcorrection leads to increased shear by tensile forces or compressive forces. This also threatens fixation.

Manufacture

The single radius of curvature is easier to manufacture, and therefore, those condylar units with complex geometry are harder to finish, but the devices with multiple radii of curvature have definite structural advantages.

Devices Currently in Use

Constrained Devices

The two currently most popular constrained devices are the Waldius hinged prosthesis, which was the first of the hinged devices, and the Guepar prosthesis, which is the most commonly used hinged device in the United States today. The problem with all hinged devices is that they cannot reproduce normal physiologic movement, and this is especially true in rotation. Torsion is therefore increased, threatening the implant fixation. No more than 3 cm of bony stock should be removed, because more than that would make an arthrodesis quite impossible. A total knee should allow flexion of more than 90 degrees. However, the Waldius hinge, because of its construction, tends to compress the posterior soft parts and allows only 90 degrees of flexion. The Guepar hinge is smaller and provides a larger range of motion, because its center is closer to the theoretical center of rotation of the knee. This is because the femoral stem is placed behind the tibial stem. Because of the size of these prostheses, the load transmission is over as wide an area as possible, but the torque is considerable. For instance, at 0 degrees of extension, the torque is 960 foot-pounds, and as the knee flexes to 45 degrees, the torque approaches 1600 foot-pounds. Wilson studied 54 knees and noted that 63 percent of those knees that have Waldius prostheses implanted showed some settling and loosening of the prosthesis. Cement fixation has not improved the situation. Interestingly, the pain was worse when the prosthesis was cemented. When cemented endoprostheses become infected, they are harder to remove and harder to fuse, and this may eventually lead to amputation. Regardless of whether cement is used or not, all constrained prostheses may settle into bone, and all show evidence of wear debris after 80 days. In spite of these disadvantages, the hinged prosthesis is still probably the best prosthesis to use with a significant bone loss, deformity, or ligamentous instability.

Semiconstrained Prostheses

This group includes the geometric prosthesis, which is probably the most commonly used device, and the duocondylar, the Eftekhar (Fig. 4), and Spirocentric knees. The anterior cruciate ligaments are sacrificed, and the joint motion is not exactly anatomical. The problem with the early geometric knee was loosening of the medial tibial component

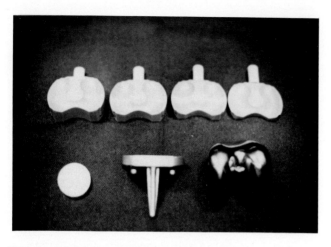

FIG. 4. Efetkhar total knee components.

because of failure to correct varus and valgus malalignment. There was progressive sinking of the medial tibial block, increasing varus with loosening of the prosthesis, and fatigue fracture of the plateaus. This led to dislocation of the prosthesis. This is because the geometric design is semiconstrained and is best used in knees that have only some bone loss and some ligamentous stability. Failures result because of poor selection of patient. Both the duocondylar and the Eftekhar design present a large tibial surface of the prosthesis, and both have a central peg to enhance tibial fixation.

Resurfacing Prostheses

Gunston's original prosthesis was a resurfacing device, as is Charnley's load inlay and Waugh's UCI device. These prostheses depend upon minimal bone loss and stable ligaments. The problem with Gunston's original design was that the runners were too small and did not allow for varus or valgus correction. Also, the femoral head had to be placed in 30 degrees of retroflexion, which is hard to estimate on the operating table, and was also placed in 5 to 8 degrees of valgus, which is similarly hard to do. Waugh's prosthesis has corrected this by making the bony cuts 90 degrees, and this is much easier to estimate at the operating table. All resurfacing prostheses have the same limitations, i.e., the collateral ligaments must be stable and posterior cruciate ligaments must be intact. A fixed flexion deformity of greater than 60 degrees or a varus or valgus deformity of greater than 20 degrees cannot be corrected. None of these prostheses are suitable for a joint previously fused, previously infected, or a neuropathic joint. All suffer from the same problems when the patient's flexion is poor, the prostheses tend to sublux and the knee is unstable, and the tibial component is deformed and loosened. Also, from a practical viewpoint, these prostheses should not be placed in a patient who weighs more than 200 pounds. Currently there are about ten designs in use throughout the country, and they fall into the three groups mentioned. The resurfacing prostheses are suitable for patients who really have minimal deformity. The semiconstrained prostheses are suitable for patients who have moderate deformity and malalignment but still retain ligamentous stability. Finally, the constrained or hinged prostheses are used for those patients who have considerable bone loss, deformity of ligamentous instability. Currently there is no one prosthesis that fits all patients best.

Bibliography

Morrison JB: Bioengineering analysis of force actions transmitted by the knee joint. Biomed Eng 3:164, 1968

Arthrography of the Postsurgical Hip

R. H. Freiberger

Arthrocentesis of the postsurgical hip to obtain fluid for laboratory examination can be a technically difficult procedure and the arthrogram is useful in determining the exact location of the needle tip. When fluid cannot be obtained, the injection of a few drops of contrast agent will reveal whether the needle tip has been placed intra-articularly or is in an extra-articular location. With intra-articular needle placement, saline without preservative can be injected and aspirated for laboratory analysis. The subsequent arthrogram can also supply useful information. It can show the tight but intact capsule, or it may opacify, communicating abscesses or sinuses.

When a prosthesis has been embedded with acrylic cement, the arthrogram can show loosening by demonstrating contrast agent in the bone cement interface (Fig. 1). Subtraction studies have proven useful when barium opacified cement has been used, since it is difficult to distinguish radiopaque contrast agent from radiopaque cement on routine films. Contrast agent becomes visible after "subtraction" of the pre-injection roentgenogram from the post-injection roentgenogram.

Arthrography of the Hip in Children

Purposes

1. Verification of the intra-articular placement of the needle tip in joint aspirations, particularly when no fluid can be obtained.
2. Evaluation of the shape and congruity of the largely or completely cartilaginous hip joint.
3. Detection and localization of loose cartilaginous or osteocartilaginous bodies.

Anatomy

The hip is a ball and socket joint with a congruous fit of a spherical femoral head in the acetabulum. In the child, the articular cartilages are very thick, and prior to approximately age 4 months, the femoral head is totally cartilaginous and cannot be seen on conventional roentgenograms. The articular cartilage of the acetabulum extends beyond the subchondral bone and forms the labrum or limbus which constitutes a large portion of the roof of the acetabulum. A space between the capsule and the limbus opacifies by intra-articularly injected contrast agent and is called the limbus thorn. The capsule inserts on the femur near the intertrochanteric ridge. Fibers constricting the capsule at the midportion of the femoral neck form the zona orbicularis. A notch at the inferior margin of the acetabulum is bridged by the transverse ligaments where the ligamentum teres takes its origin to insert into the fovea of the femoral head. The ligamentum teres is not visible on the arthrogram of a normal hip. The articular surface of the acetabulum is horseshoe shaped with the center of the acetabulum occupied by a fat pad.

Congenital Hip Dislocation and Dysplasia (CDH)

Arthrography is not used to make the diagnosis of CDH in the neonate. This is done by clinical examination. Arthrography is used later in infancy and childhood to evaluate the exact position of the femoral head with respect to the limbus. If the head lies above the limbus, a frank dislocation is present; if it is below the limbus, but elevated from its normal

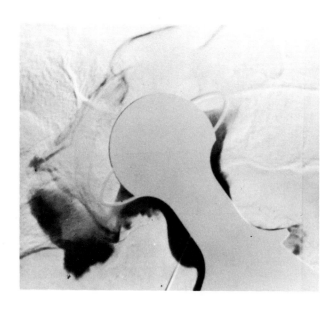

FIG. 1. Loose Charnly Miller prosthesis. A subtraction arthrogram shows the injected contrast agent in black. A thin layer of contrast agent outlines the cement bone interface of the acetabular component indicating loosening of the acetabular part of the prosthesis.

position, a subluxation is present. The size and shape of the limbus can be evaluated. Deformities of the largely or completely cartilaginous femoral head become evident as does incongruity between the femoral head and acetabulum. Thus, adequacy of reduction can be evaluated and the need for possible surgical treatment can be determined.

Septic Hip of Neonates and Children

Septic arthritis of the hip of the neonate is not uncommon. It is a disastrous occurrence which, if not detected and treated immediately, can lead to total destruction of the hip joint. Clinical symptoms may be

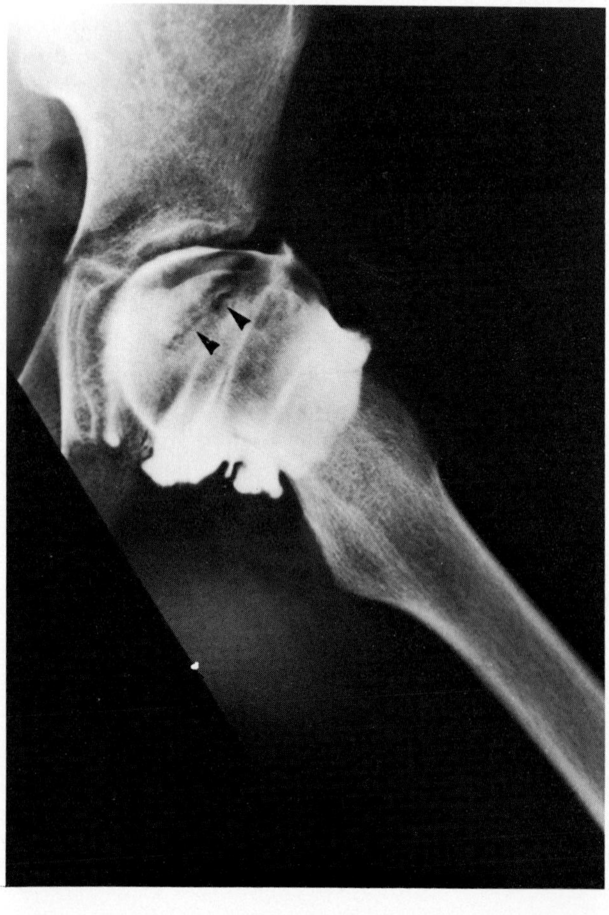

FIG. 2. An arthrogram of a hip with Legg-Perthes' disease. The fragmented capital apophysis can be seen through the contrast covered femoral head (arrows). The femoral head is mildly deformed and incongruity between the femoral head and acetabulum is shown by the thick layer of contrast agent supramedially. Evaluation of femoral deformity and incongruity of the joint is important when treatment by osteotomy is being considered.

minimal and systemic symptoms absent. A definitive diagnosis is made by aspiration, smear, and culture. The arthrogram serves to verify intra-articular needle placement, particularly when fluid cannot be obtained easily. The arthrogram will also depict the integrity of the capsule or show para-capsular abcesses or sinus formation. Destructive changes of the cartilages or incongruity of the joint can be demonstrated. By arthrogram, it can also be determined whether a dislocation of the cartilaginous femoral head is present or whether there is an epiphyseal separation displacing the shaft of the femur.

Legg Calve Perthes Disease

The arthrogram will show the deformity of the femoral head and incongruity of the hip joint (Fig. 2). This information, particularly in the older child, may be used to make a decision as to whether an iliac osteotomy or intertrochanteric osteotomy should be performed, or whether conservative treatment should be used.

Growth Disturbances of the Femoral Head

Growth disturbances such as multiple epiphyseal dysplasia and spondylo epiphyseal dysplasia, delay ossification of the femoral capital epiphysis and cause deformities of these ossification centers. The true deformites of the cartilaginous femoral head can be made visible by the arthrogram. The presence or absence of a femoral head can be determined in cases of proximal focal femoral deficiency.

Loose Osseous Bodies

Osteochondritis dissecans-like changes can occur in the femoral head as a result of previous Legg Perthes disease or post-traumatic osteonecrosis. The arthrogram can determine whether an osseous body is completely subchondral or whether it is loose and surgical removal should be considered.

Index

Abduction injuries, 332-333
Absorptiometry, photon, 186
Achondroplasia, 76
Acromioclavicular dislocations, 318, 319-320
Actinomycin D (Cosmogen, Dactinomycin), 232
Adduction injuries, 333, 335
ADI, see Atlantal-dens interval
Adjuvant chemotherapy, 234-235
Adriamycin, 232-233, 234
Alkaline phosphatase in Paget's disease, 140, 143, 144, 145
Alkeran, see Melphalan
Alkylating agents, 232
Allograft, 247
  replacement, technique of, 248-249
Alpha heavy-chain disease, 198
Amethopterin, see Methotrexate
Amyloid deposition, 203-205
Amyloidosis, 202-206
Anemia
  aplastic, 60, 61-62
  Cooley's, 53-54
  hemolytic, 53-55
  in macroglobulinemia, 195
  in myeloma, 191-192
  sickle-cell, 53, 54
Aneurysmal bone cyst, 276-277
Angiography
  bone tumors and, 243-246
  spleen, 325-326
  of trauma, 321-326
Angiomas, 7
  cervical or mediastinal, 18
  incidence of vertebral, 8-9
Angiomata of skin, 18
Angiomatosis, 7
  diagnosis of, 21
  of hands and feet, 20
  multiple, 17-22
  of skeleton, 7-27
Ankle
  arthrography, 351-356
  fractures, 329-336

  in adults, 330-331
  in children, 329-330
  injuries, patterns of, 331-336
  ligaments of, 351-353
  radiography, 329
Ankylosing spondylitis, 100, 109-114
Ankylosis
  of apophyseal joints, 91-92, 93
  carpal, 92
  cervical spine, 94
Antibiotics, 232
Antigens, transplantation, 109
Aortic injury, 321-323
Aortography, thoracic, 321
Aplastic anemia, 60, 61-62
Apophyseal joints, ankylosis of, 91-92, 93
Arnold-Chiari malformation, 85
Arteriography, 323
Arthritides, 89-138
  complications and unusual features of, 99-100
Arthritis
  acute pyogenic, 129-130
  amyloid, 204
  HLA-B27-associated, 109-118
  psoriatic, 115-116, 117
  in Reiter's syndrome, 114
  rheumatoid, 99-100
    juvenile (JRA), 89-96
  septic, of hip, 378-379
Arthrocentesis, 377
Arthrography
  ankle, 351-356
  elbow, 353-358
  knee, 371-372
  postsurgical hip, 377-379
  shoulder, 363-369
  wrist, 357-360
Arthropathy, neurotrophic, 127-128
Arthrosis, 124
Articular erosion, 89, 91
Astrocytoma, cervicomedullary, 81
Atlantas-dens interval (ADI), 304-305, 309

Atlantoaxial
  dislocations, 99-100
  rotary
    dislocation, 308
    displacement, 308-309
  subluxation, 92
Atlantooccipital dislocation, 307
Atlas fractures, 309, 311
Attenuation coefficient, mass (MAC), 177, 179, 180
Autograft, 247
Axis development, 306-307

B cell system, 201
Battered children, 343
Bence Jones proteinuria, 190, 192-194, 195
BMC, see Bone mineral content
Bone
  abnormal structure, modeling, and density of, 1-6
  abscess, localized, 131
  absorption, spontaneous, 24
  benign vascular lesions involving, 8
  calcium and, 153
  cell types, 1
  cortical, osteoporosis and, 157-158
  cyst, aneurysmal, 276-277
  disease, 139-188
    disappearing, 24-27
    metabolic, 154
  dysplasia, hereditary, with hyperphosphatasemia, 32, 34
  erosion, 135
  formation, 1
    new, 130
  giant cell tumor of, 141
  hemangioma of, see Hemangioma of bone
  infections of, 129-132
  living, changes in old, 130-131
  long, 129
    healing of, 347
    hemangiomas of, 14-15
  loss, osteoporosis and, 157-158
  marrow
    activity, peripheral extension of, 69
    imaging with radioiron, 57-69
    myeloma and, 191
    reserve, evaluation of, 64
    tumors involving, 64
  mineral
    analysis, computed tomography and, 183-187
    content (BMC), 175-181
    density, 175-181
  necrotic, changes in, 130
  neoplasms, see Neoplasms of bone
  Paget's disease of, see Paget's disease of bone
  subperiosteal resorption of, 165
  transition from woven to lamellar, 1
  trauma, scintigraphy and, 339-344
  tumor embolization, 244
  tumors
    angiography and, 243-246
    chemotherapy of malignant, 231-235
    giant cell, 141
    limb-saving resections of malignant, 247-251
    malignant, 189-301
    nuclear medicine and, 237-240
    of spine, 78-81
Bowel disease, inflammatory, 117-118
Brain, Paget's disease and, 141
Brown tumors, 165, 166

C1-C2
  ligament disruption, 308
  subluxation, 112
Calcification, periarticular, 89, 90
Calcitonin
  excess state, 31
  in Paget's disease, 141-142
  radioimmunoassay and, 173-174
Calcium, bone and, 153
Calvé's disease, 268
Cameron scanning, 178, 186
Capsulitis, adhesive, 366
Carpal
  ankylosis, 92
  tunnel syndrome, 204
Cartilage
  failure of osteocytic resorption of, 31
  space narrowing, 89
CCT, see Cortical thickness, combined
CDH, see Hip dislocation, congenital
Chemotherapy
  adjuvant, 234-235
  of Ewing's sarcoma, 235
  of malignant bone tumors, 231-235
  of osteogenic sarcoma, 234
Chondroblastoma, 292-294, 295
  benign, 277
Chondrocalcinosis, 165
Chondroitin sulfate, 121
Chondrolysis, osteocytic, 32
Chondroma
  chondroid, 225
  juxtacortical, 291-292
  periosteal, 225, 291-292
Chondromatosis, synovial, 225
Chondromyxoid fibroma, 294-300
Chondrosarcoma
  clear cell, 226
  dedifferentiated, 222
  of hands and feet, 225
  mesenchymal, 224
  newer variants of, 221, 223-226
  periosteal, 223
  radiology, 227-230
  secondary, 224
Clavicle fractures, 313-315
Compression
  deformities in osteoporosis, 155-156
  fractures, 11, 156
  injuries, vertical, 333-336
Compton scattering, 179-180

Index

Connective tissue disease, mixed (MCTD), 103, 106
Cooley's anemia, 53-54
Coracoid process, fractures of, 315
Correlation coefficient, significance of, 181
Cortical thickness, combined (CCT), 158
Cosmogen, see Actinomycin D
Cranial stenosis, 83
Cranium
   lesions of, 83-85
   neoplasms of, 84
CREST syndrome, 103
CRST syndrome, 103, 105
CT, see Tomography, computed
Cushing's syndrome, 156-157
Cyclophosphamide (Cytoxan, Endoxan), 232
Cysts, giant rheumatoid, 100
Cytoxan, see Cyclophosphamide

Dactinomycin, see Actinomycin D
Daunomycin, 233
Densitometry, film, 177
Dermatomyositis, 102, 103, 104
Desmoid, extraabdominal, 41
Diaphysis, normal, 4
Diastematomyelia, 77, 79, 85
Dimethyl triazeno imidazole carboxamide, see DTIC
Diphosphonates in Paget's disease, 142-144
Disk herniation, 77, 78
Dislocations
   acromioclavicular, 318, 319-320
   atlantooccipital, 307
   fracture, 319
   of glenohumeral joint, 316
   shoulder, 315-320
      posterior, 316-318
   sternoclavicular, 320
DTIC (dimethyl triazeno imidazole carboxamide), 233-234
Dupruytren fracture dislocation, 332-333, 334
Dyscrasias, plasma cell, see Plasma cell dyscrasias
Dysplasia epiphysealis hemimelica, 290-291
Dysplasias, 31-34
Dysraphism, 85

EHDP in Paget's disease, 142-144
Elbow arthrography, 353-358
Embolization, bone tumor, 244
Enchondromas, solitary, 287-289
Enchondromatosis, multiple, 289-290
Endoxan, see Cyclophosphamide
Eosinophilia, differentiation of benign from malignant, 62-63
Epidermoids, 13
Epiphyseal
   closure, early, 93
   development, 305-306
   enlargement, 95
   injury types, 329-330
Epiphyses, angular, 96

Erythrocytosis, 64-65
Erythropoietic tissue, detection of extramedullary, 63, 64
Erythropoietin assays, usefulness of, 65
Ewing's sarcoma, 211-212, 213
   chemotherapy of, 235
   radiology, 215-218

Facial bones, lesions of, 85
Fanconi's syndrome, 205
Fat, retrobulbar, 84
Feet
   angiomatosis of, 20
   chondrosarcoma of, 225
   hemangiomas of, 14
Femoral head, growth disturbances of, 379
Femurs, fetal, 3
Fibromas
   chondromyxoid, 294-300
   of gastrointestinal tract, 42
   nonosteogenic, 277
Fibromatosis
   congenital
      generalized, 40, 43
      generalized or disseminated, 35
      localized, 35
      multiple, 35, 36
      trauma and, 37
   juvenile, 35, 37, 39
   latent form of, 37
   multiple, 35
      skeletal, 38
   plantar, 37
   progressive and aggressive, 37
   skeleton and, 35-44
Fibrous dysplasia, 278
Fingers
   hemangioma of, 15
   osteoarthritis of, 122
Foam cells, 21, 260
Fracture dislocations, 319
Fractures, 345
   ankle, see Ankle fractures
   of atlas, 309, 311
   of clavicle, 313-315
   compression, 11, 156
   of coracoid process, 315
   growth and, 347
   Jefferson, 309, 311
   Malgaigne's, 334, 336
   odontoid, 311, 312
   of proximal humerus, 313, 314
   ring, 309, 311
   of scapula, 315
   of shoulder, 313-315
   supramalleolar, 334, 336
   treatment of, 346
   types of, 346

Gamma heavy-chain disease, 198

Gammopathy, monoclonal, 189
   essential or benign, 198
Gastrointestinal tract, fibromas of, 42
Gaucher's disease, 54, 55
Genitourinary disease, lateral spine films and, 71-74
Giant-cell
   granuloma, 277
   lesion, 278
   tumor, 275-278
Glenohumeral joint, dislocations of, 316
Gliomas, optic nerve, 84
β-Glucuronidase deficiency, 48
Gout, 100
Granuloma
   eosinophilic, 259, 261-262, 264-267, 269
   giant-cell, 277
   plasma cell, 214
Graves' disease, 84
Growth, fractures and, 347

Hamartomas, 39-40
   neoplasms versus, 18
Hamartomatosis, generalized, 41
Hand-Schüller-Christian disease, 257, 259-261, 262-263
Hands
   angiomatosis of, 20
   chondrosarcoma of, 225
   hemangiomas of, 14
Heberden's nodes, 122
Hemangiolymphangiomatosis, congenital, 9
Hemangiomas
   benign metastasizing, 7
   of bone, solitary primary, 8-16
   of finger, 15
   frontal bone, 13
   of hands and feet, 14
   intraosseous, 10
   of long bones, 14-15
   of lumbar spine, 12
   multiple, 14
   primary calvarial, 13
   of rib, 14, 15
   of skull, 11, 13
   soft-tissue, 14
   of spine, 10
Hemangiomatoses, multiple, 16, 17
Hemangiomatosis, 7-8
   of bone, 138
   diffuse, 17
Hematomas, pelvic, 324
Hemiatrophy, 83
Hemolytic anemias, 53-55
Hemophilia, pseudotumor of, 279
Hemorrhagic infarct, 147
Hepatic trauma, 326
Herniation, disk, 77, 78
Heterograft, 247
Hip
   amyloidosis of, 205
   dislocation, congenital (CHD), 377-378
   osteoarthritis of, 122, 123
   postsurgical, arthrography of, 377-379
   septic, 378-379
Hiss-Sachs deformity, 316, 317
Histiocytes, 212-213
Histiocytosis X, 213, 268, 271
   pathology, 257-258
   radiology, 259-272
Histocompatibility antigens, 109
HLA-B27-associated arthritis, 109-118
Homograft, 247
Hunter-Hurler syndrome, 47-50
Hunter syndrome, 48, 50
Hurler's syndrome, 47-48
Hydromelia of spinal cord, 85-86
Hydroxyproline excretion, 139-140, 143
Hypercalcemia, 193
Hyperparathyroidism, 163-166, 277
   primary, 164, 165, 172
      renal osteodystrophy versus, 167
   secondary, 164, 166
   tertiary, 164
Hypertrophy, skeletal, 14
Hypervascular lesions, 243
Hyperviscosity syndrome, 206
Hypovascular lesions, 243

I cell disease, 51
IgM globulins, 195-196
Ilium, myeloma of, 214
Imaging, scintillation, *see* Scintillation imaging
Involucrum, 132
Isograft, 247

Jaws, osteosarcomas of, 221
Jefferson fracture, 309, 311
Joint
   disease, degenerative, 121-124, 125
   infection, 129-132
   neurotrophic, 127-128
   subluxation, 128
JRA, *see* Arthritis, rheumatoid, juvenile

Kidney, myeloma, 193
Knee
   arthrography, 371-372
   biomechanics of, 373-375
   osteoarthritis of, 122-123
   prostheses, 375-376
   replacement, 373-376

Lamina dura, absorption of, 165
Langerhans granules, 260-261
Larynx, cartilaginous tumors of, 225
Legg-Calvé-Perthes disease, 379
Leiomyosarcoma, 244

Lesions
- ankylosing spondylitis, 111-112
- cartilaginous, see Tumors, cartilaginous
- categories of, 75
- of cranium, 83-85
- giant-cell, 278
- myeloma, 190-191
- of paranasal sinuses and facial bones, 85
- round cell, see Round cell lesions
- skeletal, simulating malignancy, 279-280

Letterer-Siwe disease, 257, 259-261, 263
Leukemia, monocytic, after myeloma, 199
Ligamentous ossification, 112-113
Ligaments of ankle, 351-353
Light chain deposition, systemic, 206
Limb-saving
- procedures, results of, 251
- resections of malignant bone tumors, 247-251

Lipomas of spinal cord, 85-86
L-PAM, see Melphalan
Lupus erythematosus, systemic (SLE), 101, 102
Lymphangiography, 24
Lymphangiomas, soft-tissue, 22
Lymphangiomatosis, 7-8
- skeletal, 22-24

Lymphoma, 237
- malignant, 212-213

M-type proteins, 190-194
MAC, see Attenuation coefficient, mass
Macroglobulinemia, 195-196
- Waldenstrom's, 206

Magnification, clinical applications
- of direct, 134-136
- for optical, 134
- in skeletal radiography, 133-137

Malgaigne's fracture, 334, 336
Malignancy, skeletal lesions simulating, 279-280
Maroteaux-Lamy syndrome, 48, 51
Marrow, bone, see Bone marrow
MCTD, see Connective tissue disease, mixed
Melphalan (alkeran, L-PAM, phenylalanine mustard, sarcolysin), 232
Meningiomas, optic nerve, 84
Meningocele-lipoma complex, anterior sacral, 71
Meningoceles
- lateral, 77, 80
- sacral, 85, 86

Meniscal anatomy, 371-372
Meniscus, cystic and discoid lateral, 372
Metaphysis, normal tapering of, 5
Methotrexate (amethopterin), 233
Metrizamide, 75
Mithramycin in Paget's disease, 144-145
MMM, see Myelofibrosis with myeloid metaplasia
Monoclonal disorder, 201
Morquio syndrome, 48, 50
MPS, see Mucopolysaccharidoses
Mucolipidosis GM I, 51
Mucopolysaccharidoses (MPS), 47-51

Musculoskeletal applications of computed tomography, 253-256
Myelofibrosis with myeloid metaplasia (MMM), 59-61
Myelography, role of, 75-81
Myeloid metaplasia
- myelofibrosis with (MMM), 59-61
- splenectomy in, 58-59

Myeloma, 213
- anemia in, 191-192
- bone marrow and, 191
- clinical manifestations of, 193
- kidney, 193
- lesions, 190-191
- management and therapy of multiple, 193
- monocytic leukemia after, 199
- multiple, 189-194, 202
- presymptomatic, 190
- solitary, 213-214
- tissue, 217

Myelomatosis, 216-218
- sclerotic forms of, 217

Myositis ossificans, scintography and, 344

Neoplasms
- of bone
  - benign, 239-240
  - primary malignant, 237-239
- of cranium, 84
- hamartomas versus, 18
- of spine, 87

Neuroblastoma, 239
Neurofibromatosis, 40, 43, 77, 80
Neurolemmomas, 77
Nuclear medicine in bone tumors, 237-240

Odontoid, 306
- fractures, 311, 312
- hypoplasia, 307, 308
- shining, 112
- trauma, 311, 312

Oncovin, see Vincristine
Orbit, CT examination of, 84
Os odontoideum, 307, 308
Osseous bodies, loose, 379
Ossification
- ligamentous, 112-113
- secondary centers of, 93, 346

Osteitis
- condensans ilii, 109-110
- deformans, 139
- fibrosa cystica, 163
- pubis, 137

Osteoarthritis, 121-124
- erosive, 125
- scintigraphy and, 343

Osteoarthrosis, 121-124
Osteoblast, 1
Osteoblastoma, benign, 277
Osteochondritis vertebralis, 268

Osteochondroma
  pedunculated, 283
  solitary, 281-284
Osteochondromatosis, multiple, 284-287
Osteoclast, 1
  activating factor, 208
  in Paget's disease, 139
Osteocyte, 1, 32
Osteocytic resorption of cartilage, failure of, 31
Osteodystrophy, renal, 166-167
  primary hyperparathyroidism versus, 167
Osteogenesis imperfecta, 31
Osteoid tissue, neoplastic, 227
Osteolysis, massive, 24-27
Osteoma, osteoid, 240
Osteomalacia, 31, 33, 158-163
  adult, 160-163, 164
  due to target cell abnormality, 160
  senile osteoporosis versus, 161
Osteomyelitis, 135
  localized form of, 132
  pathogenesis of, 130-131
  streptococcal, 131
Osteopathies, pediatric, 1-69
Osteopathy, parathyroid, 278
Osteopenia, 33, 153
Osteopetrosis, 31-32
Osteophytes, 100
  absence of, 155
Osteophytosis, 124
Osteoporosis, 153-158
  causes of, 154
  cortical bone and, 157-158
  disuse, 135
  senile, versus osteomalacia, 161
Osteosarcoma, 277
  hemorrhagic, 222
  intraosseous, 223
  multicentric, 222
  newer variants of, 221-223
  in Paget's disease, 221
  parosteal, 223
  periosteal, 223
  postradiation, 222
  radiology, 227-229
  telangiectatic, 222-223
  varieties of, 221

Paget's disease of bone, 11, 139-145, 237, 239
  osteolytic, 147-151
  osteosarcoma in, 221
Pantopaque, 75
Paranasal sinuses, lesions of, 85
Parathormone (parathyroid hormone) (PTH), 159, 163, 171-173
  structure of, 172
Parathyroid osteopathy, 278
PCDs, see Plasma cell dyscrasias
PCDUS, see Plasma cell dyscrasias of unknown significance

Pelvic hematomas, 324
Periosteal reaction, 89, 90, 266
Phenylalanine mustard, see Melphalan
Phleboliths, 14, 16
Photodensitometry, 185
Photon
  absorptiometry, 186
  energy effect on BMC measurements, 178-179
  scattered, 179-180
Plasma cell
  dyscrasias (PCDs), 189-199
    interrelationships of, 207
    progression of, 203
    radiology of, 201-208
    of unknown significance (PCDUS), 198-199
  granuloma, 214
  products, biologic properties of, 208
Plasma cells, 201
Plasmacytoma, 81, 204, 216-218
Polycythemia vera, staging of, 57-59
Polymyositis, 102
Posterior arch fracture, 309
Premyeloma, 190, 198
Prevertebral soft tissue, 304
Proteinuria, Bence Jones, 190, 192-194, 195
Proximal humerus fractures, 313, 314
Pseudocallus, 157
Pseudofractures, 161-162, 163
Pseudohyperparathyroidism, 164
Pseudotumor of hemophilia, 279
Pseudowidening, 109
Psoriatic
  arthritis, 115-116, 117
  spondylitis, 117, 118
PSS, see Sclerosis, progressive systemic
PTH, see Parathormone

Radiogallium, 342
Radiography
  ankle, 329
  magnification in skeletal, 133-137
  scintigraphy versus, 340, 341, 343
Radioimmunoassay (RIA), 171
Radioiron
  marrow imaging with, 57-69
  radioindium and radiocolloid versus, 66-67
  thrombocythemia vera and, 67-69
Radiology of plasma cell dyscrasias, 201-208
Reiter's syndrome, 114-115
Renal trauma, 325
Reticulum
  cell sarcoma, 212-213
    of bone, primary, 215-216
  cells, 212-213
Retrobulbar fat, 84
Retropharyngeal soft tissue, 304
Rheumatoid arthritis, see Arthritis, rheumatoid
RIA, see Radioimmunoassay
Rib, hemangioma of, 14, 15

Index

Rickets
  oncogenous, 160
  renal, 167
  vitamin D-dependent, 160
Ring fracture, 309, 311
Rotation injuries, ankle, 331-332
Rotator cuff, complete tear of, 365-368
Round cell lesions
  pathology, 211-214
  radiology, 215-218

Sacroiliitis, 116
Sacrum
  partial absence of, 71, 72
  scimitar defect in, 73
Sanfilippo syndrome, 48, 50
Sarcolysin, see Melphalan
Sarcoma
  bone and cartilagenous, 231
  osteogenic
    chemotherapy of, 234
    sclerosing, 238
  soft tissue, 231
Scapula fractures, 315
Scattering, coherent, 180
Scheie syndrome, 48, 50
Schlesinger-Taveras syndrome, 75-76
Scintigraphy
  bone trauma and, 339-344
  radiography versus, 340, 341, 343
Scintillation imaging, 237
Scleroderma, 103, 104, 105
Sclerosis
  progressive systemic (PSS), 103
  of terminal phalanges, 101
Sclerotic forms of myelomatosis, 217
Scoliosis, 77
Segmentation, anomalies of, 85
Sella turcica, 84
Sequestra, 132
Shoulder
  amyloid of, 205
  arthrography, 363-369
  dislocations, 315-320
    posterior, 316-318
  fractures, 313-315
Sickle-cell anemia, 53, 54
Skeletal enlargement, 14
Skeleton
  angiomatosis and, 7-27
  fibromatosis and, 35-44
  growing, trauma to, 345-349
Skin, angiomata of, 18
Skull
  computed tomography of, 83-85
  hemangiomas of, 11, 13
  osteoporosis of, 157
  Paget's disease of, 140
  trauma, 83-84
SLE, see Lupus erythematosus, systemic

Spinal
  canal
    diameter of, 305
    narrow, 75-76
  cord, hydromyelia and lipomas of, 85-86
  stenosis, 86
Spine
  bone tumors of, 78, 81
  cervical
    ankylosis, 94
    anomalies of, 307
    in child, 305-307
    trauma to upper, 303-312
  computed tomography of, 85-87
  congenital anomalies of, 85-86
  degenerative diseases of, 87
  films, genitourinary disease and, 71-74
  hemangiomas of, 10
  lumbar, hemangioma of, 12
  neoplasms of, 87
  osteoarthritis of, 123-124
  pediatric, 71-74
  spur formation of, 100
  trauma of, 86-87
Spleen angiography, 325-326
Splenectomy in myeloid metaplasia, 58-59
Spondylitis
  ankylosing, 100, 109-114
  cervical, 92
  psoriatic, 117, 118
Spondylosis, 76-77
Spongiosa, primary, 2
Spur formation of spine, 100
Squaring of vertebral bodies, 110
Sternoclavicular dislocations, 320
Subluxation
  atlantoaxial, 92
  C1-C2, 112
  joint, 128
Supramalleolar fractures, 334, 336
Symphysis pubis, widening of, 137
Syndesmophytes, 100, 110-111
Syringomyelia, 127

Terminal tuft resorption, 101
Thalassemia major, 54, 55
Thoracic aortography, 321
Thrombocythemia vera, 67-69
Tibia, normal proximal, 4
Tomography, computed (CT)
  advantages, 185-187
  bone mineral analysis and, 183-187
  musculoskeletal applications of, 253-256
  numbers, 183-184
  scanning, 180-181
  of skull, 83-85
  of spine, 85-87
Torticollis, 309, 310
Trabeculae, metaphyseal, 2
  normal, 4

Transmission, dual energy, 179
Transplantation antigens, 109
Trauma, 303-349
  angiography of, 321-326
  bone, see Bone trauma
  congenital fibromatoses and, 37
  to growing skeleton, 345-349
  hepatic, 326
  iatrogenic, 341-342
  indirect, 343
  odontoid, 311, 312
  renal, 325
  skull, 83-84
  of spine, 86-87
  to upper cervical spine, 303-312
  vascular, 321-323
  visceral, 325-326
Tumors
  bone, see Bone tumors
  brown, 165, 166
  cartilaginous, 281-300
    of larynx, 225
      location of, 281
    nomenclature of, 282
  giant-cell, 275-278
  involving bone marrow, 64

Vascular
  lesions, benign, involving bone, 8
  trauma, 321-323
Vertebra plana, 267-268
Vertebrae, osteoporotic, 155
Vincristine (Oncovin), 233
Visceral trauma, 325-326
Vitamin D metabolism, 158-160
Von Recklinghausen's disease, 77

Whiskering, 112, 113
Wrist arthrograms, 357-360

Xanthoma cells, 260
Xenograft, 247
X-ray transmission, single energy, 177